21st International Conference on Strangeness in Quark Matter (SQM 2024)

AA003583

EPJ Web of Conferences Volume 316 (2025)

Strasbourg, France
3-7 June 2024

Editors:

Cvetan Cheshkov
Rachid Guernane
Antonin Maire

ISBN: 979-8-3313-1323-4

Printed from e-media with permission by:

Curran Associates, Inc.
57 Morehouse Lane
Red Hook, NY 12571

Some format issues inherent in the e-media version may also appear in this print version.

This work is licensed under a Creative Commons Attribution 4.0 International License. License details: http://creativecommons.org/licenses/by/4.0/.

No changes have been made to the content of these proceedings. There may be changes to pagination and minor adjustments for aesthetics.

Printed with permission by Curran Associates, Inc. (2025)

For additional information, please contact EDP Sciences – Web of Conferences
at the address below.

EDP Sciences – Web of Conferences
17, Avenue du Hoggar
Parc d'Activité de Courtabœuf
BP 112
F-91944 Les Ulis Cedex A
France

Phone: +33 (0) 1 69 18 75 75
Fax: +33 (0) 1 69 28 84 91

contact-edps@webofconferences.org

Additional copies of this publication are available from:

Curran Associates, Inc.
57 Morehouse Lane
Red Hook, NY 12571 USA
Phone: 845-758-0400
Fax: 845-758-2633
Email: curran@proceedings.com
Web: www.proceedings.com

TABLE OF CONTENTS

Statement of Peer review

In submitting conference proceedings to EPJ Web of Conferences, I certify to the Publisher that I adhere to the **Policy on Publishing Integrity** of the journal in order to safeguard good scientific practice in publishing.

1. All articles have been subjected to peer review administered by the proceedings editors.
2. Reviews have been conducted by expert referees, who have been requested to provide unbiased and constructive comments aimed, whenever possible, at improving the work.
3. Proceedings editors have taken all reasonable steps to ensure the quality of the materials they publish and their decision to accept or reject a paper for publication has been based only on the merits of the work and the relevance to the journal.

Title, date and place of the conference

21st international conference on Strangeness in Quark Matter (SQM2024), 3-7 June 2024, Strasbourg, France

Proceedings editor(s):

Cvetan Cheshkov (Institut de Physique des deux Infinis, Lyon)
Rachid Guernane (Laboratoire de physique subatomique et de cosmologie, Grenoble)
Antonin Maire (Institut Pluridisciplinaire Hubert Curien, Strasbourg)

Date and editor's signature

December the 6th, 2024

Rachid Guernane Cvetan Cheshkov Antonin MAIRE

Starting SQM 2024: Theory state-of-the-art

Jacquelyn Noronha-Hostler[1,*]

[1]Illinois Center for Advanced Studies of the Universe, Department of Physics, University of Illinois at Urbana-Champaign, 1110 W. Green St., Urbana IL 61801-3080, USA

Abstract. I discuss the theoretical developments related to Strangeness in Quark Matter (SQM) leading up to the SQM2024 conference. These advances include mapping out the Quantum Chromodynamics phase diagram; puzzles that exist in hadron physics from light to heavy particles; and relativistic hydrodynamics with the inclusion of spin and magnetic fields.

1 Introduction

I will summarize recent advances in high-energy nuclear theory, pressing questions within the field, and make connections to astrophysics, gravity, nuclear structure, and cold atoms.

2 QCD Phase Diagram

The QCD phase diagram cannot be systematically studied in a single experiment or astronomical observation, rather it requires input from heavy-ion collisions [HIC] (high temperatures T, low to intermediate net baryon density n_B, range of electric charge to baryon number $Y_Q = Z/A = n_Q/n_B = [0.38, 0.5]$), cold neutron stars [NS] ($T = 0$, large n_B, $Y_Q \lesssim 0.1$), binary [NS] mergers ($T > 0$, large n_B, $Y_Q \sim [0.01, 0.2]$), and low energy nuclear physics ($T \sim 0$, nuclear saturation density $n_B = n_{sat}$, $Y_Q \sim 0.5$). This range of measurements leave gaps in the QCD phase diagram that may be difficult to interpret. Theoretical input is important e.g. causality and stability constraints $0 \leq c_s^2 \leq 1$, first principle lattice or perturbative QCD (pQCD) calculations, and well-defined effective field theories (EFT) such as chiral EFT. Constraints are summarized in [1]. *Key questions:* How to interpret new Beam Energy Scan II data? How does QCD matter move at $n_B \neq 0$? How does [HIC]/QCD constrain the [NS] equation of state (EOS)? Can out-of-equilibrium effects shed light on the core of [NS]?

[HIC]/critical point The (potential) QCD critical point [2] separates the boundary between quarks/gluons and hadrons. While current first principle calculations cannot determine its existence/location [3], recent calculations in other theoretical frameworks [4–9] indicate a critical point at $\mu_B/T \sim 4 - 8$. Preliminary measurements exist for net-proton fluctuations κ_4/κ_1 across $\sqrt{s_{NN}}$ [10] appearing flat. Factorial cumulants show qualitative features predicted from critical point models [11]. Viscous effects may smear out or alter critical fluctuations [12] and the critical region can either cause critical lensing or minimize the effect [12–15]. Theorists are developing frameworks (initial state, fluid dynamics, hadron interactions) that dynamically study $n_B \neq 0$ (and strangeness, $n_S \neq 0$, and electric charge, $n_Q \neq 0$).

*e-mail: jnorhos@illinois.edu

© The Authors, published by EDP Sciences. This is an open access article distributed under the terms of the Creative Commons Attribution License 4.0 (https://creativecommons.org/licenses/by/4.0/).

The Taylor series EOS from lattice is only valid up to $\mu_B/T \sim [2-3]$ [3], which does not cover the entire EOS needed for Beam Energy Scan II. An alternative expansion was developed that improves the reach in μ_B/T [16] and is coupled to 3D Ising [17]. In hydro simulations, every fluid cell must have a valid EOS. In [18] a solution to a limited EOS was developed in 4-dimensions (T, μ_B, μ_S, μ_Q) to smoothly map to a conformal EOS for out-of-bound fluid cells. The MUSES collaboration provides open-source EOS across the phase diagram [19].

Initial conditions include baryon stopping through strings [20], a 3-fluid model [21], color glass condensate [22, 23], or a hadron transport [24]. It is also possible to study BSQ (baron, strangeness, electric charge) fluctuations where net-$n_B = 0$ through gluon splittings into $q\bar{q}$ pairs [25]. These initial states may be far-from-equilibrium and require a pre-equilibrium state incorporating BSQ charges [26, 27]. They are fed into relativistic viscous BSQ hydrodynamic codes, existing open-source codes are `CCAKE` [18], `MUSIC` [28], and `vHLLE` [29], each have different equations of motion, numerical algorithms, and EOS solvers. Stability [30] provides further constraints on BSQ hydro. Phase transitions [31] or critical fluctuations [32, 33] incorporation in relativistic hydrodynamics is underway but not yet implemented in dynamical frameworks. Hadronization also requires criticality [34]. For low $\sqrt{s_{NN}}$ the influence of the QGP may be so short lived such that it is negligible [35]. An alternative is to incorporate potentials within a hadron transport to simulation phase transitions [36].

[NS] Scientists can now extract posteriors of the [NS] EOS ($T = 0$, large n_B, at β-equilibrium). Different approaches are used: functional forms, physics based EOS (often multiple stitched together in different ranges of n_B), or a mixture. There is strong evidence that the speed of sound squared surpasses the conformal limit i.e. $c_s^2 > 1/3$ to support heavy [NS] [37] and may near the causal limit if ultra heavy [NS] exist [38, 39]. A bump in c_s^2 could indicate a cross-over into quarks [40] or a phase transition into hyperons [41]. Functional forms should capture features in c_s^2 that indicate phase transitions/new degrees of freedom, which is not possible with polytropes/spectral EOS [38, 42]. Non-parametric approaches can reproduce these features like Gaussian processes [43], which can also be explicitly included in EOS [44]. Strong phase transitions could be measured via mass twins [45] or the binary love relation [46]. We cannot calculate the EOS directly at large n_B. However, lattice QCD calculations are possible at $T = 0$ with isospin asymmetry, leading to $c_s^2 > 1/3$ [47]. pQCD calculations are relevant at $n_B \gtrsim 40\, n_{sat}$. Using causality and stability, one can obtain bounds that reach to smaller n_B to further constrain the EOS [48]. Bulk viscosity within [NS] mergers arises from weak interactions, affecting different stages of the merger [49], and is influenced by the degrees of freedom [50] and Y_Q [51]. In [52] the first constraints on bulk viscosity in the inspiral of [NS] mergers was extracted, making tests possible with further data.

Groups use low-energy [HIC] flow data to extract the low $T \geq 0$ EOS for $Y_Q \sim 0.5$. It is important to consider structure in c_s^2, as was done first in [53], indicating a large bump in c_s^2 around $[2, 3]n_{sat}$. It is possible to convert a given [NS] EOS into [HIC] using the symmetry energy expansion (and applying saturation and causality/stability constraints) [54], which indicated ultraheavy [NS] are consistent with [HIC]. Next is to expand cold EOS into finite T, which is possible using a T/μ_B expansion [55] or an effective mass (npe matter) [56].

3 Bulk Dynamics

The QGP is well described by a relativistic viscous fluid with small viscosity [57], leading to developments of relativistic *viscous* fluids. At $n_B = 0$ and large systems, the properties can be exacted from a Bayesian analysis [58–60]. However, the QGP medium may begin extremely far-from-equilibrium such that even causality and stability break down [61], which affects the extraction of bulk viscosity [62]. Collisions of light nuclei push the boundaries of fluid dynamics even further [63], spin hydrodynamics is relevant for Λ polarization results

© The Authors, published by EDP Sciences. This is an open access article distributed under the terms of the Creative Commons Attribution License 4.0 (https://creativecommons.org/licenses/by/4.0/).

[64, 65], and magnetohydrodynamics is needed to explore large magnetic fields [66]. *Key Questions:* Does [HIC] knowledge of relativistic viscous fluids affect other fields? Could hydrodynamics be relevant for the EIC? What role does spin and magnetic fields play?

Advances in fluid dynamics Predictions between $\sqrt{s_{NN}}$ [67, 68] were confirmed [69], so one could extract nuclear structure information from collisions of deformed ions [70]. One can extract the neutron skin from ^{208}Pb [71] from [HIC], which was more consistent with ab initio [72] than the PREXII data [73]. Collaborations with nuclear structure physicists have led to suggestions of new data to be collected to shed light on α clustering [74] or interesting deformations [75, 76]. Fundamental questions still remain for small systems and peripheral collisions. Correlations between v_2 and $\langle p_T \rangle$ are hard to reproduce [77]. In extremely small systems i.e. a vector meson-ion collisions (ultra peripheral collisions) there are hints of collective behavior [78], which has interesting implications for the Electron Ion Collider. These advances have led to cold atoms connections [79].

Spin and Magnetic Fields [HIC] have extremely large, short lived magnetic fields [80]. Groups are working on developing relativistic magnetohydrodyanmics (MHD) codes for [HIC]. Since ideal MHD is standard in astrophysics, at least two astro codes have been converted to [HIC] [81, 82]. [HIC] physicists and astrophysicists incorporated Israel-Stewart equations of motion into astro MHD codes [83]. Non-linear stability and causality techniques from [HIC] [84], have been extended to MHD coupled to general relativity for accretion disks around black holes [85]. Similar constraints are also available for spin/chiral relativistic hydrodynamics [86, 87]. Λ polarization can further constrain hydrodynamic parameters [88].

4 Hadron Physics

[HIC] cannot directly probe quarks and gluons, rather detectors measure charged hadrons. Understanding their properties provides insight into freeze-out, hadronization, and Brownian motion. *Key Questions:* How is strangeness produced? What causes isospin breaking? Can we understand heavy flavor across system size? Could charm quarks be thermalized?

Puzzles from hadrons The hadron resonance gas (HRG) model helped understand hadronization. HRG is used to extract freeze-out information from thermal fits [89] and fluctuations [90, 91] (c.f. in lattice QCD [92]), although tensions remain between light and strange hadrons [93]. Increasing the number of resonances might have solved the puzzle [94], but the tension remained even with new states [91, 95]. Either there are separate freeze-out temperatures [96] or an S-matrix approach is required [97]. Extensions to the HRG include magnetic fields [98] or surface tension [99]. Experiments found isospin symmetry breaking of kaons across $\sqrt{s_{NN}}$ [100]. One explanation is a disorientated isospin condensate [101].

Puzzles from heavy flavor Heavy flavor provides an interesting probe of the QGP that is normally modeled through a Langevin equation at low p_T, or an energy loss model at high p_T. Most dynamics are described using more weakly coupled approaches [102], but others have used charm as a conserved charge within the fluid [103, 104]. Coalescence is important at low p_T to correctly capture the hadronization process (see [105] for hadronic rescattering). While it was thought heavy flavor models lead to disparate results for the nuclear modification factors R_{AA}, much could be attributed to different medium effects since the same medium led to very similar results [106]. After the first event-by-event studies with heavy flavor [107] and jets [108], soft-heavy [109, 110] and soft-hard [111, 112] correlations advanced. These correlations can constrain the T dependence of diffusion [113]. ALICE experimental data showed that the heavy flavor elliptical flow, v_2, scales with the soft v_2 [110]. Charmonium flow may be more complicated due to its internal structure [114]. Jets' sensitivity to event-by-event fluctuations can be seen by experimental measurements of jet $v_3 \neq 0$ [115]. Groups are trying to unify the soft-heavy [116] and soft-hard [117] frameworks.

© The Authors, published by EDP Sciences. This is an open access article distributed under the terms of the Creative Commons Attribution License 4.0 (https://creativecommons.org/licenses/by/4.0/).

Small systems have perplexing behavior for heavy flavor. $R_{AA} \sim 1$, while pPb D meson v_2 is large [118] but charmonium [119] and bottomonium $v_2 \sim 0$ [120] and in pp D mesons $v_2 > 0$ and B mesons $v_2 < 0$ [121]. A heavy flavor system size scan was proposed [122] with multicharm hadron predictions in [123] . Corrections to the short path length may be required [124]. Heavy flavor requires longer formation times [125] due to their masses. The interplay between the early far-from-equilibrium stage and heavy flavor is intriguing, such as how $c\bar{c}$ and $b\bar{b}$ pairs disassociate in the glasma [126].

Acknowledgments J.N.H thanks SQM24 for providing child care and acknowledges support from DOE Grant No. DE-SC0023861, Saturated Glue (SURGE) Topical Theory Collaboration, and NSF Grant No. OAC-2103680 within the framework of MUSES collaboration.

References

[1] R. Kumar et al. (MUSES), Living Rev. Rel. **27**, 3 (2024), `2303.17021`.
[2] M.A. Stephanov, K. Rajagopal, E.V. Shuryak, Phys. Rev. Lett. **81**, 4816 (1998).
[3] D. Bollweg, et al. (HotQCD), Phys. Rev. D **105**, 074511 (2022).
[4] W.j. Fu, et al., Phys. Rev. D **101**, 054032 (2020).
[5] P.J. Gunkel, C.S. Fischer, Phys. Rev. D **104**, 054022 (2021), `2106.08356`.
[6] G. Basar, Phys. Rev. Lett. **127**, 171603 (2021), `2105.08080`.
[7] P. Dimopoulos, et al., Phys. Rev. D **105**, 034513 (2022).
[8] M. Hippert, et al (2023), `2309.00579`.
[9] A. Sorensen, P. Sorensen (2024), `2405.10278`.
[10] Y. ZHANG, This SQM2024 proceedings (2024)
[11] A. Bzdak, et al, Phys. Rept. **853**, 1 (2020).
[12] T. Dore, et al, Phys. Rev. D **106**, 094024 (2022).
[13] C. Nonaka, M. Asakawa, Phys. Rev. C **71**, 044904 (2005), `nucl-th/0410078`.
[14] M.S. Pradeep, N. Sogabe, M. Stephanov, H.U. Yee, Phys. Rev. C **109**, 064905 (2024).
[15] J.M. Karthein, V. Koch, C. Ratti (2024), `2409.13961`.
[16] S. Borsányi, et al, Phys. Rev. Lett. **126**, 232001 (2021).
[17] M. Kahangirwe, et al, Phys. Rev. D **109**, 094046 (2024).
[18] C. Plumberg et al. (2024), `2405.09648`.
[19] MUSES calculation enginge (2024), `https://musesframework.io/`
[20] C. Shen, B. Schenke, Phys. Rev. C **97**, 024907 (2018), `1710.00881`.
[21] J. Cimerman, et al, Phys. Rev. C **107**, 044902 (2023).
[22] O. Garcia-Montero, et al, Phys. Rev. C **109**, 044916 (2024).
[23] O. Garcia-Montero, S. Schlichting (2024), `2409.06788`.
[24] A. Schäfer, et al (SMASH), Eur. Phys. J. A **58**, 230 (2022).
[25] P. Carzon, et al, Phys. Rev. C **105**, 034908 (2022).
[26] P. Carzon, et al, Phys. Rev. C **108**, 064905 (2023).
[27] T. Dore, This SQM2024 proceedings (2024)
[28] A. Monnai, G. Pihan, B. Schenke, C. Shen (2024), `2406.11610`.
[29] I. Karpenko, et al, Comput. Phys. Commun. **185**, 3016 (2014).
[30] D. Almaalol, T. Dore, J. Noronha-Hostler (2022), `2209.11210`.
[31] J. Kapusta, M. Singh, T. Welle (2024), `2409.16525`
[32] J.M. Karthein, et al (2024), `2409.16249`
[33] X. An, G. Basar, M. Stephanov, H.U. Yee, Phys. Rev. C **108**, 034910 (2023).
[34] M.S. Pradeep, M. Stephanov, Phys. Rev. Lett. **130**, 162301 (2023), `2211.09142`.

© The Authors, published by EDP Sciences. This is an open access article distributed under the terms of the Creative Commons Attribution License 4.0 (https://creativecommons.org/licenses/by/4.0/).

[35] G. Inghirami, H. Elfner, Eur. Phys. J. C **82**, 796 (2022), 2201.05934.
[36] A. Sorensen, V. Koch, Phys. Rev. C **104**, 034904 (2021), 2011.06635.
[37] P. Bedaque, A.W. Steiner, Phys. Rev. Lett. **114**, 031103 (2015), 1408.5116.
[38] H. Tan, et al, Phys. Rev. Lett. **125**, 261104 (2020), 2006.16296.
[39] T. Kojo, AAPPS Bull. **31**, 11 (2021), 2011.10940.
[40] L. McLerran, S. Reddy, Phys. Rev. Lett. **122**, 122701 (2019), 1811.12503.
[41] N. Cruz-Camacho, et al (2024).
[42] H. Tan, et al, Phys. Rev. D **105**, 023018 (2022).
[43] R. Essick, et al, Phys. Rev. D **108**, 043013 (2023).
[44] D. Mroczek, M.C. Miller, J. Noronha-Hostler, N. Yunes (2023), 2309.02345.
[45] M.G. Alford, S. Han, M. Prakash, Phys. Rev. D **88**, 083013 (2013), 1302.4732.
[46] H. Tan, et al, Phys. Rev. Lett. **128**, 161101 (2022).
[47] B.B. Brandt, F. Cuteri, G. Endrodi, JHEP **07**, 055 (2023), 2212.14016.
[48] O. Komoltsev, A. Kurkela, Phys. Rev. Lett. **128**, 202701 (2022), 2111.05350.
[49] E.R. Most, et al, Mon. Not. Roy. Astron. Soc. **509**, 1096 (2021).
[50] M.G. Alford, A. Haber, Phys. Rev. C **103**, 045810 (2021), 2009.05181.
[51] Y. Yang, M. Hippert, E. Speranza, J. Noronha, Phys. Rev. C **109**, 015805 (2024).
[52] J.L. Ripley, A. Hegade K. R., R.S. Chandramouli, N. Yunes (2023), 2312.11659.
[53] D. Oliinychenko, et al, Phys. Rev. C **108**, 034908 (2023).
[54] N. Yao, et al, Phys. Rev. C **109**, 065803 (2024).
[55] D. Mroczek, et al (2024), 2404.01658.
[56] C.A. Raithel, F. Ozel, D. Psaltis, Astrophys. J. **875**, 12 (2019), 1902.10735.
[57] U. Heinz, R. Snellings, Ann. Rev. Nucl. Part. Sci. **63**, 123 (2013), 1301.2826.
[58] J.E. Bernhard, J.S. Moreland, S.A. Bass, Nature Phys. **15**, 1113 (2019).
[59] D. Everett et al. (JETSCAPE), Phys. Rev. C **103**, 054904 (2021), 2011.01430.
[60] G. Nijs, et al, Phys. Rev. C **103**, 054909 (2021).
[61] C. Plumberg, et al, Phys. Rev. C **105**, L061901 (2022).
[62] T.S. Domingues, et al (2024).
[63] C. Aidala et al. (PHENIX), Nature Phys. **15**, 214 (2019), 1805.02973.
[64] L. Adamczyk et al. (STAR), Nature **548**, 62 (2017), 1701.06657.
[65] W. Florkowski, B. Friman, A. Jaiswal, E. Speranza, Phys. Rev. C **97**, 041901 (2018).
[66] U. Gursoy, D. Kharzeev, K. Rajagopal, Phys. Rev. C **89**, 054905 (2014), 1401.3805.
[67] H. Niemi, K.J. Eskola, R. Paatelainen, K. Tuominen, Phys. Rev. C **93**, 014912 (2016).
[68] J. Noronha-Hostler, M. Luzum, J.Y. Ollitrault, Phys. Rev. C **93**, 034912 (2016).
[69] J. Adam et al. (ALICE), Phys. Rev. Lett. **116**, 132302 (2016), 1602.01119.
[70] B. Bally et al. (2022), 2209.11042.
[71] G. Giacalone, G. Nijs, W. van der Schee, Phys. Rev. Lett. **131**, 202302 (2023).
[72] B. Hu et al., Nature Phys. **18**, 1196 (2022), 2112.01125.
[73] D. Adhikari et al. (PREX), Phys. Rev. Lett. **126**, 172502 (2021), 2102.10767.
[74] N. Summerfield, et al, Phys. Rev. C **104**, L041901 (2021).
[75] G. Giacalone et al. (2024), 2402.05995.
[76] G. Giacalone et al. (2024), 2405.20210.
[77] ALICE, Ali-prel-491854
[78] W. Zhao, C. Shen, B. Schenke, Phys. Rev. Lett. **129**, 252302 (2022), 2203.06094.
[79] S. Floerchinger, et al, Phys. Rev. C **105**, 044908 (2022).
[80] G. Inghirami, et al, Eur. Phys. J. C **80**, 293 (2020).

© The Authors, published by EDP Sciences. This is an open access article distributed under the terms of the Creative Commons Attribution License 4.0 (https://creativecommons.org/licenses/by/4.0/).

[81] G. Inghirami, et al, Eur. Phys. J. C **76**, 659 (2016).
[82] M. Mayer, L. Rezzolla, H. Elfner, G. Inghirami, D.H. Rischke (2024), 2403.08669.
[83] E.R. Most, J. Noronha, Phys. Rev. D **104**, 103028 (2021), 2109.02796.
[84] F.S. Bemfica, et al, Phys. Rev. Lett. **126**, 222301 (2021).
[85] I. Cordeiro, et al, Phys. Rev. Lett. **133**, 091401 (2024).
[86] N. Abboud, E. Speranza, J. Noronha, Phys. Rev. D **109**, 094007 (2024), 2308.02928.
[87] A. Daher, et al, Phys. Rev. D **109**, 114001 (2024).
[88] A. Palermo, This SQM2024 proceedings (2024)
[89] A. Andronic, et al, Nature **561**, 321 (2018).
[90] R. Bellwied, et al, Phys. Rev. C **99**, 034912 (2019).
[91] P. Alba, et al, Phys. Rev. C **101**, 054905 (2020).
[92] S. Borsanyi, et al, Phys. Rev. Lett. **113**, 052301 (2014).
[93] M. Floris, Nucl. Phys. A **931**, 103 (2014), 1408.6403.
[94] A. Bazavov et al., Phys. Rev. Lett. **113**, 072001 (2014), 1404.6511.
[95] J.S. San Martin, et al (2023), 2309.01737.
[96] R. Bellwied, et al, Phys. Rev. Lett. **111**, 202302 (2013).
[97] A. Andronic, et al, Phys. Lett. B **792**, 304 (2019).
[98] V. Vovchenko, Phys. Rev. C **110**, 034914 (2024), 2405.16306.
[99] E. Zherebtsova, This SQM2024 proceedings (2024)
[100] W. Brylinski, et al (2023).
[101] J.I. Kapusta, S. Pratt, M. Singh, Phys. Rev. C **109**, L031902 (2024), 2306.13280.
[102] S. Bass, This SQM2024 proceedings (2024)
[103] F. Capellino, et al, Phys. Rev. D **106**, 034021 (2022).
[104] F. Capellino, et al, Phys. Rev. D **108**, 116011 (2023).
[105] R. Hirayama, This SQM2024 proceedings (2024)
[106] S. Cao et al., Phys. Rev. C **99**, 054907 (2019), 1809.07894.
[107] M. Nahrgang, et al, Phys. Rev. C **91**, 014904 (2015).
[108] J. Noronha-Hostler, et al, Phys. Rev. Lett. **116**, 252301 (2016).
[109] C.A.G. Prado, et al, Phys. Rev. C **96**, 064903 (2017).
[110] M.L. Sambataro, et al, Eur. Phys. J. C **82**, 833 (2022).
[111] L. Barreto, et al (2022), 2208.02061.
[112] A. Holtermann, et al, Phys. Rev. C **108**, 064901 (2023).
[113] M.L. Sambataro, V. Minissale, S. Plumari, V. Greco, Phys. Lett. B **849**, 138480 (2024).
[114] S. Cho, Phys. Rev. C **109**, 054904 (2024), 2307.14765.
[115] G. Aad et al. (ATLAS), Phys. Rev. C **105**, 064903 (2022), 2111.06606.
[116] J. Zhao, et al, Phys. Rev. C **110**, 024909 (2024).
[117] B. Karmakar, et al (2024), 2403.17817.
[118] A.M. Sirunyan et al. (CMS), Phys. Rev. Lett. **121**, 082301 (2018), 1804.09767.
[119] A.M. Sirunyan et al. (CMS), Phys. Lett. B **791**, 172 (2019), 1810.01473.
[120] A. Tumasyan et al. (CMS), Phys. Lett. B **850**, 138518 (2024), 2310.03233.
[121] G. Aad et al. (ATLAS), Phys. Rev. Lett. **132**, 202301 (2024), 2308.16652.
[122] R. Katz, et al, Phys. Rev. C **102**, 024906 (2020).
[123] V. Minissale, et al, Eur. Phys. J. C **84**, 228 (2024).
[124] C. Faraday, A. Grindrod, W.A. Horowitz, Eur. Phys. J. C **83**, 1060 (2023).
[125] G.D. Moore, D. Teaney, Phys. Rev. C **71**, 064904 (2005), hep-ph/0412346.
[126] Pooja, M.Y. Jamal, P.P. Bhaduri, M. Ruggieri, S.K. Das (2024), 2404.05315.

© The Authors, published by EDP Sciences. This is an open access article distributed under the terms of the Creative Commons Attribution License 4.0 (https://creativecommons.org/licenses/by/4.0/).

EPJ Web of Conferences **316**, 01002 (2025) https://doi.org/10.1051/epjconf/202531601002
SQM 2024

Experimental state-of-the-art before SQM 2024

Maximiliano Puccio[1,*]

[1]European Organization for Nuclear Research (CERN), Geneva, Switzerland

Abstract. Large conferences like SQM often drive rapid advancements in physics, and this talk was intended to recap the status of the main topics discussed during the event. In these proceedings, the content has been contextualized in light of the discussions and new results presented at SQM 2024.

1 Introduction

Among all the topics discussed in the Strangeness in Quark Matter conference series, this overview is focused on three main subjects: the (circumstantial?) evidence we have of thermalised QCD matter phenomena in small collision systems, the recent headline studies on the properties of the quark gluon plasma (QGP) and of the QCD phase diagram, and the recent advancement on how we can use the "heavy-ion measurement toolkit" to study hadronic physics.

2 Thermalised QCD matter in small systems?

Since the discovery of the quark-gluon plasma in heavy-ions collisions, the question of how small a droplet of QGP can be produced has become a major subject of investigation. This has led to detailed studies of smaller collision systems like proton-proton (pp) and proton-nucleus (p-A) collisions (see [1] for a review). Recently, the field has taken an even further step: investigating whether some fundamental characteristics of a thermalized QGP might already exist in elementary collisions, such as photonuclear collisions, electron-positron (e^-e^+) collisions, or in pp collisions with the lowest multiplicities.

For example, recent results from the ATLAS Collaboration [2] shown in the left panel of Figure 1 indicate that the v_2 of charged particles in high-multiplicity γ-Pb collisions is comparable to that measured in high-multiplicity pp collisions. While this finding is surprising, one could argue that collective behavior naturally emerges in any system when the initial energy transfer is sufficient to produce a large number of charged particles.

However, two recent analyses of the charged particle yields in the "ridge" region of the two-particle angular correlation measurements in pp and e^+e^- collisions present a different perspective, as shown in the right panel of Figure 1. In pp collisions, a significant associated yield is consistently observed, regardless of the multiplicity [3]. In contrast, e^+e^- collisions do not exhibit any significant associated yield [4].

These two results are quite impactful, as they suggest that the presence of a reservoir of color charges – such as a hadron involved in the collision – is necessary to develop signs of collectivity in the form of collective flow.

*e-mail: maximiliano.puccio@cern.ch

© The Authors, published by EDP Sciences. This is an open access article distributed under the terms of the Creative Commons Attribution License 4.0 (https://creativecommons.org/licenses/by/4.0/).

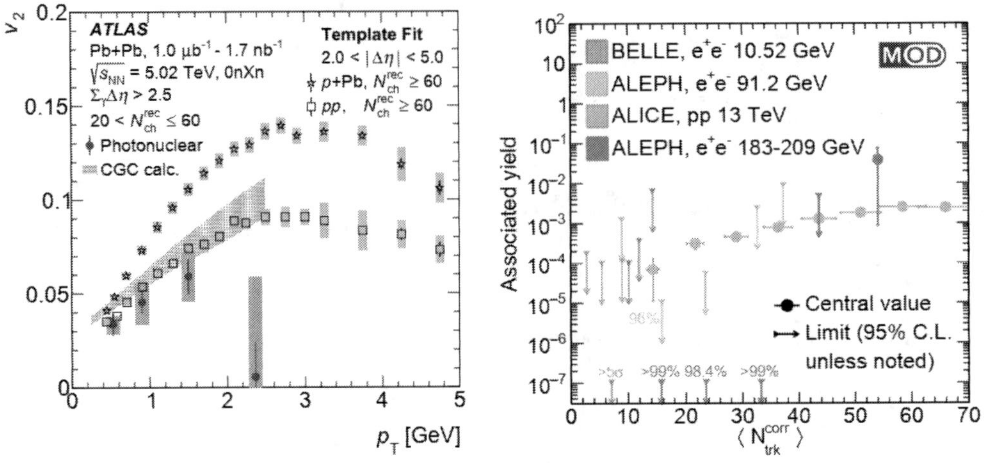

Figure 1. On the left: v_2 for charged particles in γ–Pb (red), pp (black squares), and p–Pb (black stars) collisions measured by the ATLAS Collaboration. The color glass condensate expectation are shown with a green band. Figure from [2]. On the right: charged particles yields associated to the near side peak of two-particle correlations in electron-positron [4] (coloured points) and pp [3] (grey) collisions.

The presence of collective flow is only one piece of evidence pointing to the possible existence of a thermalized, expanding QCD matter droplet in hadronic collisions. If thermalization is achieved, then the particle yields should also distribute according to the predictions of the Statistical Hadronization Model (SHM). It has been observed that the canonical formulation of the SHM describes the yields of light-flavored particles—up to multi-strange baryons—even in low-multiplicity pp collisions within 10–15% accuracy, albeit with a strangeness saturation parameter lower than unity. At the same time, recent versions of PYTHIA that include new hadronization modes with color ropes have achieved similar success in reproducing light-flavor particle yields in small collision systems.

Surprisingly, the SHM predictions are also consistent with the yield ratios of charmed hadrons measured by the ALICE Collaboration [5], as shown in the left panel of Figure 2. Considering that the first measurement of the interaction between D mesons and charged pions – performed by the ALICE Collaboration using femtoscopic techniques [6] – reveals scattering parameters compatible with zero (Figure 2, right panel), and given that any hadronic phase in pp collisions must be extremely short-lived, this suggests that if charm hadrons exhibit thermal yields, these are set at the time of hadronization rather than through subsequent interactions. However, also in the case of the charmed hadrons, modern versions of PYTHIA with new color reconnection schemes are able to describe the measured particle ratios [5].

Measured average particle yields alone are insufficient to distinguish between the string fragmentation and the SHM hadronization scenarios. To differentiate these two distinct hadronization mechanisms, experimental Collaborations have begun measuring various multi-differential observables. Among the most promising ones are the measurements of higher-order moments of the distribution of the number of produced hadrons. Recently, the ALICE Collaboration presented the first measurement of the second-order moment of number of produced multistrange baryon in pp, p–Pb, and Pb–Pb collisions [7]. The second-order moment of the net-Ξ number distribution, normalized by the average yield of Ξ baryons, effectively distinguishes between the PYTHIA and SHM predictions, with the experimental data favoring the SHM scenario across all collision systems, as shown in the left panel of

© The Authors, published by EDP Sciences. This is an open access article distributed under the terms of the Creative Commons Attribution License 4.0 (https://creativecommons.org/licenses/by/4.0/).

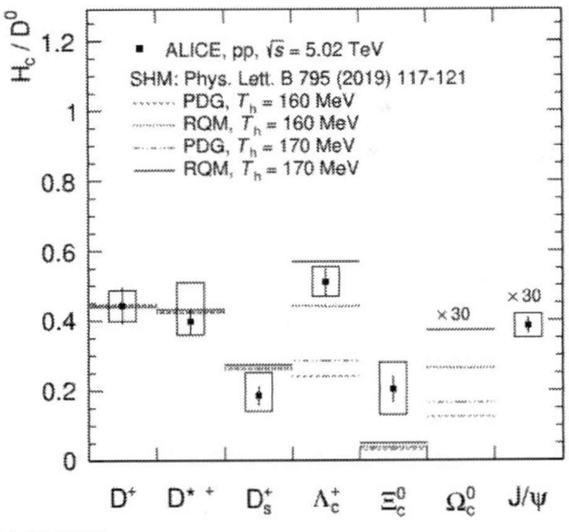

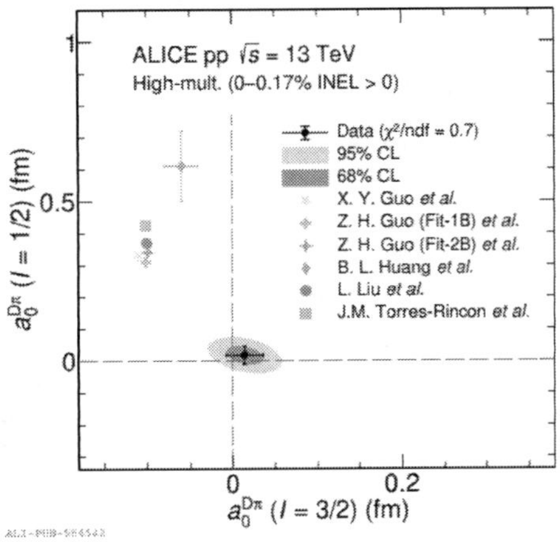

Figure 2. On the left: ratio of the production yield of charm hadrons over neutral D meson yield in pp collisions at $\sqrt{s}$ = 5 TeV. Different implementations of the SHM are shown as colored lines. Figure taken from [5]. On the right: different isospin components of the D-meson pion scattering parameter measured by the ALICE Collaboration [6] and compared to different theoretical calculations. Figure taken from [6].

Figure 3. Similarly, a study by the STAR Collaboration [8] on the higher-order cumulants of the net-proton number distribution in pp collisions at $\sqrt{s}$ = 200 GeV comes to the same conclusion. Additionally, measuring cumulants of the fourth order (C_4) and higher allowed the STAR Collaboration to directly compare their results with Lattice QCD predictions, which are expected to differ from those of the Statistical Hadronization Model (SHM) for cumulants of order higher than the fourth. However, such a comparison with LQCD is complicated by the fact that LQCD only provides predictions for the net-baryon number, whereas experiments measure only the net-proton number. Nonetheless, in both measurements at the LHC and RHIC, continuous trends as a function of multiplicity are observed, and the thermalized system baseline provided by the SHM better describes the experimental data.

3 Exploring the QCD phase diagram

While event-by-event fluctuations of the net-proton number are currently being used to understand if a thermalized QCD medium is created in pp collisions and what is the hadronisation mechanism, over the past decade they have been considered the "holy grail" observable to pinpoint the position of the QCD critical point. A couple of years ago, the STAR Collaboration released results of measurements of higher-order cumulants of the net-proton number as a function of the Au–Au collision energy [9]. These results are shown in the left panel of Figure 4. All theoretical models that do not include a critical point predict a smooth evolution of the normalized fourth-order cumulant of the net-proton number with collision energy. During the conference, an updated measurement was shown, revealing a rather intriguing difference between the C_4/C_2 measurements in central and peripheral collisions in the energy range around $\sqrt{s_{NN}}$ = 20 GeV. How to model the change of baryon number transport at midrapidity and its influence on the fluctuations of the measured number of baryons as a function of collision energy – and how this translates to the reliable definition of a non-critical baseline – remain open points of discussion in this long-awaited search.

© The Authors, published by EDP Sciences. This is an open access article distributed under the terms of the Creative Commons Attribution License 4.0 (https://creativecommons.org/licenses/by/4.0/).

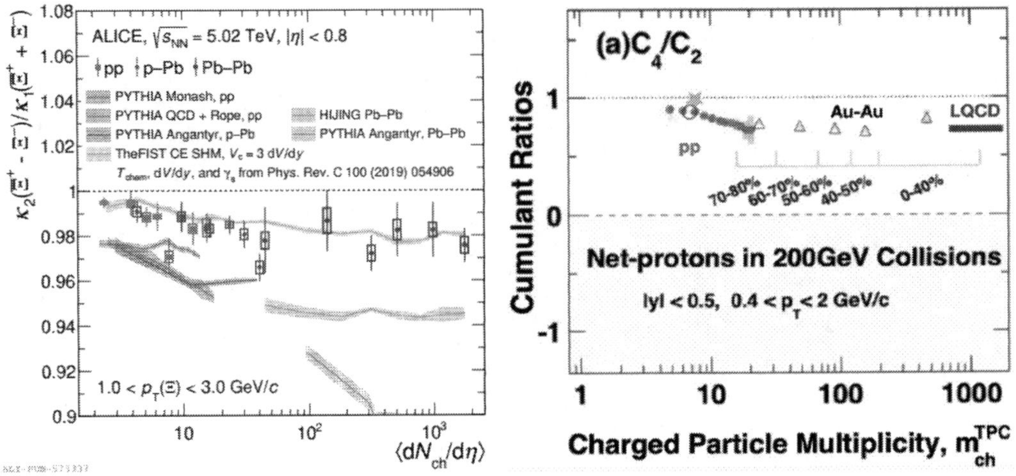

Figure 3. On the left: normalised second-order cumulant of the net-Ξ number as a function of the charged particle multiplicity in pp, p–Pb, and Pb–Pb collisions at the LHC. Figure taken from [7]. On the right: normalised fourth-order cumulant of the net-proton number as a function of charged particle multiplicity in pp and Au–Au collisions at RHIC. Figure adapted from [8].

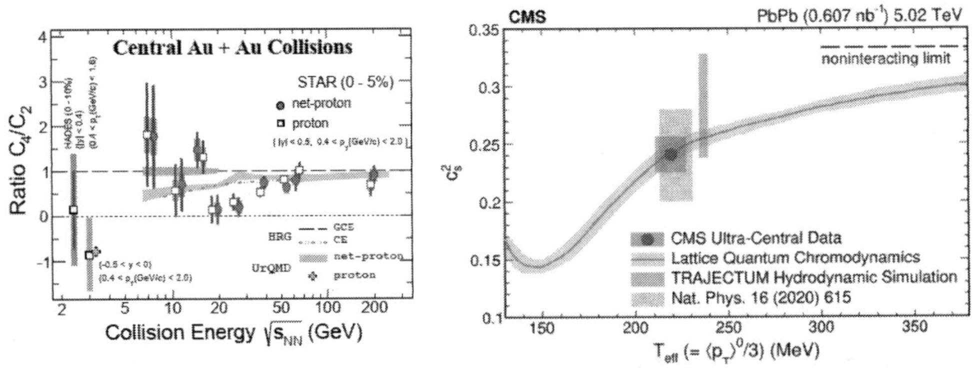

Figure 4. On the left: normalised fourth-order cumulant of the net-proton number as a function of the Au–Au collision energy at RHIC. Figure taken from [9]. On the right: speed of sound as a function of the effective temperature of the expanding QGP. The red point represent the measurement from the CMS Collaboration, while the different coloured bands are model predictions. Figure taken from [10].

While the search for the critical point is the main focus at RHIC, the characterization of the properties of the quark-gluon plasma (QGP) is the primary focus of research in heavy-ion collisions at the LHC. Recently, inspired by a seminal paper [11] on how to relate measured bulk properties of heavy-ion collisions to fundamental properties of the QGP, the CMS Collaboration released the first measurement of the speed of sound in the QGP. This measurement focuses on ultra-central collisions, where fluctuations in the number of partonic interactions lead to an increase in the system's energy density. In such collisions, the average transverse momentum ($\langle p_T \rangle$) of the produced particles and the number of produced charged hadrons (N_{ch}) act as proxies for the temperature and the entropy density, respectively. By utilizing these two proxies, the speed of sound in the QGP has been extracted from the fit to the distribution of $\langle p_T \rangle$ as a function of N_{ch}. The resulting value, which is close to half

© The Authors, published by EDP Sciences. This is an open access article distributed under the terms of the Creative Commons Attribution License 4.0 (https://creativecommons.org/licenses/by/4.0/).

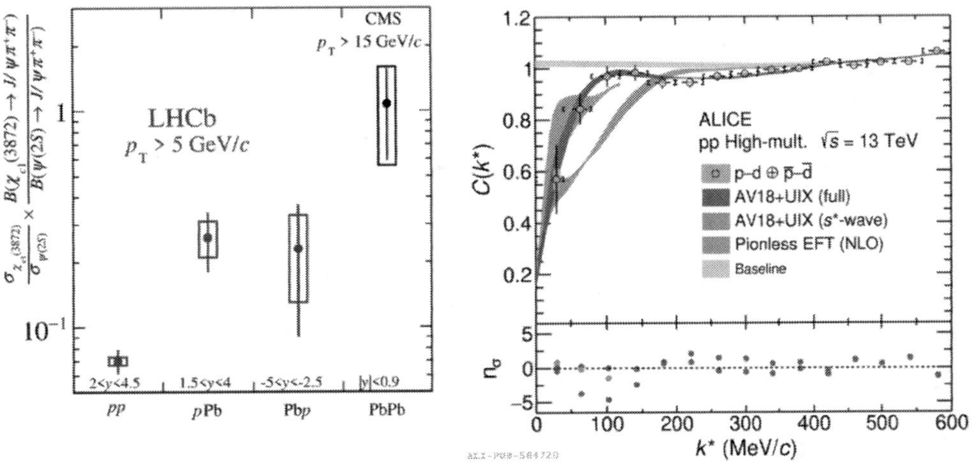

Figure 5. On the left: production cross section of the $\chi_{c1}(3872)$ normalised on the production cross section of the $\psi(2S)$ as a function of system energy. Figure taken from [12]. On the right: femtoscopic correlation function between proton and deuterons. The red-colored band represent the fit with a three-body system calculation. Figure taken from [13].

the speed of light, along with comparisons to theoretical models, is shown in the right panel of Figure 4. During the conference, the ALICE Collaboration also presented a systematic study on the extraction of this quantity, highlighting how the measurement can be influenced by different definitions of centrality estimators and their correlations with the measured $\langle p_T \rangle$ and N_{ch}. However, even in light of this study, no major issues have been found in the CMS measurement.

4 Hadronic structure and interactions

This last section could deserve dedicated proceedings due to the incredible expansion in the number and quality of studies on hadronic structure and hadronic interactions using heavy-ion and heavy-ion-inspired techniques.

In recent years, the study of the production of nuclei and hypernuclei has made headlines, showcasing how we can use their production cross sections in hadronic collisions to study their structure. This was particularly striking when the large wave function of the hypertriton was revealed by measuring its production in small collision systems [14]. Drawing an analogy to nuclear bound states, it was argued almost ten years ago [15] that one could study the structure of the $\chi_{c1}(3872)$ by examining its production in different colliding systems. In the context of these efforts, the LHCb Collaboration recently published the first measurement of the $\chi_{c1}(3872)$ in p–Pb collisions, as shown in the left panel of Figure 5. This measurement indicates that the production of $\chi_{c1}(3872)$ in p–Pb collisions relative to that of the $\psi(2S)$ is more abundant than in pp collisions, following the trend already established by the first measurement of the $\chi_{c1}(3872)$ in Pb–Pb collisions by the CMS Collaboration [16]. Such an enhancement is very similar to that observed for light nuclei; however, it is not easy to reconcile with the measurement of the $\chi_{c1}(3872)/\psi(2S)$ ratio as a function of charged particle multiplicity in pp collisions [17], which is instead decreasing. Future measurements and theoretical developments will be needed to clarify this discrepancy.

© The Authors, published by EDP Sciences. This is an open access article distributed under the terms of the Creative Commons Attribution License 4.0 (https://creativecommons.org/licenses/by/4.0/).

However, while the jury is still out in the case of the $\chi_{c1}(3872)$, the CMS Collaboration has established that the $f_0(980)$ is an ordinary meson by measuring its flow in p–Pb collisions and comparing it to that of standard mesons and baryons [18].

Finally, in the study of hadronic interactions, femtoscopic correlations are paving the way to investigate increasingly complex particle systems. For instance, the recent measurement of the correlation between protons and deuterons by the ALICE Collaboration in pp collisions [13] shows that this two-particle correlation cannot be simply explained in terms of a two-body system. Since pp collisions are small systems, the proton interacts with the components of the deuteron, creating a genuine three-body system whose modeling is required to explain the proton-deuteron correlation function, as shown in the right panel of Figure 5. This opens up the possibility of inferring properties of multi-body systems by simply measure two-particle correlations of their constituents. Studying such correlations in large collision systems can be also advantageous, as some interactions may be suppressed, making modeling easier. A first attempt in this direction was made by the STAR Collaboration, which determined the Λ–deuteron scattering parameters by measuring the corresponding two-particle femtoscopic correlation function, as discussed at this conference.

5 Summary: physics at the extremes

In heavy-ion physics, the most extreme hadron collisions ever created by human means are studied. However, we are now pushing further by studying both the extremely high multiplicity region in heavy-ion collisions and the extremely low multiplicity region in pp collisions. Not only that, but we are also pushing observables to their limits. While experimental biases and misleading assumptions can undermine such extreme studies, our field demonstrates a stunning resilience to overcome them and, as demonstrated at this conference, to make lasting contributions to the understanding of QCD matter.

References

[1] M. Arslandok et al., Hot QCD White Paper (2023), 2303.17254.

[2] G. Aad et al. (ATLAS), Two-particle azimuthal correlations in photonuclear ultraperipheral Pb+Pb collisions at 5.02 TeV with ATLAS, Phys. Rev. C **104**, 014903 (2021), 2101.10771. 10.1103/PhysRevC.104.014903

[3] S. Acharya et al. (ALICE), Emergence of Long-Range Angular Correlations in Low-Multiplicity Proton-Proton Collisions, Phys. Rev. Lett. **132**, 172302 (2024), 2311.14357. 10.1103/PhysRevLett.132.172302

[4] Y.C. Chen et al., Long-range near-side correlation in e+e− collisions at 183-209 GeV with ALEPH archived data, Phys. Lett. B **856**, 138957 (2024), 2312.05084. 10.1016/j.physletb.2024.138957

[5] S. Acharya et al. (ALICE), Charm-quark fragmentation fractions and production cross section at midrapidity in pp collisions at the LHC, Phys. Rev. D **105**, L011103 (2022), 2105.06335. 10.1103/PhysRevD.105.L011103

[6] S. Acharya et al. (ALICE), Studying the interaction between charm and light-flavor mesons, Phys. Rev. D **110**, 032004 (2024), 2401.13541. 10.1103/PhysRevD.110.032004

[7] S. Acharya et al. (ALICE), Probing strangeness hadronization with event-by-event production of multistrange hadrons (2024), 2405.19890.

[8] M. Abdulhamid et al. (STAR), Measurements of charged-particle multiplicity dependence of higher-order net-proton cumulants in pp collisions at $\sqrt{s}$=200

© The Authors, published by EDP Sciences. This is an open access article distributed under the terms of the Creative Commons Attribution License 4.0 (https://creativecommons.org/licenses/by/4.0/).

GeV from STAR at RHIC, Phys. Lett. B **857**, 138966 (2024), 2311.00934. 10.1016/j.physletb.2024.138966

[9] M.S. Abdallah et al. (STAR), Measurements of Proton High Order Cumulants in $\sqrt{s_{NN}}$ = 3 GeV Au+Au Collisions and Implications for the QCD Critical Point, Phys. Rev. Lett. **128**, 202303 (2022), 2112.00240. 10.1103/PhysRevLett.128.202303

[10] A. Hayrapetyan et al. (CMS), Extracting the speed of sound in quark–gluon plasma with ultrarelativistic lead–lead collisions at the LHC, Rept. Prog. Phys. **87**, 077801 (2024), 2401.06896. 10.1088/1361-6633/ad4b9b

[11] F.G. Gardim, G. Giacalone, M. Luzum, J.Y. Ollitrault, Thermodynamics of hot strong-interaction matter from ultrarelativistic nuclear collisions, Nature Phys. **16**, 615 (2020), 1908.09728. 10.1038/s41567-020-0846-4

[12] R. Aaij et al. (LHCb), Modification of χc1(3872) and ψ(2S) Production in pPb Collisions at $\sqrt{s_{NN}}$ =8.16 TeV, Phys. Rev. Lett. **132**, 242301 (2024), 2402.14975. 10.1103/PhysRevLett.132.242301

[13] S. Acharya et al. (ALICE), Exploring the Strong Interaction of Three-Body Systems at the LHC, Phys. Rev. X **14**, 031051 (2024), 2308.16120. 10.1103/PhysRevX.14.031051

[14] S. Acharya et al. (A Large Ion Collider Experiment, ALICE), Hypertriton Production in p-Pb Collisions at $\sqrt{s_{NN}}$=5.02 TeV, Phys. Rev. Lett. **128**, 252003 (2022), 2107.10627. 10.1103/PhysRevLett.128.252003

[15] A. Esposito, A.L. Guerrieri, L. Maiani, F. Piccinini, A. Pilloni, A.D. Polosa, V. Riquer, Observation of light nuclei at ALICE and the X(3872) conundrum, Phys. Rev. D **92**, 034028 (2015), 1508.00295. 10.1103/PhysRevD.92.034028

[16] A.M. Sirunyan et al. (CMS), Evidence for X(3872) in Pb-Pb Collisions and Studies of its Prompt Production at $\sqrt{s_{NN}}$=5.02 TeV, Phys. Rev. Lett. **128**, 032001 (2022), 2102.13048. 10.1103/PhysRevLett.128.032001

[17] R. Aaij et al. (LHCb), Observation of Multiplicity Dependent Prompt χ_{c1}(3872) and $\psi(2S)$ Production in pp Collisions, Phys. Rev. Lett. **126**, 092001 (2021), 2009.06619. 10.1103/PhysRevLett.126.092001

[18] A. Hayrapetyan et al. (CMS), Elliptic anisotropy measurement of the f_0(980) hadron in proton-lead collisions and evidence for its quark-antiquark composition (2023), 2312.17092.

© The Authors, published by EDP Sciences. This is an open access article distributed under the terms of the Creative Commons Attribution License 4.0 (https://creativecommons.org/licenses/by/4.0/).

Recent results on heavy flavours and quarkonia from ALICE

Fiorella Fionda[1,*] on behalf of the ALICE Collaboration

[1]University and INFN, Cagliari (Italy)

Abstract. Heavy-flavour hadrons, containing at least one charm or beauty quark, are excellent probes of the deconfined medium created in ultra-relativistic heavy-ion collisions, known as quark–gluon plasma. Results in smaller collision systems, such as proton–proton and p–Pb collisions, besides representing an important baseline for interpreting heavy-ion measurements, are crucial to test perturbative QCD calculations and hadronisation mechanisms in the absence of hot medium effects, as well as to search for commonalities with heavy-ion systems. Recently, measurements in proton–proton and p–Pb collisions have revealed unforeseen features with respect to the expectations based on previous results from e^+e^- and ep collisions, showing that fragmentation fractions of heavy quarks are not universal. In this contribution, an overview of the most recent ALICE heavy-flavour measurements, along with the comparison to available calculations, will be discussed.

1 Introduction

Quarkonia, bound states of a $c\bar{c}$ or a $b\bar{b}$ pair, and open heavy-flavour (HF) hadrons, which have non-zero charm or beauty quantum numbers, are effective probes of several aspects of quantum chromodynamics (QCD). Comprehensive measurements of quarkonia and open HF hadrons in Pb–Pb collisions allow for testing the mechanisms of heavy-quark transport, energy loss, and coalescence effects during hadronisation in the presence of the quark–gluon plasma (QGP). In order to correctly interpret results in Pb–Pb collisions, measurements in proton–proton (pp) and p–Pb collisions are fundamental baselines. In particular, pp collisions provide details on the vacuum production, and studies in p–Pb collisions are useful to examine the so-called cold nuclear matter effects. Besides serving as reference, HF measurements in hadronic collisions provide a stringent test of QCD since the large quark masses imply high momentum transfer processes that make perturbative QCD (pQCD) calculations applicable to a large extent. According to the factorisation approach [1], the production cross section of HF hadrons in hadronic collisions can be described as the convolution of three terms: the parton distribution functions (PDFs) of the colliding nucleons; the partonic cross-section responsible for the production of the heavy-quark pair, which can be computed using pQCD calculations; and the non-perturbative evolution into the bound states. Concerning the latter contribution, in the case of quarkonia different semi-phenomenological approaches, such as those implemented in the Color Evaporation Model (CEM) and Non-Relativistic QCD (NRQCD), are employed (see Ref. [2] for a comprehensive review). For open HF hadrons, fragmentation functions, which describe the fraction of the heavy-quark momentum carried

*e-mail: fiorella.fionda@cern.ch

© The Authors, published by EDP Sciences. This is an open access article distributed under the terms of the Creative Commons Attribution License 4.0 (https://creativecommons.org/licenses/by/4.0/).

by the HF hadron, are considered. In the factorisation approach, the hadronisation process is taken to be independent of the collision system, and this assumption implies identical fragmentation fractions and fragmentation functions for e^+e^-, ep and pp collisions. However, in the last decade a large wealth of results related to HF baryon production have disproved the assumed universality of the fragmentation fractions. In particular, an enhancement was observed for the first time by the ALICE Collaboration at midrapidity in pp collisions for the prompt Λ_c^+/D^0 production yield ratio compared to similar results in leptonic collisions [3], and this is even more pronounced for charm-strange baryon states production relative to D^0 mesons [4]. In the beauty sector, the measurement of the Λ_b^0 baryon production relative to that of beauty mesons at forward rapidity from the LHCb Collaboration [5] shows an enhancement of the corresponding fragmentation fraction in pp collisions similar to that observed for charm quarks at midrapidity. As a consequence, state-of-the-art calculations employing a model of quark fragmentation constrained by leptonic measurements, such as those considered in PYTHIA 8 with the default tune (Monash) [6], are challenged. Several models attempt to reproduce the observed baryon enhancement assuming either modified hadronisation in a parton-rich environment with respect to in-vacuum fragmentation via different mechanisms, such as colour reconnection beyond the leading-colour approximation (CR-BLC) [7] in PYTHIA 8 and coalescence [8, 9], or enhanced baryon production originating from the decay of unobserved charm-baryon resonant states [10].
The ALICE detector has excellent capabilities for reconstructing a large variety of HF hadrons in a wide rapidity window and down to very low transverse momentum (p_T), exploiting both inclusive and exclusive decay channels. It went through a major upgrade during Long Shutdown 2 (2019-2021) [11] of the LHC and restarted data taking at the beginning of LHC Run 3 in 2022. The main upgraded detector subsystems in the central barrel are the Time Projection Chamber (TPC) and the Inner Tracking System (ITS). In particular, the upgraded readout system of the TPC allows it to cope with a continuous readout mode in Run 3, while the upgraded ITS provides an approximately 3 (6) times improvement in pointing resolution in the transverse (longitudinal) direction. In addition, the Muon Forward Tracker was installed at forward rapidity, enabling the reconstruction of secondary vertices from charm- and beauty-hadron decays in the acceptance of the muon spectrometer.

2 Highlights from pp collisions

Figure 1 shows the new preliminary measurement of the inclusive J/ψ cross section reconstructed at midrapidity in pp collisions at $\sqrt{s}$ = 13.6 TeV. In the left panel, the results are compared with similar midrapidity Run 2 measurements from ALICE and ATLAS (see references on the left panel), the latter available for $p_T > 7$ GeV/c. The comparison indicates a good agreement between Run 2 and Run 3 results, as well as a significantly improved granularity with respect to previous ALICE measurements. The cross section is described within uncertainties by NRQCD-based models [12–14] and an improved version of the CEM [15], both coupled to FONLL [16] to account for the contribution of J/ψ originating from beauty hadron decays, as shown in the right panel of Fig. 1. The large uncertainties of the model calculations do not allow for discriminating among different models, however these measurements represent an important input for tuning such models, especially in the low-p_T region.
Preliminary measurements in the open HF sector based on Run 3 data are also carried out. The prompt D_s^+/D^+ production yield ratio at midrapidity in pp collisions at $\sqrt{s}$ = 13.6 TeV is depicted in Fig. 2. The comparison to previous Run 2 results in the left panel clearly shows that preliminary Run 3 measurements are compatible with published results at lower centre-of-mass energies. Furthermore, they have significantly increased granularity and are extended down to a lower p_T (0.5 GeV/c) compared to previous measurements. In the right

© The Authors, published by EDP Sciences. This is an open access article distributed under the terms of the Creative Commons Attribution License 4.0 (https://creativecommons.org/licenses/by/4.0/).

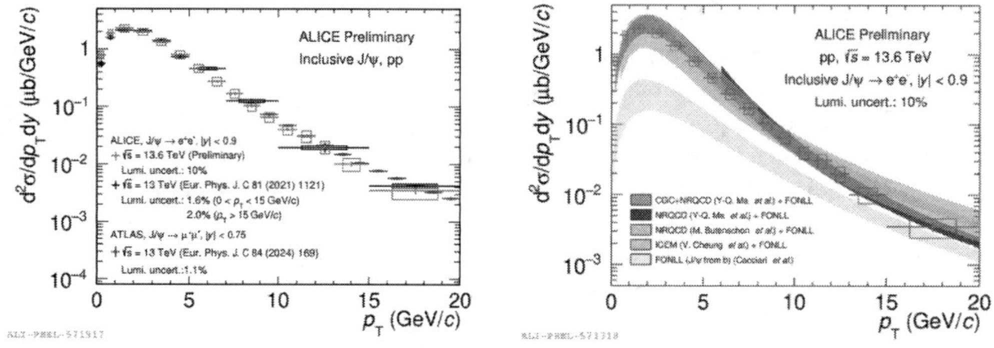

Figure 1. Inclusive J/ψ cross section measured at midrapidity in pp collisions at $\sqrt{s}$ = 13.6 TeV compared to similar measurements from ALICE and ATLAS at $\sqrt{s}$ = 13 TeV in the left panel (see references on the plot) and with several charmonium production models [12–15] coupled to FONLL [16] in the right panel.

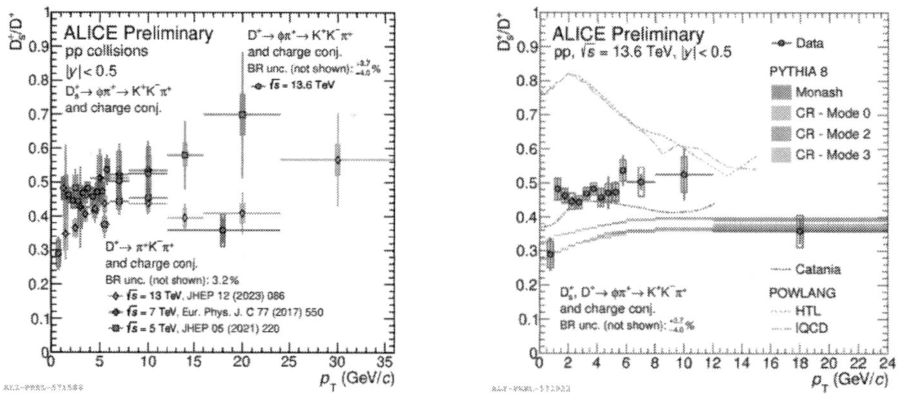

Figure 2. Prompt D_s^+/D^+ production yield ratio measured at midrapidity in pp collisions at $\sqrt{s}$ = 13.6 TeV compared to similar measurements at lower centre-of-mass energies from Run 2 in the left panel (see references on the plot) and with model calculations [6–9] in the right panel.

panel of Fig. 2, the prompt D_s^+/D^+ production yield ratio is compared to model calculations. PYTHIA 8 calculations with Monash [6] and CR-BLC [7] tunes, the latter introduced to improve the description for the baryons, underpredict the data, while POWLANG [9], which assumes an expanding fireball in pp collisions and hadronisation occurring via coalescence, overshoots the measured ratio. The Catania [8] model, which considers both coalescence and fragmentation for the charm quark hadronisation, provides a description closer to the data.
Moving to the HF baryon sector, the ALICE Collaboration measured at midrapidity, for the first time at the LHC, the relative production of the $\Sigma_c^{0,++}(2520)$ baryon to the ground state $\Sigma_c^{0,++}(2455)$. In the left panel of Fig. 3, the ratio is compared to several model calculations [6, 7, 10, 17] and to similar measurements in e^+e^- collisions at $\sqrt{s}$ = 10.52 GeV from the Belle Collaboration (see reference on the figure). No evidence of an enhancement with respect to e^+e^- collisions is observed considering the current uncertainties. Models struggle to reproduce the data; in particular, PYTHIA 8 with the Monash tune overestimates the data, while the CR-BLC tune, which describes the prompt Λ_c^+/D^0 production yield ratio

© The Authors, published by EDP Sciences. This is an open access article distributed under the terms of the Creative Commons Attribution License 4.0 (https://creativecommons.org/licenses/by/4.0/).

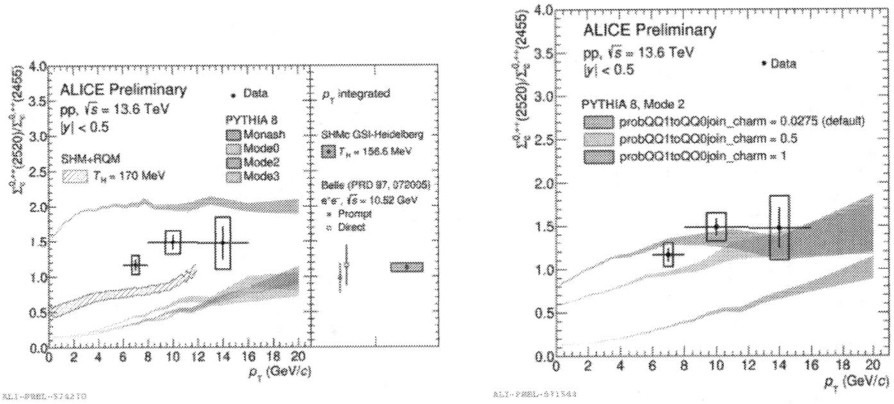

Figure 3. Relative production of the $\Sigma_c^{0,++}(2520)$ baryon to the ground state $\Sigma_c^{0,++}(2455)$ measured at midrapidity in pp collisions at $\sqrt{s}$ = 13.6 TeV. In the left panel, it is compared with model calculations [6, 7, 10, 17] and to lower energy results from Belle Collaboration (see reference on the plot). In the right panel, the ratio is compared with PYTHIA 8 CR-BLC calculations [7] after tuning one parameter of the model (see text for details).

fairly well, significantly underestimates the $\Sigma_c^{0,++}(2520)/\Sigma_c^{0,++}(2455)$ yield ratio. The statistical hadronisation model (SHMc) [17], which assumes that the abundances of all particles, including open and hidden charm hadrons, are determined by thermal weights at the chemical freeze-out, provides the p_T-integrated $\Sigma_c^{0,++}(2520)/\Sigma_c^{0,++}(2455)$ ratio which is consistent with the p_T-differential measurement from ALICE. The statistical hadronisation model coupled with the Relativistic Quark Model (SHM+RQM) [10], for which an augmented baryon production is obtained from the significant feed-down from excited charm baryon states beyond those listed in the PDG, predicts lower values compared to the measurement. An interesting example of how new measurements can be used to provide constraints for model tuning is shown in the right panel of Fig. 3. In particular, the $\Sigma_c^{0,++}(2520)/\Sigma_c^{0,++}(2455)$ production yield ratio can be reproduced by PYTHIA 8 calculations with the CR-BLC tune after the parameter that suppresses spin-one with respect to spin-zero charm-light diquark systems is tuned on the measured $\Lambda_c^+(\leftarrow \Sigma_c^{0,++})/\Lambda_c^+$ production yield ratio (see details in Ref. [18]).

3 Highlights from p–Pb and Pb–Pb collisions

In the left panel of Fig. 4, the prompt Ξ_c^0 p_T-differential cross section in p–Pb collisions at $\sqrt{s_{NN}}$ = 5.02 TeV [19] is compared with POWHEG+PYTHIA 6 simulations coupled to the EPPS16 nPDFs (see references in [19]). These predictions, which incorporate fragmentation fractions from e^+e^- collisions, significantly underestimate the prompt Ξ_c^0 production. The agreement improves with the predictions from Quark Combination Model (QCM), which assumes the coalescence of charm- and light-flavour quarks close in the momentum space, without considering the coordinate space. However, also in this case the prompt Ξ_c^0 cross section remains underestimated by a factor of 2. The measurement of the Ξ_c^0 production combined with the others already available for charm mesons and baryons in p–Pb collisions at $\sqrt{s_{NN}}$ = 5.02 TeV, allowed for measuring fragmentation fractions of charm hadrons in this collision system. They are shown in the right panel of Fig. 4 where they are compared with results from pp collisions at $\sqrt{s}$ = 5.02 TeV as well as results from e^+e^- and ep collisions at lower $\sqrt{s}$ (see [20] and references therein). The fragmentation fractions in p–Pb collisions are

© The Authors, published by EDP Sciences. This is an open access article distributed under the terms of the Creative Commons Attribution License 4.0 (https://creativecommons.org/licenses/by/4.0/).

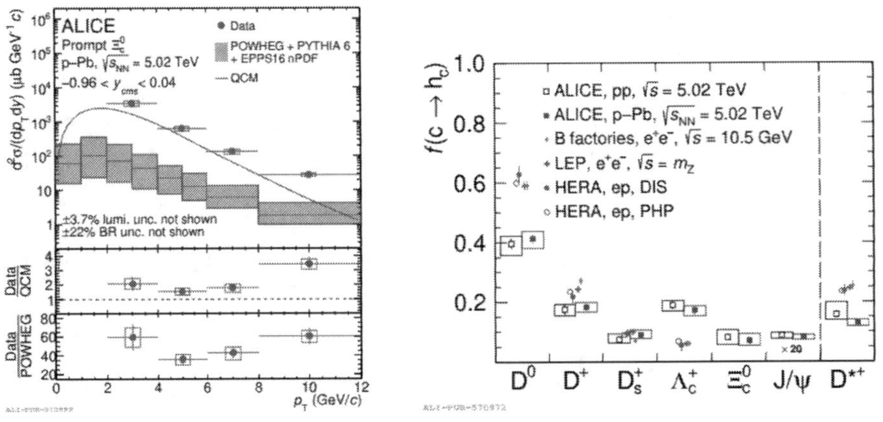

Figure 4. Left panel: prompt Ξ_c^0 p_T-differential cross section in p–Pb collisions at $\sqrt{s_{NN}}$ = 5.02 TeV compared with POWHEG+PYTHIA 6 simulations coupled to the EPPS16 nPDF parametrisation and to QCM (see [19] and references therein). Right panel: fragmentation fractions for charm hadrons in p–Pb collisions at $\sqrt{s_{NN}}$ = 5.02 TeV, compared to results from pp collisions at $\sqrt{s}$ = 5.02 TeV and from e^+e^- and ep collisions at lower energies (see [20] and references therein).

consistent with those in pp collisions, showing no dependence on the system size at the LHC. Furthermore, they are significantly enhanced (diminished) compared to those from leptonic collisions in the case of baryons (mesons), confirming the hypothesis of differing hadronisation in a parton-rich environment as already argued from previous pp measurements.

In the charmonium sector, midrapidity production measurements of prompt and non-prompt J/ψ in Pb–Pb collisions at $\sqrt{s_{NN}}$ = 5.02 were recently published [21]. In the left panel of Fig. 5, the p_T-differential prompt J/ψ nuclear modification factor R_{AA}, measured in the centrality class 0–10%, is compared to similar measurements from the CMS and ATLAS Collaborations, as well as with several model calculations (see references in [21]). Results from ALICE nicely extend high-p_T measurements down to 1.5 GeV/c, and are in good agreement with CMS and ATLAS measurements in the overlapping p_T intervals. The comparison with calculations suggests that models which include re-combination either at the chemical freeze-out (SHMc) or during the QGP phase (Boltzmann Transport, BT) are able to describe the rising trend of the R_{AA} down to low p_T. In the right panel of Fig. 5, the p_T-integrated ($1.5 < p_T < 10$ GeV/c) prompt J/ψ R_{AA} is shown as a function of centrality (expressed by the average number of participants, $\langle N_{part} \rangle$), where it is compared to the BT model. Experimental data show a rising trend with increasing centrality, as expected from the re-generation scenario, and are well described within uncertainties by the BT model above $\langle N_{part} \rangle \approx 50$.

4 Conclusions and outlook

The ALICE Collaboration produced a large collection of new HF physics results based on Run 2 and fresh Run 3 data, the latter collected using the upgraded ALICE detector. In these proceedings, a few selected highlights from different collision systems are presented. Run 3 data taking is efficiently progressing, and will continue until the end of 2025. Several upgrade projects are expected to be installed during Long Shutdown 3 (2026-2028). Among others, the ITS3 [22] project foresees to replace the innermost 3 layers of the current ITS with truly cylindrical silicon layers, providing an improvement of the pointing resolution, crucial for the reconstruction of HF hadrons, by a factor 2 compared to the current ITS. These

© The Authors, published by EDP Sciences. This is an open access article distributed under the terms of the Creative Commons Attribution License 4.0 (https://creativecommons.org/licenses/by/4.0/).

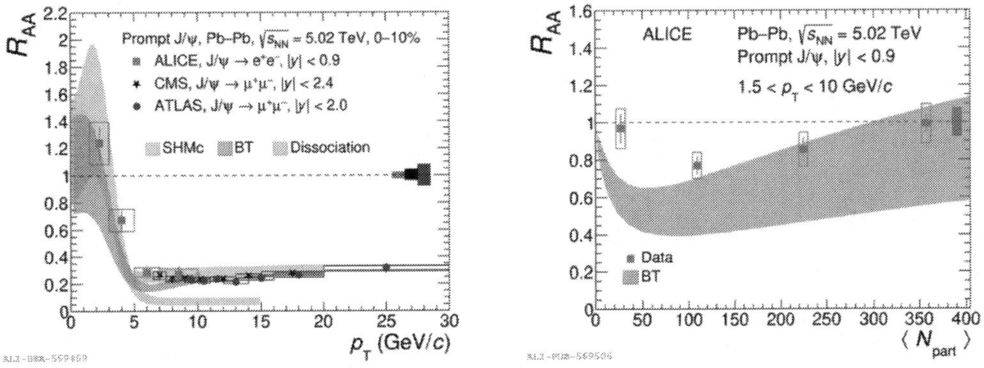

Figure 5. Left panel: p_T-differential prompt J/ψ R_{AA} measured at midrapidity in Pb–Pb collisions (0–10% centrality class) at $\sqrt{s_{NN}}$ = 5.02 TeV compared to similar measurements from other experiments and with theoretical models (see [21] and references therein). Right panel: prompt J/ψ R_{AA} for $1.5 < p_T < 10$ GeV/c as a function of $\langle N_{part} \rangle$ compared with BT calculations (see [21] and references therein).

improvements, as well as the increase in terms of amount of collected data, which by the end of Run 4 is expected to be about 100 (1000) times larger for Pb–Pb (pp and p–Pb) compared to the total recorded during the Runs 1 and 2, will allow for high-precision HF measurements, including direct beauty reconstruction and the production of multi-charm baryons.

References

[1] Collins, J. C. *et al.*, Adv. Ser. Direct. High Energy Phys. **5**, 1–91 (1989).
[2] Andronic, A. *et al.*, Eur. Phys. J. C **76**, 3 (2016).
[3] Acharya, S. *et al.*, ALICE Collaboration, Phys. Rev. C **107** 064901 (2023).
[4] Acharya, S. *et al.*, ALICE Collaboration, JHEP **10** 159 (2021).
[5] Aaij, R. *et al.*, LHCb Collaboration, Phys. Rev. D **100**, 031102 (2019).
[6] Skands, P. *et al.*, Eur. Phys. J. C **74** 3024 (2014).
[7] Christiansen J. R. and Skands P., JHEP **08** 003 (2015).
[8] Greco, V. *et al.*, Phys. Lett. B **821** 136622 (2021).
[9] Beraudo, A. *et al.*, arXiv:2306.02152.
[10] He M. and Rapp R., Phys. Lett. B **795** 117 (2019).
[11] Acharya, S. *et al.*, ALICE Collaboration, JINST **19** P05062 (2024).
[12] Butenschön M. and Kniehl B. A., Phys. Rev. Lett. **106** 022003 (2011).
[13] Ma, Y.-Q. *et al.*, Phys. Rev. Lett. **106** 042002 (2011).
[14] Ma Y.-Q. and Venugopalan R., Phys. Rev. Lett. **113** 192301 (2014).
[15] Cheung V. and Vogt R., Phys. Rev. D **98** 114029 (2018).
[16] Cacciari, M. *et al.*, JHEP **05** 007 (1998).
[17] Andronic, A. *et al.*, Phys. Lett. B **797** 134836 (2019).
[18] Altmann J. *et al.*, arXiv:2405.19137v2.
[19] Acharya, S. *et al.*, ALICE Collaboration, arXiv:2405.14538.
[20] Acharya, S. *et al.*, ALICE Collaboration, arXiv:2405.14571.
[21] Acharya, S. *et al.*, ALICE Collaboration, JHEP **02** 066 (2024).
[22] Acharya, S. *et al.*, ALICE Collaboration, CERN-LHCC-2024-003 (2024).

© The Authors, published by EDP Sciences. This is an open access article distributed under the terms of the Creative Commons Attribution License 4.0 (https://creativecommons.org/licenses/by/4.0/).

Recent results from LHCb

Thomas Boettcher[1,*] on behalf of the LHCb collaboration

[1]University of Cincinnati

Abstract. The LHCb experiment is a fully instrumented detector in the forward region at the Large Hadron Collider. LHCb's forward acceptance and unique detector capabilities enable a broad heavy-ion physics program. LHCb has studied colliding systems from pp to PbPb, as well as fixed-target collisions with a wide variety of target species. This contribution presents recent results from the LHCb experiment.

1 Introduction

The LHCb detector is a forward spectrometer at the Large Hadron Collider [1]. The LHCb detector is fully instrumented, consisting of tracking, calorimetry, and particle identification systems. In particular, the LHCb tracking system includes a Vertex Locator (VELO), a silicon strip detector close to the beamline that provides transverse vertex resolution of $10 - 50\ \mu$m. In addition, LHCb also includes two Ring Imaging Cherenkov (RICH) detectors, which allow for charged hadron identification. As a result, the LHCb detector can reconstruct and identify charged hadrons, photons, leptons, and light nuclei produced in high-energy hadron collisions. These capabilities, along with the detector's forward acceptance, allow LHCb to perform unique studies of heavy-ion collisions in unexplored kinematic regions.

In addition to operating as a collider detector, LHCb can also operate as a fixed-target experiment by injecting gas into the VELO region using the System for Measuring the Overlap with Gas (SMOG) [2]. Consequently, LHCb is the highest energy fixed-target experiment ever conducted. With SMOG, LHCb has performed studies of both proton-gas and lead-gas collisions using a wide variety of noble gas targets.

2 Open heavy-flavor production and hadronization

LHCb's excellent vertex resolution and particle identification capabilities make it an ideal detector for studying heavy-flavor production. The LHCb experiment has used these capabilities to study bottom quark hadronization in high-multiplicity pp collisions [3]. LHCb measured the Λ_b^0/B^0 production ratio as a function of multiplicity in pp collisions at $\sqrt{s} = 13$ TeV. This measurement uses the similar $\Lambda_b^0 \to J/\psi p K^-$ and $B^0 \to J/\psi \pi^- K^+$ final states to maximize the cancellation of systematic uncertainties in the baryon-to-meson ratio. As a result, this measurement requires precise identification of both charged hadrons and leptons. The resulting baryon-to-meson ratio is shown as a function of multiplicity in Fig. 1. This ratio agrees with results from e^+e^- collisions at low multiplicity [4], but grows as multiplicity increases.

*e-mail: boettcts@ucmail.uc.edu

© The Authors, published by EDP Sciences. This is an open access article distributed under the terms of the Creative Commons Attribution License 4.0 (https://creativecommons.org/licenses/by/4.0/).

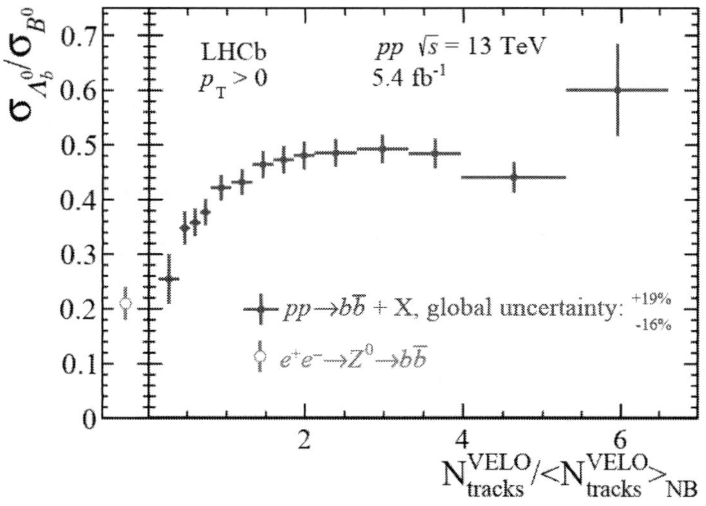

Figure 1. Ratio of Λ_b^0 to B^0 cross section as a function of the track multiplicity in the VELO detector normalized to the average multiplicity in minimum bias pp events.

This result indicates that the hadronization of b quarks is modified in high-multiplicity environments.

LHCb has also studied the modification of charm quark hadronization by studying $D_s^{\pm}$ production relative to $D^{\pm}$ production as a function of multiplicity in pPb collisions at $\sqrt{s_{NN}}$ = 8.16 TeV [5]. The resulting $D_s^{\pm}/D^{\pm}$ ratio is shown in Fig. 2. The measured ratio shows an enhancement of D_s^+ in high-multiplicity events, independent of rapidity. These results agree with previous results from both the LHCb and ALICE experiments [6–9]. The data are better described by models including final state effects such as coalescence or color reconnection, although no model provides and accurate description of the data over the entire kinematic range of the measurement.

3 Conventional and exotic charmonium production

Charmonium suppression could be a signal of color screening by the Quark-Gluon Plasma (QGP), which would cause the charmonium to disassociate. Charmonium states with different binding energies will disassociate at different energies. As a result, studying the suppression of different charmonium states can reveal the temperature of the medium produced in high-energy hadron collisions. LHCb recently measured χ_c production relative to J/ψ production in both pPb and pp collisions, where χ_c refers to the combination of χ_{c1} and χ_{c2} states [10]. The double ratio of production ratios of different quarkonium states is shown in Fig. 3. The LHCb results show no suppression of the χ_c states. Because their binding energies are just above the expected QGP freeze-out temperature, these results indicate that charmonium suppression in pp and pPb collisions is not a product of QGP formation.

LHCb also recently performed measurements of charmonium production in pp collisions as a function of multiplicity [11]. LHCb measured the $\psi(2S)/J/\psi$ production ratio, and the results are shown in Fig. 4. The LHCb measurement shows suppression at high multiplicity for promptly produced particles and no suppression for charmonia from b hadron decays.

© The Authors, published by EDP Sciences. This is an open access article distributed under the terms of the Creative Commons Attribution License 4.0 (https://creativecommons.org/licenses/by/4.0/).

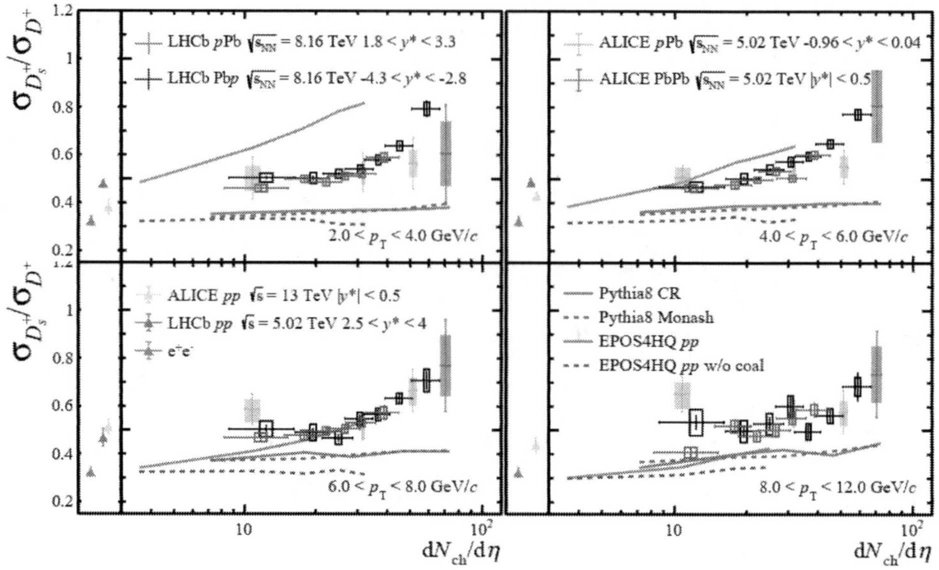

Figure 2. Ratio of D_s^+ to D^+ cross sections as a function of the charged particle pseudorapidity density in different transverse momentum regions. Statistical uncertainties are shown as error bars and systematic uncertainties are shown as boxes.

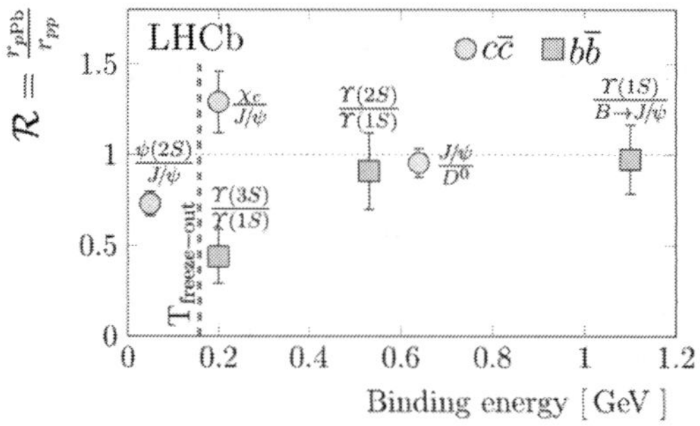

Figure 3. Ratio of production ratios in pPb to pp collisions. The particle species used in each production ratio are shown in the data point labels. The double ratios are shown as a function of the binding energy of the quarkonium state. The dashed line shows the estimated QGP freeze-out temperature.

The results agree with predictions from a co-mover model, which describes how charmonium disassociates by interacting with nearby hadrons. This ratio was also measured in pPb collisions, and the results also agree with co-mover predictions.

In addition to conventional charmonium states, the LHCb experiment has also studied exotic charmonium production in heavy-ion collisions. LHCb measured $\chi_{c1}(3872)$ production relative to $\psi(2S)$ production in pp and pPb collisions [12]. LHCb measured the nuclear modification factor of the $\chi_{c1}(3872)$ meson at $\sqrt{s_{NN}} = 8.16$ TeV. The measured nuclear modification factor is shown in Fig. 5. The resulting modification is larger than that of the $\psi(2S)$, although the uncertainties are large. These results indicate that conventional charmonium states are modified differently than exotic states.

© The Authors, published by EDP Sciences. This is an open access article distributed under the terms of the Creative Commons Attribution License 4.0 (https://creativecommons.org/licenses/by/4.0/).

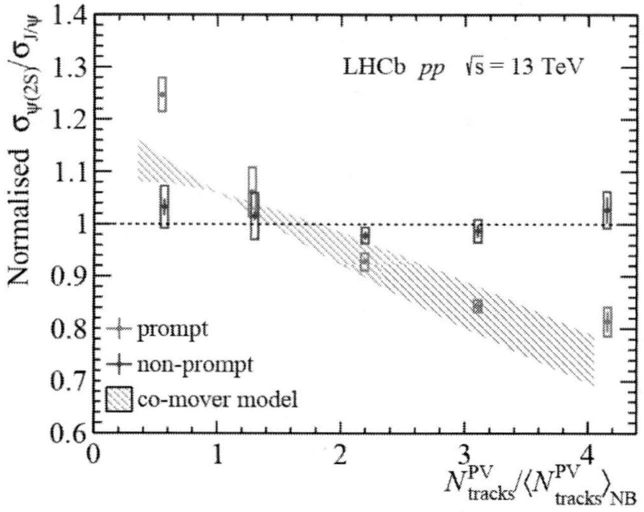

Figure 4. The ratio of $\psi(2S)$ to J/ψ production cross sections in pp collisions at $\sqrt{s}$ = 13 TeV. The ratio is normalized such that the mean value is unity. The ratio is shown as a function of the multiplicity of tracks originating from a collision point normalized to the mean value in minimum bias pp collisions.

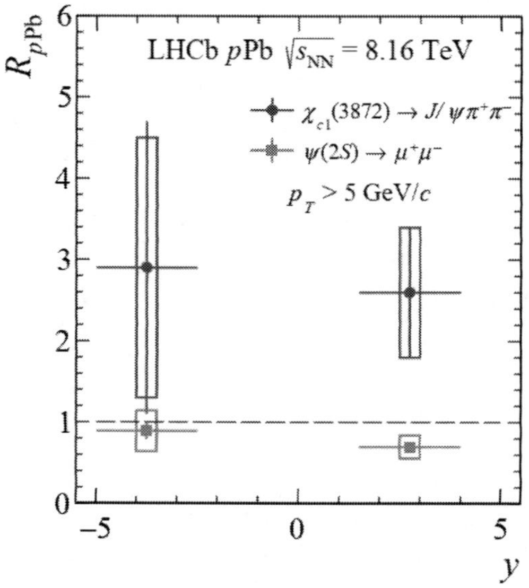

Figure 5. Nuclear modification factor $R_{p\mathrm{Pb}}$ of $\chi_{c1}(3872)$ and $\psi(2S)$ mesons at $\sqrt{s_{\mathrm{NN}}}$ = 8.16 TeV. The statistical uncertainties are shown as error bars, while the systematic uncertainties are shown as boxes.

© The Authors, published by EDP Sciences. This is an open access article distributed under the terms of the Creative Commons Attribution License 4.0 (https://creativecommons.org/licenses/by/4.0/).

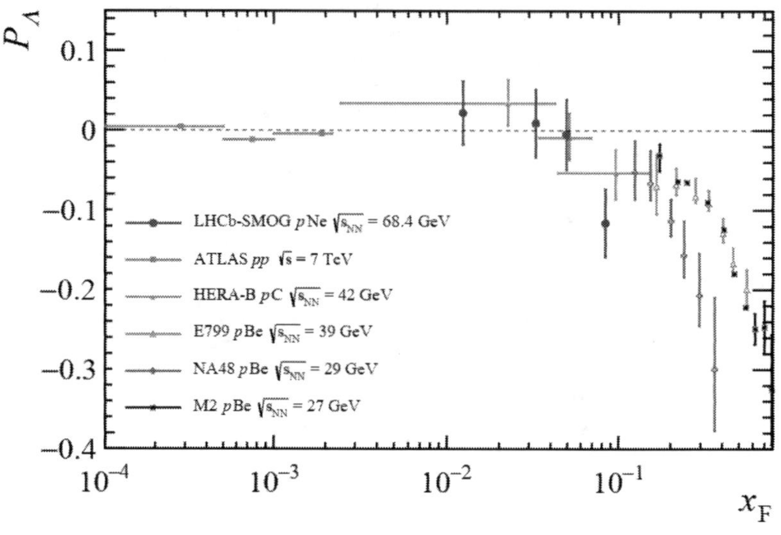

Figure 6. Transverse polarisation of Λ baryons as a function of x_F for various collision systems and center-of-mass energies.

4 Fixed-target physics

The LHCb experiment operates in fixed-target mode using the SMOG system. The fixed-target configuration provides complementary kinematic coverage to other hadron collider experiments. LHCb recently measured the transverse polarization of Λ baryons produced in fixed-target *p*Ne collisions with $\sqrt{s_{NN}}$ = 68.4 GeV [14]. This measurement probes transverse-momentum dependent fragmentation functions at larger Feynman *x* (x_F) than that accessible to other LHC experiments. Results are shown in Fig. 6. The measured Λ polarization is in broad agreement with measurements across a wide range of collision systems and center-of-mass energies.

5 LHCb in Run 3

The LHCb detector recently underwent a major upgrade which included replacement of the entire tracking system, as well as a new software-only trigger system [15]. The new detector is capable of studying more central PbPb collisions than the original LHCb detector, reaching centralities of 30%. Additionally, a new gas target cell called SMOG2 was installed. This dedicated gas cell is located upstream of the VELO and allows for a two-order-of-magnitude increase in gas pressure. The upgraded detector is capable of collecting data in both collider and fixed-target modes simultaneously and has done so throughout 2024. As a result, this upgrade will enable the expansion of LHCb's already rapidly growing heavy-ion physics program.

6 Summary

The LHCb experiment has a broad and growing heavy-ion physics program. Its forward acceptance, vertexing and particle identification capabilities, and gas injection system allow LHCb to play a unique role in the heavy-ion physics landscape. Recent upgrades will further

© The Authors, published by EDP Sciences. This is an open access article distributed under the terms of the Creative Commons Attribution License 4.0 (https://creativecommons.org/licenses/by/4.0/).

expand this program and allow LHCb to make major contributions to high-energy nuclear physics for the foreseeable future.

References

[1] A. A. Alves, Jr. *et al.* [LHCb], JINST **3** (2008), S08005 doi:10.1088/1748-0221/3/08/S08005
[2] R. Aaij *et al.* [LHCb], JINST **9** (2014) no.12, P12005 doi:10.1088/1748-0221/9/12/P12005 [arXiv:1410.0149 [hep-ex]].
[3] R. Aaij *et al.* [LHCb], Phys. Rev. Lett. **132** (2024) no.8, 081901 doi:10.1103/PhysRevLett.132.081901 [arXiv:2310.12278 [hep-ex]].
[4] Y. S. Amhis *et al.* [HFLAV], Eur. Phys. J. C **81** (2021) no.3, 226 doi:10.1140/epjc/s10052-020-8156-7 [arXiv:1909.12524 [hep-ex]].
[5] R. Aaij *et al.* [LHCb], Phys. Rev. D **110** (2024) no.3, L031105 doi:10.1103/PhysRevD.110.L031105 [arXiv:2311.08490 [hep-ex]].
[6] R. Aaij *et al.* [LHCb], JHEP **06** (2017), 147 doi:10.1007/JHEP06(2017)147 [arXiv:1610.02230 [hep-ex]].
[7] S. Acharya *et al.* [ALICE], JHEP **12** (2023), 086 doi:10.1007/JHEP12(2023)086 [arXiv:2308.04877 [hep-ex]].
[8] S. Acharya *et al.* [ALICE], JHEP **12** (2019), 092 doi:10.1007/JHEP12(2019)092 [arXiv:1906.03425 [nucl-ex]].
[9] S. Acharya *et al.* [ALICE], JHEP **10** (2018), 174 doi:10.1007/JHEP10(2018)174 [arXiv:1804.09083 [nucl-ex]].
[10] R. Aaij *et al.* [LHCb], Phys. Rev. Lett. **132** (2024) no.10, 102302 doi:10.1103/PhysRevLett.132.102302 [arXiv:2311.01562 [hep-ex]].
[11] R. Aaij *et al.* [LHCb], JHEP **05** (2024), 243 doi:10.1007/JHEP05(2024)243 [arXiv:2312.15201 [hep-ex]].
[12] R. Aaij *et al.* [LHCb], Phys. Rev. Lett. **132** (2024) no.24, 242301 doi:10.1103/PhysRevLett.132.242301 [arXiv:2402.14975 [hep-ex]].
[13] A. M. Sirunyan *et al.* [CMS], Phys. Rev. Lett. **128** (2022) no.3, 032001 doi:10.1103/PhysRevLett.128.032001 [arXiv:2102.13048 [hep-ex]].
[14] R. Aaij *et al.* [LHCb], [arXiv:2405.11324 [hep-ex]].
[15] R. Aaij *et al.* [LHCb], JINST **19** (2024) no.05, P05065 doi:10.1088/1748-0221/19/05/P05065 [arXiv:2305.10515 [hep-ex]].

© The Authors, published by EDP Sciences. This is an open access article distributed under the terms of the Creative Commons Attribution License 4.0 (https://creativecommons.org/licenses/by/4.0/).

EPJ Web of Conferences **316**, 01005 (2025) https://doi.org/10.1051/epjconf/202531601005
SQM 2024

ATLAS highlights: Recent results from ATLAS

Petr Balek[1,*], on behalf of the ATLAS Collaboration

[1]AGH University of Krakow, al. Adama Mickiewicza 30, 30-059 Kraków, Poland

Abstract. The ATLAS experiment at the Large Hadron Collider encompasses a wide-ranging physics programme that includes various collision systems and centre-of-mass energies. These proceedings provide an overview of the recent ATLAS measurements regarding heavy ion collisions, heavy flavour production, and collective behaviour. The data utilized in the presented results were recorded during LHC Run 2 (2015–2018). The measurements can help to understand various physics phenomena from hard scatterings to soft processes, and from large systems to small systems.

1 Introduction

The heavy-ion programme of the ATLAS experiment at the Large Hadron Collider [1] is rich and encompasses collision systems of different sizes. In the large systems, Pb+Pb and Xe+Xe collisions, it explores the phenomenon of jet quenching in quark-gluon plasma (QGP) and its dependency on the sub-structure, colour charge, or quark flavour of jets, as well as collective effects in QGP such as the azimuthal anisotropy or transverse momentum (p_T) fluctuations. In the small collision systems, pp and p+Pb, the programme focuses on the origin of collectivity, the formation of quarkonia, and the mechanism behind the jet modification. Furthermore, $\gamma+\gamma$ and γ+Pb collisions are studied using intense photon fluxes accompanying Pb+Pb collisions.

2 ATLAS detector

The ATLAS experiment [1] at the LHC is a multipurpose particle detector with a forward–backward symmetric cylindrical geometry and a near 4π coverage in solid angle. It consists of an inner tracking detector surrounded by a thin superconducting solenoid providing a 2 T axial magnetic field, electromagnetic and hadronic calorimeters, and a muon spectrometer. A two-level trigger system is used to select events.

Heavy-ion events are categorized into centrality intervals using the Glauber model parameterization of the ΣE_T distribution in the forward calorimeter [2]. Each centrality interval represents a range in ΣE_T, starting at 0% for the most central collisions.

*e-mail: petr.balek@cern.ch

© The Authors, published by EDP Sciences. This is an open access article distributed under the terms of the Creative Commons Attribution License 4.0 (https://creativecommons.org/licenses/by/4.0/).

3 Recent ATLAS analyses

3.1 Production of $t\bar{t}$ pairs in p+Pb collisions

The measurement of top-quark production in p+Pb collision may provide new constraints for nuclear parton distribution functions (nPDFs) in Bjorken-x about 0.005–0.05 and in Q^2 about $m_t^2 \approx 3 \times 10^4\,\mathrm{GeV}^2$, where m_t denotes top quark mass. Information about gluon nPDF may also be extracted. Anti-shadowing and EMC effects are expected to be responsible for an enhancement of ~10% in the $t\bar{t}$ production in p+Pb with respect to pp collisions.

The analysis of $t\bar{t}$ pairs production [3] uses 165 nb^{-1} of p+Pb data at $\sqrt{s_{\mathrm{NN}}} = 8.16\,\mathrm{TeV}$ and focuses on two decay channels. In the single-lepton channel, it selects events with the decay mode $t\bar{t} \to W^-\bar{b} + W^+b \to \ell^-\bar{\nu}_\ell + \bar{b} + q\bar{q} + b$ and analogical process for W^+. The events are required to have one electron or muon and at least four jets; out of these jets, at least one jet has to be b-tagged [4]. In the di-lepton channel, both W-bosons have to decay leptonically: $t\bar{t} \to W^-\bar{b} + W^+b \to \ell^-\bar{\nu}_\ell + \bar{b} + \ell^+\nu_\ell + b$. Thus, the events are required to have two oppositely charged leptons and at least two jets while at least one jet has to be b-tagged. Events with the same lepton-flavour pairs with the invariant mass $80 < m_{\ell\ell} < 100\,\mathrm{GeV}$ are rejected to exclude Z candidates. The background comes from W and Z bosons produced in association with jets, from single t production, and from di-boson production.

The signal events are simulated using next-to-leading-order Monte Carlo (MC) event generator Powheg Box v2 [5], while Pythia generator [6] is used for the modelling of parton shower and hadronisation. The background Z+jets, W+jets, and di-bosons+jets events are simulated with Sherpa [7]; the background t and tW events are simulated with Powheg Box v2. The signal and background samples are produced in pp and pn configurations, and then embedded to the real p+Pb events. This way, their underlying-event contribution is the same as in the data.

The largest systematic uncertainties arise from the jets reconstruction, signal modelling, background modelling, luminosity measurement, and other smaller sources, in this order.

The signal and background events are studied as a function of $H_{\mathrm{T}}^{\ell,\mathrm{j}}$, a scalar sum of p_{T} of all leptons and jets within the event. The signal strength is then extracted using profile-likelihood fit. The post-fit $H_{\mathrm{T}}^{\ell,\mathrm{j}}$ distributions are shown in Figure 1 for both single- and di-lepton channels. The value of measured inclusive cross-section of $t\bar{t}$ pair production in p+Pb collisions is $\sigma_{t\bar{t}} = 58.1 \pm 2.0\,(\mathrm{stat.})\,^{+4.8}_{-4.4}\,(\mathrm{syst.})$.

Figure 2 (left) shows the measured cross-section in comparison with several MC models, CMS p+Pb measurement [8], and ATLAS+CMS pp measurement [9]. Figure 2 (right) shows the nuclear modification factor, defined as the ratio of p+Pb and pp cross sections, measured at the same $\sqrt{s_{\mathrm{NN}}}$ energy, scaled by the number of nucleons in the Pb nucleus. For this comparison, pp cross section is extrapolated from $\sqrt{s} = 8\,\mathrm{TeV}$ to $\sqrt{s} = 8.16\,\mathrm{TeV}$. The nuclear modification factor is $R_{\mathrm{pA}} = 1.090 \pm 0.039\,(\mathrm{stat.})\,^{+0.094}_{-0.087}\,(\mathrm{syst.})$.

3.2 Azimuthal correlations of muon pairs in Pb+Pb collisions

Since heavy-flavour quarks cannot be produced thermally in QGP, they are mostly produced in the hard scattering at the early stages of the collisions and thus experience the whole QGP evolution. The quarks may suffer energy loss due to their interactions with the QGP. Additionally, anisotropy of the heavy-flavour quarks, or their decay products, is usually present and can be described by Fourier coefficients v_n.

The analysis of muon pairs originating in the decays of charm and bottom quarks [10] uses Pb+Pb and pp data at $\sqrt{s_{\mathrm{NN}}} = 5.02\,\mathrm{TeV}$ with integrated luminosities of 1.94 nb^{-1} and 0.26 fb^{-1}, respectively. Muons considered for this analysis are required to have $p_{\mathrm{T}} > 4\,\mathrm{GeV}$

© The Authors, published by EDP Sciences. This is an open access article distributed under the terms of the Creative Commons Attribution License 4.0 (https://creativecommons.org/licenses/by/4.0/).

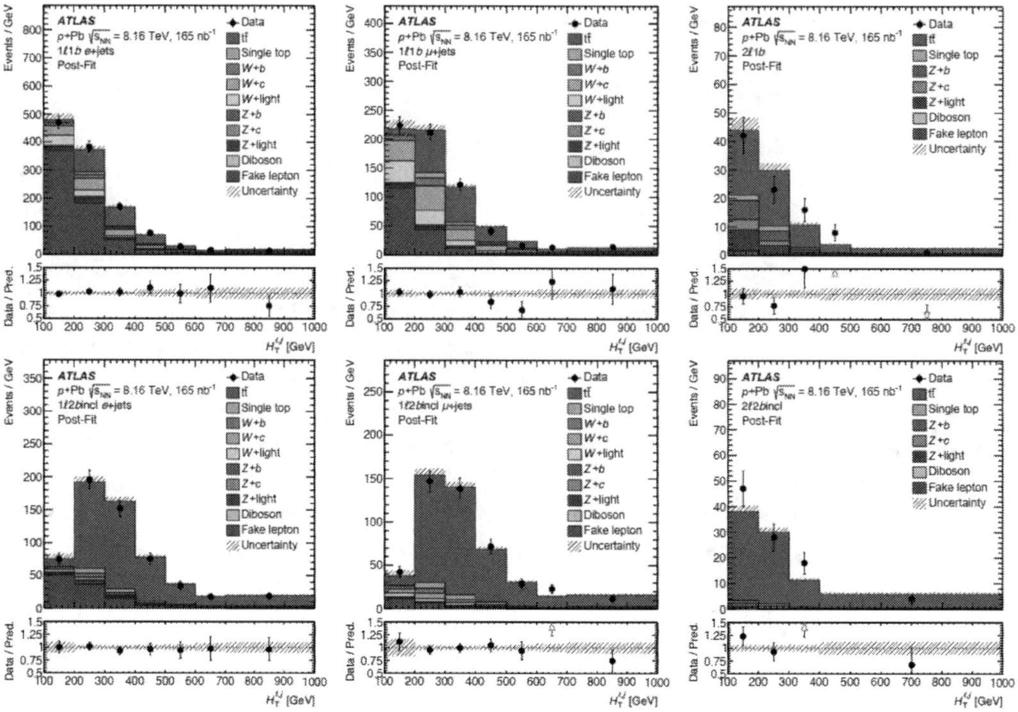

Figure 1: Post-fit distributions of $H_{\mathrm{T}}^{\ell,\mathrm{j}}$ [3]. The distributions are shown for 1 b-tagged jet (top row) and 2 b-tagged jets (bottom row); for single-electron channel (left column), single-muon channel (middle column), and two-lepton channels (right column). Data are shown with black markers, signal with red histogram, and background with histograms of other colours.

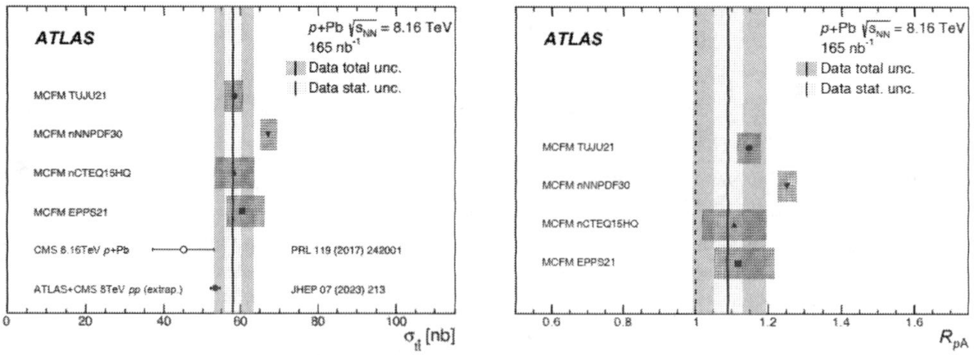

Figure 2: **Left:** Measured cross section of $t\bar{t}$ production in p+Pb collisions [3] compared to MC predictions as well as previous CMS and ATLAS results [8, 9]. **Right:** Nuclear modification factor R_{pA} compared to several MC predictions [3].

and $|\eta| < 2.4$. To suppress pairs with low trigger efficiency, the average p_{T} of the pair has to be above 5 GeV. To suppress contributions from quarkonia decays and from the production within jets, the muons distance in pseudorapidity has to be more than 0.8; this eliminates decay products of light vector mesons and J/ψ. Opposite-sign pairs with invariant mass $9.2 < m_{\mu\mu} < 10.4$ GeV or $70 < m_{\mu\mu} < 110$ GeV are excluded to eliminate contributions from Υ

© The Authors, published by EDP Sciences. This is an open access article distributed under the terms of the Creative Commons Attribution License 4.0 (https://creativecommons.org/licenses/by/4.0/).

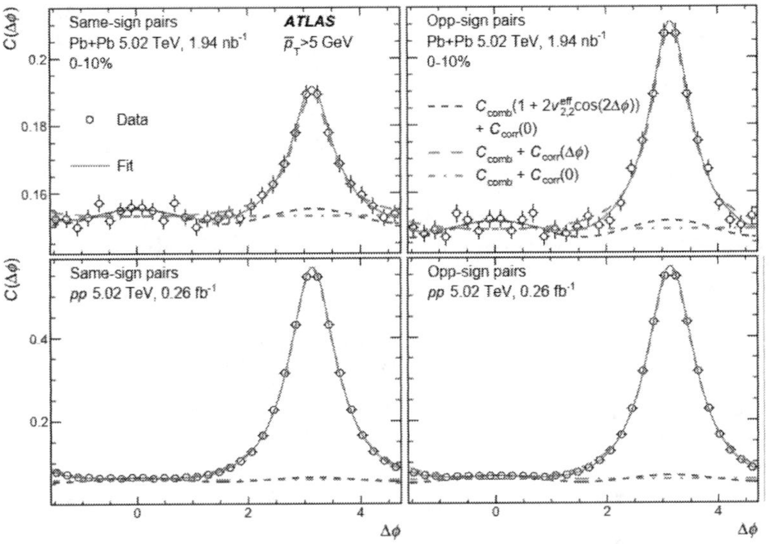

Figure 3. Distributions of azimuthal angle separation, $\Delta\phi$, for same-sign (left column) and opposite-sign (right column) muons pairs in central Pb+Pb (top row) and *pp* (bottom row) collisions [10]. The red solid line shows the full template fit of the data; the dashed magenta, green, and blue lines show the contribution of pedestal, pedestal with Breit-Wigner distribution, and pedestal with elliptic flow modulation, respectively.

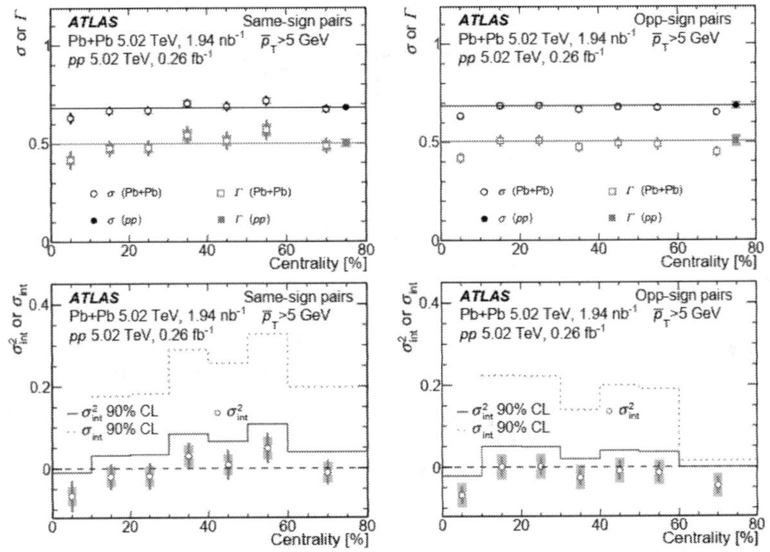

Figure 4. Top: Centrality dependence of Γ and σ widths for same-sign (left) and opposite-sign (right) muon pairs [10]. **Bottom:** Centrality dependence of additional angular deflection introduced by the QGP for same-sign (left) and opposite-sign (right) muon pairs and their 90% confidence levels (CL) [10].

and Z, respectively. In Pb+Pb, di-muons originating in $\gamma+\gamma$ scattering are also rejected [11]. To correct for trigger and reconstruction inefficiencies, as well as for the exclusion of the opposite-sign pairs mentioned above, a weight is assigned to each pair [11].

The systematic uncertainties arise from the muon quality selection, trigger and reconstruction efficiency corrections, correction for the $m_{\mu\mu}$ exclusion, background parametrization, and the method used to extract the observables.

The muon pairs are studied in terms of their azimuthal angle separation, $\Delta\phi$, shown in Figure 3. All the correlation functions, $C(\Delta\phi)$, show a clear enhancement at $\Delta\phi = \pi$ on top of a pedestal. The pedestal is significantly higher for Pb+Pb collisions. The correlation functions are fitted with a sum of the flat pedestal, elliptic flow modulation, and Breit-Wigner distribution. The former two terms are from a combinatorial contribution while the third one

© The Authors, published by EDP Sciences. This is an open access article distributed under the terms of the Creative Commons Attribution License 4.0 (https://creativecommons.org/licenses/by/4.0/).

is from a correlated contribution:

$$C_{fit}(\Delta\phi) = C_{comb}\left[1 + 2\,v_2 \cos(2\Delta\phi)\right] + C_{corr,max}\,\Gamma^2 / \left[(\Delta\phi - \pi)^2 + \Gamma^2\right]$$

Figure 4 (top) shows centrality dependence of Γ and σ, a standard deviation of $C(\Delta\phi)$ in $0 < \Delta\phi < 2\pi$. There is almost no variation of either with centrality; they both exhibit similar behaviour for same-sign and opposite-sign pairs. Figure 4 (bottom) shows centrality dependence of additional angular deflection due to the presence of QGP, $\sigma^2_{int} = \sigma^2_{PbPb} - \sigma^2_{pp}$. The 90% confidence levels for σ^2_{int} and σ_{int} are evaluated using combined statistical and systematic uncertainties, providing a model-independent constraint on the stochastic deflection in the QGP.

3.3 Speed of sound and p_T fluctuations in Xe+Xe and Pb+Pb collisions

The hydrodynamic expansion of the QGP is not only responsible for the anisotropic flow but also induces a significant boost to the transverse momentum of the particles. This translates to the variation in the mean p_T in each event, $[p_T]$, which is a sensitive probe of QGP properties, such as the equation of state and the associated speed-of-sound, c_s, in the plasma.

The study of momentum fluctuations [12] uses Pb+Pb collisions at $\sqrt{s_{NN}} = 5.02$ TeV and Xe+Xe collisions at $\sqrt{s_{NN}} = 5.44$ TeV with luminosities of $470\,\mu b^{-1}$ and $3\,\mu b^{-1}$, respectively. Events with pileup, i.e. multiple inelastic collisions, are rejected. Tracks considered in this analysis have $p_T > 5$ GeV and $|\eta| < 2.5$. The track reconstruction efficiency and the fake rate are estimated using events generated by Hijing [13]. A weight, w_i, is applied to each particle in an event to compensate for the fake rate and the reconstruction inefficiency. The charged particle multiplicity corrected for detector effects is denoted by $N_{ch} = \Sigma_i\, w_i$.

Systematic uncertainties arise from the track selection, detector material modelling, residual pileup, centrality definition, and MC closure check.

The $[p_T]$, dimensionless normalized variance, k_2, and skewness, k_3, are calculated as:

$$[p_T] = \frac{\Sigma_i\, w_i\, p_{T,i}}{\Sigma_i\, w_i}, \qquad k_2 = \frac{\Sigma_{i\neq j}\, w_i\, w_j\, \delta p_{T,i}\, \delta p_{T,j}}{\langle [p_T]\rangle^2\, \Sigma_{i\neq j}\, w_i\, w_j}, \qquad k_3 = \frac{\Sigma_{i\neq j\neq k}\, w_i\, w_j\, w_k\, \delta p_{T,i}\, \delta p_{T,j}\, \delta p_{T,k}}{\langle [p_T]\rangle^3\, \Sigma_{i\neq j\neq k}\, w_i\, w_j\, w_k},$$

where summing is done over all tracks in the event and (for k_2 and k_3) over non-repeating unique sets of indices; $\delta p_{T,i} \equiv p_{T,i} - \langle [p_T]\rangle$; and angle brackets, $\langle \ldots \rangle$, denote the average over all events in the ensemble. In the scenario of independent particle production from participating nucleons, N_{ch} is roughly proportional to N_{part}. Consequently, it is expected that $k_2 \approx N_{ch}^{-1}$ and $k_3 \approx N_{ch}^{-2}$.

Figure 5 (left) presents k_2 and k_3 scaled by N_{ch} and N_{ch}^2, respectively. Both values rise up to $N_{ch} \approx 1500$ for both systems. Then, only for Pb+Pb collisions, they show only a little dependence on N_{ch} in the range $2000 \lesssim N_{ch} \lesssim 4000$. In even more central collisions, $N_{ch}k_2$ displays a sharp decrease, while $N_{ch}^2 k_3$ shows an abrupt increase followed by a sharp decrease; this is true for both Pb+Pb and Xe+Xe. Such behaviour can be described by models incorporating both geometrical and intrinsic fluctuations of the particle production sources [14].

The increase of $[p_T]$ is related to the speed of sound in the QGP:

$$c_s^2(T_{eff}) \propto \frac{d\,\ln(\langle [p_T]\rangle)}{d\,\ln(N_{ch})} \approx \frac{\Delta p_T / \langle [p_T]\rangle}{\Delta N_{ch} / \langle N_{ch}\rangle}$$

where T_{eff} is the effective temperature of QGP. The fraction can be understood as the slope in Figure 5 (right). An agreement with Music model [14] is achieved for $c_s^2 \approx 0.23$ and $T_{eff} = 222$ MeV. Nevertheless, these values depend on several choices of the analysis, such as the selected p_T range of the particles.

© The Authors, published by EDP Sciences. This is an open access article distributed under the terms of the Creative Commons Attribution License 4.0 (https://creativecommons.org/licenses/by/4.0/).

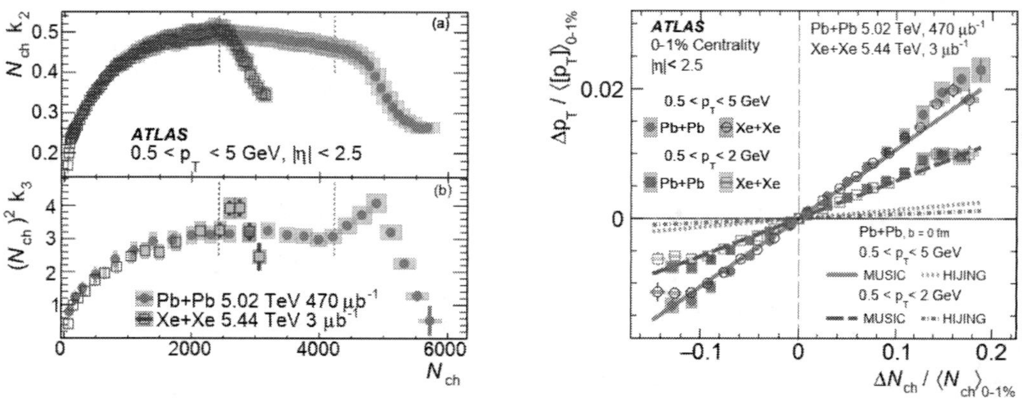

Figure 5: **Left:** Dimensionless normalized variance k_2 scaled by N_{ch} (top) and dimensionless normalized skewness k_3 scaled by N_{ch}^2 (bottom) as a function of N_{ch} in Pb+Pb and Xe+Xe collisions [12]. The vertical dashed lines indicate the N_{ch} values corresponding to 5% centrality in both systems. **Right:** Correlation between $\Delta p_{\mathrm{T}}/\langle[p_{\mathrm{T}}]\rangle$ and $\Delta N_{\mathrm{ch}}/\langle N_{\mathrm{ch}}\rangle$ in the 0-1% most central Pb+Pb and Xe+Xe collisions [12]. Data are compared to calculations of the Music and Hijing models [13, 14]. Both are presented for two p_{T} ranges of the particles.

4 Summary & Acknowledgements

In the 21st International Conference on Strangeness in Quark Matter (SQM 2024), the ATLAS experiment presented several new results, utilizing data recorded during Run 2 of the LHC. This included the production of $t\bar{t}$ pairs in p+Pb collisions, azimuthal correlations of muons in Pb+Pb collisions, and the study p_{T} fluctuations with its implications to the speed of sound in QGP in Xe+Xe and Pb+Pb collisions.

This research project is supported by the programme "Excellence Initiative – Research University" for the AGH University in Krakow, application #9041.

References

[1] ATLAS Collaboration, JINST **3**, S08003 (2008).
[2] ATLAS Collaboration, JHEP **01**, 051 (2020), arXiv:1904.04808.
[3] ATLAS Collaboration (2024), arXiv:2405.05078.
[4] ATLAS Collaboration, Eur. Phys. J. C **79**, 970 (2019), arXiv:1907.05120.
[5] S. Alioli, P. Nason, C. Oleari, E. Re, JHEP **06**, 043 (2010), arXiv:1002.2581.
[6] T. Sjöstrand et al., Comput. Phys. Commun. **191**, 159 (2015), arXiv:1410.3012.
[7] E. Bothmann et al., SciPost Phys. **7**, 034 (2019), arXiv:1905.09127.
[8] CMS Collaboration, Phys. Rev. Lett. **119**, 242001 (2017), arXiv:1709.07411.
[9] ATLAS and CMS Collaborations, JHEP **07**, 213 (2023), arXiv:2205.13830.
[10] ATLAS Collaboration, Phys. Rev. Lett. **132**, 202301 (2024), arXiv:2308.16652.
[11] ATLAS Collaboration, Phys. Rev. C **107**, 054907 (2023), arXiv:2206.12594.
[12] ATLAS Collaboration (2024), arXiv:2407.06413.
[13] X.N. Wang, M. Gyulassy, Phys. Rev. D **44**, 3501 (1991).
[14] B. Schenke et al., Phys. Rev. C **82**, 014903 (2010), arXiv:1004.1408.

© The Authors, published by EDP Sciences. This is an open access article distributed under the terms of the Creative Commons Attribution License 4.0 (https://creativecommons.org/licenses/by/4.0/).

EPJ Web of Conferences **316**, 01006 (2025) https://doi.org/10.1051/epjconf/202531601006
SQM 2024

PHENIX highlights: Recent results from PHENIX

Maya Shimomura[1,2,*] for the PHENIX Collaboration

[1]Department of Mathematical and Physical Sciences, Nara Women's University
[2]RIKEN Nishina Center

Abstract. The Relativistic Heavy Ion Collider (RHIC) PHENIX experiment has provided new insights into the behavior of quark-gluon plasma (QGP) under extreme conditions. While PHENIX concluded data collection in 2016, the collaboration remains actively engaged in analyzing the rich dataset yielding the following three important results recently. (1) Heavy flavor (HF) with charm and bottom quark separation indicates a clear mass dependence observed in central collisions. Additionally, forward rapidity J/ψ and v_2 measurements show a result consistent with zero, in contrast to LHC findings. (2) Precise direct photon yields obtained via large statistical datasets, spanning a wide range of transverse momentum (p_T) and centrality bins have the scaling behavior across various large systems, and non-prompt direct photon results reveal temperature (T_{eff}) dependence on p_T. (3) Charged hadron production was examined in collision systems of varying sizes, revealing that R_{AA} is primarily dependent on the overlap volume (N_{part}). Moreover, R_{dAu} obtained by experimental N_{coll} suggests the clear yield suppression in central collisions.

1 Introduction

Quark-gluon plasma (QGP) is a state of matter in which quarks and gluons are no longer confined within hadrons. The study of this extreme state of matter has been one of the major topics in high-energy nuclear physics for several decades. The Relativistic Heavy Ion Collider (RHIC) at Brookhaven National Laboratory (BNL) in the United States has played a central role in deepening our understanding of this phase. The RHIC PHENIX (Pioneering High Energy Nuclear Interaction Experiment) experiment has obtained accumulated extensive data for nine collision species and nine collision energies, providing crucial experimental insights into the properties of QGP under temperatures and energy densities similar to those found in the early universe.

Since its first operation in 2000, PHENIX has produced numerous significant experimental results. For instance, the discovery of strong elliptic flow (v_2) in non-central collisions provided important evidence for the collective motion of the QGP, suggesting that its viscosity is extremely low, consistent with nearly perfect fluid behavior. Furthermore, the observation of yield suppression at high transverse momentum, commonly referred to as jet quenching, revealed that QGP strongly interacts with high transverse momentum partons produced only in the initial collision stages, leading to significant energy loss.

Although data collection concluded in 2016, the PHENIX collaboration remains highly active, continuing to produce new results based on the extensive data set. Recent studies

*e-mail: maya@cc.nara-wu.ac.jp

© The Authors, published by EDP Sciences. This is an open access article distributed under the terms of the Creative Commons Attribution License 4.0 (https://creativecommons.org/licenses/by/4.0/).

have reported new findings on the interactions of heavy-flavor quarks, the production of direct photons, and charged hadron production in various collision systems. These analyses represent important steps toward a deeper understanding of QGP properties and dynamics, contributing not only to mapping the QCD phase diagram but also to research on the boundary conditions for QGP production and comparisons with results from the LHC. This paper focuses on reporting these recent findings.

2 Results from heavy flavor production

The first topic is the heavy flavor at mid and forward rapidity measured by PHENIX. The details of this study are discussed here [1] also. Heavy-flavor quarks (HQ), such as charm (c) and bottom (b), provide a valuable probe for studying the properties of the quark-gluon plasma (QGP) due to their large mass. At RHIC energies, heavy-flavor quarks are primarily produced during the initial hard scattering phase of the collision. Their production can be accurately calculated using perturbative QCD, as their large mass restricts their creation to the early stages of the collision. Once produced, heavy-flavor quarks do not disappear but pass through the QGP, experiencing energy loss through interactions with the QGP medium. This interaction, along with their contribution to the collective motion of the QGP (flow), modifies heavy-flavor production a sensitive indicator of QGP properties.

In PHENIX, the charm and bottom quarks were successfully separated and measured. The transverse momentum (p_T) spectra of charm hadrons and bottom hadrons were obtained for several centrality bins [2]. These results allow for a detailed study of the mass dependence of heavy-flavor interactions with the QGP.

We also calculated the nuclear modification factor (R_{AA}) for the various centralities. The left panel in Figure 1 shows the results for central collisions (0-10%), while the right panel presents the results for peripheral collisions (40-60%) as a function of the electron transverse momentum (p_T). In the 0-10% centrality range, a clear suppression of both bottom and charm quarks is observed, with a distinct difference in the degree of suppression at mid p_T. However, in the 40-60% peripheral collisions, the suppression of both bottom and charm quarks is much smaller and their suppression patterns are similar. This indicates a clear centrality dependence as well as mass ordering in the suppression.

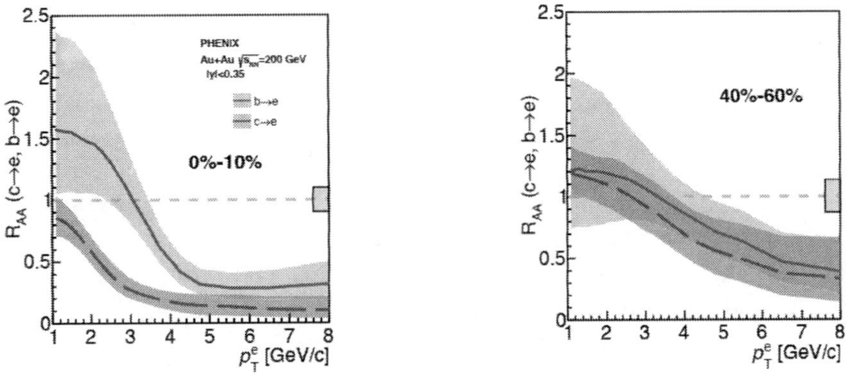

Figure 1. Central (Left) and peripheral (Right) R_{AA} for b and c decayed electron separately in Au + Au. [2]

In addition to these results, we measured the elliptic flow (v_2) of heavy-flavor muons at forward rapidity and compared it to what is obtained at mid-rapidity. The v_2 of heavy-flavor

© The Authors, published by EDP Sciences. This is an open access article distributed under the terms of the Creative Commons Attribution License 4.0 (https://creativecommons.org/licenses/by/4.0/).

quarks is positive at both forward and mid-rapidity, and the results are mostly consistent. In contrast, the v_2 of charged hadrons, which are predominantly composed of light quarks, is larger than that of heavy-flavor quarks. At mid-rapidity, we further separated the v_2 measurements for the charm and bottom quarks. The left panel in Figure 2 shows the v_2 for charm quarks, while the middle panel shows the v_2 for bottom quarks, compared with the v_2 of charged hadrons. As expected from mass order, the v_2 of charm quarks is larger than that of bottom quarks, while charged hadrons exhibit the largest v_2.

We also measured the v_2 of J/ψ at forward rapidity and found that it is consistent with zero, as shown in the right panel in Figure 2. The red points represent the PHENIX results, while the blue points correspond to the LHC-ALICE data, which are not consistent with zero. The R_{AA} for central collisions at RHIC and LHC energies have also significant differences [6]. These results are consistent with the regeneration scenario of charm and anti-charm quarks at LHC energies.

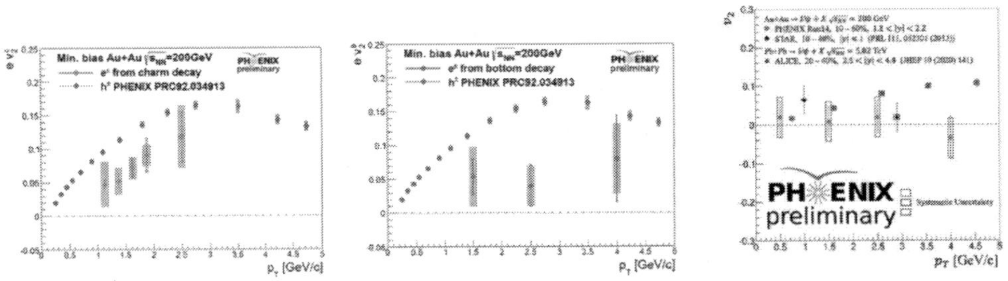

Figure 2. v_2 for b (left) and c (right) decayed electron at mid-rapidity and for J/ψ at forward in Au + Au.

3 The results of direct photon measurements

PHENIX measures direct photon production with large statistics. Figure 3 shows direct photon R_γ vs. p_T, with different centrality intervals for each panel [3]. The different panel shows different centrality [3]. External conversion method and large statistics give a precise measurement for wider p_T ranges for all measured centrality bins. Red points are new results that are consistent with previous measurements shown here but have much less uncertainty. $dN_{\text{direct photon}}/d\eta$ as a function of the $dN_{\text{ch}}/d\eta$ at mid-rapidity. The scaling of the yields holds for various large systems and is different from the p+p scaled by N_{coll} as expected.

A significant enhancement is observed at low p_T when comparing scaled p+p spectra to the central Au+Au one. Same comparisons but different system sizes are also obtained such as Au+Au peripheral, p+Au central, p+Au minimum-bias, and p+p. It seems that this enhancement is related to QGP size [3].

Non-prompt photons were extracted to gain more insight. The left panel in Figure 4 is the ratio of the spectra data to exponential fit where the effective temperature is fixed to a certain value. The ratio has clear p_T dependence but not centrality. Then we applied two different p_T ranges for fitting where the effective temperature is a free parameter and obtained the right figure in Figure 4. The result shows the effective temperature is also p_T dependent, but not dependent on the multiplicity. More details are discussed at [3].

© The Authors, published by EDP Sciences. This is an open access article distributed under the terms of the Creative Commons Attribution License 4.0 (https://creativecommons.org/licenses/by/4.0/).

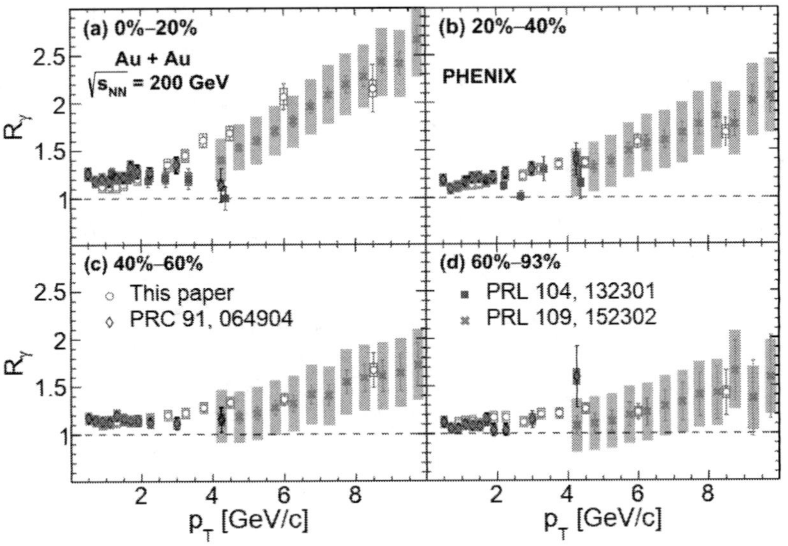

Figure 3. R_γ for direct photon measurement vs. p_T and increasing centrality intervals (panels (a) to (d)).

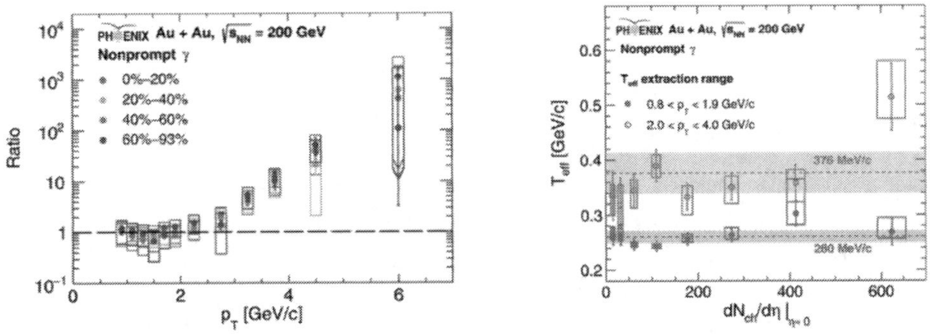

Figure 4. The left panel is the ratio of the spectra data to exponential fit where the effective temperature is fixed to a certain value in Au + Au while the right panel shows the effective temperature vs. $dN/d\eta$ for two indicated p_T ranges [3].

4 Results from PID charged hadron production

Systematics studies of hadron production are performed with PHENIX [4]. Similarly to inclusive charged hadrons, R_{AA} for PID hadrons in both large and small systems depend on collision overlap size (N_{part}) but not collision systems. Figure 5 shows R_{dA} for ϕ, π, K and proton for small colliding systems ($p + Al$, ${}^3He + Au$) as a function of p_T. It shows that ϕ meson has a similar trend to other mesons such as π and K and these are consistent with unity at high p_T.

For d+Au we developed the new method to calculate N_{coll} with direct photon measurement [7]. The left figures in Figure 6 are p_T spectra. π^0, direct photons spectra and their ratio as a function of p_T are shown in the top, middle, and bottom panels respectively. Since the direct photon at high p_T is not suppressed, N_{coll} can be redefined by the direct photon ratio of

© The Authors, published by EDP Sciences. This is an open access article distributed under the terms of the Creative Commons Attribution License 4.0 (https://creativecommons.org/licenses/by/4.0/).

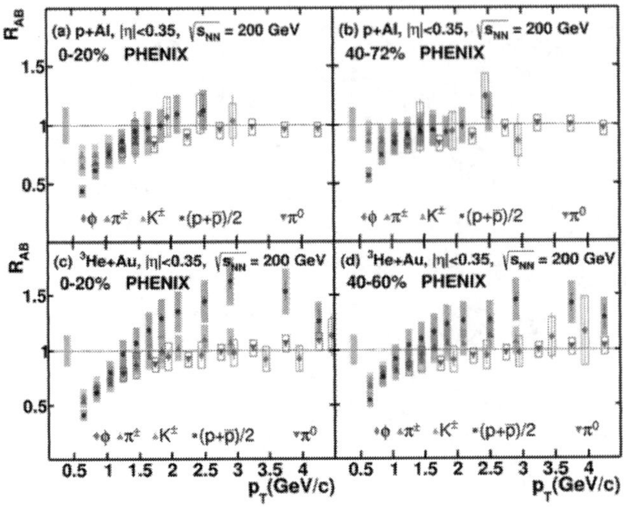

Figure 5. PID R_{AB} of ϕ, π, K and $p + \overline{p}$ for small colliding systems as a function of p_{T}. ϕ meson has a similar trend to other mesons such as π and K.

d+Au to p+p experimentally as following this equation.

$$N_{\mathrm{coll}}^{\mathrm{EXP}}(p_{\mathrm{T}}) = \frac{Y_{d\mathrm{Au}}^{\gamma^{\mathrm{dir}}}(p_{\mathrm{T}})}{Y_{pp}^{\gamma^{\mathrm{dir}}}(p_{\mathrm{T}})}. \tag{1}$$

The right figures in Figure 6 are the calculated experimental N_{coll} compared with Glauber MC N_{coll}. You can see for 0-100% centrality, N_{coll} Glauber and N_{coll} experiment agree well. However, if we divide into certain centrality bins, these are different. To avoid the N_{coll} Glauber bias at small system, we use this experimental N_{coll} to re-calculate R_{AA}.

Figure 7 shows the re-defined R_{dA} with experimental N_{coll} for different centrality as a function of p_{T}, and the average of these R_{dA} is plotted at right figure as a function of the N_{coll}. The clear suppression can be seen at central in d+Au while it's consistent with unity in peripheral d+Au. Therefore, we could conclude there is a suppression at the central [7].

5 Summary

Recent PHENIX results are summarized. c/b separated heavy flavor R_{AA} and v_2 are successfully measured at various centrality bins and clear mass dependence can be seen at central. Measurement of the forward J/ψ v_2 is also performed and it's consistent with zero, and unlike LHC results, it is consistent with the regeneration scenario of charm and anti-charm at LHC energy. External conversion method and large statistics give precise measurement for wide p_{T} ranges and all centrality bins, and the scaling of yields holds for various large systems. Non-prompt direct photons are extracted and show the T_{eff} has the dependence of the p_{T}. PHENIX measured charged hadron production at small to large collision systems and found mostly R_{AA} only depends on overlap volume (N_{part}). Experimental N_{coll} gives non-GL biased R_{dAu}, and indicates the clear yield suppression for central collisions.

© The Authors, published by EDP Sciences. This is an open access article distributed under the terms of the Creative Commons Attribution License 4.0 (https://creativecommons.org/licenses/by/4.0/).

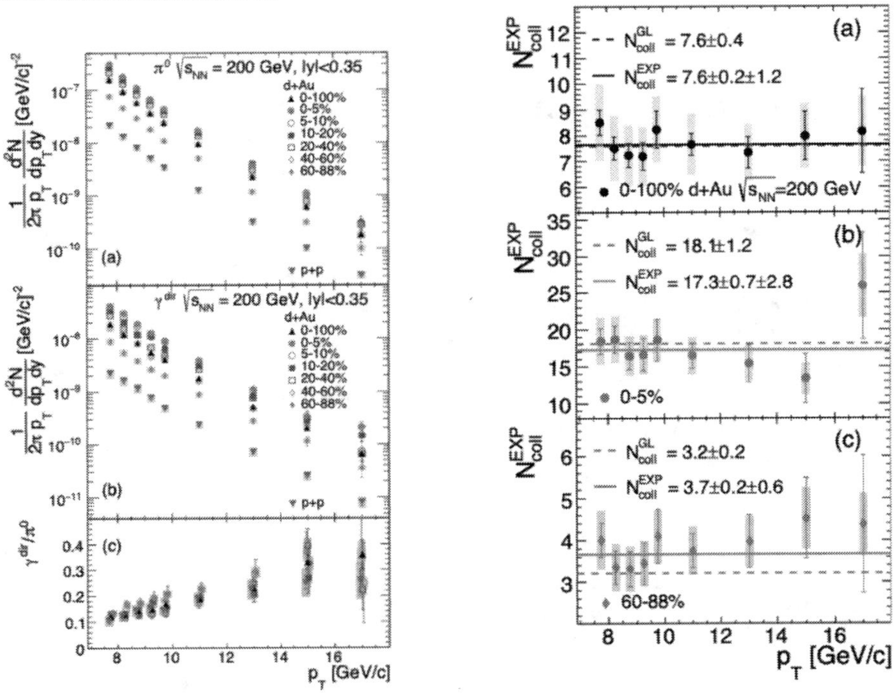

Figure 6. Left panels are p_T spectra for π^0 (top), direct photons (middle), and the ratio (bottom). The right panels show the calculated experimental N_{coll} compared with Glauber MC N_{coll} [7].

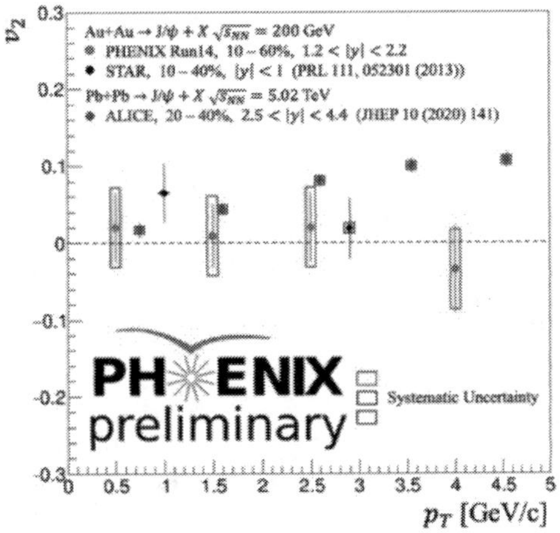

Figure 7. Nuclear modification factor for π^0 with the experimental N_{coll}.

References

[1] Luis Bichon, these proceedings.
[2] PHENIX Collaboration, Phys Rev C **109**, 044907 (2024).
[3] PHENIX Collaboration, Phys Rev C **109**, 044912 (2024).
[4] PHENIX Collaboration, Phys Rev C **109**, 054910 (2024).
[5] PHENIX Collaboration, Phys Rev C **107**, 014907 (2023).
[6] PHENIX Collaboration, Phys Rev C **84**, 054912 (2011).
[7] PHENIX Collaboration, https://doi.org/10.48550/arXiv.2303.12899

© The Authors, published by EDP Sciences. This is an open access article distributed under the terms of the Creative Commons Attribution License 4.0 (https://creativecommons.org/licenses/by/4.0/).

EPJ Web of Conferences **316**, 01007 (2025) https://doi.org/10.1051/epjconf/202531601007
SQM 2024

sPHENIX Highlights: First Results from sPHENIX at RHIC

Rachid Nouicer[1,*] (for the sPHENIX Collaboration)

[1]Physics Department, Brookhaven National Laboratory, Upton, New York 11973, United States

Abstract. First results from the sPHENIX experiment on the π^0 v_2 and $dE_T/d\eta$ in Au+Au collisions at $\sqrt{s_{NN}}$ = 200 GeV using detector commissioning data during the RHIC 2023 Run are presented. These results are shown across a large centrality range, and compared to previous PHENIX and STAR results. These measurements demonstrate the sPHENIX capabilities towards accomplishing the sPHENIX jet physics program. From the ongoing RHIC 2024 Run, $p + p$ collisions at $\sqrt{s}$ = 200 GeV, a first glimpse of the measurements K^0_S and Λ^0 obtained by the sPHENIX Time Projection Chamber (TPC) detector are shown.

1 Introduction

sPHENIX is a new experiment currently operating at the Relativistic Heavy Ion Collider (RHIC) at Brookhaven National Laboratory (BNL). The sPHENIX detector was commissioned and took first Au + Au data during the RHIC 2023 Run [1, 2]. The sPHENIX experiment provides state-of-art capabilities for studies of the strongly interacting quark-gluon plasma using jet and heavy-flavor observables. The goal of sPHENIX is to understand the microscopic structure of the plasma and reveal how its strongly interacting nature arises from the underlying interactions of quarks and gluons described by quantum chromodynamics [3].

This paper presents first sPHENIX measurements on the π^0 v_2 and $dE_T/d\eta$ in Au+Au collisions at $\sqrt{s_{NN}}$ = 200 GeV obtained during detector commissioning in the RHIC 2023 Run. In addition, a first look at the K^0_S and Λ^0 measured by the sPHENIX TPC in $p + p$ collisions at $\sqrt{s}$ = 200 GeV from the ongoing RHIC 2024 Run are presented.

2 sPHENIX Detector

The sPHENIX detector is a cylindrical spectrometer covering $|\eta| \leq 1.1$ and the full azimuth. Figure 1 shows a schematic of the sPHENIX detector components during RHIC 2023 and 2024 Runs. It uses the former 1.4 Tesla BaBar superconducting solenoid magnet to contain an inner tracking system followed by electromagnetic and hadronic calorimeters. The sPHENIX subdetectors are as follows.

1) Tracking system: the tracking system from inside-out radially consists of three components:

- **MAPS Vertex Tracker (MVTX):** the MVTX sensors are a Monolithic Active Pixel Sensors with pixel pitch of 27 μm. It is a copy of the inner three layers of the ALICE Inner Tracking System (ITS) with layers at radial positions of 2.5, 3.2, and 4.0 cm. The MVTX three-layer silicon pixel detector are located just outside the sPHENIX beryllium beam-pipe. The MVTX measures the collision vertex, the displaced vertex of charged tracks, and contributes to the overall tracking resolution of sPHENIX.

*e-mail: nouicer@bnl.gov

© The Authors, published by EDP Sciences. This is an open access article distributed under the terms of the Creative Commons Attribution License 4.0 (https://creativecommons.org/licenses/by/4.0/).

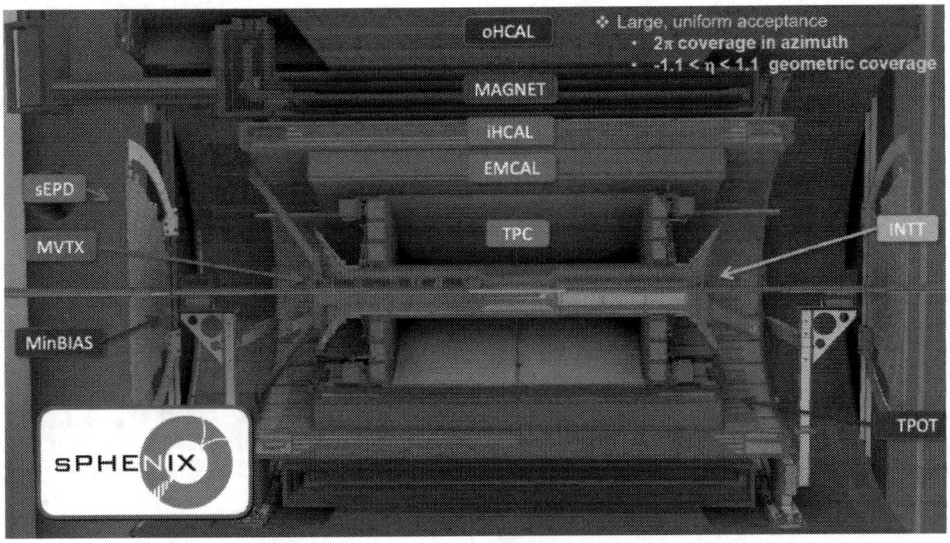

Figure 1. A Schematic of side view of the sPHENIX detector during RHIC 2023 and 2024 Runs.

- **Silicon Strip INTermediate Tracker (INTT)**: the INTT is a silicon strip detector with strip pitch 78 μm in ϕ, consisting of two hermetic cylindrical barrels. The INTT plays key roles in the sPHENIX experiment; 1) it provides seeds and enhances the resolution of the track reconstruction of the charged particles, and 2) it has 60 ns time resolution allowing it to read out collision data from each single RHIC beam bunch crossing and suppress event pileup background.
- **Time Projection Chamber (TPC)**: the TPC is a compact cylindrical tracking detector, 2.1 m in length with an outer radius of 80 cm and a total gas volume of ≈ 4 m^3 . The TPC operates with an Ar-CF_4-Isobutane 75/20/5 gas mixture. The TPC field cage is constructed from carbon fiber/honeycomb for mechanical strength and use 18 layers of 3 mil kapton to insulate high voltage (45 kV) from ground. The device is read out on each end with quad-GEM modules each containing four layers of GEM foils plus a cathode pad plane. There are 36 quad-GEM modules on each end of the TPC. The TPC makes high precision measurements (space point resolution < 200 μm) of the momenta of all charged tracks passing through sPHENIX at mid-rapidity. Figure 2 shows a first look at the K_S^0 and Λ^0 peaks in $p+p$ collisions at $\sqrt{s} = 200$ GeV from the ongoing RHIC 2024 Run. These measurements were obtained using only the TPC reconstructed tracks passing a loose selection cuts with no corrections to the measured track trajectory.
- **Time Projection Outer Tracker (TPOT)**: the TPOT consists of eight identical Micromegas modules, grouped in three sectors [4]. The three sectors are mounted to the Electromagnetic Calorimeter (EmCal) at the bottom facing TPC. Each TPOT module measures a space point of charged tracks passing through TPC. Its function is to provide tracking distortion correction-information for the TPC.

2) **Electromagnetic Calorimeter (EmCal)**: just outside the TPC/TPOT, the EmCal detector is an electromagnetic calorimeter with solid angle coverage of $|\eta| < 1.1$ and 2π in ϕ. The EmCal is 20.1 radiation lengths deep and is designed to measure photons, electrons and positrons via electromagnetic showers. The sPHENIX EmCal is constructed from a tungsten powder absorber with embedded scintillating fibers to make towers of size $\Delta\eta \times \Delta\phi = 0.024 \times 0.024$. The light from the scintillating fibers is collected by a light guide onto four gain-matched silicon photomultipliers (SiPMs). The EmCal is 0.83 hadronic interaction lengths deep and therefore sees a significant amount of hadronic shower energy [5].

© The Authors, published by EDP Sciences. This is an open access article distributed under the terms of the Creative Commons Attribution License 4.0 (https://creativecommons.org/licenses/by/4.0/).

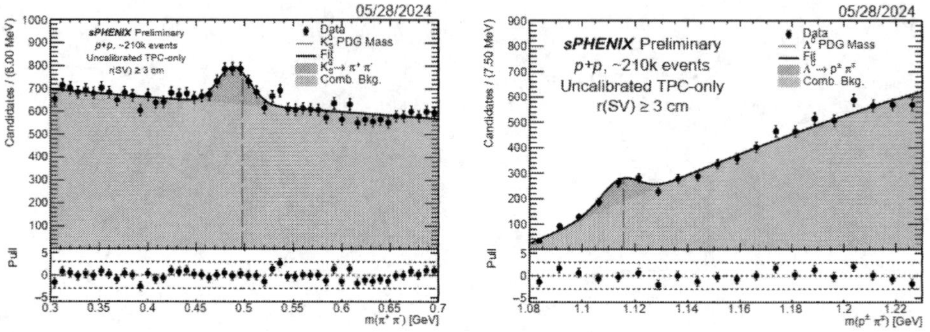

(a) K_S^0 peak measurement in $p + p$ at 200 GeV

(b) Λ^0 peak measurement in $p + p$ at 200 GeV

Figure 2. Measured invariant mass spectra by the sPHENIX-TPC detector: (a) $\pi^+\pi^-$, and (b) $p^\pm\ \pi^\mp$ obtained in $p + p$ collisions at $\sqrt{s}$ = 200 GeV from RHIC 2024-Run. The combinatorial background and signals are shown in gray and blue, respectively.

3) **Hadronic Calorimeter (HCal)**: the HCal consists of two components; Inner Hadronic Calorimeter (IHCal) located inside the magnet cryostat (just after EmCal), and Outer Hadronic Calorimeter (OHCal) located outside the magnet. Both IHCal and OHCal are cylindrical. The HCal is 4.9 hadronic interaction lengths thick. Both the IHCal and OHCal are sampling calorimeters consisting of aluminum (inner)/steel (outer) absorbing plates and scintillating tiles with tower size $\Delta\eta \times \Delta\phi = 0.1 \times 0.1$. Both IHCal and OHCal scintillating tiles are set at an angle offset to the transverse direction to reduce channeling particles that do not interact with the active volumes of these calorimeters. The light from these tiles is collected into SiPMs.

4) **Global Detectors** are comprised of following elements:

- **Minimum Bias Trigger Detector (MBD)** consists of 64 photomultiplier arrays with quartz radiator (PMTs) on each of the two arms covering the very forward region in $3.61 < |\eta| < 4.51$. The PMTs are arranged in 3 concentric rings around the beam-pipe, covering 2π in azimuth. The MBD system timing resolution is 50 ps. The MBD is used for minimum bias, centrality-selected event triggering.
- **sPHENIX Event Plane Detection (sEPD)**: the sEPD is a pair of disks of scintillating tiles finely segmented in ϕ with readout into SiPMs located at forward pseudorapidity $2 < |\eta| < 4.9$. Its role is high-resolution event plane determination.
- **Zero Degree Calorimeters (ZDC)**: the ZDC are small transverse area hadron calorimeters located downstream of the DX dipole magnets in each side of the sPHENIX experiment about 18 m from the interaction point. It is interleaved layers of tungsten and PMMA optical fibers readout with photomultiplier tubes. The ZDC measure neutral energy within a 2 mrad cone about the beam direction.

3 First Results

3.1 Measurement π^0 v_2 in Au + Au Collisions at $\sqrt{s_{NN}}$ = 200 GeV From RHIC 2023 Run

The data used for this analysis was taken with the EmCal, and MBD detectors. The EmCal is used to measure photons to reconstruct π^0's, and the MBD is used for minimum bias,

© The Authors, published by EDP Sciences. This is an open access article distributed under the terms of the Creative Commons Attribution License 4.0 (https://creativecommons.org/licenses/by/4.0/).

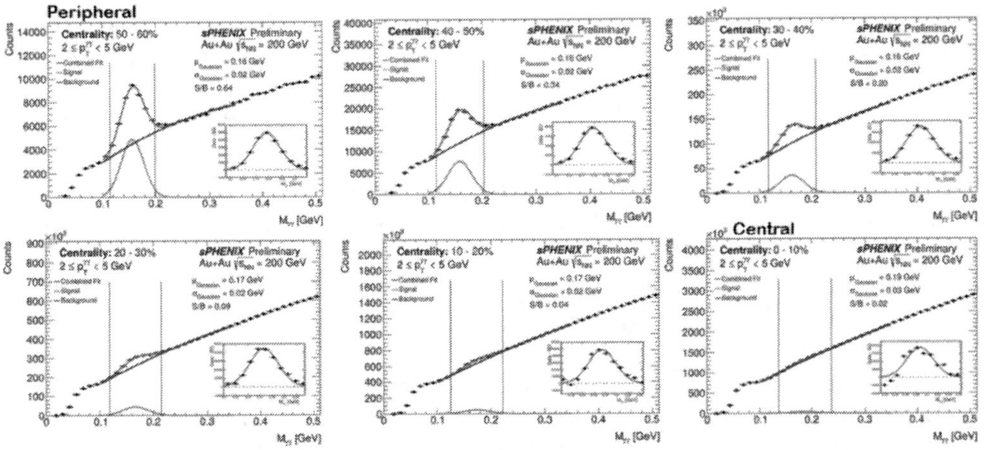

Figure 3. Diphoton invariant mass $M_{\gamma\gamma}$ spectrum for each centrality bin measured by sPHENIX in Au + Au at $\sqrt{s_{NN}} = 200$ GeV during detector commissioning in the RHIC 2023 Run. In each panel, the red curve is the total fit, the orange curve is the Gaussian used to fit the signal, and the blue curve is the polynomial used to fit the combinatorial background. The insets show the fitted mass peak after background subtraction. The uncertainties in these invariant mass distributions are statistical only.

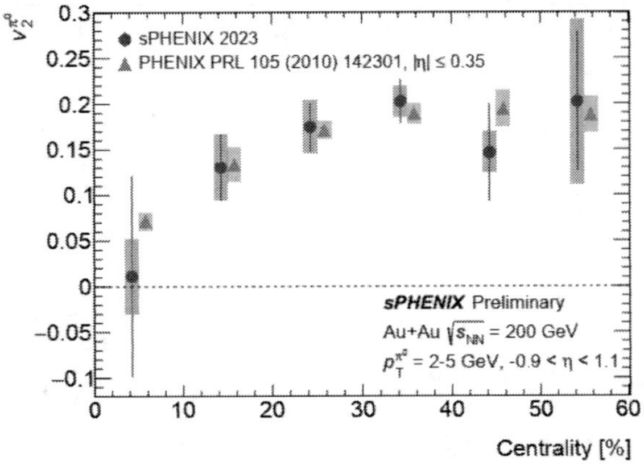

Figure 4. Elliptic flow $v_2^{\pi^0}$ as a function of centrality measured by sPHENIX in Au + Au at $\sqrt{s_{NN}} = 200$ GeV during detector commissioning in the RHIC 2023 Run. These measurements are integrated over the range of $2 \leq p_T \leq 5$ GeV. The results, shown in blue with both statistical and systematic uncertainties, are offset to the left, while PHENIX data (integrated over the same p_T range as the sPHENIX points) [7], shown in red, is displaced to the right in each centrality bin for visibility.

centrality-selected event triggering, and event centrality determination [6]. In addition, signals from the MBD PMTs are used within this analysis to calculate the flow reaction plane in collision events. During the 2023 commissioning run, subsystems were in partial operation, and the EmCal was read out in the region $-0.9 < \eta < 1.1$. The EmCal data analysis starts by doing a tower-by-tower energy calibration and making the tower response uniform in ϕ. Pairs of clusters are then found and the invariant mass of the pair is determined. η-dependent calibration constants which move the π^0 mass peak to the expected value are extracted based on the η location of the most energetic tower within the higher energy cluster of the pair. This procedure is done iteratively until the calibration is stable for all η rings. In reconstructing the π^0 candidates, each pair of clusters in the event that satisfies cuts of $\chi^2 < 4$ and $E_{\text{cluster}} \geq 1$ GeV is considered. The $E_{\text{cluster}} \geq 1$ GeV requirement is applied to exclude clusters that originate from noise in the EmCal. After applying the selection criteria, in-

© The Authors, published by EDP Sciences. This is an open access article distributed under the terms of the Creative Commons Attribution License 4.0 (https://creativecommons.org/licenses/by/4.0/).

variant mass ($M_{\gamma\gamma}$) distributions of π^0 candidates with $2 \leq p_T \leq 5$ GeV are generated for six centrality selections from 0–60%. The distributions are fit with a combined Gaussian and second-order polynomial function.

Figure 3 shows the diphoton invariant mass $M_{\gamma\gamma}$ spectrum for each centrality bin measured by sPHENIX in Au + Au at $\sqrt{s_{NN}}$ = 200 GeV during detector commissioning in the RHIC 2023 Run. The vertical black lines mark the signal bounds, calculated in the range $\mu_{\text{Gauss}} \pm 2\sigma_{\text{Gauss}}$. Here, μ_{Gauss} denotes the Gaussian mean (peak position), and σ_{Gauss} the standard deviation. The fit is performed over the range $0.1 < M_{\gamma\gamma} < 0.35$ GeV, with the upper bound optimized to adequately cover the background while excluding the influence of the η meson (with $M_{\gamma\gamma} = 0.55$ GeV), and the lower bound is set to exclude the region $M_{\gamma\gamma} < 0.1$ GeV [6].

Figure 4 shows the results of elliptic flow, $v_2^{\pi^0}$, integrated over the range of $2 \leq p_T \leq 5$ GeV as a function of centrality bins. The results are overlaid with p_T-integrated data from PHENIX measurement [7]. Good agreement is observed between the two measurements across the full centrality range studied.

3.2 Measurement of $dE_T/d\eta$ in Au + Au Collisions at $\sqrt{s_{NN}}$ = 200 GeV From RHIC 2023 Run

To obtain $dE_T/d\eta$ measurements, we use data from all three calorimeters, EmCal, IHCal, and OHCal. In addition, the MBD detector is used for determination of the collision vertex and the centrality of the event triggering.

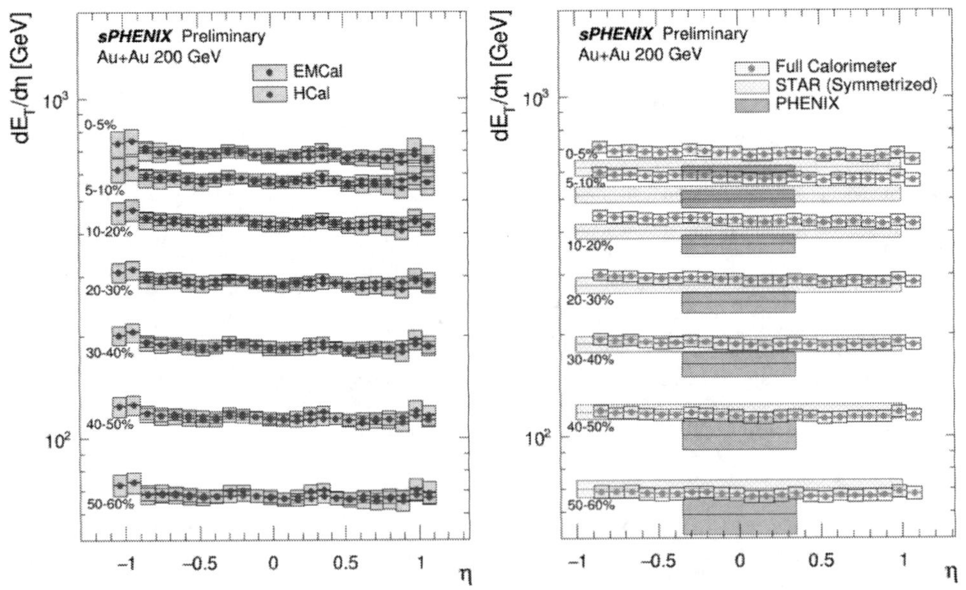

(a) $dE_T/d\eta$ measurements EmCal vs HCal

(b) $dE_T/d\eta$ measurements from sPHENIX, PHENIX, and STAR

Figure 5. Measured $dE_T/d\eta$ for different centrality bins obtained in Au + Au at $\sqrt{s_{NN}}$ = 200 GeV using detector commissioning data from the RHIC 2023 Run. (a) comparison of $dE_T/d\eta$ measurements for EmCal-only results ($-0.9 < \eta < 1.1$) and HCal-only results ($-1.1 < \eta < 1.1$). (b) Comparison of $dE_T/d\eta$ measurements for different centrality bins obtained by sPHENIX, PHENIX, and STAR experiments.

© The Authors, published by EDP Sciences. This is an open access article distributed under the terms of the Creative Commons Attribution License 4.0 (https://creativecommons.org/licenses/by/4.0/).

In the present analysis [6], the absolute energy scale of the EmCal is established from an η-dependent calibration of the π^0 meson peak in data to the same position as in simulation. The η region for the EmCal is limited to $-0.9 < \eta < 1.1$ because the EmCal readout electronics were only partially instrumented during this stage of commissioning. It should be noted that the calibration of the EmCal is the same as in the π^0 v_2 analysis presented in the previous section. The IHCal and OHCal absolute EM energy scale calibration is performed using the minimum ionizing particle energy depositions from cosmic ray muons from data taken in early 2024. Temperature-dependent corrections to the detector energy scale are applied to these cosmics calibrations to account for gain variations from detector conditions during collision data-taking. The uncorrected $dE_T/d\eta$ is calculated for a given centrality class as the sum of calorimeter tower E_T as a function of η. The correction factors for each calorimeter sub-system are fairly constant with centrality.

In Fig. 5(a), the EmCal-only and HCal-only $dE_T/d\eta$ measurements are overlaid to highlight their agreement. This is a particularly encouraging result as the EmCal and HCal see different contributions of the collision energy. Further, for all calorimeter measurements, $dE_T/d\eta$ at positive η and negative η are compatible within uncertainties. The uncertainty contributions to the measurement include the calorimeter energy response, the application of the particle spectra reweighing in simulation, calorimeter noise, detector acceptance, and z-vertex resolution effects.

Figure 5(b) shows comparison of the present $dE_T/d\eta$ measurements from sPHENIX to published results from the PHENIX [8] and STAR [9] experiments. The sPHENIX results are consistently higher than the results from PHENIX for all centrality bins but agree within uncertainties for mid-central bins 30-60%; the sPHENIX results are above the STAR results in the centrality range of 0-10% but are in agreement in other centrality intervals. Presently, these sPHENIX results use a preliminary centrality calculation for Run 2023 data, which is expected to be updated before publication.

4 Summary

sPHENIX measured π^0 v_2 and $dE_T/d\eta$ in Au + Au at $\sqrt{s_{NN}}$ = 200 GeV obtained during detector commissioning in the RHIC 2023 Run. These results are measured with the calorimeter system, including the hadronic calorimeter, which is used for the first time for such observables at RHIC. Results are shown as a function of η across a large centrality range, and compared to previous PHENIX and STAR results demonstrating good understanding of the sPHENIX calorimeter response, and laying the path towards accomplishing the sPHENIX jet and heavy flavor physics program at RHIC.

References

[1] A. Adare et al. (sPHENIX Collaboration), arXiv:1501.06197
[2] E. O'Brien (sPHENIX Collaboration) EPJ Web of Conf. **296**, 01007 (2024)
[3] R. Belmont et al. Nucl. Phys. **A 1043**, 122821 (2024)
[4] S. Aune et al., Nuclear Inst. and Methods in Physics Research, **A 1066**, 169615 (2024)
[5] C. A. Aidala et al., IEEE Transactions on Nuclear Science, **vol. 68**, no. 2 , 173 (2021)
[6] E. McLaughlin (sPHENIX Collaboration), this conference proceeding
[7] A. Adare et al. (PHENIX Collaboration), Phys. Rev. Lett. **105**, 142301 (2010)
[8] A. Adare et al. (PHENIX Collaboration), Phys. Rev. C **93**, 024901 (2016)
[9] J. Adams et al. (STAR Collaboration), Phys. Rev. C **70**, 054907 (2004)

© The Authors, published by EDP Sciences. This is an open access article distributed under the terms of the Creative Commons Attribution License 4.0 (https://creativecommons.org/licenses/by/4.0/).

EPJ Web of Conferences **316**, 01008 (2025) https://doi.org/10.1051/epjconf/202531601008
SQM 2024

Recent results from NA61/SHINE

Andrzej Rybicki[1,*] for the NA61/SHINE Collaboration

[1]Institute of Nuclear Physics, Polish Academy of Sciences, Radzikowskiego 152, 31-342 Kraków, Poland

Abstract. The NA61/SHINE experiment at the CERN SPS is a multipurpose fixed-target spectrometer for charged and neutral hadron measurements. Its research program includes studies of strong interactions as well as reference measurements for neutrino and cosmic-ray physics. A significant advantage of NA61/SHINE over collider experiments is its extended coverage of phase space available for hadron production. This includes the nearly entire forward hemisphere for charged hadrons and additionally, a large part of the backward hemisphere for specific neutrals. This paper summarizes a selected set of new results, obtained by NA61/SHINE since the last SQM conference (Busan, 2022). Particular attention is devoted to (1) the first-ever direct measurement of open charm production in nucleus-nucleus collisions at SPS energies (2) the difference observed between charged and neutral meson production in Ar+Sc reactions, up to now not understood by existing models, and (3) the importance of baseline effects in the search for the critical point of strongly interacting matter.

1 Introduction

NA61/SHINE [1] is the second largest fixed target experiment at CERN. As a direct, upgraded continuation of its predecessor NA49, the detector offers a nearly continuous experimental data flow for three decades. Its research program includes heavy ion physics as well as measurements for neutrino (J-PARC, Fermilab) and cosmic-ray studies (Pierre-Auger, KASCADE, satellite experiments). In this paper, only a selected set of most recent results related to the NA61/SHINE strong interaction program will be addressed.

2 Open charm physics at the SPS

The first-ever direct measurement of open charm production in nucleus-nucleus collisions at CERN SPS energies has recently been performed by NA61/SHINE. This result, obtained in Xe+La collisions at $\sqrt{s_{NN}}$=16.8 GeV, is described in detail in Ref. [2]. The $D^0 + \overline{D}^0$ yield visible in acceptance was extrapolated to the total 4π yield using three different models to estimate the corresponding model-related uncertainty. As apparent in Fig. 1, the NA61/SHINE measurement offers sufficient discrimination power to bring a decisive improvement to existing theoretical models of open charm production, which in their present version differ by as much as three orders of magnitude at this collision energy.

The second direct measurement in A+A reactions, and the first-ever for open charm in heavy ion collisions at the SPS, is also in progress in the framework of the NA61/SHINE open

*e-mail: andrzej.rybicki@ifj.edu.pl

© The Authors, published by EDP Sciences. This is an open access article distributed under the terms of the Creative Commons Attribution License 4.0 (https://creativecommons.org/licenses/by/4.0/).

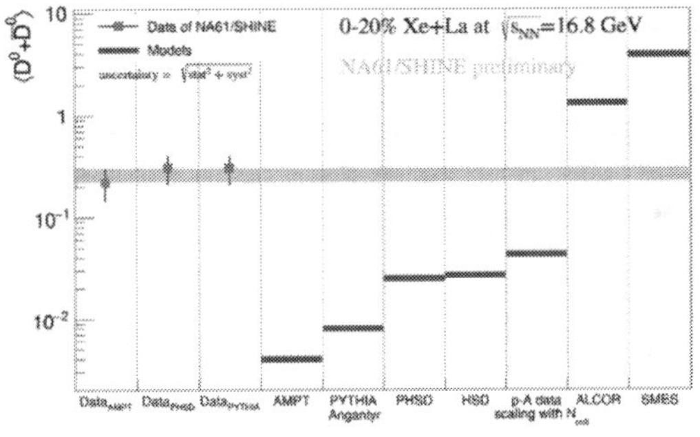

Figure 1. The open charm 4π yield $\langle D^0 + \overline{D}^0\rangle$ in central Xe+La collisions at $\sqrt{s_{NN}}$=16.8 GeV, put in comparison to predictions of theoretical models. The red band gives the theoretical (model-related) uncertainty. The figure comes from Ref. [2] (see therein for more details).

charm program with the upgraded detector. The corresponding data-taking of Pb+Pb collisions started in 2022 and is planned to be completed in 2025. The accumulated statistics will allow for the first measurements of open charm as a function of Pb+Pb collision centrality, rapidity, or transverse momentum ever made at SPS energies. This will further significantly improve the theoretical knowledge of charm production mechanisms close to threshold. A continuation of charm studies at the SPS with still higher statistics was also proposed by the NA60+ Collaboration, in a Letter of Intent at the end of 2022 [3].

3 Unexpected excess of charged over neutral kaons

A simultaneous, large-acceptance measurement of K^+, K^-, and K^0_S production in Ar+Sc collisions at $\sqrt{s_{NN}}$=11.9 GeV has recently been performed [4, 5] (see also a detailed account in Ref. [6]). A comparison of averaged charged to neutral kaon rapidity spectra is shown in Fig. 2 (a). A clear excess of charged kaon production is apparent. Isospin symmetry between u and d quarks, which can be regarded as a specific case of the flavour symmetry of QCD in the massless quark limit, predicts the equality $(K^+ + K^-)/2 = (K^0 + \overline{K}^0)/2 = K^0_S$ for „charge-symmetric" nuclei ($Z = N \equiv A-Z$) once we neglect the very small effect of CP violation. The observed inequality of charged and neutral kaon yields does not follow this prediction. As

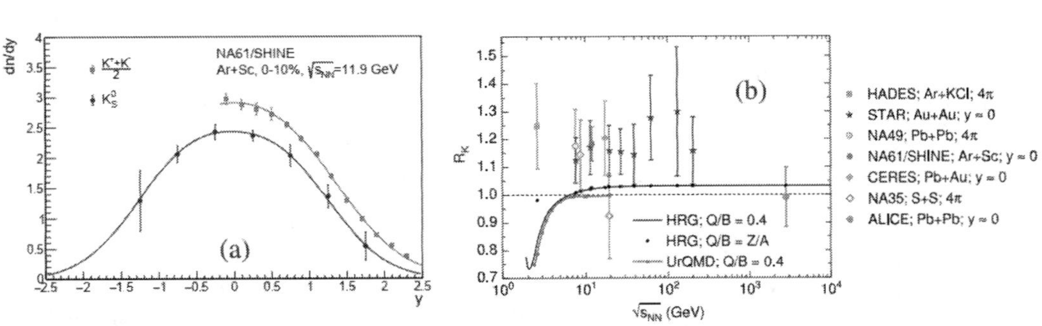

Figure 2. (a) Comparison of rapidity spectrum of neutral kaons (K^0_S) with the averaged spectrum of charged kaons $(K^+ + K^-)/2$ in Ar+Sc collisions. From Ref. [5]. (b) Ratio $R_K = (K^+ + K^-)/(2K^0_S)$, drawn as a function of $\sqrt{s_{NN}}$ for NA61/SHINE data and the compilation of results from earlier experiments. The compilation includes ratios of mid-rapidity densities as well as 4π yields, see Ref. [5] for a full bibliography. Experimental data is compared to predictions of UrQMD and HRG models (Q/B is the assumed charge-to-baryon number ratio). From Ref. [7], updated. In both plots total uncertainties are drawn.

© The Authors, published by EDP Sciences. This is an open access article distributed under the terms of the Creative Commons Attribution License 4.0 (https://creativecommons.org/licenses/by/4.0/).

such, it is to be concluded that this constitutes an evidence of an unexpectedly large violation of isospin symmetry, or flavour symmetry, in the high-energy Ar+Sc reaction.[1]

A comparison of the NA61/SHINE result to a compilation performed on the basis of other existing data is shown in Fig. 2 (b). Although with large uncertainties for the individual measurements, the latter consistently support the Ar+Sc result in the range $2.6 < \sqrt{s_{NN}} < 200$ GeV. The observed effect cannot be understood on the basis of UrQMD and HRG models, including known effects of isospin symmetry violation (see also Ref. [7]).

An unexpected effect of strong violation of isospin symmetry in multi-particle production can be of fundamental importance for the understanding of the strong force. This calls for a concentrated experimental and theoretical effort.

On the experimental side, it is interesting to note that the experimental uncertainties of the earlier measurements shown in Fig. 2 (b) appear larger than those of NA61/SHINE. This is in spite of the larger statistics of data published by collider experiments. Evidently, this question requires clarification from the respective Collaborations (the discussion of NA61/SHINE uncertainties can be found in Refs. [4, 5]). We note that a part of this problem could originate from the fact that collider measurements of dn/dy yields at mid-rapidity can never be performed in the complete kinematic range due to their inherent p_T−cut off, which will increase their uncertainty. The ability to perform measurements starting at $p_T = 0$ (sharp) is a clear advantage of NA61/SHINE.

Finally, we note that the excess of charged over neutral kaons has also been observed by NA61/SHINE in π^-+C collisions. This effect cannot be explained by microscopic models (for more details, see Ref. [8]). Therefore, in October 2024 the experiment will perform a simultaneous measurement of a charge-symmetric ensemble of π^++C and π^-+C reactions [9]. This will allow to clarify whether the unexpected violation of flavour symmetry is unique to nucleus-nucleus collisions or is a more general property of strong interactions.

4 Energy dependence of strangeness production

The SPS regime has a unique position in the energy and system size dependence of strangeness production. This is perfectly apparent in Fig. 3 (a), which shows the evolution of the K^+/π^+ ratio as a function of $\sqrt{s_{NN}}$ for different systems measured by NA61/SHINE with addition of other experiments, including our new preliminary data on Xe+La collisions [10]. With increasing system size, the K^+/π^+ ratio remains similar for $p+p$ and Be+Be reactions but rapidly enhances for Ar+Sc collisions, where it approaches Xe+La and Pb+Pb data at top SPS energy. The non-monotonic "horn" in strangeness production is, at present, seen only in Pb+Pb and Au+Au reactions.

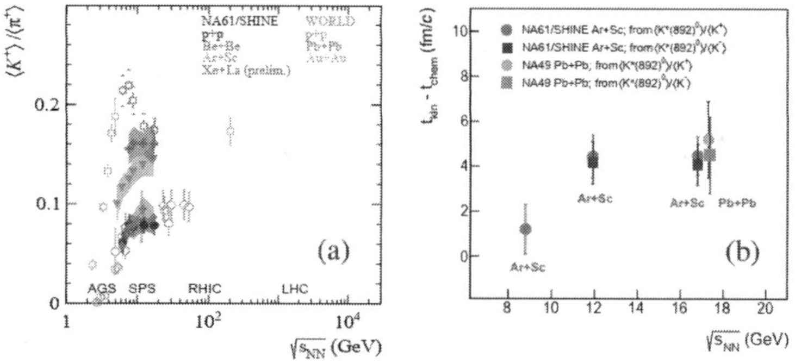

Figure 3. (a) Energy dependence of the K^+/π^+ ratio of 4π multiplicities in different systems (b) energy dependence of the time between two freeze-outs as discussed in the text.

[1]The Ar and Sc nuclei are not exactly charge-symmetric. The moderate excess of neutrons over protons is expected to decrease charged kaon compared to neutral kaon production, contrary to the result shown in Fig. 2 (a).

© The Authors, published by EDP Sciences. This is an open access article distributed under the terms of the Creative Commons Attribution License 4.0 (https://creativecommons.org/licenses/by/4.0/).

Complementary, Fig. 3 (b) shows the energy dependence of the estimated time between chemical and kinetic freeze-outs in Ar+Sc reactions ($t_{kin} - t_{chem}$), also compared to Pb+Pb collisions at top SPS energy. The latter estimate is evaluated on the basis of the suppression of $K^*(892)/K^{+,-}$ ratios between p+p and central Ar+Sc (Pb+Pb) collisions, using the same methodology as in Ref. [11]. This is possible thanks to the new preliminary data on $K^*(892)$ production in Ar+Sc reactions from NA61/SHINE [12]. A very peculiar pattern emerges: at $\sqrt{s_{NN}} \gtrsim 12$ GeV the time between the two freeze-outs exceeds 4 fm/c, but it is consistent with zero at $\sqrt{s_{NN}}$=8.8 GeV (no suppression of the K^*/K ratio in Ar+Sc compared to p+p collisions indicates a shorter hadronic phase at this collision energy [13]).

These apparently rapid changes of system properties with system size and energy will be further quantified by NA61/SHINE, including a new energy scan of light, charge-symmetric nuclei (B+B, O+O, Mg+Mg) after LS3 [14].[2]

5 Importance of non-critical effects in the search for the CP

Comprehensive theoretical reviews of the status of the search for the hypothetical critical point of the phase diagram of strongly interacting matter have been given at this Conference [15, 16]. NA61/SHINE contributes to this search by a versatile program of experimental studies bringing an altogether negative result, but at the same time demonstrating the crucial importance of a properly complete understanding of the non-critical „baseline", that is, of spurious effects that could possibly be erroneously interpreted as indications for the system freezing out at the critical point.

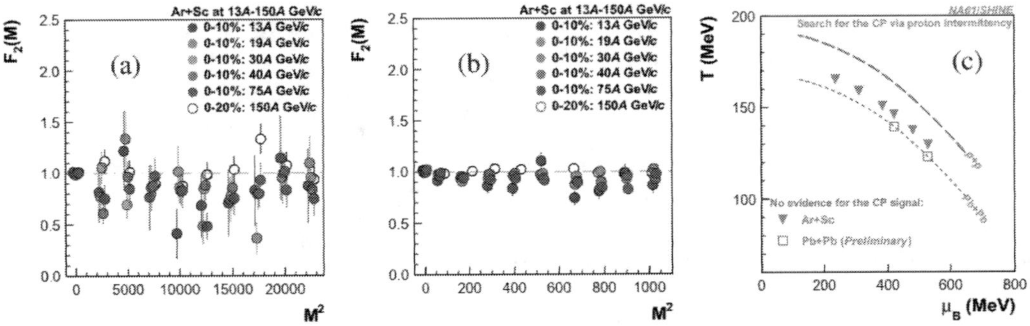

Figure 4. Proton intermittency studies in NA61/SHINE. Scaled Factorial Moments F_2 of proton multiplicity, drawn as a function of the number M^2 of cells in the cumulative transverse momentum plane for (a) $1^2 \leq M^2 \leq 150^2$, and (b) $1^2 \leq M^2 \leq 32^2$. Points for different energies are slightly shifted horizontally. (c) Diagram of freeze-out temperature and baryochemical potential, marking reactions where no evidence for the CP was found. All plots come from Ref. [17]. Preliminary Pb+Pb data from Ref. [18].

In Figs. 4 (a)-(b), new results from NA61/SHINE proton intermittency studies in Ar+Sc collisions are shown. The data comes from Refs. [17, 19] where all methodological details and definitions can be found. The second-order Scaled Factorial Moments F_2 show no apparent power-law scaling with the number M^2 of cells in the transverse momentum plane, which could have been an indication of the CP. The studied reactions with no evidence for the CP cover an appreciable part of the phase diagram, see Fig. 4 (c). As described in the cited papers, this analysis employs dedicated procedures aimed at elimination of spurious, non-critical effects. This includes, among others, a cumulative transformation of protons' (p_x, p_y)

[2] For a full bibliography of NA61/SHINE and other data discussed in this section, see Refs. [4, 12].

© The Authors, published by EDP Sciences. This is an open access article distributed under the terms of the Creative Commons Attribution License 4.0 (https://creativecommons.org/licenses/by/4.0/).

vectors which eliminates the non-scale invariant correlations but preserves the scale-invariant power-law behaviour [20, 21].

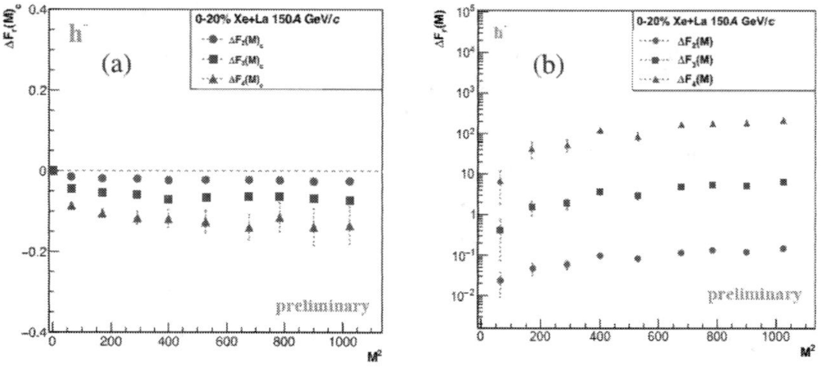

Figure 5. Intermittency studies of negatively charged hadrons in NA61/SHINE. The results obtained for Subtracted Scaled Factorial Moments of hadron multiplicity ΔF_r are shown up to the fourth order, for the case where the cumulative transformation of particles' (p_x, p_y) vectors (a) was applied, and (b) was not applied to the analysis. From Ref. [21].

A similar study of Subtracted Scaled Factorial Moments $\Delta F_r(M)_c$ for negatively charged hadrons produced in central Xe+La collisions is shown in Fig. 5 (a), with no indication of non-trivial power-law scaling. For comparison, the same analysis made with no cumulative transformation that would eliminate spurious effects is shown in Fig. 5 (b). A characteristic hierarchical structure of increase of $\Delta F_r(M)$ with M and moment order r emerges. This structure is qualitatively similar to that appearing in charged hadron results from STAR [22], also discussed at this Conference in the experimental CP review talk [23]. The fact that the dedicated procedure of cumulative transformation, Fig. 5 (a), eliminates completely the characteristic structure from Fig. 5 (b), has been investigated by NA61/SHINE. It has been argued [21, 24] that this structure can be explained by short-range (HBT) correlations and consequently, that the latter could also be at the source of the cited STAR result.[3]

These findings clearly call for further development of experimental and theoretical methodologies for elimination of spurious (non-critical) effects in the CP search, and strongly confirm the need for a good baseline model emphasized at this Conference [16]. For the time being, numerous models available on the market still have difficulties in description of various non-critical phenomena (see, *e.g.*, section 3 for HRG and UrQMD models). This may appear as a serious obstacle in present and future efforts to establish the presence of the critical point.

6 Summary

The unique capabilities of the multipurpose NA61/SHINE spectrometer make it a good tool for exploitation of the versatile scientific potential of the SPS. Our most recent findings include the first-ever direct measurement of open charm production in nucleus-nucleus collisions at SPS energies, to be followed soon by our new series of first direct open charm measurements in a heavy ion system at the SPS. This is supplemented by the observation of

[3]Note: the subtraction of uncorrelated background for ΔF_r is made differently for the cumulative and non-cumulative analyses. The non-cumulative approach relies on event-mixing, $\Delta F_r(M) = F_r^{data}(M) - F_r^{mix}(M)$ [23] while for the cumulative procedure, the subtraction reduces to $\Delta F_r(M)_c = F_r^{data}(M) - F_r^{data}(1)$ [21].

© The Authors, published by EDP Sciences. This is an open access article distributed under the terms of the Creative Commons Attribution License 4.0 (https://creativecommons.org/licenses/by/4.0/).

an unexpected excess of charged over neutral kaon production in agreement with data from earlier experiments but with better precision. This gives evidence for violation of isospin (flavour) symmetry and calls for a concentrated experimental and theoretical effort. Findings from NA61/SHINE give no indication of the critical point, but point at the crucial importance of a proper understanding of non-critical (baseline) effects in the CP search.

We thank all the Organizers for their effort invested in the realization of a very successful and interesting SQM Conference (merci!). This work was supported by the Polish Ministry of Science and Higher Education under contract No. 2021/WK/10.

References

[1] N. Abgrall *et al.* [NA61/SHINE], JINST **9**, P06005 (2014). https://doi.org/10.1088/1748-0221/9/06/P06005
[2] A. Merzlaya [NA61/SHINE], this Proceedings
[3] C. Ahdida *et al.* [NA60+], arXiv:2212.14452 [nucl-ex]
[4] H. Adhikary *et al.* [NA61/SHINE], Eur. Phys. J. C **84**, no.4, 416 (2024). https://doi.org/10.1140/epjc/s10052-024-12602-2, and references therein
[5] H. Adhikary *et al.* [NA61/SHINE], arXiv:2312.06572 [nucl-ex], and references therein
[6] T. Šuša [NA61/SHINE], this Proceedings
[7] W. Bryliński *et al.*, arXiv:2312.07176 [nucl-th], and references therein
[8] H. Adhikary *et al.* [NA61/SHINE], Phys. Rev. D **107**, no.6, 062004 (2023). https://doi.org/10.1103/PhysRevD.107.062004
[9] S. Kowalski *et al.* [NA61/SHINE], CERN-SPSC-2024-022
[10] O. Panova [NA61/SHINE], Poster at this Conference
[11] S. Acharya *et al.* [ALICE], Phys. Lett. B **802**, 135225 (2020). https://doi.org/10.1016/j.physletb.2020.135225
[12] B. Kozłowski [NA61/SHINE], Poster at this Conference
[13] A. Knospe, Talk at this Conference
[14] M. Maćkowiak-Pawłowska *et al.* [NA61/SHINE], CERN-SPSC-2023-022
[15] M. Stephanov, Talk at this Conference
[16] A. Bzdak, Talk at this Conference
[17] H. Adhikary *et al.* [NA61/SHINE], Eur. Phys. J. C **84**, no.7, 741 (2024). https://doi.org/10.1140/epjc/s10052-024-13012-0
[18] H. Adhikary [NA61/SHINE], EPJ Web Conf. **274**, 06008 (2022). https://doi.org/10.1051/epjconf/202227406008
[19] H. Adhikary *et al.* [NA61/SHINE], Eur. Phys. J. C **83**, no.9, 881 (2023). https://doi.org/10.1140/epjc/s10052-023-11942-9
[20] A. Białas and M. Gazdzicki, Phys. Lett. B **252**, 483-486 (1990). https://doi.org/10.1016/0370-2693(90)90575-Q
[21] V. Z. Reyna Ortiz [NA61/SHINE], Talk presented at New Trends in High-Energy and Low-x Physics, 1-5 Sept. 2024, Sfantu Gheorghe, Romania
[22] M. Abdulhamid *et al.* [STAR], Phys. Lett. B **845**, 138165 (2023). https://doi.org/10.1016/j.physletb.2023.138165
[23] Y. Zhang, Talk at this Conference
[24] V. Z. Reyna Ortiz [NA61/SHINE], Talk presented at CPOD 2024 - 15th Workshop on Critical Point and Onset of Deconfinement, 20-24 May 2024, Berkeley, California

© The Authors, published by EDP Sciences. This is an open access article distributed under the terms of the Creative Commons Attribution License 4.0 (https://creativecommons.org/licenses/by/4.0/).

Strangeness in Astrophysics

Laura Tolos[1,2,3,*]

[1]Institute of Space Sciences (ICE, CSIC), Campus UAB, Carrer de Can Magrans, 08193, Barcelona, Spain
[2]Institut d'Estudis Espacials de Catalunya (IEEC), 08860 Castelldefels (Barcelona), Spain
[3]Frankfurt Institute for Advanced Studies, University of Frankfurt, Ruth-Moufang-Str. 1, 60438 Frankfurt am Main, Germany

Abstract. In this contribution the role of strangeness in astrophysics is discussed and, more precisely, strange hadronic matter in the interior of neutron stars. A special attention is payed to certain phenomena involving strange hadronic matter, such as the hyperon puzzle, kaon condensation and the thermal behaviour of hyperons in neutron star mergers.

1 Introduction

Compact stars and, more precisely, neutron stars (NSs) are an excellent laboratory to study matter under extreme conditions of density, isospin asymmetry and temperature [1, 2]. NSs result from core-collapse supernovae and are in hydrostatic equilibrium with the gravitational collapse counterbalanced by the pressure inside NSs. NSs usually have masses around 1-2 $M_\odot$ and radii about 10-12 km, with an onion-like configuration. The largest region of an NS is the inner region, whose composition is not known. Thus, several hypothesis have been put forward to describe the NS interior.

Among others, hadronic matter made of strange baryons (also called hyperons) and/or strange mesons (mainly antikaons) has been postulated as possible scenarios in the inner core of NSs. In this contribution we aim at reviewing some specific phenomena related to the presence of strange hadronic matter inside NSs, such as the hyperon puzzle, the possible existence of kaon condensation as well as the thermal behaviour of hyperons as a possible indicator of their presence in NS mergers.

2 The hyperon puzzle

Hyperons are baryons with one or more strange quarks, such as Λ, Σ and Ξ. Since the seminal work of Ref. [3], hyperons have been thoroughly studied in the inner core of NS (for recent reviews see [1, 2, 4, 5]). NSs are charged neutral objects equilibrated by weak interactions involving the different species inside the core (β-equilibrium). The equilibrium can be expressed in terms of the chemical potentials of the different components. In particular, hyperons are expected to appear in the interior of NSs at densities of $\approx$ 2-3ρ_0, as the nucleon chemical potential could be so large at these densities that is energetically more favourable

*e-mail: tolos@ice.csic.es

© The Authors, published by EDP Sciences. This is an open access article distributed under the terms of the Creative Commons Attribution License 4.0 (https://creativecommons.org/licenses/by/4.0/).

to have hyperons rather than highly energetic nucleons at the top of the Fermi seas. As a consequence, matter becomes softer when only nucleons are present as the system relieves Fermi pressure. Hence, the less pressure there is inside an NS, the less mass it can sustain. The softening of the equation of state (EoS) of matter could then lead to maximum masses not compatible with $2M_{\odot}$ measurements, such as the masses of the PSR J1614-2230 [6, 7], PSR J0348+0432 [8] and PSR J0740+6620 [9]. This fact is usually denoted as *the hyperon puzzle*. There is, however, no solution yet to this puzzle, although several answers have been put forward in order to have hyperons in the interior of $2M_{\odot}$ NSs.

Among them, one solution considers stiff hyperon-nucleon and hyperon-hyperon interactions so as to compensate for the softening of the EoS (see, for example, [10–13]). Another way for stiffening the EoS relies on the three-body forces [14–22]. Another solution pushes hyperons to larger densities by the occurrence of new species, such as Δ baryons [23–25] or a kaon condensate (as discussed in the following section). Also, solutions based on non-hadronic degrees of freedom have been considered, such as an early phase transition to quark matter below the hyperon onset (see Refs. [26–30] for recent papers). And, finally, modified gravity theories allowing for hyperons inside $2M_{\odot}$ stars have been also described in the literature [31].

3 Kaon condensation in neutron stars

Another possibility for strange hadronic matter is to consider strange mesons, such as antikaons in the core of NSs. As mentioned before, the composition of matter in NSs is determined by demanding β-equilibrium. In the particular case of matter made of neutrons, protons and electrons, the chemical potential of the electron substantially increases with density. Thus, antikaons could be produced if their effective mass gets smaller than the electron chemical potential at a given density inside NSs. In that case, antikaons would appear forming a condensate, hence leading to the phenomenon of *kaon condensation*.

Since the pioneering work of Ref. [32], this phenomenon has been widely discussed. The question is whether the mass of antikaons could be largely modified in nuclear matter. Whereas this is the case for some phenomenological models (see the recent results in Refs. [33–36]), microscopic unitarized schemes do not obtain such large modifications in the antikaon mass (see, for example, Refs. [37–44]).

4 Thermal behaviour of hyperons in neutron star mergers

In spite of the previous scenarios where strange hadronic matter might be present inside an NS, there is still the open question whether there is a clear signal of the presence of strange hadronic matter, and in particular hyperons, in NSs. For example, by comparing the mass-radius relations coming from EoSs with only nucleons or with nucleons and hyperons, it is not straightforward to tell from a measured mass-radius relation if the underlying EoS contains hyperons. Therefore, it is very much welcome to identify NS features that can be clearly linked to the appearance of hyperons in NS.

Recently, the *thermal behavior of hyperons* has been proposed as a potential indicator for the presence of hyperons in NS mergers [45]. The approach is based on the fact that the hyperonic EoSs produce lower average thermal indeces in matter than the nucleonic ones [46, 47]. The influence of the different thermal treatment on the dominant postmerger GW frequency is then systematically investigated using numerical simulations [45].

First, this frequency is obtained by running simulations using the full temperature- and composition-dependent nucleonic and hyperonic EoSs. Then, in another set of simulations,

© The Authors, published by EDP Sciences. This is an open access article distributed under the terms of the Creative Commons Attribution License 4.0 (https://creativecommons.org/licenses/by/4.0/).

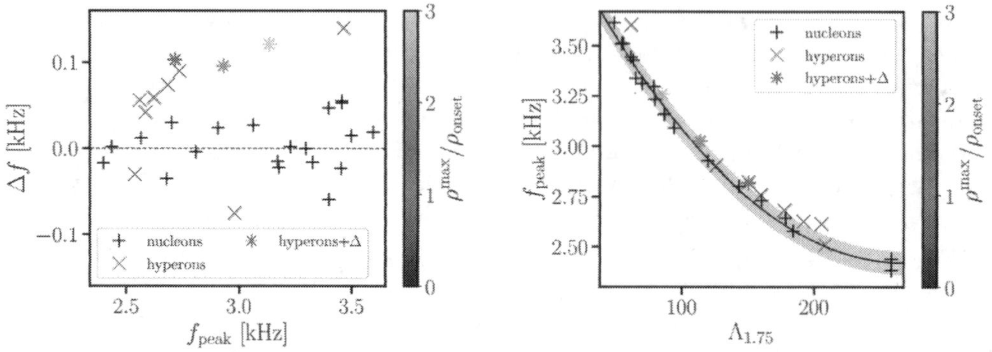

Figure 1. Left plot: $\Delta f = f_{\rm peak} - f_{\rm peak}^{1.75}$ as a function of $f_{\rm peak}$. Right plot: $f_{\rm peak}$ of 1.4-1.4 $M_\odot$ mergers as function of tidal deformability of a 1.75 $M_\odot$ NS. Black symbols depict purely nucleonic models, whereas crosses display hyperonic models and asterisks refer to models which include Δs. The coloring shows the ratio between the maximum rest-mass density in the postmerger remnant and the hyperon onset rest-mass density at zero temperature. Black curve on the right plot shows least-squares fit to purely nucleonic models, while grey band displays maximum residual of purely nucleonic models from the fit. Plots taken from [45].

all EoSs at zero temperature in neutrinoless β-equilibrium are used and supplemented with the ideal-gas treatment of thermal pressure with $\Gamma_{\rm th} = 1.75$, mimicking the thermal behavior of purely nucleonic matter. And, finally, the dominant postmerger GW frequencies of the full EoS models, $f_{\rm peak}$, with the frequencies $f_{\rm peak}^{1.75}$ obtained from the $\Gamma_{\rm th} = 1.75$ calculations are compared, so that $\Delta f \equiv f_{\rm peak} - f_{\rm peak}^{1.75}$ measures the deviation from an idealized nucleonic thermal behavior. The results are shown in the l.h.s of Fig. 1, distinguishing purely nucleonic and hyperonic models. A characteristic increase of the dominant postmerger gravitational-wave frequency by up to $\sim$ 150 Hz is observed with the hyperonic EoS models (with a sufficient amount of hyperons) compared to purely nucleonic ones, thus linking this effect to the occurrence of hyperons in NSs.

Also, regarding the identification of hyperons by a frequency shift within a more concrete detection scenario, it is interesting to relate $f_{\rm peak}$ from the simulations with the fully temperature-dependent EoSs as function of the tidal deformability of 1.75 $M_\odot$ NSs, as done in the r.h.s of Fig. 1.The dominant postmerger frequency of hyperonic models is characteristically increased compared to the nucleonic models, so that the presence of hyperons may be deduced by an increased postmerger frequency that is not compatible with a nucleonic EoS.

Some caveats from the previous analysis need, however, to be mentioned, such as the necessity of enough statistical power in GW measurements, the dependence on the abundance of hyperons and hyperon threshold, or the fact that other exotic degrees of freedom might lead to a similar frequency shift.

5 Conclusions

In this contribution certain phenomena associated to the occurrence of strange hadronic matter in the inner core of NSs have been presented. More precisely, we have reviewed the hyperon puzzle and the possible solutions as well as the potential existence of a kaon condensed phase in NSs. Furthermore, the thermal behaviour of hyperons in NS mergers has been discussed as a feasible indicator of strangeness in the inner core of NSs.

© The Authors, published by EDP Sciences. This is an open access article distributed under the terms of the Creative Commons Attribution License 4.0 (https://creativecommons.org/licenses/by/4.0/).

Acknowledgments

L.T. thanks H. Kochankovski and A. Ramos for the careful reading of this contribution. She also acknowledges support from CEX2020-001058-M (Unidad de Excelencia "María de Maeztu") and PID2022-139427NB-I00 financed by the Spanish MCIN/AEI/10.13039/501100011033/FEDER,UE as well as from the Generalitat de Catalunya under contract 2021 SGR 171, by the EU STRONG-2020 project, under the program H2020-INFRAIA-2018-1 grant agreement no. 824093, and by the CRC-TR 211 'Strong-interaction matter under extreme conditions'- project Nr. 315477589 - TRR 211.

References

[1] L. Tolos, L. Fabbietti, Strangeness in Nuclei and Neutron Stars, Prog. Part. Nucl. Phys. **112**, 103770 (2020), 2002.09223. 10.1016/j.ppnp.2020.103770

[2] G.F. Burgio, H.J. Schulze, I. Vidana, J.B. Wei, Neutron stars and the nuclear equation of state, Prog. Part. Nucl. Phys. **120**, 103879 (2021), 2105.03747. 10.1016/j.ppnp.2021.103879

[3] V.A. Ambartsumyan, G.S. Saakyan, The Degenerate Superdense Gas of Elementary Particles, Soviet Astronomy **4**, 187 (1960).

[4] I. Vidaña, Hyperons: the strange ingredients of the nuclear equation of state, Proc. Roy. Soc. Lond. **A474**, 0145 (2018), 1803.00504. 10.1098/rspa.2018.0145

[5] D. Chatterjee, I. Vidaña, Do hyperons exist in the interior of neutron stars?, Eur. Phys. J. A **52**, 29 (2016), 1510.06306. 10.1140/epja/i2016-16029-x

[6] P. Demorest, T. Pennucci, S. Ransom, M. Roberts, J. Hessels, Shapiro Delay Measurement of A Two Solar Mass Neutron Star, Nature **467**, 1081 (2010), 1010.5788. 10.1038/nature09466

[7] E. Fonseca et al., The NANOGrav Nine-year Data Set: Mass and Geometric Measurements of Binary Millisecond Pulsars, Astrophys. J. **832**, 167 (2016), 1603.00545. 10.3847/0004-637X/832/2/167

[8] J. Antoniadis et al., A Massive Pulsar in a Compact Relativistic Binary, Science **340**, 6131 (2013), 1304.6875. 10.1126/science.1233232

[9] H.T. Cromartie et al., Relativistic Shapiro delay measurements of an extremely massive millisecond pulsar, Nature Astronomy p. 439 (2019), 1904.06759. 10.1038/s41550-019-0880-2

[10] I. Bednarek, P. Haensel, J.L. Zdunik, M. Bejger, R. Manka, Hyperons in neutron-star cores and two-solar-mass pulsar, Astron. Astrophys. **543**, A157 (2012), 1111.6942. 10.1051/0004-6361/201118560

[11] S. Weissenborn, D. Chatterjee, J. Schaffner-Bielich, Hyperons and massive neutron stars: vector repulsion and SU(3) symmetry, Phys. Rev. **C85**, 065802 (2012), [Erratum: Phys. Rev.C90,no.1,019904(2014)], 1112.0234. 10.1103/PhysRevC.85.065802, 10.1103/PhysRevC.90.019904

[12] M. Oertel, C. Providência, F. Gulminelli, A.R. Raduta, Hyperons in neutron star matter within relativistic mean-field models, J. Phys. **G42**, 075202 (2015), 1412.4545. 10.1088/0954-3899/42/7/075202

[13] K.A. Maslov, E.E. Kolomeitsev, D.N. Voskresensky, Solution of the Hyperon Puzzle within a Relativistic Mean-Field Model, Phys. Lett. **B748**, 369 (2015), 1504.02915. 10.1016/j.physletb.2015.07.032

© The Authors, published by EDP Sciences. This is an open access article distributed under the terms of the Creative Commons Attribution License 4.0 (https://creativecommons.org/licenses/by/4.0/).

[14] J. Haidenbauer, U.G. Meißner, N. Kaiser, W. Weise, Lambda-nuclear interactions and hyperon puzzle in neutron stars, Eur. Phys. J. **A53**, 121 (2017), 1612.03758. 10.1140/epja/i2017-12316-4

[15] D. Logoteta, I. Vidaña, I. Bombaci, Impact of chiral hyperonic three-body forces on neutron stars, Eur. Phys. J. **A55**, 207 (2019), 1906.11722. 10.1140/epja/i2019-12909-9

[16] D. Gerstung, N. Kaiser, W. Weise, Hyperon–nucleon three-body forces and strangeness in neutron stars, Eur. Phys. J. A **56**, 175 (2020), 2001.10563. 10.1140/epja/s10050-020-00180-2

[17] T. Takatsuka, S. Nishizaki, Y. Yamamoto, Necessity of extra repulsion in hypernuclear systems: Suggestion from neutron stars, The European Physical Journal A - Hadrons and Nuclei **13**, 213 (2002). 10.1140/epja1339-35

[18] T. Takatsuka, Hyperon-mixed neutron stars, Prog. Theor. Phys. Suppl. **156**, 84 (2004). 10.1143/PTPS.156.84

[19] I. Vidaña, D. Logoteta, C. Providencia, A. Polls, I. Bombaci, Estimation of the effect of hyperonic three-body forces on the maximum mass of neutron stars, EPL **94**, 11002 (2011), 1006.5660. 10.1209/0295-5075/94/11002

[20] Y. Yamamoto, T. Furumoto, N. Yasutake, T.A. Rijken, Multi-pomeron repulsion and the Neutron-star mass, Phys. Rev. **C88**, 022801 (2013), 1308.2130. 10.1103/PhysRevC.88.022801

[21] Y. Yamamoto, T. Furumoto, N. Yasutake, T.A. Rijken, Hyperon mixing and universal many-body repulsion in neutron stars, Phys. Rev. **C90**, 045805 (2014), 1406.4332. 10.1103/PhysRevC.90.045805

[22] D. Lonardoni, A. Lovato, S. Gandolfi, F. Pederiva, Hyperon Puzzle: Hints from Quantum Monte Carlo Calculations, Phys. Rev. Lett. **114**, 092301 (2015), 1407.4448. 10.1103/PhysRevLett.114.092301

[23] T. Schuerhoff, S. Schramm, V. Dexheimer, Neutron stars with small radii – the role of delta resonances, Astrophys. J. **724**, L74 (2010), 1008.0957. 10.1088/2041-8205/724/1/L74

[24] A. Drago, A. Lavagno, G. Pagliara, D. Pigato, Early appearance of Δ isobars in neutron stars, Phys. Rev. **C90**, 065809 (2014), 1407.2843. 10.1103/PhysRevC.90.065809

[25] P. Ribes, A. Ramos, L. Tolos, C. Gonzalez-Boquera, M. Centelles, Interplay between Δ Particles and Hyperons in Neutron Stars, Astrophys. J. **883**, 168 (2019), 1907.08583. 10.3847/1538-4357/ab3a93

[26] S. Weissenborn, I. Sagert, G. Pagliara, M. Hempel, J. Schaffner-Bielich, Quark Matter In Massive Neutron Stars, Astrophys. J. **740**, L14 (2011), 1102.2869. 10.1088/2041-8205/740/1/L14

[27] L. Bonanno, A. Sedrakian, Composition and stability of hybrid stars with hyperons and quark color-superconductivity, Astron. Astrophys. **539**, A16 (2012), 1108.0559. 10.1051/0004-6361/201117832

[28] T. Klaehn, R. Lastowiecki, D.B. Blaschke, Implications of the measurement of pulsars with two solar masses for quark matter in compact stars and heavy-ion collisions: A Nambu–Jona-Lasinio model case study, Phys. Rev. **D88**, 085001 (2013), 1307.6996. 10.1103/PhysRevD.88.085001

[29] R. Lastowiecki, D. Blaschke, H. Grigorian, S. Typel, Strangeness in the cores of neutron stars, Acta Phys. Polon. Supp. **5**, 535 (2012), 1112.6430. 10.5506/APhysPolBSupp.5.535

© The Authors, published by EDP Sciences. This is an open access article distributed under the terms of the Creative Commons Attribution License 4.0 (https://creativecommons.org/licenses/by/4.0/).

[30] J.L. Zdunik, P. Haensel, Maximum mass of neutron stars and strange neutron-star cores, Astron. Astrophys. **551**, A61 (2013), 1211.1231. 10.1051/0004-6361/201220697

[31] A.V. Astashenok, S. Capozziello, S.D. Odintsov, Maximal neutron star mass and the resolution of the hyperon puzzle in modified gravity, Phys. Rev. **D89**, 103509 (2014), 1401.4546. 10.1103/PhysRevD.89.103509

[32] D.B. Kaplan, A.E. Nelson, Strange Goings on in Dense Nucleonic Matter, Phys. Lett. **B175**, 57 (1986). 10.1016/0370-2693(86)90331-X

[33] N. Gupta, P. Arumugam, Impact of hyperons and antikaons in an extended relativistic mean-field description of neutron stars, Phys. Rev. C **88**, 015803 (2013). 10.1103/PhysRevC.88.015803

[34] V.B. Thapa, M. Sinha, J.J. Li, A. Sedrakian, Massive Δ-resonance admixed hypernuclear stars with antikaon condensations, Phys. Rev. D **103**, 063004 (2021), 2102.08787. 10.1103/PhysRevD.103.063004

[35] T. Malik, S. Banik, D. Bandyopadhyay, Equation-of-state Table with Hyperon and Antikaon for Supernova and Neutron Star Merger, Astrophys. J. **910**, 96 (2021), 2104.00775. 10.3847/1538-4357/abe860

[36] T. Muto, T. Maruyama, T. Tatsumi, Effects of three-baryon forces on kaon condensation in hyperon-mixed matter, Phys. Lett. B **820**, 136587 (2021), 2106.03449. 10.1016/j.physletb.2021.136587

[37] M. Lutz, Nuclear kaon dynamics, Phys. Lett. **B426**, 12 (1998), nucl-th/9709073. 10.1016/S0370-2693(98)00299-8

[38] A. Ramos, E. Oset, The Properties of $\bar{K}$ in the nuclear medium, Nucl. Phys. **A671**, 481 (2000), nucl-th/9906016. 10.1016/S0375-9474(99)00846-5

[39] L. Tolos, A. Ramos, A. Polls, T.T.S. Kuo, Partial wave contributions to the antikaon potential at finite momentum, Nucl. Phys. **A690**, 547 (2001), nucl-th/0007042. 10.1016/S0375-9474(00)00711-9

[40] L. Tolos, A. Ramos, A. Polls, The Anti-kaon nuclear potential in hot and dense matter, Phys. Rev. **C65**, 054907 (2002), nucl-th/0202057. 10.1103/PhysRevC.65.054907

[41] L. Tolos, A. Ramos, E. Oset, Chiral approach to antikaon s and p-wave interactions in dense nuclear matter, Phys. Rev. **C74**, 015203 (2006), nucl-th/0603033. 10.1103/PhysRevC.74.015203

[42] L. Tolos, D. Cabrera, A. Ramos, Strange mesons in nuclear matter at finite temperature, Phys. Rev. **C78**, 045205 (2008), 0807.2947. 10.1103/PhysRevC.78.045205

[43] D. Cabrera, L. Tolos, J. Aichelin, E. Bratkovskaya, Antistrange meson-baryon interaction in hot and dense nuclear matter, Phys. Rev. **C90**, 055207 (2014), 1406.2570. 10.1103/PhysRevC.90.055207

[44] M.F.M. Lutz, C.L. Korpa, M. Moller, Antikaons and hyperons in nuclear matter with saturation, Nucl. Phys. **A808**, 124 (2008), 0707.1283. 10.1016/j.nuclphysa.2008.05.008

[45] S. Blacker, H. Kochankovski, A. Bauswein, A. Ramos, L. Tolos, Thermal behavior as indicator for hyperons in binary neutron star merger remnants, Phys. Rev. D **109**, 043015 (2024), 2307.03710. 10.1103/PhysRevD.109.043015

[46] H. Kochankovski, A. Ramos, L. Tolos, Equation of state for hot hyperonic neutron star matter, Mon. Not. Roy. Astron. Soc. **517**, 507 (2022), [Erratum: Mon.Not.Roy.Astron.Soc. 518, 6376], 2206.11266. 10.1093/mnras/stac2671

[47] A.R. Raduta, Equations of state for hot neutron stars-II. The role of exotic particle degrees of freedom, Eur. Phys. J. A **58**, 115 (2022), 2205.03177. 10.1140/epja/s10050-022-00772-0

© The Authors, published by EDP Sciences. This is an open access article distributed under the terms of the Creative Commons Attribution License 4.0 (https://creativecommons.org/licenses/by/4.0/).

EPJ Web of Conferences **316**, 01010 (2025) https://doi.org/10.1051/epjconf/202531601010
SQM 2024

Strange Resonances and Exotic States

Anders G. Knospe[1,*]

[1]Department of Physics, Lehigh University, Bethlehem, Pennsylvania, USA

Abstract. In these proceedings, I discuss how strange resonances and exotic states are used as unique probes in the field of ultrarelativistic heavy-ion physics. These particles can be used to study the properties of the hadronic phase of ion-ion collisions, the hadrochemistry of the system, flow and vorticity in the hot and dense medium, and the structures of the hadrons themselves. I provide an overview of recent results to illustrate these concepts.

1 Introduction

Strange hadronic resonances and exotic states provide a variety of ways to learn about the physics of heavy-ion collisions and about hadron structure. In these proceedings, I discuss a selection of experimental measurements and theoretical calculations, with an emphasis on recent results. The reader is invited to consult the cited presentations, proceedings, and papers for figures illustrating these results.

2 Resonance-to-Stable Yield Ratios & Hadronic Phase Properties

Measurements of resonance yields provide a way to study the properties of the later, post-hadronization phase of an ion-ion collision. Between chemical and kinetic freeze out, re-scattering and regeneration may modify the measurable yields of resonances; the strengths of these processes will be influenced by the properties of the hadronic phase (the temperatures at chemical and kinetic freeze out and the time interval between them), the lifetime of the resonance, and the scattering cross sections of its decay products. The K^* and ϕ mesons are easily reconstructed resonances that have been used frequently to study these effects. The K^*/K yield ratio in central nucleus-nucleus collisions is observed to be suppressed with respect to the yields in peripheral A+A collisions, smaller collision systems (*e.g.* p+p and p+A), and statistical thermal model calculations. [1, 2] This suggests that re-scattering is dominant over regeneration, leading to suppression of the K^* yield. This suppression has been observed at SPS energies, at RHIC energies (including in the Beam Energy Scan program) [3], and at the LHC. However, this suppression may vanish at low energies: NA61/SHINE observes no suppression of the K^{*0}/K ratio in Ar+Sc collisions at $\sqrt{s_{NN}}$ = 8.8 GeV (in comparison to p+p collisions at the same energy) [4]. They do, however observe suppression at higher collision energies; the lowest energy collisions may have a shorter-lived hadronic phase, which would result in less influence of re-scattering on the K^{*0} yield. In contrast, no suppression of the ϕ/K yield ratio is observed in central A+A collisions. This is likely due to fact that the

*e-mail: knospe@lehigh.edu

© The Authors, published by EDP Sciences. This is an open access article distributed under the terms of the Creative Commons Attribution License 4.0 (https://creativecommons.org/licenses/by/4.0/).

lifetime of the ϕ meson ($c\tau$ = 46.4 fm/c) is an order of magnitude greater than that of the K^{*0} meson ($c\tau$ = 4.17 fm/c). The ϕ lives longer than typical estimates of the hadronic phase lifetime, so its yield should not be affected strongly by re-scattering or regeneration. The ϕ/K ratio appears to be independent of the colliding ions and of collision energy from $\gtrsim$ 10 GeV up to the top LHC energies. The K^*/K ratio depends on the system size, parameterized by the number of wounded nucleons, the number of participant nucleons, or the charged-particle multiplicity. When examined as a function of these quantities, the K^*/K and ϕ/K ratios do not seem to depend on collision species or energy (*i.e.*, the K^*/K ratio at a given event multiplicity is quite similar for different collision energies or species) [3, 5]. This is true even when the ratios are measured as a function of multiplicity in p+p and p+Pb collisions. The resonance yields seem to controlled primarily the system size or multiplicity, an observation that also applies to long-lived hadron yields and strangeness enhancement. Theoretical models have had varying degrees of success at describing this behavior. EPOS uses UrQMD to model hadronic-phase effects and is able to qualitatively describe the observed suppression trends [6]. PHSD includes re-scattering and absorption of decay products in hadronic phase; it can describe the observed K^*/K suppression trend [7]. MUSIC with the SMASH afterburner to describe the hadronic phase gives a K^*/K ratio that decreases with increasing collision centrality (although that decrease is less pronounced than seen experimentally), but MUSIC without SMASH does not exhibit such suppression [8]. The γ_S Canonical Statistical Model predicts only a slow decrease in the K^*/K ratio [9], while the Hadron Resonance Gas in Partial Chemical Equilibrium provides a good agreement with ALICE measurements of the K^*/K ratio. [10]

Under the assumption of simultaneous chemical freeze out of all long-lived hadron species and negligible regeneration of resonances, K^*/K ratio measurements can be used to estimate the time between chemical and kinetic freeze out. The lower limit on the hadronic phase lifetime $\tau_{\rm low}$ is extracted from the following expression:

$$\left(\frac{K^*}{K}\right)_{\rm kinetic} = \left(\frac{K^*}{K}\right)_{\rm chemical} \times \exp\left(-\frac{\tau_{\rm low}}{\tau_{K^*}}\right). \tag{1}$$

The yield ratio in p+p collisions can be used as a proxy for the ratio at chemical freeze out, while the final measured yield ratio in a given centrality/multiplicity class is taken to be the ratio at kinetic freeze out. The extracted value for $\tau_{\rm low}$ is corrected by a Lorentz factor based on the mean momentum of the K^* mesons. Using this method, ALICE has estimated $\tau_{\rm low}$ values consistent with 0 for low-multiplicity p+p and p+Pb collisions, ranging up to approximately 6 fm/c in central Pb+Pb collisions [2]. This lifetime appears to depend on the charged-particle multiplicity, but not on the species of the colliding ions (for a given event multiplicity). ALICE has also performed a similar study using $\Lambda(1520)$ baryons and ρ^0 mesons. NA61/SHINE employed the same technique using K^* mesons to estimate the hadronic phase lifetime, which they found to be near 0 for $\sqrt{s_{NN}}$ = 8.8 GeV [4].

Other resonance-to-stable particle yield ratios can also be measured to shed light on the properties of the hadronic phase. Figure 1 (an updated version of a figure in [6]) is a summary of such measurements (see [1, 5] and the references listed in [6]). For clarity, only a selection of results are presented: the figure mostly shows ALICE p+p, p+Pb, and Pb+Pb results, supplemented with analogous STAR results when necessary. These results are compared to EPOS calculations. Each resonance yield is divided by the yield of a long-lived hadron with similar quark content in order to cancel out strangeness enhancement or similar effects. The resonance lifetimes decrease from the top to the bottom of the figure. The Ξ^{*0} baryon lives about half as long as the ϕ meson and may be expected to be somewhat more affected by re-scattering and/or regeneration. ALICE observes possible weak suppression of the Ξ^{*0}/Ξ^-

© The Authors, published by EDP Sciences. This is an open access article distributed under the terms of the Creative Commons Attribution License 4.0 (https://creativecommons.org/licenses/by/4.0/).

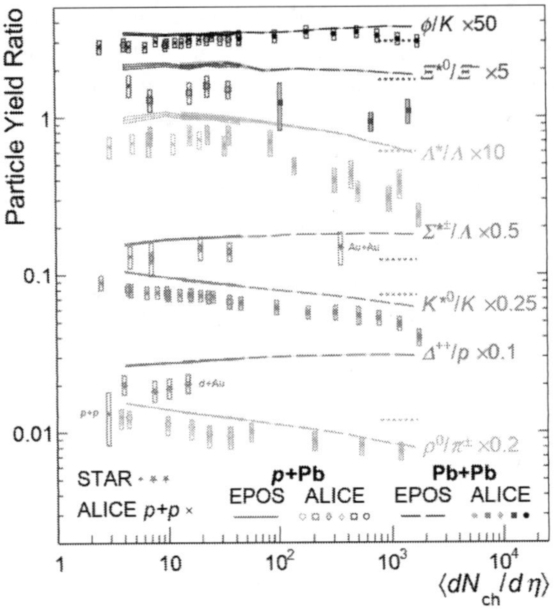

Figure 1. Summary plot showing a collection of particle yield ratios that illustrate the suppression trends for various resonances. Results from ALICE in small and large collision systems (supplemented by STAR data when necessary) are plotted as functions of the mid-rapidity charged-particle density and compared to EPOS calculations [6]. The ratios are scaled for visibility. The resonance lifetime decreases going from the top to the bottom of the plot. Bars represent statistical uncertainties, open boxes represent total systematic uncertainties, and shaded boxes represent systematic uncertainties that are uncorrelated between multiplicity/centrality classes. The experimental results come from [1, 5] and the references listed in [6].

ratio in (mid-)central A+A collisions. The $\Lambda(1520)$ baryon has a lifetime about 3 times shorter than the ϕ and 3 times longer than the K^{*0}. The ρ^0 meson has a significantly shorter lifetime than K^{*0}. Both of these exhibit clear suppression trends with increasing centrailty in A+A collisions, qualitatively similar to the trend seen for K^*/K. The suppression or non-suppression trends of the resonances mentioned so far can be at least qualitatively described by EPOS. However, resonance suppression may depend on properties other than the lifetime. The Σ^* and Δ baryons have lifetimes similar to or smaller than the K^* meson. Despite these short lifetimes, EPOS predicts no suppression (and, in fact, a small enhancement) of the Σ^*/Λ and Δ/p ratios with increasing system size. Measurements of these resonances are unfortunately difficult, particularly in A+A collisions. The STAR measurement of the Σ^*/Λ ratio exhibits no suppression trend. ALICE sees hints of possible weak suppression of the $\Sigma^*/\pi^\pm$ yield ratio in A+A collisions [11], but ALICE has not published final measurements of the Σ^*/Λ ratio. Publication of that result, measurements of the Δ baryon in A+A collisions, or measurements of other resonance species could help us improve our picture of the hadronic phase.

3 Hadrochemistry

Resonances including the ϕ meson can also be used to study other aspects of the hadrochemistry of the collision system, shedding light on how hadron yields are affected by the properties of the medium and the hadron. ALICE recently released a comprehensive study of hadron production as a function of transverse spherocity [12]. For the 0–10% highest multiplicity p+p collisions at $\sqrt{s}$ = 13 TeV, ALICE split the data sample into spherocity classes, ranging from very jet-like to very isotropic. ALICE observes that the production of hadrons with open strangeness, like Ξ, is suppressed in jet-like collisions. This suggests that strangeness enhancement is connected to the properties of the underlying even and soft processes. However, ϕ meson production is observed to be independent of spherocity in this study, suggesting that the ϕ behaves like a particle that contains no strange quarks. This is in contrast to studies of the evolution of the ϕ/π yield ratio with event multiplicity. The ϕ/π ratio is observed to increase with multiplicity [13]; this increase appears to be more pronounced than the increases of the K/π and Λ/π ratios, but smaller in magnitude than the increase in the

© The Authors, published by EDP Sciences. This is an open access article distributed under the terms of the Creative Commons Attribution License 4.0 (https://creativecommons.org/licenses/by/4.0/).

Ξ/π ratio. This suggests that the ϕ meson may behave as if it has an effective strangeness content between 1 and 2 units. Further theoretical investigations are likely needed to reconcile this with the aforementioned spherocity study.

Other interesting results come from studies of ϕ meson production at low energies. NA49 and NA61/SHINE observe that the ϕ/π yield ratio increases with collision energy over the range $5 \lesssim \sqrt{s_{NN}} \lesssim 20$ GeV [14]. This increase is seen in large and small collisions and may be attributable to strangeness enhancement. The ϕ/K yield ratio has also been measured in A+A collisions at energies $\sqrt{s_{NN}} \lesssim 3$ GeV, below the ϕ production threshold for p+p collisions. While the ϕ/K yield ratio is essentially constant from top SPS energies up to LHC energies, HADES, FOPI, and STAR observe that the ϕ/K ratio is higher in this sub-threshold regime than it is at larger energies [15]. This is likely due to the canonical suppression of the kaon yields. These measurements can be used to estimate the correlation radius R_C in canonical statistical models, although there is currently a discrepancy in the R_C values extracted from the ϕ/K^- and ϕ/Ξ yield ratios. Theoretical frameworks including UrQMD and PHSD have had some success in describing this behavior.

4 Hadron Structure

Recent measurements of f and a mesons and excited kaons have the potential to provide information about the structure of those hadrons. These mesons may have simple $q\bar{q}$ structures or they may have more exotic structures. Among the possibilities are hybrid mesons (containing valence gluons and quarks), glueballs, meson-meson molecules, and other tetraquark configurations. ALICE has measured the $f_0(980)/\pi$ yield ratio in p+p collisions [16]. This measurement is compared to two different implementations of the canonical statistical model [9], one assuming that the $f_0(980)$ is a $s\bar{s}$ state (the $|S| = 2$ hypothesis) and the other assuming that $f_0(980)$ is a $q\bar{q}$ meson with no $s\bar{s}$ content (the $|S| = 0$ hypothesis). The measured $f_0(980)/\pi$ ratio is in better agreement with the $|S| = 0$ hypothesis. ALICE also observes that the $f_0(980)/K^{*0}$ ratio decreases with increasing multiplicity in p+Pb collisions [17]. This suppression is described by the γ_S Canonical Statistical Model with the $|S| = 0$ hypothesis; the $|S| = 2$ hypothesis predicts a slow increase in the $f_0(980)/K^{*0}$ ratio with multiplicity. ALICE measurements of the $f_0(980)$ nuclear modification factor in p+Pb collisions do not exhibit Cronin enhancement, which is taken to support the hypothesis that this meson has a simple quark-antiquark structure. CMS has measured the v_2 elliptic flow parameter of $f_0(980)$ in p+Pb collisions [18]. By performing n_q scaling with a variable number of constituent partons, they found that the preferred n_q value is near 2, strongly favoring a simple quark-antiquark structure for $f_0(980)$.

ALICE has measured the $f_1(1285)$ meson in p+p collisions [19]. The measured $f_1(1285)/\pi$ ratio is better described by the γ_S Canonical Statistical Model with the $|S| = 0$ hypothesis [9], suggesting little hidden strangeness content for this meson. ALICE has measured the correlation functions for $K^0_S K^0_S$ pairs (which can originate from $f_0(980)$ decays) and $K^0_S K^\pm$ pairs (which can originate from $a_0(980)^\pm$ decays) [20]. While the correlation radius R is the same for both types of pairs, the correlation strength λ is significantly smaller (by about a factor of 2) for $K^0_S K^\pm$ pairs, suggesting that the $a_0(980)^\pm$ may be a tetraquark. ALICE has also measured the correlation function for $K^0_S \pi^\pm$ pairs, which may originate from $K^*_0(700)^\pm$ decays [21]. The correlation strength λ is observed to increase as a function of the correlation radius R, which is consistent with a tetraquark configuration.

© The Authors, published by EDP Sciences. This is an open access article distributed under the terms of the Creative Commons Attribution License 4.0 (https://creativecommons.org/licenses/by/4.0/).

5 Flow & Vorticity

K^* and ϕ can also be used to explore collective flow and vorticity in nucleus-nucleus collisions. PHENIX has reported measurements of the v_2 of ϕ mesons in Cu+Au and U+U collisions [22]. The $v_2(p_T)$ trends in the 0-20% most central Cu+Au collisions are consistent with those for the 0-30% most central Au+Au collisions, while $v_2(p_T)$ for the 0-50% most central U+U collisions is consistent with mid-central (30-80%) Au+Au collisions. However, these various trends can be collapsed to essentially a single $v_2(p_T)$ by scaling by the empirical factor $\varepsilon N_{\rm part}^{1/3}$, where ε is the eccentricity of the nuclear overlap region.

Non-central A+A collisions have a large orbital angular momentum perpendicular to the reaction plane. This leads to vorticity in the QGP produced in the collision, which induces the global polarization of quarks in that QGP, which may be observable as a global polarization of the hadrons containing those quarks, including Λ baryons and K^* and ϕ vector mesons. This polarization can be measured by finding the angular distributions of K^* and ϕ decay products with respect to a suitable quantization axis (*e.g.*, the direction of the orbital angular momentum). From this, the elements of the spin-density matrix can be extracted. Deviations of the ρ_{00} spin-density matrix element from the unpolarized value of $\frac{1}{3}$ indicated a net polarization of these mesons. STAR recently published such a study [23]. They observe ρ_{00} values consistent with $\frac{1}{3}$ for Au+Au collisions at $\sqrt{s_{NN}}$ = 200 GeV. However, measurements for $11.5 \leq \sqrt{s_{NN}} \leq 62$ GeV have $\rho_{00} > \frac{1}{3}$ at the 7.4σ level for the ϕ meson. This result is supported by new STAR preliminary results from the second RHIC Beam Energy Scan [24]. It is also observed that ρ_{00} becomes larger and deviates even farther from $\frac{1}{3}$. This is consistent with expectations, since particles moving with larger rapidities encounter larger magnetic field fluctuations in the direction perpendicular to their motion. Earlier measurements by ALICE [25] indicate that at LHC energies, $\rho_{00} < \frac{1}{3}$ for the K^*, using a lower p_T range than the STAR measurement. ALICE observes ρ_{00} consistent with the unpolarized value for vector mesons in p+p collisions and for K^0_S mesons (scalar particles) in A+A collisions.

I wish to thank the organizers for inviting me to give a presentation on this topic, and the many experimentalists and theorists whose work I have discussed here.

References

[1] S. Acharya et al. (ALICE), Evidence of rescattering effect in Pb–Pb collisions at the LHC through production of $K^*(829)^0$ and $\phi(1020)$ mesons, Phys. Lett. B **802**, 135225 (2020). 10.1016/j.physletb.2020.135225

[2] S. Acharya et al. (ALICE), System-size dependence of the hadronic rescattering effect at energies available at the CERN Large Hadron Collider, Phys. Rev. C **109**, 014911 (2024). 10.1103/PhysRevC.109.014911

[3] S. Singha (STAR), K^{*0} production in Ru+Ru and Zr+Zr collisions at $\sqrt{s_{NN}}$ = 200 GeV at RHIC, these proceedings (2024).

[4] B. Kozłowski (NA61/SHINE), K^*/K ratio and the time between freeze-outs for intermediate-mass Ar+Sc system at the SPS energy range, these proceedings (2024).

[5] S. Acharya et al. (ALICE), Multiplicity dependence of $K^*(892)^0$ and $\phi(1020)$ production in pp collisions at $\sqrt{s}$ = 13 TeV, Phys. Lett. B **807**, 135501 (2020). 10.1016/j.physletb.2020.135501

[6] A.G. Knospe et al., Hadronic resonance production and interaction in p–Pb collisions at LHC energies in EPOS3, Phys. Rev. C **104**, 054907 (2021). 10.1103/PhysRevC.104.054907

[7] A. Ilner et al., Probing hot and dense nuclear matter with K^*, $\overline{K}^*$ vector mesons, Phys. Rev. C **99**, 024914 (2019). 10.1103/PhysRevC.99.024914

© The Authors, published by EDP Sciences. This is an open access article distributed under the terms of the Creative Commons Attribution License 4.0 (https://creativecommons.org/licenses/by/4.0/).

[8] D. Oliinychenko, C. Shen, Resonance production in PbPb collisions at 5.02 TeV via hydrodynamics and hadronic afterburner, arXiv:2105.07539 (2021). 10.48550/arXiv.2105.07539

[9] V. Vovchenko, B. Dönigus, H. Stoecker, Canonical statistical model analysis of p–p, p-Pb, and Pb-Pb collisions at energies available at the CERN Large Hadron Collider, Phys. Rev. C **100**, 054906 (2019). 10.1103/PhysRevC.100.054906

[10] A. Motornenko et al., Kinetic freeze-out temperature from yields of short-lived resonances, Phys. Rev. C **102**, 024909 (2020). 10.1103/PhysRevC.102.024909

[11] S. Acharya et al. (ALICE), $\Sigma(1385)^{\pm}$ resonance production in Pb–Pb collisions at $\sqrt{s_{\rm NN}} = 5.02$ TeV, Eur. Phys. J. C **83**, 351 (2023). 10.1140/epjc/s10052-023-11475-1

[12] S. Acharya et al. (ALICE), Light-flavor particle production in high-multiplicity pp collisions at $\sqrt{s}$ = 13 TeV as a function of transverse spherocity, J. High Energy Phys. **05**, 184 (2024). 10.1007/JHEP05(2024)184

[13] S. Acharya et al. (ALICE), Multiplicity dependence of light-flavor hadron production in pp collisions at $\sqrt{s}$ = 7 TeV, Phys. Rev. C **99**, 024906 (2019). 10.1103/PhysRevC.99.024906

[14] L. Rozpłochowski (NA61/SHINE), Energy dependence of $\phi(1020)$ meson production in nucleus-nucleus collisions at the CERN SPS, these proceedings (2024).

[15] M. Lorenz (HADES), HADES highlights: Recent results from HADES, these proceedings (2024).

[16] S. Acharya et al. (ALICE), $f_0(980)$ production in inelastic pp collisions at $\sqrt{s}$ = 5.02 TeV, Phys. Lett. B **846**, 137644 (2023). 10.1016/j.physletb.2022.137644

[17] S. Acharya et al. (ALICE), Observation of abnormal suppression of $f_0(980)$ production in p–Pb collisions at $\sqrt{s_{\rm NN}}$ = 5.02 TeV, Phys. Lett. B **853**, 138665 (2024). 10.1016/j.physletb.2024.138665

[18] A. Hayrapetyan et al. (CMS), Elliptic anisotropy measurement of the $f_0(980)$ hadron in proton-lead collisions and evidence for its quark-antiquark composition, arXiv:2312.17092 (2023). 10.48550/arXiv.2312.17092

[19] P. Das (ALICE), Investigating the hidden strangeness content of exotic resonance with ALICE, these proceedings (2024).

[20] S. Acharya et al. (ALICE), $\rm K^0_S K^0_S$ and $\rm K^0_S K^{\pm}$ femtoscopy in pp collisions at $\sqrt{s}$ = 5.02 and 13 TeV, Phys. Lett. B **833**, 137335 (2022). 10.1016/j.physletb.2022.137335

[21] S. Acharya et al. (ALICE), Investigating the composition of the $\rm K^*_0(700)$ state with $\pi^{\pm}\rm K^0_S$ correlations at the LHC, Phys. Lett. B **856**, 138915 (2024). 10.1016/j.physletb.2024.138915

[22] N.J. Abdulameer et al. (PHENIX), Measurement of ϕ-meson production in Cu+Au collisions at $\sqrt{s_{NN}}$ = 200 GeV and U+U collisions at $\sqrt{s_{NN}}$ = 193 GeV, Phys. Rev. C **107**, 014907 (2023). 10.1103/PhysRevC.107.014907

[23] M.S. Abdallah et al. (STAR), Pattern of global spin alignment of ϕ and K^{*0} mesons in heavy-ion collisions, Nature **614**, 244 (2023). 10.1038/s41586-022-05557-5

[24] G. Wilks (STAR), Differential measurements of ϕ-meson global spin alignment in Au+Au collisions at STAR, these proceedings (2024).

[25] S. Acharya et al. (ALICE), Evidence of spin-orbital angular momentum interactions in relativistic heavy-ion collisions, Phys. Rev. Lett. **125**, 012301 (2020). 10.1103/PhysRevLett.125.012301

© The Authors, published by EDP Sciences. This is an open access article distributed under the terms of the Creative Commons Attribution License 4.0 (https://creativecommons.org/licenses/by/4.0/).

Experimental Search on the QCD Critical Point

Yu Zhang[1,*]

[1]Guangxi Normal University, Guilin, 541006, China

Abstract. The QCD critical point is a landmark on the QCD phase diagram, attracting much effort to confirm its location. In these proceedings, we present recent experimental progress in the search for the QCD critical point. The results, comparisons with models, and implications for the QCD critical point will also be discussed.

1 Introduction

The QCD critical point (CEP) is predicted to be the end point of the first-order phase transition boundary between the Quark-Gluon Plasma (QGP) and hadron gas on the QCD phase diagram. Much effort has been made to look for it from both theory and experiment. In these proceedings, we show recent progress from experiment on the search for the QCD critical point.

2 Intermittency Analysis

Intermittency is used to look for the critical behavior induced by the critical point in heavy-ion collisions [1]. In the vicinity of CEP, the correlation length of the system diverges, and the system turns to be self-similar. Based on the 3D-Ising universality class, the density-density correlation function for small momentum transfer has a power-law structure, which leads to large density fluctuations. Intermittency analysis can capture these fluctuations using scaled factorial moments in transverse momentum space. qth-order scaled factorial moments are defined by Eq. (1) where M^D is the number of cells in D-dimensional phase space and n_i is the measured multiplicity of a given event in the ith cell. Two types of power-law behavior are shown as $F_q(M) \propto (M^D)^{\phi_q}$ and $F_q(M) \propto F_2(M)^{\beta_q}$, separately, where ϕ_q and β_q are called intermittency index and scaling index. According to Ginzburg-Landau theory, the β_q follows $\beta_q \propto (q-1)^{\nu}$ and is predicted to be 1.034, which is 1.0 in 2D-Ising model calculation.

$$F_q(M) = \frac{\langle \frac{1}{M^D} \sum_{i=1}^{M^D} n_i(n_i-1)\cdots(n_i-q+1)\rangle}{\langle \frac{1}{M^D} \sum_{i=1}^{M^D} n_i \rangle^q} \quad (1)$$

Figure 1 shows $\Delta F_q(M)$ ($q > 2$, $F_q(M)$ after background subtraction) up to sixth-order of identified charged hadrons as a function of second-order $\Delta F_2(M)$ in 0-5% central Au+Au collisions at $\sqrt{s_{NN}}$ = 7.7 - 200 GeV [2]. The black solid lines indicate linear fitting to data. The power-law structure of $F_q(M) \propto F_2(M)^{\beta_q}$ are clearly seen in each energy. The power

*e-mail: zhangyu@gxnu.edu.cn

© The Authors, published by EDP Sciences. This is an open access article distributed under the terms of the Creative Commons Attribution License 4.0 (https://creativecommons.org/licenses/by/4.0/).

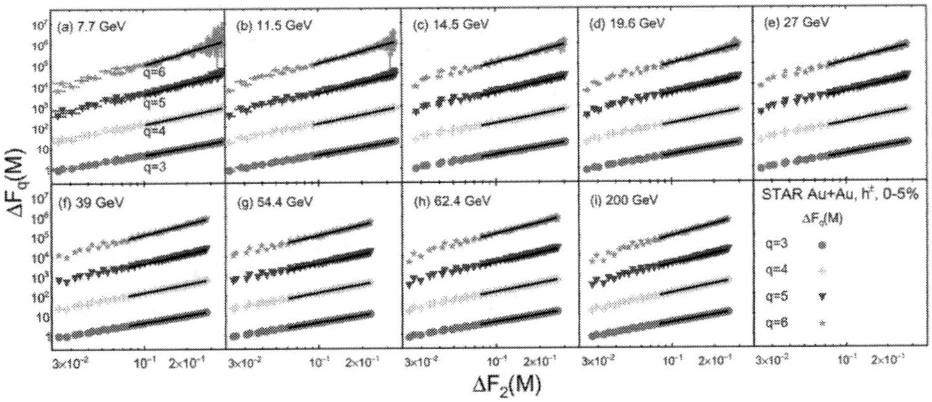

Figure 1. $\Delta F_q(M)$ as a function of $\Delta F_2(M)$ in most central Au+Au collisions at $\sqrt{s_{NN}}$ = 7.7-200 GeV. The solid black lines represent the best power-law fit of $\Delta F_q(M) \propto \Delta F_2(M)^{\beta_q}$.

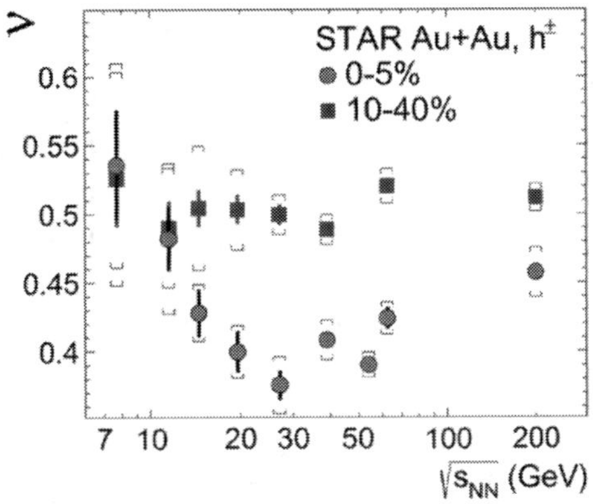

Figure 2. Energy dependence of the scaling exponent (ν) for identified charged hadrons ($h^{\pm}$) in Au+Au collisions at $\sqrt{s_{NN}}$ = 7.7-200 GeV.

series β_q is then extracted from each energy. In agreement with theoretical expectation, β_q obeys a power-law behavior with $q-1$. The scaling exponent ν can be obtained through a power-law fitting on $\beta_q \propto (q-1)^{\nu}$. Figure 2 shows the energy dependence of β_q for identified charged hadrons in 0-5% central and 10-40% mid-central Au+Au collisions. In 5% most central collisions, ν exhibits a non-monotonic behavior as a function of collision energy and reaches a minimum at around $\sqrt{s_{NN}} = 27$ GeV. In contrast, ν shows flat energy dependence in 10-40% mid-central collisions. The observed non-monotonic energy dependence of ν in 0-5% most central collisions could be due to the signal of density fluctuations induced by the QCD critical point. The measured ν in Fig. 2 is smaller than the theoretical calculation of the critical $\nu = 1.30$ by the Ginzburg-Landau theory. However, the theoretical calculation is performed over an entire phase space while the data is measured within limited η and p_T acceptance. Theoretical calculations implemented in a limited transverse momentum phase space and equivalent experimental acceptance, are necessary to understand the measurements.

© The Authors, published by EDP Sciences. This is an open access article distributed under the terms of the Creative Commons Attribution License 4.0 (https://creativecommons.org/licenses/by/4.0/).

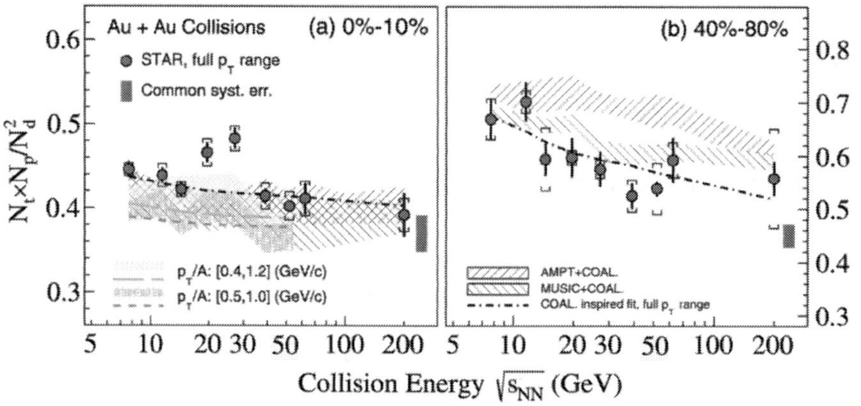

Figure 3. Collision energy, centrality, and p_T dependence of the yield ratio $N_t \times N_p/D_d^2$ in Au+Au collisions at RHIC. Solid circles are the results from 0–10% central (left panel) and 40–80% peripheral (right panel) collisions.

3 Light Nuclei Yield Ratio

The production of light nuclei like deuteron, triton, and helium-3 provides important information on nuclear matter at high baryon density and temperature. It is proposed by the coalescence model [3] that the compound yield ratio $N_t \times N_p/N_d^2$ of tritons (N_t), deuterons (N_d), and protons (N_p) is sensitive to neutron density fluctuations, making it a promising observable for the QCD critical point. Below we show recent progress from RHIC-STAR experiment.

Figure 3 shows the energy dependence of the yield ratio $N_t \times N_p/N_d^2$ at mid-rapidity in 0-10% central and 40-80% peripheral Au+Au collisions at $\sqrt{s_{NN}}$ = 7.7 - 200 GeV [4]. For comparisons, the calculations of AMPT, MUSIC+UrQMD hybrid models are shown in shaded areas. For 0-10% most central Au+Au collisions, the yield ratios are consistent with the coalescence baseline and model calculation, except for the enhancements of the yield ratios to the coalescence baseline with a significance of 2.3σ and 3.4σ observed at $\sqrt{s_{NN}} = 19.6$ and 27 GeV, respectively. In peripheral collisions, similar to data, model calculations have a smooth decreasing trend as a function of collision energy. If the enhancements in 0-10% central collisions are due to large baryon density fluctuations near the critical point, further theoretical studies are needed, especially the dynamical modeling of heavy-ion collisions with a realistic equation of state.

4 Baryon-strangeness Correlation

Baryon-strangeness correlation is proposed to be a diagnostic of the degree of freedom and correlations of strongly interacting matter. It could be used to identify the onset of deconfinement [5]. The definition of the baryon-strangeness correlation is $C_{BS} = -3\frac{\langle BS\rangle - \langle B\rangle\langle S\rangle}{\langle S^2\rangle - \langle S\rangle^2}$, where B and S indicate number of baryon and strangeness, respectively. For QGP, if quark flavors are uncorrelated, then the C_{BS} ratio is unity. While for hadron gas, the C_{BS} depends on hadron environment. The estimation of C_{BS} in the statistical model for a hadron resonance

© The Authors, published by EDP Sciences. This is an open access article distributed under the terms of the Creative Commons Attribution License 4.0 (https://creativecommons.org/licenses/by/4.0/).

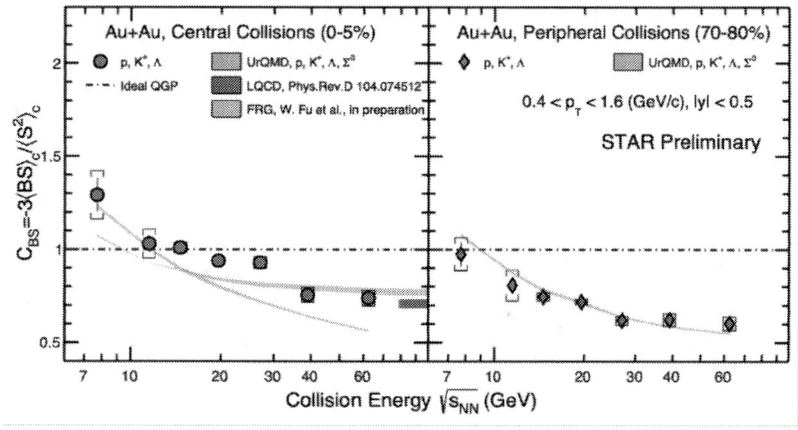

Figure 4. Collision energy dependence of the baryon-strangeness correlation $C_{\rm BS}$ in 0-5% and 70-80% Au+Au collisions.

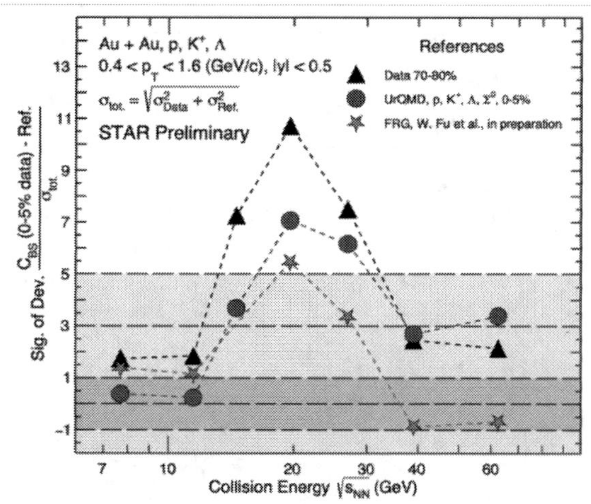

Figure 5. Collision energy dependence of deviation of $C_{\rm BS}$ ratio from peripheral data, UrQMD and FRG model calculations.

gas model at zero baryon chemical potential Here, we show recent measurements on baryon-strangeness correlation in Au+Au collisions at $\sqrt{s_{NN}}$ = 7.7 - 62.4 GeV from RHIC-STAR. Particles in measurements include (anti)proton, $K^{\pm}$, and Λ ($\bar{\Lambda}$).

Figure 4 shows the energy dependence of C_{BS} ratio in Au+Au collisions at $\sqrt{s_{NN}}$ = 7.7 - 62.4 GeV for 0-5% and 70-80% centrality. The data points are shown with red circles, while colored bands represent model calculations from UrQMD, LQCD, and FRG. The experimental measurements are done within $|y| < 0.5$ and $0.4 < p_T < 1.6$ GeV/c. The data in peripheral collisions are qualitatively consistent with calculation from the UrQMD model, while in 0-5% central collisions, the data show deviations from various model calculations. The deviations as a function of collision energy between data from central collisions and peripheral collisions and model calculations are shown in Fig. 5. It can be seen that the deviation reaches a maximum at 19.6 GeV. The physics behind this discrepancy needs further investigation.

© The Authors, published by EDP Sciences. This is an open access article distributed under the terms of the Creative Commons Attribution License 4.0 (https://creativecommons.org/licenses/by/4.0/).

5 Net-proton Cumulants

Cumulants of conserved quantities like net-baryon, net-charge, and net-strangeness number are proposed as sensitive observables [6] to the QCD critical point and have been studied extensively both in experiment and theory.

Recently, preliminary results of net-proton cumulants on STAR BES-II datasets are available. This measurement is performed in Au+Au collisions at $\sqrt{s_{NN}}$ = 7.7 - 27 GeV. With upgrades on STAR detector like iTPC, higher-quality data with extended particle acceptance are collected. Figure 6 shows collision energy dependence of net-proton C_4/C_2 ($\kappa\sigma^2$) in Au+Au collisions at STAR BES energies. In the left panel of Fig. 6, new measurements are shown with red circles and red diamonds for 0-5% and 70-80% collision centrality, respectively. The measurements on BES-I datasets are shown with black open circles and diamonds [7]. Calculations from non-critical models like hydrodynamic model, HRG with a canonical ensemble, and the UrQMD model are also shown. Compared to model calculations, it can be seen that the trend of data in 0-5% central collisions can be qualitatively described by model calculations while there are quantitive differences. The right panel of Fig. 6 shows the deviations of net-proton C_4/C_2 in 0-5% central collisions as a function of collision energy from each model. It can be seen that the deviation from non-critical models reaches a maximum at $\sqrt{s_{NN}}$ = 19.6 GeV for both three models with a significance level of 4σ. The result could be the sign of the QCD critical point, but there is still a gap at low energies $3 < \sqrt{s_{NN}} < 7.7$ GeV. The measurements at these lower energies will be crucial to confirm the existence of the QCD critical point. Currently, the analysis at STAR fixed-target energies focusing on $\sqrt{s_{NN}} = 3.0 - 4.5$ GeV is ongoing. Hopefully, this measurement as well as measurements from other heavy-ion experiments at high baryon density will give a confirmative answer.

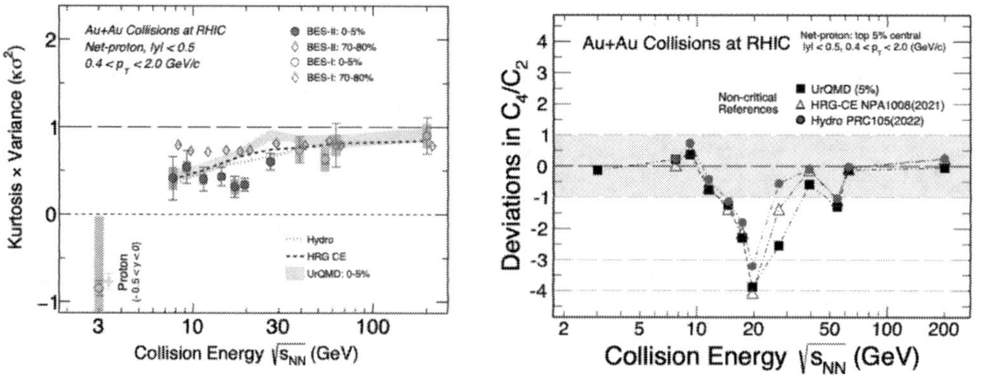

Figure 6. Left: Net-proton $\kappa\sigma^2$ as a function of collision energy ($\sqrt{s_{NN}}$) in 0-5% and 70-80% collision centrality. Right: Deviations of net-proton $\kappa\sigma^2$ from various model calculations as a function of collision energy ($\sqrt{s_{NN}}$).

6 Summary

In these proceedings, I show a brief introduction to the recent progress from experiments on the search for the QCD critical point. From various measurements, we see a maximum deviation from non-critical baselines at around 20 GeV. There could be important physics at these energies which require efforts from both experimental and theoretical sides. More important results are not shown here due to limited space.

© The Authors, published by EDP Sciences. This is an open access article distributed under the terms of the Creative Commons Attribution License 4.0 (https://creativecommons.org/licenses/by/4.0/).

Acknowledgements

The National Natural Science Foundation of China supported the author under No. 12305141.

References

[1] R.C. Hwa, M.T. Nazirov, Intermittency in second order phase transition, Phys. Rev. Lett. **69**, 741 (1992). 10.1103/PhysRevLett.69.741

[2] M. Abdulhamid et al. (STAR), Energy dependence of intermittency for charged hadrons in Au+Au collisions at RHIC, Phys. Lett. B **845**, 138165 (2023), 2301.11062. 10.1016/j.physletb.2023.138165

[3] K.J. Sun, L.W. Chen, C.M. Ko, J. Pu, Z. Xu, Light nuclei production as a probe of the QCD phase diagram, Phys. Lett. B **781**, 499 (2018), 1801.09382. 10.1016/j.physletb.2018.04.035

[4] M. Abdulhamid et al. (STAR), Beam Energy Dependence of Triton Production and Yield Ratio ($N_t \times N_p/N_d^2$) in Au+Au Collisions at RHIC, Phys. Rev. Lett. **130**, 202301 (2023), 2209.08058. 10.1103/PhysRevLett.130.202301

[5] V. Koch, A. Majumder, J. Randrup, Baryon-strangeness correlations: A Diagnostic of strongly interacting matter, Phys. Rev. Lett. **95**, 182301 (2005), nucl-th/0505052. 10.1103/PhysRevLett.95.182301

[6] M.A. Stephanov, Non-Gaussian fluctuations near the QCD critical point, Phys. Rev. Lett. **102**, 032301 (2009), 0809.3450. 10.1103/PhysRevLett.102.032301

[7] J. Adam et al. (STAR), Nonmonotonic Energy Dependence of Net-Proton Number Fluctuations, Phys. Rev. Lett. **126**, 092301 (2021), 2001.02852. 10.1103/PhysRevLett.126.092301

© The Authors, published by EDP Sciences. This is an open access article distributed under the terms of the Creative Commons Attribution License 4.0 (https://creativecommons.org/licenses/by/4.0/).

Experimental overview on light (anti)(hyper)nuclei production at the LHC

Chiara Pinto[1,*]

[1]European Organisation for Nuclear Research (CERN), Geneva, Switzerland

Abstract. For many years, significant efforts at the LHC have been put towards the measurements of the production of (anti)(hyper)nuclei in high-energy hadronic collisions. The ALICE experiment has contributed with measurements in pp, p–Pb, Xe–Xe, and Pb–Pb collisions, at various center-of-mass energies. Recently, also the LHCb experiment has measured the production of nuclei in both collider and fixed-target modes. The highlights of recent experimental measurements are discussed in the context of the production models used to describe the hadronization mechanism.

1 Introduction

At the Large Hadron Collider, the baryochemical potential is close to zero [1] and therefore it is possible to measure the production of both nuclei and antinuclei, in several collision systems, from pp to Pb–Pb, in a range of center-of-mass energies between $\sim$ 1 and $\sim$ 13 TeV. The measurements of the production of (anti)(hyper)nuclei at accelerators allow the study of the nucleosynthesis mechanism, which is nowadays under intense debate in the scientific community. The production mechanism is usually described by three classes of phenomenological models, i.e., the statistical hadronization model (SHM), the coalescence approach, and hybrid models that include relativistic hydrodynamics with a subsequent coalescence afterburner, to describe the survival of bound states in the hadron gas phase with intense rescattering. Many observables have been pointed out to have a strong discrimination power among the model predictions, and therefore the experimental challenge today is to get the most possibly precise measurement to compare to the expectations of the models. Besides the interest in the hadronization mechanism per se, the study of the production of (anti)(hyper)nuclei at the LHC has important applications to the indirect dark matter searches and to the study of the particle interactions needed for the equation of state to describe neutron star observations.

2 Testing production models

According to SHM, light (anti)(hyper)nuclei are produced by a source in local thermal and hadrochemical equilibrium at the chemical freeze-out, with a temperature common to all light flavor species [2], of about 155 MeV. Such chemical freeze-out temperature corresponds to the phase-transition temperature, that is the same in Pb–Pb [3] and Xe–Xe [4] collisions, and corresponds to the pseudo-critical temperature for the transition from QGP to hadron

*e-mail: chiara.pinto@cern.ch

© The Authors, published by EDP Sciences. This is an open access article distributed under the terms of the Creative Commons Attribution License 4.0 (https://creativecommons.org/licenses/by/4.0/).

gas predicted by lattice QCD calculations [5]. In heavy-ion collisions, the grand-canonical ensemble of the statistical model [2, 6, 7], where all the charges are only conserved on average but fluctuate from one microscopic state to another, holds true to describe the particle yield in the case of sufficiently large reaction volumes. In small collision systems (e.g., pp and p–Pb collisions), the exact conservation of charges from one microscopic state to another plays an important role in determining the final state particle yield. Therefore, the canonical ensemble of the SHM is used, resulting in the so-called canonical suppression of the yields of particles carrying conserved charges, relative to their grand-canonical values.

In the coalescence model, light (anti)(hyper)nuclei are formed by the coalescence of nucleons that are close in phase space and with matching spin-isospin configurations, at kinetic freeze-out (occurring when the elastic interactions stop and the momentum of the particles is fixed). In the state-of-the-art implementation of the coalescence model, the formation probability is calculated by folding the phase-space distributions of the constituent nucleons with the Wigner density distribution of the bound state [8].

Recent developments of the coalescence model based on the Wigner function formalism have shown that the coalescence of deuterons is extremely sensitive to the wavefunction of the nucleus and to the size of the baryon-emitting source [7–9]. For instance, in order to reproduce the measured transverse momentum distribution of deuterons measured in the high-multiplicity sample of pp collisions at a center-of-mass energy of $\sqrt{s}$ = 13 TeV by ALICE with a coalescence model based on the Wigner function formalism [8], the realistic wavefunction of deuterons, i.e., Argonne v18 [10], is needed in addition to the measured baryon-emitting source size [11], as shown in Fig. 1.
On the contrary, the SHM predictions depend only on the mass and on the spin degeneracy factor of the nucleus. This aspect is particularly relevant in two cases: (i) in small collision systems, since the size of the nucleus (r_{deuteron} = 1.96 fm and $r_{^3\text{He}}$ = 1.76 fm [12]) and that of the baryon-emitting source ($r \sim$ 1 fm in pp collisions [13] and $r \sim$ 1.5 fm in p–Pb colli-

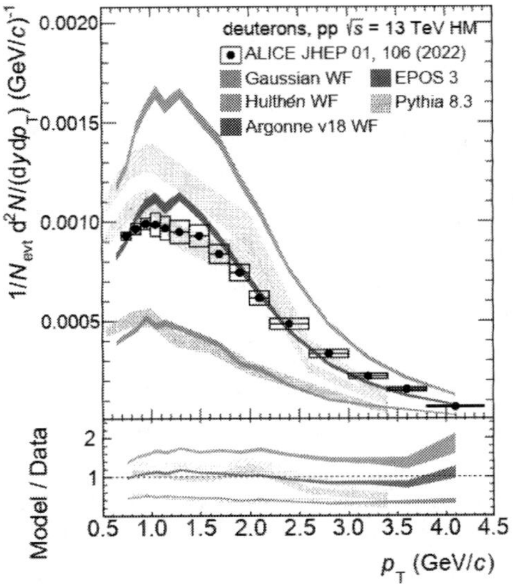

Figure 1. Transverse momentum distribution of deuterons measured in the high multiplicity class of pp collisions at $\sqrt{s}$ = 13 TeV by ALICE, compared to the expectations of a coalescence model based on the Wigner function formalism that uses several hypotheses for the wavefunction, implemented as an afterburner on top of events generated with EPOS 3 or PYTHIA 8.3.

© The Authors, published by EDP Sciences. This is an open access article distributed under the terms of the Creative Commons Attribution License 4.0 (https://creativecommons.org/licenses/by/4.0/).

sions [14]) are comparable; (ii) in the ratios of the yields of nuclei and hypernuclei with mass number $A \geq 3$.

In the left panel of Figure 2 the ratios of the integrated yields of deuterons and protons are shown as a function of the charged-particle multiplicity and compared to the expectations of the production models. The predictions of two implementations of the canonical statistical model (CSM) with different hypotheses for the freeze-out temperature and those of the coalescence model describe qualitatively the smoothly increasing trend of the deuteron-to-proton yield ratios with multiplicity and the overall yields. Notably, all the predictions give similar results for the $A = 2$ yield ratios, making it difficult to discriminate among the hadronization mechanisms. A better discrimination power is present for the ratios of ^{3}He-to-proton yields, as shown in the right panel of Fig. 2: especially at the multiplicities corresponding to heavy-ion collisions, models differ by $\sim 40\%$. However, the current experimental precision is not enough to distinguish among the predictions of the different models.

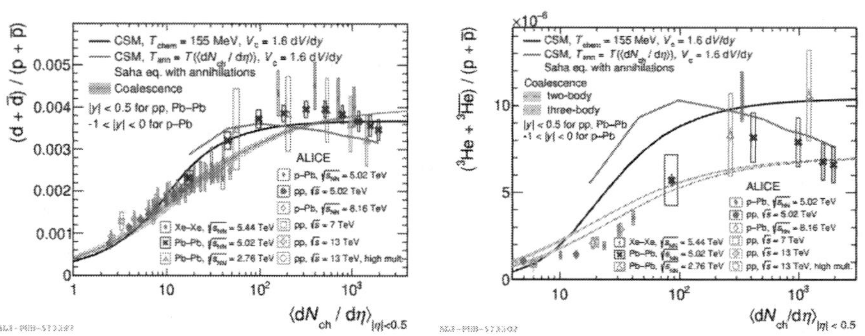

Figure 2. The ratios of the integrated yields of deuterons and protons (left panel) and ^{3}He and protons (right panel) are shown as a function of the charged-particle multiplicity and compared to the expectations of several production models.

In this scenario, hypernuclei (nuclei with at least one hyperon constituent) play an important role as they are very loosely bound states with a large spatial extension of their wavefunction. For instance, hypertriton is a hypernucleus composed of a proton, a neutron, and a Λ hyperon, with a separation energy of the Λ of about 102 keV [15] and a radius of about 10 fm [16]. The ratio of the yields of the $^3_\Lambda$H and of the ^{3}He as a function of multiplicity has a great discrimination power between the SHM and coalescence (see left panel of Fig. 3), because the two species have comparable masses ($m_{^3_\Lambda H}$=2.991 and $m_{^3He}$=2.809 GeV/c^2) but very different sizes ($r_{^3_\Lambda H}$ ~10 fm and $r_{^3He}$=1.76 fm). Hence, the SHM, which depends only on the masses of the nuclei, predicts a flat ratio as a function of multiplicity, while the coalescence model, which takes into account the interplay between the source size and the spatial extension of the nucleus wavefunction, results in an increasing trend of the predicted ratio as a function of multiplicity, that reflects the increasing size of the baryon-emitting source. Finally, another powerful tool to discriminate among the predictions of the production models is the multiplicity-dependent ratio of the yields of $^3_\Lambda$H and those of the Λ baryon. The models give very different predictions for such a ratio, especially in the low multiplicity range, corresponding to small systems. Only few measurements of this ratio are currently available from the ALICE Collaboration, as shown in the right panel of Fig. 3, but with a precision good enough to discriminate among the models. The data agrees with the predictions of the coalescence model within 1σ at $\langle dN_{ch}/d\eta \rangle < 10$, and exclude the SHM predictions by $\sim 6\sigma$.

© The Authors, published by EDP Sciences. This is an open access article distributed under the terms of the Creative Commons Attribution License 4.0 (https://creativecommons.org/licenses/by/4.0/).

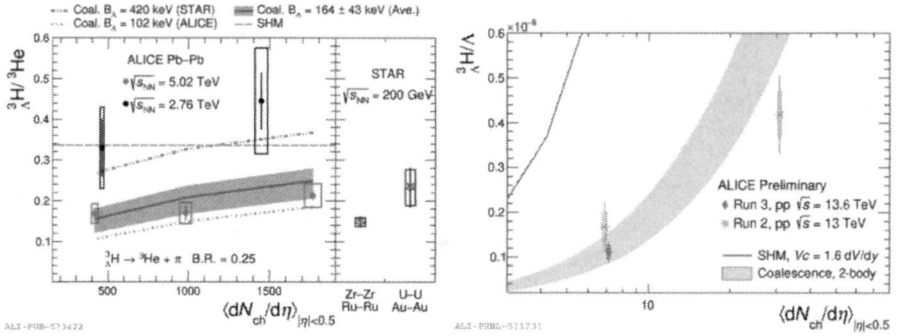

Figure 3. Ratio of the integrated yields of ${}^{3}_{\Lambda}$H and ^{3}He (left panel) and of ${}^{3}_{\Lambda}$H and Λ (right panel) as a function of the charged-particle multiplicity, compared to the expectations of different production models.

3 Impact on astrophysics

One interesting application of the measurement of the production of antinuclei in accelerator experiments concerns the search for indirect dark matter (DM). Antinuclei in the Galaxy might be produced either by reactions of primary cosmic rays (CRs) with the interstellar medium (ISM), both constituted for the 90% by protons and for the 9% by alpha nuclei, or by annihilation of dark matter candidates, or by even more exotic sources such as anticlouds or antistars [17]. Although antinuclei heavier than antiprotons still elude detection in space, many theoretical efforts are ongoing in order to model the flux of antinuclei from cosmic rays, with the ultimate goal of correctly interpret any future measurement of antinuclear CR fluxes. In this sense, information on the production of antinuclei from accelerator experiments, particularly in small collision systems, is essential for the theoretical description of the background constituted by antinuclei from cosmic-ray collisions with ISM. Notably, the production of antinuclei at the LHC is typically measured at midrapidity ($|y| < 0.5$), and then, combined with several coalescence models, employed to predict the flux of antinuclei from CR interactions at forward rapidity. However, the possible impact of a rapidity dependence of the production yield and of the coalescence probability[1] of antinuclei was suggested in Ref. [18] and recently investigated by the ALICE Collaboration in pp collisions at $\sqrt{s}$ = 13 TeV [19]. In such work, the rapidity dependence of the production yield of $\overline{\mathrm{p}}$ and $\overline{\mathrm{d}}$ and of the coalescence parameter B_2 (obtained from Eq. 1 with $A = 2$) has been investigated. The measurements have been carried out rapidity-differentially up to $|y| < 0.7$, and then, with the use of the a state-of-the-art coalescence model based on the Wigner formalism [8], the integrated yields and B_2 have been extrapolated to forward rapidity and high p_{T}, as shown in Fig. 4. These model predictions have been used as input for the calculations of the flux of antinuclei from cosmic rays in Ref. [20]. In such work, the author investigates the role of the rapidity extrapolation in the predictions of the CR flux of antinuclei assessing that about 90% of the flux is due to antideuterons with rapidity between 0.5 and 1.5. This kinematical region is the one where the CR flux is expected to be dominated by the DM signal, according

[1]The coalescence parameter is defined as the ratio between the invariant yield of the nucleus with mass number A and that of the protons, raised to the power of A, assuming protons and neutrons to have the same transverse momentum distributions as they are isospin partners, and $p_{\mathrm{T}}^{\mathrm{p}} = p_{\mathrm{T}}^{A}/A$:

$$B_A = \left(\frac{1}{2\pi p_{\mathrm{T}}^{A}}\left(\frac{\mathrm{d}^2N}{\mathrm{d}y\mathrm{d}p_{\mathrm{T}}}\right)_A\right) \Big/ \left(\frac{1}{2\pi p_{\mathrm{T}}^{\mathrm{p}}}\left(\frac{\mathrm{d}^2N}{\mathrm{d}y\mathrm{d}p_{\mathrm{T}}}\right)_{\mathrm{p}}\right)^A. \tag{1}$$

© The Authors, published by EDP Sciences. This is an open access article distributed under the terms of the Creative Commons Attribution License 4.0 (https://creativecommons.org/licenses/by/4.0/).

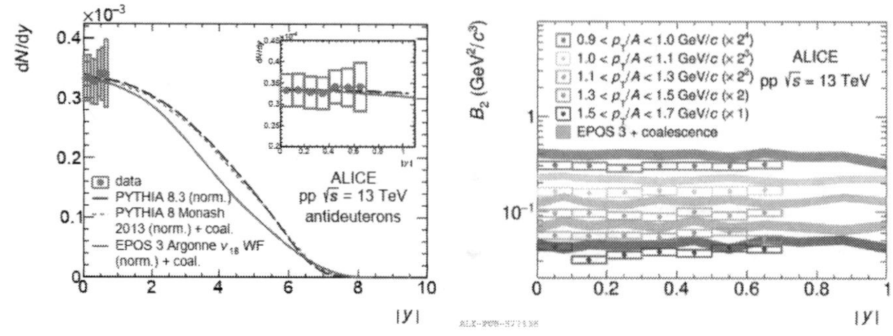

Figure 4. Deuteron yields (left panel) and coalescence parameter B_2 as a function of rapidity for fixed intervals of the p_T/A (right panel), measured in pp collisions at $\sqrt{s}$ = 13 TeV by ALICE [19]. The data are compared to the predictions of a coalescence afterburner based on the Wigner function formalism, on top of events generated with EPOS 3 or PYTHIA.

to many cosmological models [21]. The relevant rapidity region ($0.5 < |y| < 1.5$) is in reach of present or future LHC experiments, such as LHCb [22] in its fixed-target configuration, CMS [23] in its Run 4 program, and the future ALICE 3 facility [24].

Finally, one open point is still the extrapolation to the low-energy scale needed for astrophysical models: while the LHC experiments cover the TeV scale, the production of antinuclei in the Galaxy occurs at center-of-mass energies of the order of 1–10 GeV (for instance, antideuterons are produced in collisions of $\sqrt{s}$ = 17 GeV [21]). The current assumption is that the production models employed at the LHC energy scale can also be used at lower energies. The LHCb Collaboration recently measured ^{3}He nuclei [25] and successfully identified (anti)hypertritons [26]. Future (anti)deuteron and (anti)^{3}He measurements as a function of rapidity provided by LHCb would be highly useful for the cosmic-ray flux modeling, as the LHCb fixed-target mode would allow the scan of the low energy collision range (of the order of 1–10 GeV), relevant for astrophysics.

References

[1] S. Acharya et al. (ALICE), Measurements of chemical potentials in Pb-Pb collisions at $\sqrt{s_{NN}}$ = 5.02 TeV (2023), 2311.13332.

[2] A. Andronic, P. Braun-Munzinger, K. Redlich, J. Stachel, Decoding the phase structure of QCD via particle production at high energy, Nature **561**, 321 (2018), 1710.09425. 10.1038/s41586-018-0491-6

[3] A. Andronic, P. Braun-Munzinger, B. Friman, P.M. Lo, K. Redlich, J. Stachel, The thermal proton yield anomaly in Pb–Pb collisions at the LHC and its resolution, Phys. Lett. B **792**, 304 (2019), 1808.03102. 10.1016/j.physletb.2019.03.052

[4] S. Acharya et al. (ALICE), Measurement of the production and elliptic flow of (anti)nuclei in Xe-Xe collisions at $\sqrt{s_{NN}}$ = 5.44 TeV (2024), 2405.19826.

[5] A. Bazavov et al. (HotQCD), Chiral crossover in QCD at zero and non-zero chemical potentials, Phys. Lett. B **795**, 15 (2019), 1812.08235. 10.1016/j.physletb.2019.05.013

[6] F. Bellini, A.P. Kalweit, Testing production scenarios for (anti-)(hyper-)nuclei and exotica at energies available at the CERN Large Hadron Collider, Phys. Rev. C **99**, 054905 (2019), 1807.05894. 10.1103/PhysRevC.99.054905

[7] R. Scheibl, U.W. Heinz, Coalescence and flow in ultrarelativistic heavy ion collisions, Phys. Rev. **C59**, 1585 (1999), nucl-th/9809092. 10.1103/PhysRevC.59.1585

© The Authors, published by EDP Sciences. This is an open access article distributed under the terms of the Creative Commons Attribution License 4.0 (https://creativecommons.org/licenses/by/4.0/).

[8] M. Mahlein, L. Barioglio, F. Bellini, L. Fabbietti, C. Pinto, B. Singh, S. Tripathy, A realistic coalescence model for deuteron production, Eur. Phys. J. C **83**, 804 (2023), 2302.12696. 10.1140/epjc/s10052-023-11972-3

[9] H. Sato, K. Yazaki, On the coalescence model for high energy nuclear reactions, Phys. Lett. B **98**, 153 (1981). https://doi.org/10.1016/0370-2693(81)90976-X

[10] R.B. Wiringa, V.G.J. Stoks, R. Schiavilla, Accurate nucleon-nucleon potential with charge-independence breaking, Phys. Rev. C **51**, 38 (1995). 10.1103/PhysRevC.51.38

[11] S. Acharya et al. (ALICE), Search for a common baryon source in high-multiplicity pp collisions at the LHC, Phys. Lett. B **811**, 135849 (2020), 2004.08018. 10.1016/j.physletb.2020.135849

[12] G. Ropke, Light nuclei quasiparticle energy shift in hot and dense nuclear matter, Phys. Rev. C **79**, 014002 (2009), 0810.4645. 10.1103/PhysRevC.79.014002

[13] S. Acharya et al. (ALICE), p-p, p-Λ and Λ-Λ correlations studied via femtoscopy in pp reactions at $\sqrt{s} = 7$ TeV, Phys. Rev. C **99**, 024001 (2019), 1805.12455. 10.1103/PhysRevC.99.024001

[14] S. Acharya et al. (ALICE), First Observation of an Attractive Interaction between a Proton and a Cascade Baryon, Phys. Rev. Lett. **123**, 112002 (2019), 1904.12198. 10.1103/PhysRevLett.123.112002

[15] S. Acharya et al. (ALICE), Measurement of the Lifetime and Λ Separation Energy of HΛ3, Phys. Rev. Lett. **131**, 102302 (2023), 2209.07360. 10.1103/PhysRevLett.131.102302

[16] F. Hildenbrand, H.W. Hammer, Three-Body Hypernuclei in Pionless Effective Field Theory, Phys. Rev. C **100**, 034002 (2019), [Erratum: Phys.Rev.C 102, 039901 (2020)], 1904.05818. 10.1103/PhysRevC.100.034002

[17] V. Poulin, P. Salati, I. Cholis, M. Kamionkowski, J. Silk, Where do the AMS-02 antihelium events come from?, Phys. Rev. D **99**, 023016 (2019), 1808.08961. 10.1103/PhysRevD.99.023016

[18] Z. Citron et al., Report from Working Group 5: Future physics opportunities for high-density QCD at the LHC with heavy-ion and proton beams, CERN Yellow Rep. Monogr. **7**, 1159 (2019), 1812.06772. 10.23731/CYRM-2019-007.1159

[19] S. Acharya et al. (ALICE), Rapidity dependence of antideuteron coalescence in pp collisions at $\sqrt{s} = 13$ TeV with ALICE (2024), 2407.10527.

[20] K. Blum, Rapidity dependence of nuclear coalescence: Impact on cosmic ray antinuclei, Phys. Rev. C **109**, L031904 (2024), 2306.13165. 10.1103/PhysRevC.109.L031904

[21] L. Šerkšnytė et al., Reevaluation of the cosmic antideuteron flux from cosmic-ray interactions and from exotic sources, Phys. Rev. D **105**, 083021 (2022), 2201.00925. 10.1103/PhysRevD.105.083021

[22] A.A. Alves, Jr. et al. (LHCb), The LHCb Detector at the LHC, JINST **3**, S08005 (2008). 10.1088/1748-0221/3/08/S08005

[23] New opportunities of heavy ion physics with CMS-MTD at the HL-LHC (2021), `https://cds.cern.ch/record/2800541`

[24] Letter of intent for ALICE 3: A next-generation heavy-ion experiment at the LHC (2022), 2211.02491

[25] R. Aaij et al. (LHCb), Helium identification with LHCb, JINST **19**, P02010 (2024), 2310.05864. 10.1088/1748-0221/19/02/P02010

[26] H. Jage, Observation of antihelium and antihypertriton in pp collisions with LHCb, PoS **EPS-HEP2023**, 254 (2024). 10.22323/1.449.0254

© The Authors, published by EDP Sciences. This is an open access article distributed under the terms of the Creative Commons Attribution License 4.0 (https://creativecommons.org/licenses/by/4.0/).

Recent theoretical developments in quarkonia production in relativistic heavy ion collisions

Jiaxing Zhao[1,2,3,*]

[1]Helmholtz Research Academy Hesse for FAIR (HFHF), GSI Helmholtz Center for Heavy Ion Physics, Campus Frankfurt, 60438 Frankfurt, Germany
[2]Institut für Theoretische Physik, Johann Wolfgang Goethe-Universität,Max-von-Laue-Straße 1, D-60438 Frankfurt am Main, Germany
[3]SUBATECH, Nantes University, IMT Atlantique, IN2P3/CNRS 4 rue Alfred Kastler, 44307 Nantes cedex 3, France

Abstract. Quarkonia are golden probes of heavy ion collisions that have attracted much attention from both experimental and theoretical perspectives. This paper will review recent theoretical studies on quarkonium thermal properties, with a particular focus on the heavy quark finite-temperature potential obtained by Lattice QCD and other approaches. Moreover, it will examine the advancements in the real-time evolution of quarkonia in heavy ion collisions.

1 Introduction

Quarkonia, the bound state of a heavy quark (HQ) and its antiquark, are commonly referred to as charmonium and bottomonium, respectively. Due to their large mass, they are mostly produced in the early stages of heavy ion collisions and thus undergo all stages of evolution. Relativistic heavy ion collisions produce a hot and dense deconfined matter, known as the quark-gluon plasma (QGP). Accordingly, the production of quarkonium in relativistic heavy ion collisions depends on its thermal properties in the hot QGP medium and its real-time dynamical evolution framework. From a theoretical perspective, HQs with large mass and scale separation facilitate the advancement of the heavy quark effective theory. Furthermore, the QGP offers an optimal thermal environment for investigating quantum chromodynamics (QCD) at finite temperature, encompassing both perturbative and non-perturbative aspects.

2 Quarkonium thermal properties

The hierarchy of the heavy quark energy scales, namely $m_Q \gg m_Q v \gg m_Q v^2$, in vacuum allows for the quarkonia to be described by a series of effective theories based on QCD, such as, the non-relativistic QCD (NRQCD) and potential non-relativistic QCD (pNRQCD) [2]. The equation of motion in pNRQCD returns to a Schrödinger-like equation with the transition between the color singlet state and the octet state [2], where the leading order potential for the singlet state is $V(r) = -C_F\alpha_s/r$ with $C_F = (N_c^2 - 1)/(2N_c)$ and the number of colors $N_c = 3$. Moreover, phenomenological studies and non-perturbative lattice simulations indicate the

*e-mail: jzhao@itp.uni-frankfurt.de

© The Authors, published by EDP Sciences. This is an open access article distributed under the terms of the Creative Commons Attribution License 4.0 (https://creativecommons.org/licenses/by/4.0/).

presence of a linear confinement potential between a heavy quark-antiquark pair in the long range. The full range potential can be described by a Cornell form, namely, $V(r) = -\alpha/r + \sigma r + \text{constant}$. The two-body Schrödinger equation with the Cornell potential can provide a successful description of the quarkonium mass spectra, see the review pape [3].

To study the evolution and production of quarkonium in relativistic heavy ion collisions, it is essential to calculate its thermal properties in the hot QCD medium. All in-medium properties of the quarkonium are encoded in its spectral function. The quarkonium bound state manifests as a peak in the spectral function. As the temperature increases, the peak shifts in position and broadens. From the perturbative perspective, the mass shift can be attributed to the static color screening effect in the hot medium, whereas the width is a consequence of the thermal decay of the color singlet octet and the Landau damping, which correspond to gluon dissociation and inelastic scattering, respectively. Meanwhile, the in-medium properties can be incorporated into a temperature-dependent HQ potential. If the HQs interact with the medium for a very long time, the potential approaches the free energy [4]. But in heavy ion collisions, the focus is on the HQ potential over a timescale that is comparable to the internal time scale of quarkonium. In the weak coupling regime, the potential is given by the hard thermal loop (HTL) [5], which has a complex form. In the strong coupling regime, lattice QCD has been employed to simulate the Euclidean Wilson loop, which can be related to the spectral function via a Laplace transform. Nevertheless, the reconstruction of spectral functions from Euclidean correlation functions is an ill-posed inverse problem. A significant discrepancy is attributable to the extraction strategy, as shown in the recent 2+1 flavor lattice QCD study by the HotQCD Collaboration [6]! Assuming a Lorentzian form for the spectral functions, the extracted HQ potential exhibits no screening but displays a substantial imaginary component [7]. The real and imaginary parts of the potentials at T=205 MeV from the aforementioned approaches are clearly illustrated in Fig. 1.

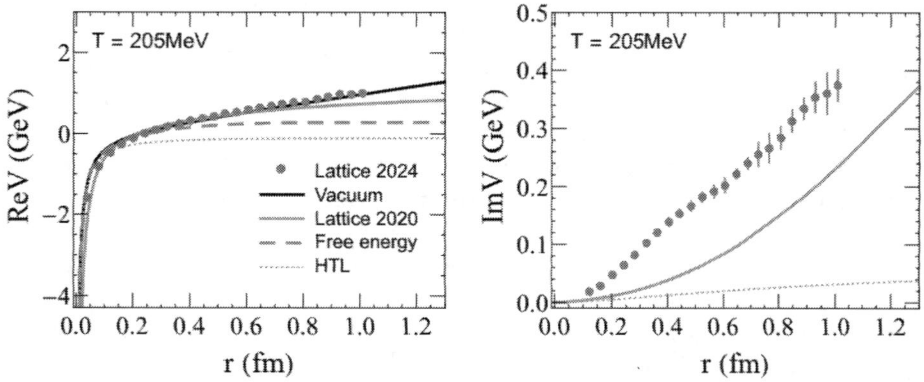

Figure 1. Comparison of the different real (left panel) and imaginary (right panel) part potentials at T=205 MeV. The results of lattice 2020 is from Ref. [8] and lattice 2024 is from Ref. [7].

This no-screening HQ potential is very different from the free energy and previous lattice results. The no-screening or less-screening potential has been demonstrated in many prior studies. For instance, it has been extracted from bottomonium mass and width [9], from bottomonium experimental observables [10], and by fitting the Wilson line correlators and EOS in the T-matrix approach [11]. Furthermore, the HQ potential has been studied with the HTL resumed perturbation method within the Gribov-Zwanziger approach, which enables the study of confinement at the perturbative level [12]. This "new" HQ potential suggests

© The Authors, published by EDP Sciences. This is an open access article distributed under the terms of the Creative Commons Attribution License 4.0 (https://creativecommons.org/licenses/by/4.0/).

a markedly different picture of quarkonium melting in the QGP, with dynamic dissociation playing a dominant role rather than the color screening that has been recognized for many years [1]. The question of how to detect this potential in experiments and which observable it is sensitive to remains unanswered. What is needed is a real-time study of the evolution of quarkonium in the QGP.

3 Real-time evolution in hot QCD medium

The question can be posed as to whether a quarkonium can be defined as a classical particle or a quantum object. If a quarkonium is a very tightly bound state with a narrow width, it can be treated as a classical point particle. Moreover, if one assumes that the quarkonium is entirely dissociated in the QGP and is produced exclusively at the QCD phase boundary, The statistical hadronization model provides a description of the yield of quarkonium. A core-corona approach has recently been employed to study the transverse momentum spectra and the collective flow of quarkonium [13]. This framework provides a comprehensive description of the spectra and collective flow of quarkonium, particularly in the low p_T region [13]. Nevertheless, the tightly bound quarkonium with a high dissociation temperature can survive in the QGP medium. Therefore, the real dynamics of the quarkonium is only partially dissociated and accompanied by the regeneration within the QGP. In this case, the classical transport equation is employed to describe the quarkonium real-time evolution, such as the THU model (Boltzmann equation) [14] and the TAMU model (rate equation) [15], which have successful explained the quarkonium observables from RHIC to LHC energies. Recently, the non-thermal charm distribution and the space-momentum correlation have been considered in the context of regeneration, which appears to significantly enhance the v_2 of J/ψ in the intermediate p_T region. These classical transport models have been extended to study the directed flow v_1 and the triangle flow v_3, the quarkonium polarization, and also the quarkonium-like states such as B_c and $X(3872)$.

It is therefore pertinent to inquire as to the significance of quantum effects in quarkonium transport, and to elucidate the nature of these effects. Two principal quantum effects have been identified. One such effect is that of quantum coherence and decoherence. In the classical transport model, it is typically assumed that the quarkonium is produced before it enters the hot medium. In the quantum view, however, the initially produced state is a compact wave packet that can be treated as a superposition of different quarkonium eigenstates. It takes time for this compact wave packet to decohere into quarkonium eigenstates, which is called "formation time" and is sometimes included in a phenomenological parameter in the classical transport approach. Nevertheless, further investigation into the manner in which the system decoheres within a hot medium would be beneficial. Another quantum effect is the decay process, which results in a significant width compared to that of a classical particle

Assuming that the quarkonium is a quantum wavefunction, its real-time evolution can be described by the time-dependent Schrödinger equation with a temperature-dependent complex HQ potential. This framework has been employed to elucidate the evolution and production of bottomonium in heavy ion collisions [16–18]. Nevertheless, the wavefunction is employed solely for the pure state ψ_i. In order to address the mixed state (color octet and color singlet state), the density matrix is introduced. The total density operator for the system is given by $\rho_{tot} = \sum_i p_i|\psi_i >< \psi_i|$. The evolution of the density operator is governed by the von Neumann equation, $\frac{d\hat{\rho}_{tot}}{dt} = -i[\hat{H}_{tot}, \hat{\rho}_{tot}]$. A heavy quark pair $Q\bar{Q}$ in a hot QCD medium can be treated as an open quantum system (OQS). The total Hamiltonian $\hat{H}_{tot}$ can be separated into the environment part, the $Q\bar{Q}$ subsystem part, and the interaction part. The system exhibits three principal time scales:

© The Authors, published by EDP Sciences. This is an open access article distributed under the terms of the Creative Commons Attribution License 4.0 (https://creativecommons.org/licenses/by/4.0/).

- Environment relaxation time scale: $\tau_e \sim 1/(\pi T)$;
- Intrinsic time scale of subsystem: $\tau_s \sim 1/E_{\rm bind}$;
- Subsystem relaxation time scale: $\tau_r \sim 1/\eta \approx m_Q/T^2$.

We can see that these time scales are related to the temperature of the QGP. The very high temperature regime, where $T \gg m_Q$, is called the non-Markovian regime. In the relativistic heavy ion collisions, the typical temperature of the QGP medium is much smaller than the heavy quark mass, i.e. the relaxation time of the environment is much smaller than the relaxation time of the subsystem, $\tau_e \ll \tau_r$. In this instance, it is possible to trace the degrees of freedom of the environment and simplify the evolution equation of the density operator to the Lindblad form. This approximation is also referred to as the Markovian approximation as illustrated in Fig. 2.

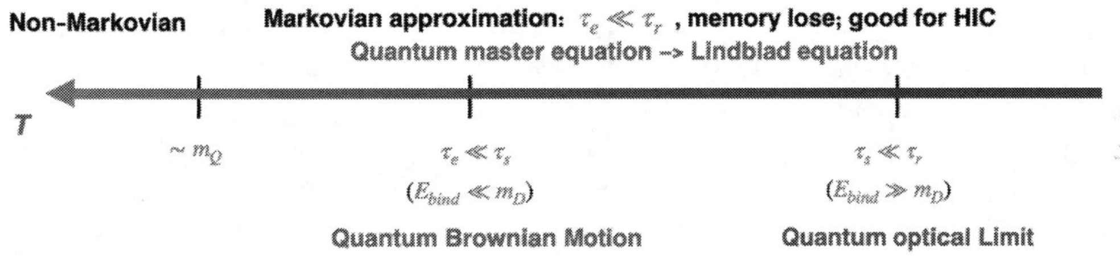

Figure 2. Sketch of different time scales along the temperature line.

Furthermore, if the relaxation time of the environment is significantly shorter than the intrinsic time scale of the subsystem $\tau_e \ll \tau_s$, indicating that the binding energy $E_{\rm bind}$ of the quarkonium is less than the Debye screening mass m_D, the system is in the quantum Brownian motion region. This region is characterized by high temperatures, and the quarkonium exhibits a broad peak and a short quantum decoherence time. In the low-temperature region, the quarkonium exhibits a large binding energy and a narrow width. In this region, the intrinsic time of the subsystem is much smaller than the relaxation time of the subsystem, $\tau_s \ll \tau_r$, which is typically referred to as the Quantum Optical Limit. In recent years, significant theoretical advances have been made in these two regions, based on the first-principle QCD and OQS approachs, see recent review papers [19–21]. I will show them separately below.

- *Duke approach.* It works in the Quantum Optical Limit and based on the pNRQCD which means with the dipole approxiamtion. Following a semi-classical expansion, it was demonstrated that the Lindblad equation is equivalent to the classical Boltzamann equation [22]. This establishes a connection between the OQS and the classical transport equation. This approach has been employed to elucidate the production of bottomonium in heavy ion collisions.
- *Munich-Kent approach.* It is based on the pNRQCD as well but works in the Quantum Brownian Motion regime with the scale hierarchy, $m_Q \geq 1/a_0 \gg \pi T \sim m_D \gg E_{\rm bind}$. The Lindblad operators can be obtained by expanding the self-energy and transition function at the leading order or the next-to-leading order of $E_{\rm bind}/T$ [23, 24]. In this instance, the Lindblad operators are associated with two coefficients: κ, which is the heavy quark momentum diffusion coefficient, and γ, which is the dispersive quantity. κ and γ are respectively related to the thermal width and mass shift of the bottomonium. Recently, the researchers have implemented the quantum jumps and utilized the novel framework to investigate the bottomonium production in heavy ion collisions. The results demonstrated that the bot-

© The Authors, published by EDP Sciences. This is an open access article distributed under the terms of the Creative Commons Attribution License 4.0 (https://creativecommons.org/licenses/by/4.0/).

tomonium R_{AA} and the double ratio can be well described with quantum jumps and without color screening [25]. This result may support the aforementioned "new" potential.

- *Nantes approach.* It works in Quantum Brownian Motion regime but based on NRQCD with the scale, $m_Q \gg T \sim m_D \geq E_{\rm bind}$, that means beyond the dipole approxiamtion. The Lindblad operators $\mathcal{L}_i$ are given by the expansion of the τ_e/τ_s [26, 27]. A numerical solution of the quantum master equations in one dimension has recently been carried out using different initial states and medium configurations. Furthermore, the contributions of the various operators that govern the evolution are also examined [28].
- *Osaka approach.* It works in Quantum Brownian Motion regime as well and based on NRQCD. In the weak coupling, the Lindblad equation is governed by two quantities $V(r)$ and $D(r)$, which are defined in terms of gluon two-point functions. In the HTL approximation, the explicit forms of $V(r)$ and $D(r)$ correspond to the real and imaginary part potentials, respectively [29]. Beyond the weak coupling, the assumption is made that the real and imaginary potentials can be derived from lattice QCD. The related study and the rationale behind the dipole approximation were recently investigated [30].

In the heavy ion collisions, the created QGP expands rapidly and the temperature drops gradually. The system transitions from the quantum Brownian motion regime at high temperatures to the quantum optical limit region at low temperatures. It is therefore necessary to construct a framework that connects these two regimes. Moreover, these studies concentrate on a single pair of heavy quarks, which is likely a reasonable approximation for the production of bottomonium. For charmonium, regeneration from many uncorrelated charm quarks plays an important role at top RHIC and LHC energies. Consider these two requirements, A microscopic model based on the density matrix formalism has been developed to describe both charmonium [31, 32] and bottomonium [33] production in heavy ion collisions. In this framework, a quarkonium is represented as a Wigner density, which is a fully quantum object. The only approximation is to simulate the N-body Wigner density as a classical phase-space distribution. A unified framework for both open heavy and quarkonium production, based on EPOS4, will be available in the near future.

4 Summary

In conclusion, the recent advancements in the theoretical understanding of the thermal properties of quarkonium and its real-time evolution in the hot QCD medium are reviewed. The recent lattice QCD along with many other research efforts, indicate that the HQ in-medium potential exhibits either no color screening effect or a diminished one, accompanied by a pronounced imaginary component. This may challenge our current understanding of quarkonium dissociation in the QGP. The dynamical scattering plays a dominant role instead of the static color screening. The classical transport models can describe the experimental data very well and have been extended to many interesting studies in recent years. The creation of a comprehensive, first-principles-based framework for the real-time evolution of quarkonium is a crucial step in advancing our understanding of QCD. Furthermore, it can more effectively address the limitations of classical transport models. Additionally, there have been significant advancements in the field of quarkonium real-time evolution, particularly in the context of the NRQCD or pNRQCD formalism and based on the the open quantum system framework.

Acknowledgement: The work is supported by the European Union's Horizon 2020 research and innovation program under grant agreement No 824093 (STRONG-2020), and the Deutsche Forschungsgemeinschaft (DFG, German Research Foundation) through the grant CRC-TR 211 'Strong-interaction matter under extreme conditions' - Project number 315477589 - TRR 211.

© The Authors, published by EDP Sciences. This is an open access article distributed under the terms of the Creative Commons Attribution License 4.0 (https://creativecommons.org/licenses/by/4.0/).

References

[1] T. Matsui and H. Satz, Phys. Lett. B **178**, 416-422 (1986).
[2] N. Brambilla, A. Pineda, J. Soto and A. Vairo, Nucl. Phys. B **566**, 275 (2000).
[3] J. Zhao, K. Zhou, S. Chen and P. Zhuang, Prog. Part. Nucl. Phys. **114**, 103801 (2020).
[4] O. Kaczmarek, F. Karsch, F. Zantow and P. Petreczky, Phys. Rev. D **70**, 074505 (2004) [erratum: Phys. Rev. D **72**, 059903 (2005)].
[5] M. Laine, O. Philipsen, P. Romatschke and M. Tassler, JHEP **03**, 054 (2007).
[6] D. Bala *et al.* [HotQCD], Phys. Rev. D **105**, no.5, 054513 (2022).
[7] A. Bazavov *et al.* [HotQCD], Phys. Rev. D **109**, no.7, 074504 (2024).
[8] D. Lafferty and A. Rothkopf, Phys. Rev. D **101**, no.5, 056010 (2020).
[9] S. Shi, K. Zhou, J. Zhao, S. Mukherjee and P. Zhuang, Phys. Rev. D **105**, no.1, 014017 (2022).
[10] X. Du, S. Y. F. Liu and R. Rapp, Phys. Lett. B **796**, 20-25 (2019).
[11] Z. Tang, S. Mukherjee, P. Petreczky and R. Rapp, Eur. Phys. J. A **60**, no.4, 92 (2024).
[12] W. Wu, G. Huang, J. Zhao and P. Zhuang, Phys. Rev. D **107**, no.11, 114033 (2023).
[13] A. Andronic, P. Braun-Munzinger, H. Brunßen, J. Crkovská, J. Stachel, V. Vislavicius and M. Völkl, [arXiv:2308.14821 [hep-ph]].
[14] L. Yan, P. Zhuang and N. Xu, Phys. Rev. Lett. **97**, 232301 (2006).
[15] X. Zhao and R. Rapp, Nucl. Phys. A **859**, 114-125 (2011).
[16] A. Islam and M. Strickland, Phys. Lett. B **811**, 135949 (2020).
[17] G. Chen, B. Chen and J. Zhao, Eur. Phys. J. C **84**, no.8, 869 (2024).
[18] V. Kumar, P. Shukla and A. Bhattacharyya, Prog. Part. Nucl. Phys. **131**, 104044 (2023).
[19] X. Yao, Int. J. Mod. Phys. A **36**, no.20, 2130010 (2021).
[20] Y. Akamatsu, Prog. Part. Nucl. Phys. **123**, 103932 (2022).
[21] A. Andronic, P. B. Gossiaux, P. Petreczky, R. Rapp, M. Strickland, J. P. Blaizot, N. Brambilla, P. Braun-Munzinger, B. Chen and S. Delorme, *et al.* Eur. Phys. J. A **60**, no.4, 88 (2024).
[22] X. Yao and T. Mehen, Phys. Rev. D **99**, no.9, 096028 (2019).
[23] N. Brambilla, M. Á. Escobedo, M. Strickland, A. Vairo, P. Vander Griend and J. H. Weber, Phys. Rev. D **104**, no.9, 094049 (2021).
[24] N. Brambilla, M. Á. Escobedo, A. Islam, M. Strickland, A. Tiwari, A. Vairo and P. Vander Griend, JHEP **08**, 303 (2022).
[25] N. Brambilla, M. Á. Escobedo, A. Islam, M. Strickland, A. Tiwari, A. Vairo and P. Vander Griend, Phys. Rev. D **108**, no.1, L011502 (2023).
[26] J. P. Blaizot and M. A. Escobedo, JHEP **06**, 034 (2018).
[27] S. Delorme, T. Gousset, R. Katz and P. B. Gossiaux, EPJ Web Conf. **258**, 05009 (2022).
[28] S. Delorme, R. Katz, T. Gousset, P. B. Gossiaux and J. P. Blaizot, JHEP **06**, 060 (2024).
[29] Y. Akamatsu, Phys. Rev. D **87**, no.4, 045016 (2013).
[30] T. Miura, Y. Akamatsu, M. Asakawa and Y. Kaida, Phys. Rev. D **106**, no.7, 074001 (2022).
[31] D. Y. A. Villar, J. Zhao, J. Aichelin and P. B. Gossiaux, Phys. Rev. C **107**, no.5, 054913 (2023).
[32] T. Song, J. Aichelin and E. Bratkovskaya, Phys. Rev. C **107**, no.5, 054906 (2023).
[33] T. Song, J. Aichelin, J. Zhao, P. B. Gossiaux and E. Bratkovskaya, Phys. Rev. C **108**, no.5, 054908 (2023).

© The Authors, published by EDP Sciences. This is an open access article distributed under the terms of the Creative Commons Attribution License 4.0 (https://creativecommons.org/licenses/by/4.0/).

Measurements of quarkonia production

Laure Massacrier[1,*]

[1]Université Paris-Saclay, CNRS/IN2P3, IJCLab, Orsay, France

Abstract. Quarkonium production in high-energy hadronic collisions is a useful tool to investigate fundamental aspects of Quantum Chromodynamics, from the proton and nucleus structure to deconfinement and the properties of the Quark Gluon Plasma (QGP). In these proceedings, emphasis is made on few recent quarkonium results from the RHIC and LHC colliders in proton-proton (pp), proton-nucleus (p-A) and nucleus-nucleus (A-A) collisions. In addition, results for some key observables are compiled to discuss the state-of-the-art in quarkonium production, with a focus on quarkonium hadroproduction.

1 Introduction

Quarkonium, a bound state made of a heavy quark and its antiquark, is an important tool to study Quantum Chromodynamics in high energy hadronic collisions [1]. Elementary pp (or e^+e^-) collisions are used to investigate quarkonium properties (production mechanisms, spectroscopy, rare decays..) and provide reference for quarkonium studies in heavy-ion collisions. Hadronic quarkonium production involves hard scales (the creation of the heavy quark-antiquark pair via hard scattering) and the subsequent soft-scale process (the binding of the pair into a colorless final state). Therefore, the quarkonium production mechanism is sensitive to the gluon content of the colliding proton and thus provides information on parton distribution functions (PDFs) [2]. In p-A collisions, the influence of a nucleus in the initial state on quarkonium production can be investigated. These effects, known as Cold Nuclear Matter effects (CNM), include among others: the modification of parton densities in nuclei (nPDFs) [3], coherent energy-loss effects involving the initial- and final-state partons [4], nuclear absorption [5] and the interaction of the quarkonium state with other particles produced in the collision (the comovers) [6]. In A-A collisions, quarkonium is a sensitive probe of the formed hot strongly interacting medium, as hard-scattering processes occur at the early stage of the collision, on a timescale that is in general shorter than the QGP lifetime. Quarkonium production is expected to be suppressed in the QGP due to static color screening [7] or dynamical dissociation [8], the suppression being sequential, according to the binding of each quarkonium state. In this simple picture, the in-medium dissociation of the quarkonium state should provide an estimate of the initial temperature reached in the collision. However, at high energy, if the number of heavy quark pairs produced is large enough, quarkonium recombination [9, 10] can take place. Evidence for such phenomenon exist at LHC energies for charmonium [11], and it directly points to the existence of a deconfined QGP. It is worth mentioning that low transverse momentum (p_{T}) heavy quarks could participate, through their interactions with the medium, in the collective expansion of the system and possibly reach

*e-mail: laure.marie.massacrier@cern.ch

© The Authors, published by EDP Sciences. This is an open access article distributed under the terms of the Creative Commons Attribution License 4.0 (https://creativecommons.org/licenses/by/4.0/).

thermal equilibrium with its constituent. Recently, several measurements in high multiplicity pp and p-Pb collisions at LHC, have revealed the presence of phenomena typically attributed to the creation of a QGP (eg. first observation of ridge in two-particle azimuthal correlations [12]). Multiplicity-dependent quarkonium studies offer a valuable testing ground for examining the onset of QGP-like effects in small systems, while also probing the interplay between soft and hard particle production and the role of multiparton interactions (MPI).

2 Quarkonium production in proton-proton collisions

Prompt quarkonium production is in general rather well described over a wide range of rapidity, p_{T} and center-of-mass (c.o.m) energy, by Non Relativistic QCD (NRQCD) [13] and Improved Color Evaporation (ICEM) [14] models. In recent years, LHC quarkonium data are becoming more precise than theoretical model uncertainties. In order to further constrain models and therefore quarkonium production mechanisms, the ALICE Collaboration has measured the ratio of J/ψ and ψ(2S) cross sections at various c.o.m energies to the ones obtained at $\sqrt{s}$ = 13 TeV [15]. These ratios benefit from the partial cancellation of several theoretical scale uncertainties and it was shown that it becomes challenging for models to reproduce quarkonium cross section ratios. The simultaneous description of quarkonium production cross section and polarization has also been a long standing puzzle for models. Indeed, S-wave quarkonia are produced almost unpolarized at LHC [1], challenging NRQCD predictions favoring sizable polarization, especially at high p_{T}. However, recent results from CMS on χ_{c1} and χ_{c2} P-wave quarkonia [16] favor a scenario where at least one of the two states is strongly polarized, as expected from NRQCD. Recent developments in NRQCD using global fit analyses were able to improve the description of quarkonium S-wave polarization results [17]. Studies at both RHIC and LHC have revealed that quarkonium production shows a noticeable dependence on the event multiplicity [18, 19], with increasing yields as multiplicity rises. This suggests that quarkonium production is influenced by the overall activity in the event, hinting at complex interactions between hard and soft processes. The enhanced quarkonium production rate with multiplicity is usually interpreted as a result of MPI. Excited to ground state quarkonium yield ratio as a function of the charged particle multiplicity can permit to further pin down final state effects on quarkonia production. In the comover scenario [6], quarkonia can be dissociated by interaction with the comoving particles in the final state, with the dissociation probability depending on the binding energy of each quarkonium state and on the density of comoving particles. Figure 1 shows recent measurements from LHCb on the self-normalized ψ(2S)-to-J/ψ cross section ratio as a function of the number of tracks considered for primary vertex reconstruction in the backward (left) and forward (right) pseudorapidity regions [20]. Results are shown separately for prompt and non-prompt charmonia. While the ratio of charmonia from b-hadrons shows no evident multiplicity dependence irrespective of the region in which the multiplicity is measured, the prompt ψ(2S)-to-J/ψ ratio exhibits a decreasing trend with multiplicity for both multiplicity estimators. The prompt ψ(2S) are more suppressed than the prompt J/ψ for high forward multiplicities[1] (i.e when the multiplicity is measured in a similar acceptance as the charmonia), in line with expectations from a comover scenario. This suggests a correlation between the observed suppression and the local particle multiplicity. On the other hand, the fact that the ψ(2S)-to-J/ψ ratio is not flat as a function of the backward multiplicity, as expected in a naive comover picture, could be attributed to the existence of strong correlations between the forward and backward multiplicities (see Fig. 4 of [20]). These studies demonstrate the

[1] Note that the muon tracks from the charmonium decays are included in the forward multiplicity estimator

© The Authors, published by EDP Sciences. This is an open access article distributed under the terms of the Creative Commons Attribution License 4.0 (https://creativecommons.org/licenses/by/4.0/).

importance of performing a systematic set of studies varying the quarkonium and multiplicity rapidity (y) windows, for proper interpretation of quarkonium versus multiplicity results.

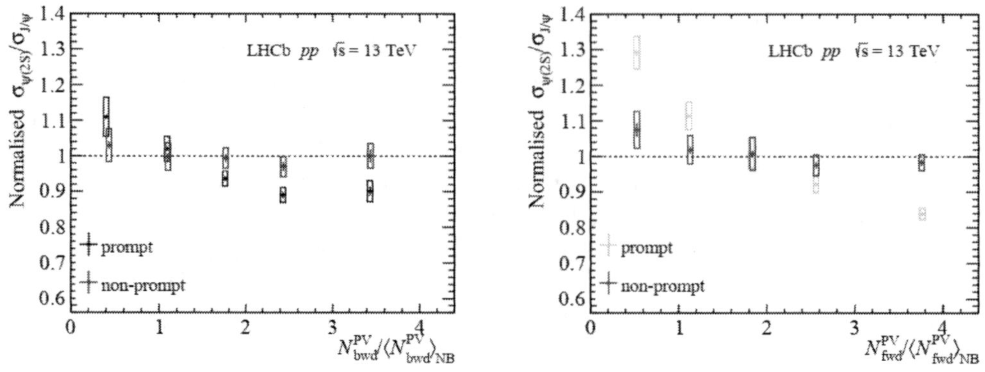

Figure 1. Self-normalized ψ(2S)-to-J/ψ cross section ratio as a function of the number of tracks (considered for primary vertex reconstruction) measured by the VELO of LHCb, in the backward (left) and forward (right) pseudorapidity regions, in pp collisions at $\sqrt{s}$ = 13 TeV. Results for (non-prompt) prompt charmonium are reported in (blue) black (left) and (blue) green (right) [20].

3 Quarkonium production in proton-nucleus collisions

The production of the J/ψ charmonium in p–A collisions has been extensively studied from RHIC to the LHC energies [1]. The J/ψ nuclear modification factor (R_{pPb}) at LHC energies exhibits almost no suppression at backward rapidity [21, 22], moderate suppression at midrapidity mainly occurring in the low-p_T region [23] and stronger suppression at forward rapidity [21, 22]. This behavior is in general well reproduced by several nPDFs sets (see reference therein), and coherent energy loss model [4]. These calculations predict similar R_{pPb} for J/ψ and ψ(2S) as they assume that the nuclear suppression is prior to the hadronisation timescale, and that the nuclear effects apply similarly to both resonances given their proximity in mass. Figure 2 left shows recent LHCb measurement of the prompt ψ(2S) R_{pPb} to J/ψ R_{pPb} ratio as a function of the c.o.m rapidity, in p–Pb collisions at $\sqrt{s_{NN}}$ = 8.16 TeV [24]. The results indicate a stronger suppression for the prompt ψ(2S) with respect to the prompt J/ψ in all y intervals, and slightly more pronounced at backward rapidity. The observed behaviour can be described by models incorporating final state interactions breaking preferentially the ψ(2S) meson compared to the J/ψ meson [6]. However, recent measurements of the excited χ_c charmonium states in p–Pb collisions could challenge these interpretations. Figure 2 right illustrates the χ_c to prompt J/ψ cross section ratio between p–Pb and pp collisions, measured by LHCb [26, 27]. The ratio is close to unity at both backward and forward rapidities, indicating no significant cold nuclear matter effects on χ_c, despite the similar binding energies of ψ(2S) and χ_c states. In the bottomonia sector, recent results from CMS at midrapidity on the Υ(nS) R_{pPb} as a function of p_T are shown in Fig. 3 (left) [28]. The Υ(nS) states follow an ordered suppression pattern, similar to the one observed in the charmonium sector for ψ(2S) with respect to J/ψ. Indeed, the measured R_{pPb}(Υ(1S)) is systematically larger than that of Υ(2S), which in turn is systematically larger than the R_{pPb}(Υ(3S)), also suggesting different level of modifications for the three states by final state effects. Measurement of quarkonium elliptic flow (v_2) in high-multiplicity p–Pb collisions are complementary to the ones for light flavor particles, in order to understand the collective behavior of particles in small collision

© The Authors, published by EDP Sciences. This is an open access article distributed under the terms of the Creative Commons Attribution License 4.0 (https://creativecommons.org/licenses/by/4.0/).

systems. Models based on the hydrodynamic expansion of a tiny QGP droplet, or alternative scenarios based on gluon saturation in the initial state have been developed in order to interpret the long-range collective azimuthal correlations observed in high multiplicity pp and p–Pb collisions [29]. Figure 3 right shows the transverse-momentum dependent elliptic flow of J/ψ in high multiplicity p–Pb collisions measured by ALICE [30] and CMS [31] in different rapidity windows. Positive J/ψ v_2 values are measured in both experiments for about $p_T > 3$ GeV/c, indicating charm quark collectivity in high multiplicity p–Pb events. The collective effects are found to be weaker for charm quarks than for light quarks [32]. Data at midrapidity are compared with a transport model calculation including path-length effects for primordial charmonium traversing the elliptic fireball, and a non-zero v_2 for regenerated charmonia [33]. The model strongly underestimates the J/ψ v_2, suggesting that the observed v_2 in data cannot come from final-state interactions alone. The fact that ALICE measures similar J/ψ v_2 at backward and forward rapidities, while the reached multiplicity is rather different in both rapidity regions, is also in line with this conclusion.

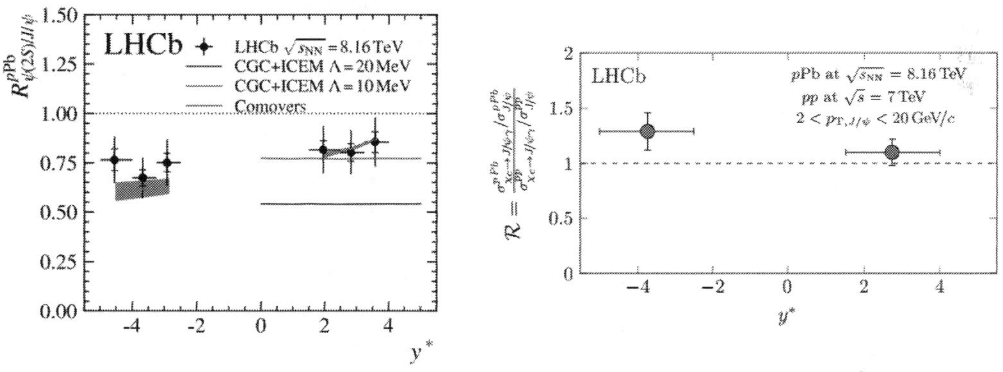

Figure 2. Left: $R^{pPb}_{\psi(2S)/J/\psi}$ for prompt charmonium, as a function of the c.o.m rapidity $y^\star$ measured by LHCb in p–Pb collisions at $\sqrt{s_{NN}} = 8.16$ TeV [24]. Data are compared to a CGC soft gluon interaction model [34] and a comover model [6]. Right: Double ratio $\sigma_{\chi_c \to J/\psi\gamma}/\sigma_{J/\psi}$ between p–Pb collisions at $\sqrt{s_{NN}} = 8.16$ TeV and pp collisions at $\sqrt{s} = 7$ TeV, as a function of $y^\star$, measured by LHCb [26, 27]. The measured χ_c cross section is the sum of χ_{c1} and χ_{c2} cross sections.

4 Quarkonium production in nucleus-nucleus collisions

Quarkonia are long standing probes of the QGP. In a hot and deconfined medium, they are suppressed by static color screening [7] or dynamical dissociation [8]. This suppression should occur sequentially, according to the binding energy of each mesons: strongly bound states such as Υ(1S) and J/ψ are expected to melt at higher temperatures with respect to the loosely bound states (e.g. χ_b, Υ(2S), Υ(3S) for the bottomonia family, ψ(2S) and χ_c for the charmonia family). However, the prediction of a sequential suppression pattern is complicated by several factors, such as feed-down (from higher mass resonances, or b-hadrons in the case of charmonium), as well as other hot nuclear matter effects (such as recombination), or cold nuclear matter effects. A comprehensive set of measurements of several quarkonium states is therefore crucial to interpret QGP effects on quarkonium production. Figure 4 left shows the J/ψ R_{AA} measured at various c.o.m energies, from SPS [35] and RHIC [43] to the LHC [37], in most central A–A collisions (Pb–Pb for SPS/LHC and Au–Au for RHIC). No significant energy dependence of the J/ψ R_{AA} is observed for $\sqrt{s_{NN}} < 200$ GeV, while

© The Authors, published by EDP Sciences. This is an open access article distributed under the terms of the Creative Commons Attribution License 4.0 (https://creativecommons.org/licenses/by/4.0/).

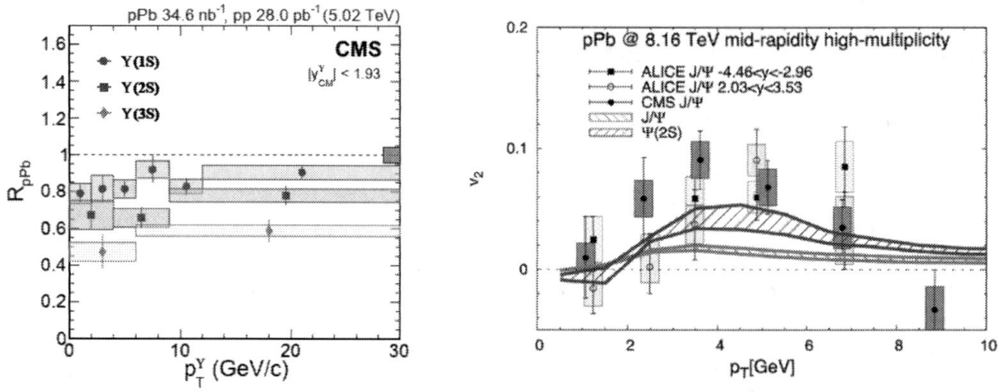

Figure 3. Left: R_{pPb} of Υ(1S) (red), Υ(2S) (blue), Υ(3S) (green) as a function of p_T measured at midrapidity by CMS in p–Pb collisions as $\sqrt{s_{NN}}$ = 5.02 TeV [28]. Right: Transverse-momentum dependent elliptic flow of J/ψ in high multiplicity p–Pb collisions at $\sqrt{s_{NN}}$ = 8.16 TeV, measured by ALICE at forward (open circle) and backward (full square) rapidities [30], and by CMS at midrapidity (full circle) [31]. Data are compared to a transport model from [33].

an important increase is seen at LHC energies. Data are compared to the transport model from Ref. [38] which includes two main components. The primordial component represents the anomalous suppression of primordial J/ψ due to CNM and color screening effects. If only CNM effects are considered, the R_{AA} is expected to be around 0.6. The regeneration component arises from the combination of uncorrelated $c\bar{c}$ pairs in medium. In this model, the feed-down from ψ(2S) and χ_c to the J/ψ are accounted for. As the QGP temperature increases, the J/ψ suppression by color screening becomes more important. At the same time, the regeneration component increases with collision energy due to the increased production of charm quark pairs, leading to a compensation of the stronger suppression. The higher J/ψ R_{AA} at LHC tends to indicate that the surviving J/ψ are mainly originating from a recombination contribution. In order to better constrain regeneration models, the comparison of the ψ(2S) and J/ψ yield modifications in A–A collisions is a golden probe. Figure 4 right shows the J/ψ and ψ(2S) R_{AA} measured by ALICE at forward rapidity [39, 40] and CMS at midrapidity [41] in Pb–Pb collisions. At high p_T, both ψ(2S) and J/ψ are strongly suppressed, the suppression being the strongest for the ψ(2S) state. The ψ(2S) R_{AA} hints at an increase towards low p_T, following a similar pattern as for the J/ψ, which is understood as a direct consequence of the regeneration process for charm and anticharm quarks. The J/ψ and ψ(2S) R_{AA} are well described as a function of p_T by the transport model from Ref. [42] which includes charmonium regeneration in the QGP phase. The bottomonium family is expected to provide additional information on the thermal properties of the deconfined medium. The Υ(1S) resonance is the most tightly bound quarkonium state and therefore has one of the highest dissociation temperature (~ $2T_c$, with T_c the critical temperature for deconfinement). In addition, recombination-like processes for bottomonia are expected to be negligible compared to charmonia at LHC energies, as the number of beauty quark pairs per event is much smaller compared to the number of charm quark pairs. Figure 5 left shows the Υ(nS) R_{AA} as a function of $\langle N_{part} \rangle$, measured at RHIC [43] and the LHC [44–47]. A gradual decrease of the Υ(nS) R_{AA} is observed with increasing $\langle N_{part} \rangle$ at LHC energies. The Υ(3S) resonance is measured for the first time in Pb–Pb collisions by CMS [44] and is found to be strongly suppressed in central Pb–Pb collisions. A sequential suppression pattern of the Υ(nS) states is observed at LHC energy, and indication for such pattern is also visible at RHIC in central

© The Authors, published by EDP Sciences. This is an open access article distributed under the terms of the Creative Commons Attribution License 4.0 (https://creativecommons.org/licenses/by/4.0/).

Au–Au collisions for the Υ(1S) and Υ(2S) states, although with a smaller significance. Remarkably, the Υ(1S) R_{AA} gets similar values at LHC and RHIC energies, despite a factor ~ 25 difference in $\sqrt{s_{NN}}$. This observation is in favor of a negligible melting of the direct Υ(1S) state at the probed energies, but might be explained by the suppression of the excited states in combination with CNM effects. Being sensitive to the dynamics of the partonic stages of heavy-ion collisions, the v_2 observable can provide complementary information to the R_{AA}, in particular on the quarkonium production mechanisms and the degree of thermalization of heavy quarks in medium. Indeed, charmonia produced via regeneration should inherit the elliptic flow of charm quarks in the QGP. On the other hand, primordial charmonia can acquire v_2 because of path-length dependent suppression through the medium. This latter contribution becomes dominant at high p_T, while the influence of a regeneration component should dominate at low p_T. Figure 5 right shows a compilation of inclusive and prompt J/ψ v_2 results at the LHC in semi-central Pb–Pb collisions [48–50], as well as the first prompt ψ(2S) v_2 measured by CMS [48]. A significant J/ψ v_2 is observed at low-p_T, stronger than in p–Pb collisions, favoring a scenario with significant regeneration of J/ψ in Pb–Pb collisions at the LHC. Given the lower binding energy of the ψ(2S) state, the recombination of ψ(2S) can occur at a later stage of the QGP evolution, leading to a larger v_2 for the regenerated ψ(2S) with respect to the regenerated J/ψ [51]. The prompt ψ(2S) v_2 measured by CMS tends to exhibit similar, and possibly larger v_2 than the J/ψ, albeit large uncertainties. More precise ψ(2S) measurements extended towards lower p_T would help probing the ψ(2S) production mechanisms in A–A collisions. It is worth mentioning that bottomonia (charmonia) measurements at LHC[52, 53] (RHIC[54]) in A–A collisions do not exhibit significant v_2 within the large experimental uncertainties.

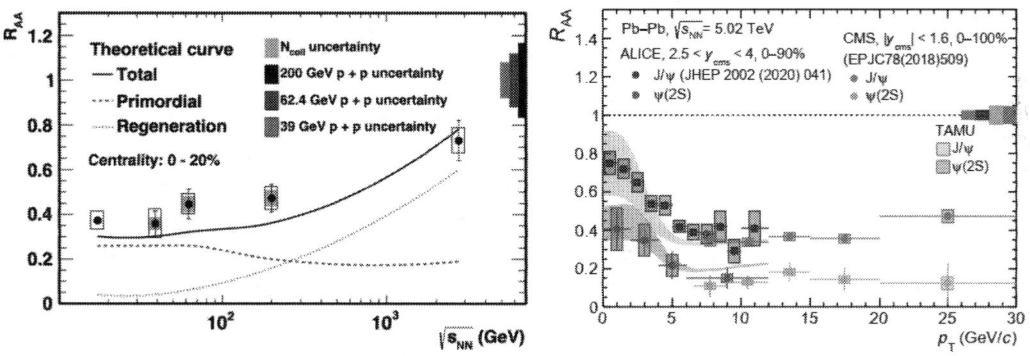

Figure 4. Left: J/ψ R_{AA} as a function of $\sqrt{s_{NN}}$ measured at SPS [35], RHIC [43] and the LHC [37] in most central A–A collisions. Data are compared to the transport model calculations from [38]. Right: J/ψ and ψ(2S) R_{AA} as a function of p_T measured by ALICE at forward rapidity [39, 40] and CMS at midrapidity [41], in Pb–Pb collisions at $\sqrt{s_{NN}}$ = 5.02 TeV. Results are compared to the transport model from [42].

5 Conclusion

A large variety of precise quarkonium cross section measurements have been performed at both RHIC and the LHC, in pp collisions, and are overall rather well described by NRQCD and ICEM calculations. Recent improvements in the simultaneous description of quarkonium production cross section and polarization have been made by the same models. Quarkonium

© The Authors, published by EDP Sciences. This is an open access article distributed under the terms of the Creative Commons Attribution License 4.0 (https://creativecommons.org/licenses/by/4.0/).

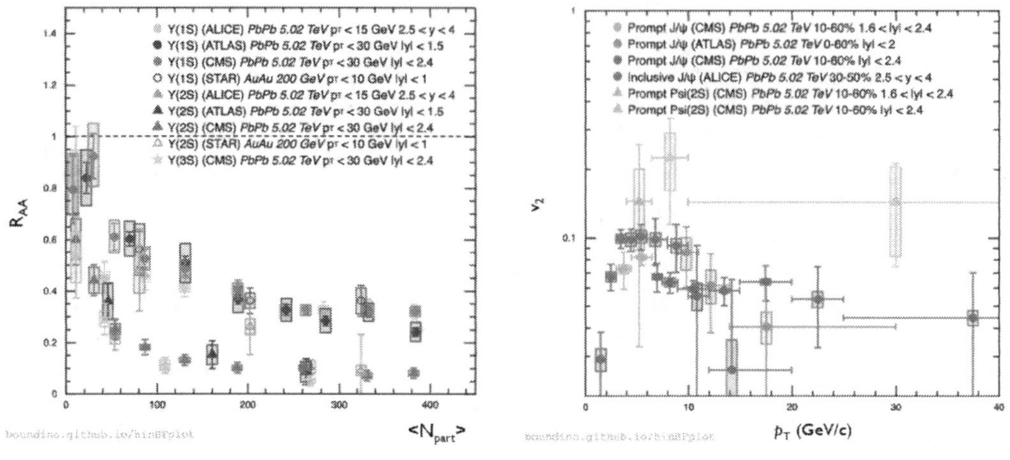

Figure 5. Left: Υ(nS) R_{AA} as a function of the mean number of participant nucleons $<N_{part}>$, measured at RHIC [43] in Au–Au collisions and the LHC [44–47] in Pb–Pb collisions. Right: Charmonium elliptic flow as a function of p_T measured by ALICE [49], ATLAS [50] and CMS [48] in semi-central Pb–Pb collisions at $\sqrt{s_{NN}} = 5.02$ TeV. These compilation plots are generated from [55].

production as a function of the event multiplicity has demonstrated to be a powerful tool to study MPI and final state interaction (including interactions with comoving particles) on quarkonium production. In pA collisions at the LHC, the systematic studies of several charmonia and bottomonia states allow for a better understanding of CNM effects (nPDFs, coherent energy loss) on quarkonium production, including the role of feed-downs. Weakly bound resonances (ψ(2S),Υ(3S)) seem to be suppressed with respect to their vector meson ground states (except for χ_c), indicating the need for final state effects to describe quarkonium production in pA. The non-zero J/ψ v_2 measured in high multiplicity pA collisions enlightens heavy quark collectivity however cannot be explained by final state interactions only. In AA collisions, charmonia follow a sequential suppression pattern at RHIC, while at LHC energies there is a clear evidence for charmonium regeneration competing with the sequential dissociation. Whether charmonium regeneration takes place throughout the QGP lifetime or at the phase transition cannot be disentangled with current data, and would require for instance more precise measurement of charmonium excited states. The elliptic flow measurement of J/ψ in AA collisions at LHC exhibits sign of charm quark thermalization in medium. In the bottomonia sector, Υ(nS) states tend to exhibit a sequential suppression pattern at both RHIC and the LHC. The combination of both measurements seems to indicate that the temperature reached in the QGP is not large enough to melt the direct Υ(1S) state, even at LHC energies.

Acknowledgments

The author acknowledges the support of the French Agence Nationale de la Recherche (ANR) under reference ANR-22-CE31-0005 MALICE.

References

[1] A. Andronic et al., Eur. Phys. J. C **76**, 107 (2016)
[2] M. Ozcelik, PoS DIS2019, 159 (2019), arXiv:1907.01400

© The Authors, published by EDP Sciences. This is an open access article distributed under the terms of the Creative Commons Attribution License 4.0 (https://creativecommons.org/licenses/by/4.0/).

[3] K. Eskola et al., JHEP **04**, 065 (2009)
[4] F. Arleo et al., JHEP **10**, 073 (2014)
[5] E. G. Ferreiro et al., Phys. Lett. Eur. Phys. J. C **73**, 2427 (2013)
[6] E. G. Ferreiro et al., Phys. Lett. **B749**, 98-103 (2015)
[7] T. Matsui et al., Phys. Lett. B. **178**, 416-422 (1986)
[8] A. Rothkopf et al., Phys. Rept. **858**, 1-117 (2020)
[9] P. Braun-Munzinger et al., Phys. Lett. B. **490**, 196-202 (2000)
[10] R. L. Thews et al., Phys. Rev. C **63**, 054905 (2001)
[11] B. Abelev et al., ALICE Collaboration, Phys. Rev. Lett. **109**, 072301 (2012)
[12] V. Khachatryan et al., CMS Collaboration, JHEP **09**, 91 (2010)
[13] G. T. Bodwin et al., Phys. Rev. D **51**, 1125-1171 (1195)
[14] V. Cheung et al., Phys. Rev. D **98**, 114029 (2018)
[15] S. Acharya et al., ALICE Collaboration, Eur. Phys. J. C **83**, 61 (2023)
[16] A. M. Syrunian et al., CMS Collaboration, Phys. Rev. Lett. **124**, 162002 (2020)
[17] P. Faccioli et al., Phys. Lett. B. **736**, 98-109 (2014)
[18] S. Acharya et al., ALICE Collaboration, arXiv:2209.04241
[19] N. J. Abdulameer et al., PHENIX Collaboration, arXiv:2409.03728
[20] R. Aaij et al., LHCb Collaboration, JHEP **05**, 243 (2024)
[21] S. Acharya et al., ALICE Collaboration, JHEP **2018**, 160 (2018)
[22] R. Aaij et al., LHCb Collaboration, Phys. Lett. B **774**, 159-178 (2017)
[23] S. Acharya et al., ALICE Collaboration, JHEP **2022**, 11 (2022)
[24] R. Aaij et al., LHCb Collaboration, JHEP **04**, 111 (2024)
[25] R. Aaij et al., LHCb Collaboration, Phys. Rev. Lett. **132**, 102302 (2024)
[26] R. Aaij et al., LHCb Collaboration, Phys. Rev. Lett. **132**, 102302 (2024)
[27] R. Aaij et al., LHCb Collaboration, LHCb-PAPER-2023-028, CERN-EP-2023-224
[28] A. Tomasyan et al., CMS Collaboration, Phys. Lett. B. **835**, 137397 (2022)
[29] K. Dusling et al., Int. J. Mod. Phys. E **25** 01, 1630004 (2016)
[30] S. Acharya et al., ALICE Collaboration, Phys. Lett. B **780**, 7 (2018)
[31] A. M. Syrunyan et al., CMS Collaboration, Phys. Lett. B. **791**, 172 (2019)
[32] A. M. Syrunyan et al., CMS Collaboration, Phys. Rev. Lett. **121**, 082301 (2018)
[33] X. Du et al., HEP **2019**, 15 (2019)
[34] Y.-Q. Ma et al., Phys. Rev. C **97**, 014909 (2018)
[35] M.C. Abreu et al., NA50 Collaboration, Phys. Lett. B **477**, 28 (2000)
[36] L. Adamczyk et al., STAR Collaboration, Phys. Lett. B **771**, 13-20 (2017)
[37] B. Abelev et al., ALICE Collaboration, Phys. Lett. B **734**, 314 (2014)
[38] X. Zhao et al., STAR Collaboration, Phys. Rev. C **82**, 064905 (2010)
[39] S. Acharya et al., ALICE Collaboration, Phys. Rev. Lett. **132**, 042301 (2024)
[40] J. Adam et al., ALICE Collaboration, Phys. Lett. B **766**, 212-224 (2017)
[41] A. M. Sirunyan et al., CMS Collaboration, Eur. Phys. J. C **78**, 509 (2018)
[42] X. Du et al., Nucl. Phys. A **943**, 147-158 (2015)
[43] B. E. Aboona et al., Phys. Rev. Lett. **130**, 112301 (2023)
[44] A. Tumasyan et al., CMS Collaboration, Phys. Rev. Lett. **133**, 022302 (2024)
[45] A. M. Sirunyan et al., CMS Collaboration, Phys. Lett. B **790**, 270 (2019)
[46] A. Acharya et al., ALICE Collaboration, Phys. Lett. B **822**, 136579 (2021)
[47] G. Aad et al., ATLAS Collaboration, Phys. Rev. C **107**, 054912 (2023)
[48] A. Tumasyan et al., CMS Collaboration, JHEP **10**, 115 (2023)

© The Authors, published by EDP Sciences. This is an open access article distributed under the terms of the Creative Commons Attribution License 4.0 (https://creativecommons.org/licenses/by/4.0/).

[49] A. Acharya et al., ALICE Collaboration, JHEP **10**, 141 (2020)
[50] M. Aaboud et al., ATLAS Collaboration, Eur. Phys. J. C **78**, 784 (2018)
[51] X. Du et al., Nucl. Phys. A **943** 147-158 (2015)
[52] A. Acharya et al., ALICE Collaboration, Phys. Rev. Lett. **123**, 192301 (2019)
[53] A. M. Sirunyan et al., CMS Collaboration, Phys. Lett. B **819**, 136385 (2021)
[54] L. Adamczyk et al., Phys. Rev. Lett. **111**, 052301 (2013)
[55] https://boundino.github.io/hinHFplot/ by J. Wang

© The Authors, published by EDP Sciences. This is an open access article distributed under the terms of the Creative Commons Attribution License 4.0 (https://creativecommons.org/licenses/by/4.0/).

EPJ Web of Conferences **316**, 01015 (2025) https://doi.org/10.1051/epjconf/202531601015
SQM 2024

Molecular states with charm: insights from vacuum and finite-temperature analyses

Juan M. Torres-Rincon[1,*]

[1]Departament de Física Quàntica i Astrofísica and Institut de Ciències del Cosmos (ICCUB), Facultat de Física, Universitat de Barcelona, Martí i Franquès 1, 08028 Barcelona, Spain

Abstract. This contribution to the SQM2024 conference covers the molecular hypothesis for the internal structure of some open and hidden charm states. The use of effective theories that incorporate heavy-quark spin symmetry, combined with unitarization techniques, has provided strong evidences supporting this interpretation. In the heavy-light sector, we discuss the double pole structure of the $D_0^*(2300)$ and the generation of the $D_{s0}^*(2317)$. In the hidden charm sector, we focus on the exotic $X(3872)$ and its heavy-quark partner, the $X(4014)$. Furthermore, we emphasize the benefits of femtoscopic measurements in $p + p$ collisions to establish the nature of these states, as well as the potential role of temperature to discern their internal structure.

1 Introduction

Since the advent of the quark model, observed hadrons were usually classified as composite objects consisting of valence quarks and antiquarks. Following the developments of QCD, the inclusion of the gluon as a possible valence component expanded the possibilities for combining these pieces to form color neutral particles. In the literature various composite structures can be classified as (for recent reviews, see [1–5]),

- Conventional hadrons: mesons ($\bar{q}q$) and baryons (qqq).
- Glueballs: gg, ggg...
- Multiquark states: tetraquarks, pentaquarks...
- Hybrids: $qqqg$, $q\bar{q}g$...
- ...

In this contribution, we focus on another interpretation known as *hadronic molecules*, in which colorless states—such as standard mesons and baryons—are bound by the residual strong force. Their interactions are typically modeled using an appropriate effective field theory (EFT), and a unitarization (or a two-body) technique.

In addition, we consider states that contain at least one charm quark, for which heavy-quark effective theory can serve as an initial starting point. We explore open-charm as well as hidden-charm (exotic) states, all of them within the molecular hypothesis. Predictions for these states from EFTs exist not only in vacuum, but also at finite temperatures, where the properties of the molecular states, such as masses and decay widths, are modified by the

*e-mail: torres@fqa.ub.edu

© The Authors, published by EDP Sciences. This is an open access article distributed under the terms of the Creative Commons Attribution License 4.0 (https://creativecommons.org/licenses/by/4.0/).

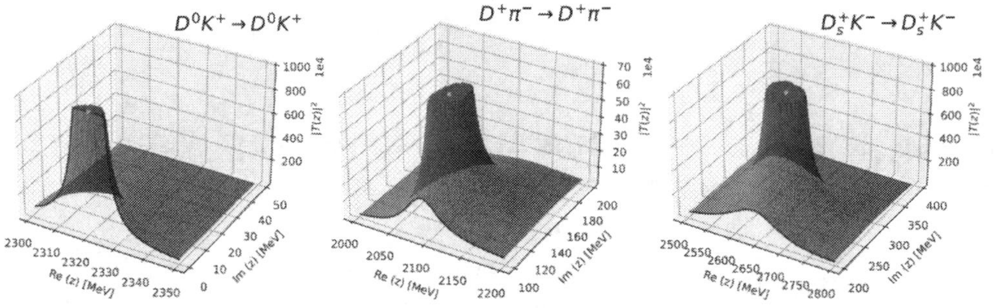

Figure 1. Left panel: Pole of the T matrix corresponding to the physical $D^*_{s0}(2317)$ bound state. Middle and right panels: two poles of the $D^*_0(2300)$ molecular state, shown in different final states of the same coupled-channel problem. Figures obtained from the results given in Ref. [19].

medium. In fact, even the nature of the states at $T = 0$ can change when immersed in the thermal bath. This is particularly relevant in the context of relativistic heavy-ion collisions (RHICs) where hadron production starts at around $T \simeq 160$ MeV, until they freeze out at a lower temperature. Therefore the properties of these states and the subsequent decay products can be affected by the temperature.

2 Heavy-light states

Consider the lightest charm meson, D, and its heavy-quark partner, D^*, forming composite states with light mesons through their attractive interactions. These can be described by an EFT that incorporates chiral and heavy-quark spin symmetries [6–11]. Other approaches extend the chiral symmetry to four flavors [12]. For a recent review, see Ref. [13].

For example, the narrow $D^*_{s0}(2317)$ [14] can be understood as a bound state of the D^0K^+ interaction (and its coupled channels) in an EFT approach [6–8] (the state contains a small width due to the isospin-breaking decay mechanism). In the left panel of Fig. 1 we show the T-matrix pole associated to this state. The $D_{s1}(2460)$ corresponds to the heavy-quark spin partner, replacing the D meson with the vector D^*. Other exotic states with similar masses have also been explained using meson-meson dynamics, such as the $X_0(2866)$, interpreted as a $D^*\bar{K}^{*-}$ molecule [15].

An interesting case arises with two rather broad states, the $D^*_0(2300)$ and $D^*_1(2430)$, which were difficult to reconcile with the results of EFTs given their experimental masses and widths. The solution of this puzzle came through the hypothesis of a "double-pole structure", where two different poles of the T matrix emerge with the same quantum numbers and interacting channels [16–18]. However, the decay probability of each pole into a particular final state depends on its coupling to that channel. In Fig. 1 we show the T-matrix element of the $D^+\pi^- \to D^+\pi^-$ reaction (middle panel), where the lower pole is identified; and the one of the $D_s^+K^- \to D_s^+K^-$ reaction (right panel), where the higher pole emerges. In Ref. [16] the two poles are separated by 250 MeV in mass, both with large decay widths (> 200 MeV).

Meson-baryon molecular candidates are also supported by several EFT calculations. Examples where the s−wave interactions produce multiparticle states include the $\Lambda_c(2595)$, $\Lambda_c(2880)$, $\Sigma_c(2800)$ and $\Xi_c(2970)$ (among many others). See for example [7, 20–22], as well as Ref. [2] for a more comprehensive review.

© The Authors, published by EDP Sciences. This is an open access article distributed under the terms of the Creative Commons Attribution License 4.0 (https://creativecommons.org/licenses/by/4.0/).

3 Heavy-heavy states

Molecular states containing double charm have also been explored within the same framework. In this case the bound states can be generated by two D/D^* mesons, or by a D/D^* meson and a $\bar{D}/\bar{D}^*$ meson, in the hidden charm case. The exotic $X(3872)$, first measured at Belle [23], has been proposed to admit different interpretations, including tetraquark, conventional $\chi_{c1}(2P)$, hybrid, and others [5]. The molecular possibility is particularly appealing since the mass of the $X(3872)$ is very close to the $D\bar{D}^*$ threshold, with an extremely small width. This argument is supported by explicit calculations using SU(4) chiral theory [24] as well as the local hidden gauge formalism [25]. Using this EFT, Ref. [26] generated both the $X(3872)$ and its heavy-quark partner, the $X(4014)$.

Many exotic states (traditionally denoted as X, Y, Z) have been discovered in the last years, with some of them lying close to open-flavor thresholds and postulated to be dominated by a molecular component. For further details on these states see Refs. [2, 5]. Some insights can also be gained from the heavy-ion program. For example, the molecular and tetraquark configurations of the $X(3872)$ would yield very different production rates as a function of the collision centrality [27]. The first measurement of this state in RHICs was reported in [28].

4 Charm femtoscopy in p-p collisions

The femtoscopy technique [29–31] has gained a lot of attention in recent years thanks to the experimental capabilities for measuring correlation functions in high-energy p+p collisions. Also the variety of hadrons that can be studied is very broad. Being sensitive to the final-state interactions of the measured pair, femtoscopy is able to help in identifying possible bound states or resonances formed by the measured hadron pair.

In the context of heavy-light systems with charm, the ALICE collaboration has published a study of the $\bar{D}N$ interaction [32] and, more recently, the femtoscopy correlation functions of $D^{(*)}\pi$, $D^{(*)}K$ and $D^{(*)}\bar{K}$ pairs [33]. Theoretical predictions incorporating heavy-light pairs have been focused on the the neutral channels [34–36], with the exception of [19] which presented predictions for the measured charged channels. While the sizable errors bars in the experimental data allow for a reasonable agreement across channels, the $D^+\pi^-$ case exhibits strong deviations between theory and experiment. The experimental results suggest little to no interaction, but the theoretical prediction indicates significant attraction, dominated by the lower pole of the $D_0^*(2300)$ state, which manifests as a shallow minimum in the correlation function. Incorporating the emission source for this broad resonance into the experimental data may improve the comparison. A proposed new channel where the molecular nature of the $D_0^*(2300)$ would more strongly influence the correlation function is the $D_s^+K^-$ pair [19].

5 Studies at finite temperature

Models that dynamically generate states from an EFT description can be formally extended to finite temperature using the imaginary-time formalism. This approach allows for the study of differences in the properties of molecular states compared to those in vacuum. The phenomenological relevance is motivated by measurements at RHICs, where a significant temperature is achieved.

In the charm sector, several works have examined—under different approximations—the thermal modification of the D-meson mass and decay width [37–41]. A summary of these results can be found in the review [13]. A recent calculation has extracted the spectral function of D and D^* mesons using a self-consistent study [42, 43]. As a byproduct, the properties of the generated molecular states $D_0^*(2300)$ and $D_{s0}^*(2317)$ can be monitored as a

© The Authors, published by EDP Sciences. This is an open access article distributed under the terms of the Creative Commons Attribution License 4.0 (https://creativecommons.org/licenses/by/4.0/).

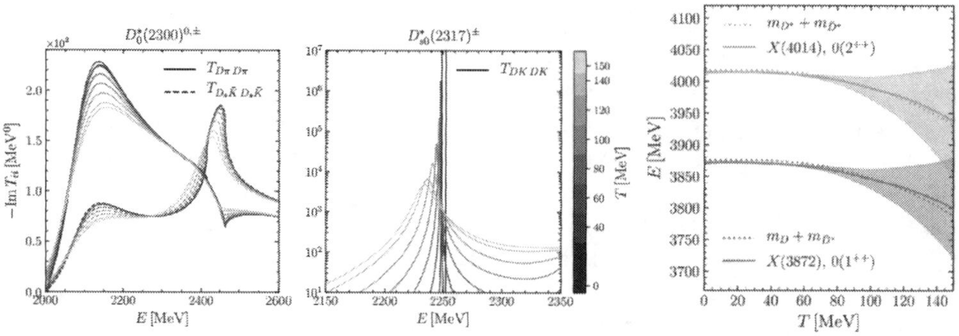

Figure 2. Left panel: Temperature dependence of the spectral shapes of the two poles of the $D_0^*(2300)$ state. Middle panel: Same for the narrow $D_{s0}^*(2317)$ state. Right panel: Temperature dependence of the masses and decay widths of the $X(3872)$ and $X(4014)$. Figures taken from Refs. [26, 42].

function of temperature. This is illustrated in the left and middle panels of Fig. 2, where the spectral shapes of the two poles of the $D_0^*(2300)$ and the $D_{s0}^*(2317)$ are shown. While all masses exhibit a small dependence on temperature up to $T = 150$ MeV ($\Delta m \simeq -20$ MeV), the thermal widths acquire sizable values, particularly for the $D_{s0}^*(2317)$, which was bound in vacuum. This change would alter their production in RHICs and enhance their decay rates.

In the hidden charm sector, molecular states such as the $X(3872)$ and the $X(4014)$ also change their nature from weakly bound states to thermal resonances as the temperature increases. As explained in [26], these states inherit the width of the D and $\bar{D}^*$ mesons, of which they are composed. As shown in the right panel of Fig. 2, the dependence of their masses and widths varies only slightly with temperature between $T = 0$ and $T \simeq 100$ MeV. However, for $T > 150$ MeV, the masses of both exotics exceed the thermal masses of their constituents, and the decay widths (represented as a broad band around the mass) increase largely due to phase-space opening. Since these are the temperatures relevant for RHICs, the resulting charm mesons from the decay of these exotics can be enhanced as compared to vacuum expectations if the molecular hypothesis is correct.

6 Conclusions

Several features of some particular hadrons containing open and hidden charm, such as their proximity to two-particle thresholds and small decay widths, suggest their possible molecular nature. Effective field theory calculations demonstrate that this possibility is well founded through the explicit generation of these states with their correct properties (without excluding that other configurations, such as tetraquarks, may also play a role). Many states found in experimental facilities, such as the $D_0^*(2300)$ and the narrow $D_{s0}^*(2317)$, can be described as heavy meson—light meson molecular states. Also, the exotic states $X(3872)$ and $X(4014)$ can be viewed as $D\bar{D}^*$ and $D^*\bar{D}^*$ bound states, respectively.

Predictions at finite temperature—relevant for RHIC phenomenology—can also help to discriminate between different configurations. Explicit finite temperature calculations using EFTs show that the thermal bath modifies the properties of the molecular states—such as providing a finite decay width to bound states, like the $D_{s0}^*(2317)$ and the $X(3872)$—leading to changes in the predicted population of their decay products. This serves as an additional test

© The Authors, published by EDP Sciences. This is an open access article distributed under the terms of the Creative Commons Attribution License 4.0 (https://creativecommons.org/licenses/by/4.0/).

for the molecular configuration hypothesis. Finally, charm femtoscopy has recently emerged as a valuable technique for studying the presence of these molecular states by analyzing specific characteristics in the correlation functions of their constituent hadrons.

7 Acknowledgments

This work has been supported by the project number CEX2019-000918-M (Unidad de Excelencia "María de Maeztu"), PID2020-118758GB-I00 and PID2023-147112NB-C21, financed by the Spanish MCIN/ AEI/10.13039/501100011033/, the EU STRONG-2020 project, under the program H2020- INFRAIA-2018-1 grant agreement no. 824093, and DFG project no. 315477589 - TRR 211 (Strong-interaction matter under extreme conditions).

References

[1] H. X. Chen *et al.*, Rept. Prog. Phys. **80** (2017) no.7, 076201 https://doi.org/10.1088/1361-6633/aa6420
[2] F. K. Guo *et al.*, Rev. Mod. Phys. **90** (2018) no.1, 015004 [erratum: Rev. Mod. Phys. **94** (2022) no.2, 029901] https://doi.org/10.1103/RevModPhys.90.015004
[3] N. Brambilla *et al.*, Phys. Rept. **873** (2020), 1-154 https://doi.org/10.1016/j.physrep.2020.05.001
[4] H. X. Chen *et al.*, Rept. Prog. Phys. **86** (2023) no.2, 026201 https://doi.org/10.1088/1361-6633/aca3b6
[5] M. Z. Liu *et al.*, https://arxiv.org/abs/2404.06399
[6] E. E. Kolomeitsev and M. F. M. Lutz, Phys. Lett. B **582** (2004), 39-48 https://doi.org/10.1016/j.physletb.2003.10.118
[7] J. Hofmann and M. F. M. Lutz, Nucl. Phys. A **733** (2004), 142-152 https://doi.org/10.1016/j.nuclphysa.2003.12.013
[8] F. K. Guo, P. N. Shen, H. C. Chiang, R. G. Ping and B. S. Zou, Phys. Lett. B **641** (2006), 278-285 https://doi.org/10.1016/j.physletb.2006.08.064
[9] F. K. Guo, C. Hanhart, S. Krewald and U. G. Meissner, Phys. Lett. B **666** (2008), 251-255 https://doi.org/10.1016/j.physletb.2008.07.060
[10] L. Tolos and J. M. Torres-Rincon, Phys. Rev. D **88** (2013), 074019 https://doi.org/10.1103/PhysRevD.88.074019
[11] M. Altenbuchinger, L. S. Geng and W. Weise, Phys. Rev. D **89** (2014) no.1, 014026 https://doi.org/10.1103/PhysRevD.89.014026
[12] D. Gamermann, E. Oset, D. Strottman and M. J. Vicente Vacas, Phys. Rev. D **76** (2007), 074016 https://doi.org/10.1103/PhysRevD.76.074016
[13] S. K. Das, J. M. Torres-Rincon and R. Rapp, https://arxiv.org/abs/2406.13286
[14] B. Aubert *et al.* [BaBar], Phys. Rev. Lett. **90**, 242001 (2003)
[15] M. Z. Liu, J. J. Xie and L. S. Geng, Phys. Rev. D **102** (2020) no.9, 091502 https://doi.org/10.1103/PhysRevD.102.091502
[16] M. Albaladejo, P. Fernandez-Soler, F. K. Guo and J. Nieves, Phys. Lett. B **767** (2017), 465-469 https://doi.org/10.1016/j.physletb.2017.02.036
[17] U. G. Meißner, Symmetry **12**, no.6, 981 (2020) https://doi.org/10.3390/sym12060981
[18] J. M. Torres-Rincon, Symmetry **13**, no.8, 1400 (2021) https://doi.org/10.3390/sym13081400
[19] J. M. Torres-Rincon, À. Ramos and L. Tolos, Phys. Rev. D **108** (2023) no.9, 096008 https://doi.org/10.1103/PhysRevD.108.096008

© The Authors, published by EDP Sciences. This is an open access article distributed under the terms of the Creative Commons Attribution License 4.0 (https://creativecommons.org/licenses/by/4.0/).

[20] T. Mizutani and A. Ramos, Phys. Rev. C **74** (2006), 065201 https://doi.org/10.1103/PhysRevC.74.065201

[21] J. He, Y. T. Ye, Z. F. Sun and X. Liu, Phys. Rev. D **82** (2010), 114029 https://doi.org/10.1103/PhysRevD.82.114029

[22] X. G. He, X. Q. Li, X. Liu and X. Q. Zeng, Eur. Phys. J. C **51** (2007), 883-889 https://doi.org/10.1140/epjc/s10052-007-0347-y

[23] S. K. Choi *et al.* [Belle], Phys. Rev. Lett. **91** (2003), 262001 https://doi.org/10.1103/PhysRevLett.91.262001

[24] D. Gamermann and E. Oset, Eur. Phys. J. A **33** (2007), 119-131 https://doi.org/10.1140/epja/i2007-10435-1

[25] R. Molina and E. Oset, Phys. Rev. D **80** (2009), 114013 https://doi.org/10.1103/PhysRevD.80.114013

[26] G. Montaña, A. Ramos, L. Tolos and J. M. Torres-Rincon, Phys. Rev. D **107** (2023) no.5, 054014 https://doi.org/10.1103/PhysRevD.107.054014

[27] H. Zhang, J. Liao, E. Wang, Q. Wang and H. Xing, Phys. Rev. Lett. **126** (2021) no.1, 012301 https://doi.org/10.1103/PhysRevLett.126.012301

[28] A. M. Sirunyan *et al.* [CMS], Phys. Rev. Lett. **128** (2022) no.3, 032001 https://doi.org/10.1103/PhysRevLett.128.032001

[29] U. W. Heinz and B. V. Jacak, Ann. Rev. Nucl. Part. Sci. **49** (1999), 529-579 https://doi.org/10.1146/annurev.nucl.49.1.529

[30] M. A. Lisa, S. Pratt, R. Soltz and U. Wiedemann, Ann. Rev. Nucl. Part. Sci. **55** (2005), 357-402 https://doi.org/10.1146/annurev.nucl.55.090704.151533

[31] L. Fabbietti, V. Mantovani Sarti and O. Vazquez Doce, Ann. Rev. Nucl. Part. Sci. **71** (2021), 377-402 https://doi.org/10.1146/annurev-nucl-102419-034438

[32] S. Acharya *et al.* [ALICE], Phys. Rev. D **106** (2022) no.5, 052010 https://doi.org/10.1103/PhysRevD.106.052010

[33] S. Acharya *et al.* [ALICE], Phys. Rev. D **110**, no.3, 032004 (2024) https://doi.org/10.1103/PhysRevD.110.032004

[34] M. Albaladejo, J. Nieves and E. Ruiz-Arriola, Phys. Rev. D **108** (2023) no.1, 014020 https://doi.org/10.1103/PhysRevD.108.014020

[35] Z. W. Liu, J. X. Lu and L. S. Geng, Phys. Rev. D **107** (2023) no.7, 074019 https://doi.org/10.1103/PhysRevD.107.074019

[36] K. P. Khemchandani, L. M. Abreu, A. Martinez Torres and F. S. Navarra, Phys. Rev. D **110**, no.3, 036008 (2024) https://doi.org/10.1103/PhysRevD.110.036008

[37] C. Fuchs, B. V. Martemyanov, A. Faessler and M. I. Krivoruchenko, Phys. Rev. C **73** (2006), 035204 https://doi.org/10.1103/PhysRevC.73.035204

[38] M. He, R. J. Fries and R. Rapp, Phys. Lett. B **701** (2011), 445-450 https://doi.org/10.1103/PhysRevC.83.018201

[39] S. Ghosh, S. Mitra and S. Sarkar, Nucl. Phys. A **917** (2013), 71-84 https://doi.org/10.1016/j.nuclphysa.2013.08.010

[40] M. Cleven, V. K. Magas and A. Ramos, Phys. Rev. C **96** (2017) no.4, 045201 https://doi.org/10.1103/PhysRevC.96.045201

[41] J. M. Torres-Rincon, G. Montaña, À. Ramos and L. Tolos, Phys. Rev. C **105** (2022) no.2, 025203 https://doi.org/10.1103/PhysRevC.105.025203

[42] G. Montaña, À. Ramos, L. Tolos and J. M. Torres-Rincon, Phys. Lett. B **806** (2020), 135464 https://doi.org/10.1016/j.physletb.2020.135464

[43] G. Montaña, À. Ramos, L. Tolos and J. M. Torres-Rincon, Phys. Rev. D **102** (2020) no.9, 096020 https://doi.org/10.1103/PhysRevD.102.096020

© The Authors, published by EDP Sciences. This is an open access article distributed under the terms of the Creative Commons Attribution License 4.0 (https://creativecommons.org/licenses/by/4.0/).

Investigation of charm-quark hadronization into baryons in hadronic collisions with ALICE

Jaeyoon Cho[1,*]

[1]Department of Physics, Inha University, Incheon 22212, Republic of Korea

Abstract. A comprehensive investigation into charm-baryon production provides insight into understanding charm-quark hadronization process. In this contribution, the charm-quark hadronization mechanism is discussed in terms of the production yield ratio between charm hadron species measured by ALICE Collaboration. Measurements of Λ_c^+, $\Sigma_c^{0,++}$, $\Xi_c^{0,+}$ and Ω_c^0 in pp collisions, Λ_c^+ and Ξ_c^0 in p–Pb collisions are presented, together with the Λ_c^+ measurement in larger collision systems, Pb–Pb collisions. Furthermore, the charm-quark hadronization mechanism is discussed based on the fragmentation fractions computed for pp and p–Pb collisions. A preliminary result on $\Sigma_c^{0,++}$ measurement with Run 3 data is also presented.

1 Introduction

In hadronic collisions, heavy-flavour (HF) hadrons are produced from the hadronization of heavy quarks generated through the hard-scattering process occurring in the initial collision stages. Typically, the production cross section of HF hadrons in pp collisions is computed based on the QCD factorization approach, as a convolution of three independent terms: i) the parton distribution functions, ii) the production cross section of the heavy quarks and iii) the fragmentation functions which model the heavy-quark hadronization. The fragmentation functions, which cannot be calculated with perturbative QCD (pQCD) framework, are constrained from the measurements performed for e^+e^- or ep collisions. This approach relies on the assumption that the fragmentation functions are universal across different collision systems.

From an experimental point of view, measuring HF hadrons in pp collisions provides a fundamental test for the validity of the pQCD calculations and their assumptions. In particular, the production yield ratio between hadron species can be a good tool to understand how the heavy quarks hadronize into given hadrons. The hadronization process in more complex environments can be studied in larger collisions systems, such as p–Pb or Pb–Pb collisions. In p–Pb collisions, the cold nuclear matter effects can be studied in detail, and the measurements in Pb–Pb collisions allow us to investigate the modification of the hadronization process in an extreme environment, in which strongly interacting QCD matter exists. Furthermore, a comprehensive study of the measurements from small to large collision systems provides insight into how the hadronization process evolves with the collision system size and on the possible onset of further hadronization mechanisms in addition to quark fragmentation, such as the coalescence.

*e-mail: jaeyoon.cho@cern.ch

© The Authors, published by EDP Sciences. This is an open access article distributed under the terms of the Creative Commons Attribution License 4.0 (https://creativecommons.org/licenses/by/4.0/).

Thanks to its excellent tracking, vertexing and particle identification capabilities, ALICE was able to measure several charm baryons, Λ_c^+, $\Sigma_c^{0,++}$, $\Xi_c^{0,+}$ and Ω_c^0, in different collision systems, using the data samples collected during the LHC Run 2 data taking.

2 Charm baryon measurement with ALICE

2.1 Non-strange charm-baryon measurements

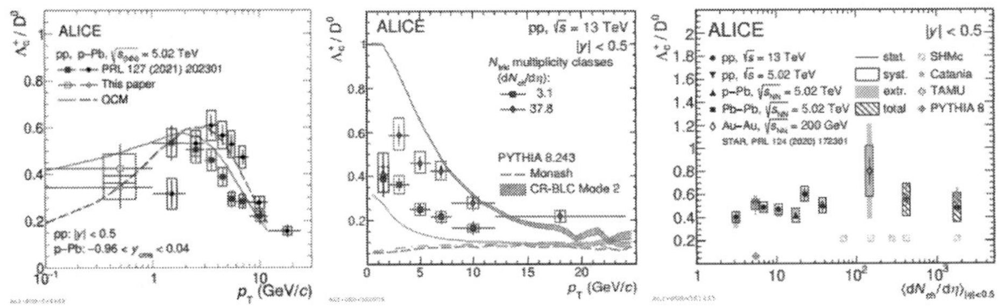

Figure 1. Left: Λ_c^+/D^0 ratios in pp and p–Pb collisions at $\sqrt{s_{NN}}$ = 5.02 TeV along with QCM predictions. Middle: Λ_c^+/D^0 ratios in pp collisions at $\sqrt{s}$ = 13 TeV for two different classes of event-multiplicity. Right: p_T-integrated Λ_c^+/D^0 ratio as a function of average event multiplicity along with theory predictions.

As shown in left panel of Figure 1, ALICE measured the Λ_c^+/D^0 ratio in pp and p–Pb collisions at $\sqrt{s_{NN}}$ = 5.02 TeV down to $p_T = 0$ [1]. The overall magnitude of the Λ_c^+/D^0 ratio is similar in the two collision systems, but in the intermediate p_T region a hint of a shift toward higher p_T is observed in p–Pb collisions with respect to pp collisions. Compared to the baryon-to-meson ratio measured in e^+e^- collisions, whose LEP average was found to be 0.113 ± 0.013 (stat.) ± 0.006 (syst.), the Λ_c^+/D^0 ratio in both hadronic collision systems shows a significant enhancement with strong p_T dependence. The QCM model [2] which includes a recombination scenario even for pp collisions, describes the measurements for both pp and p–Pb collisions.

Performing the measurement in different event-multiplicity classes provides further information on how hadronization process is influenced by the event-multiplicity. Figure 1 middle panel shows the Λ_c^+/D^0 ratio measured in low and high event-multiplicity classes in pp collisions at $\sqrt{s}$ = 13 TeV [3]. The Λ_c^+/D^0 ratio in the high event-multiplicity class is found to be 5.3σ higher than that in the low event-multiplicity class. The PYTHIA 8 event generator with Monash tune [4], using fragmentation functions tuned to the measurements from e^+e^- collisions, failed to describe both p_T and event-multiplicity dependent Λ_c^+/D^0 ratios obtained in data, as had already been observed in the comparison to the multiplicity-integrated Λ_c^+/D^0 ratio measurement. The PYTHIA 8 CR-BLC Mode 2 tune [5], which allows further combinations of partons from different Multiple Parton Interaction (MPI) introducing *junction* topologies, reproduces the data. ALICE also measured Λ_c^+/D^0 ratio in Pb–Pb collisions at $\sqrt{s_{NN}}$ = 5.02 TeV [6]. In that measurement, the Λ_c^+/D^0 ratio in central and mid-central Pb–Pb collisions was higher than that in pp collisions by 3.7σ and 2.0σ, respectively.

Based on these measurements, the p_T-integrated Λ_c^+/D^0 ratio as a function of average event-multiplicity was calculated to study the possible trend with increasing event-multiplicity, as shown in the right panel of Figure 1 [6]. No significant multiplicity dependence is observed, differently from what observed for the p_T-differential Λ_c^+/D^0 ratio for $1 < p_T < 12$ GeV/c. This feature could be explained by assuming a different redistribution of p_T for baryons and mesons across different multiplicities, rather than an overall baryon

© The Authors, published by EDP Sciences. This is an open access article distributed under the terms of the Creative Commons Attribution License 4.0 (https://creativecommons.org/licenses/by/4.0/).

production enhancement. The SHMc[7], Catania[8] and TAMU[9] describe well the data, whereas PYTHIA 8 significantly underestimates the measurements.

The $\Sigma_c^{0,++}(2455)$ baryon, which is the isospin $I = 1$ partner of the isospin singlet Λ_c^+ baryon, contributes to Λ_c^+ production via its strong decay. ALICE has measured the $\Sigma_c^{0,++}(2455)$ production using the Run 2 data sample [10], and recently performed a measurement of the production of $\Sigma_c^{0,++}(2520)$ state with the larger data samples of the LHC Run 3. Figure 2 left panel shows preliminary results for the $\Sigma_c^{0,++}(2520)/\Sigma_c^{0,++}(2455)$ ratio measured in pp collisions at $\sqrt{s} = 13.6$ TeV with Run 3 data, along with PYTHIA 8 CR-BLC Mode 2 predictions. Such predictions underestimate the $\Sigma_c^{0,++}(2520)/\Sigma_c^{0,++}(2455)$ ratio. However, as the parameter (probQQ1toQQ0join_charm), which regulates the suppression of the heavy-diquark spin $S = 1$ state with respect to $S = 0$ state is increased with respect to its default value, the predictions get closer to the measurement. This preliminary results suggest that the $S = 1$ diquarks may play a significant role in $\Sigma_c^{0,++}$ production [11]. This measurement is expected to provide valuable constraints for the theory development.

2.2 Charm-strange baryon measurements

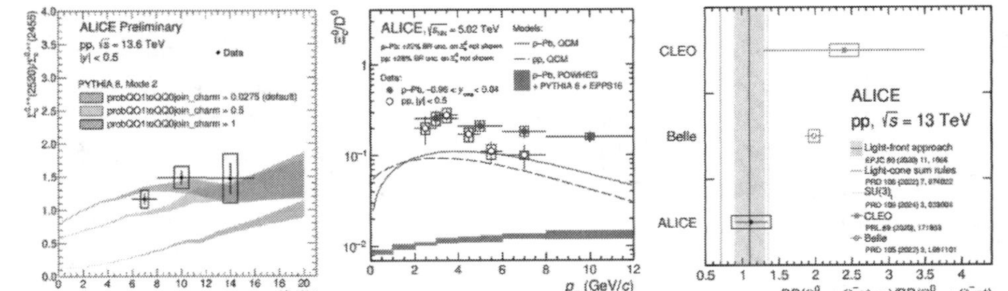

Figure 2. Left: $\Sigma_c^{0,++}(2520)/\Sigma_c^{0,++}(2455)$ ratio in pp collisions at $\sqrt{s} = 13.6$ TeV along with PYTHIA 8 CR-BLC predictions. Middle: Ξ_c^0/D^0 ratio in pp and p–Pb collisions at $\sqrt{s_{NN}} = 5.02$ TeV with model predictions. Right: $BR(\Omega_c^0 \to \Omega^- e^+ \nu_e)/BR(\Omega_c^0 \to \Omega^- \pi^+)$ compared with theory calculations.

ALICE measured multiple charm baryons containing strange quarks as well. The middle panel of Figure 2 shows the Ξ_c^0/D^0 ratio measured in pp and p–Pb collisions at $\sqrt{s_{NN}} = 5.02$ TeV [12]. The QCM model, which describes correctly the Λ_c^+/D^0 ratio, underestimates the Ξ_c^0/D^0 ratio by a factor of about 2 for both pp and p–Pb collisions collisions, and POWHEG+PYTHIA6 with fragmentation functions tuned to e^+e^- measurements, underestimates the data by a factor of about 50, implying that the baryon enhancement in charm-strange baryons is even larger than for non-strange charm baryons.

The Ω_c^0 baryon was also measured in pp collisions at $\sqrt{s} = 13$ TeV [13], but a precise comparison with theory predictions is not possible, since the branching ratio (BR) of the hadronic decay channel used for the Ω_c^0 reconstruction ($\Omega_c^0 \to \Omega^- \pi^+$) has not been expermentally measured yet, and its theory estimations are affected by large uncertainties. As shown in the right panel of Figure 2, the recent ALICE measurement of the BR ratio between hadronic and semileptonic decay channels [14] is expected to provide a constraint for the Ω_c^0 branching ratio. This measurement is consistent with that measured by the CLEO [15] and Belle [16] Collaborations within 2.3σ and 1σ respectively, and also compatible with current state-of-the-art theory calculations [17–19].

2.3 Charm-quark fragmentation fraction

Measurements of the various charm baryons provide the ingredients to calculate the fragmentation fractions of charm quarks in pp and p–Pb collisions at $\sqrt{s_{NN}} = 5.02$ TeV [20], as shown

© The Authors, published by EDP Sciences. This is an open access article distributed under the terms of the Creative Commons Attribution License 4.0 (https://creativecommons.org/licenses/by/4.0/).

in Figure 3. The fragmentation fractions measured for pp and p–Pb collisions are consistent with each other within the uncertainties. Compared to leptonic collisions, a significantly enhanced baryon production and reduced non-strange D meson production are observed. This discrepancy suggests that the hadronization process in parton-rich environment is governed by different mechanisms with respect to leptonic collisions.

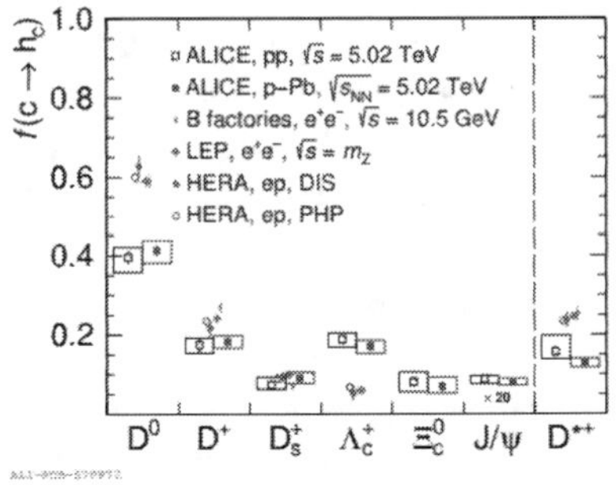

Figure 3. Fragmentation fractions of charm-quarks in hadronic collisions at $\sqrt{s_{\mathrm{NN}}}$ = 5.02 TeV [20].

3 Summary and outlook

ALICE measured charm baryons in various hadronic collision systems, and observed an enhancement of baryon-to-meson ratio with respect to that measured in leptonic collisions. The fragmentation fractions computed from these measurements clearly show that the hadron production mechanism in hadronic collisions is different from that in leptonic collisions. Models implementing baryon enhancement via modified hadronization with respect to in-vacuum fragmentation explain the charm non-strange baryon productions, while they still do not fully reproduce the phenomena obtained for charm-strange baryons. The increased size of data samples being collected at the LHC during Run 3, together with the improved detector performance after the ALICE upgrade, will allow us to reach extended p_{T} region with improved precision and shed further light on the charm-quark hadronization mechanisms.

References

[1] Acharya, S. *et al.*, ALICE Collaboration, Phys. Rev. C **107** 064901 (2023).
[2] H. Li. *et al.*, Chin. Phys. C **45** 113105 (2021).
[3] Acharya, S. *et al.*, ALICE Collaboration, Phys. Lett. B **829**, 137065 (2022).
[4] P. Skands. *et al.*, Eur. Phys. J. C **74** 3024 (2014).
[5] J.R. Christiansen. *et al.*, JHEP **08** 003 (2015).
[6] Acharya, S. *et al.*, ALICE Collaboration, Phys. Lett. B **839**, 137796 (2023).
[7] A. Andronic. *et al.*, JHEP **07** 035 (2021).
[8] S. Plumari *et al.*, Eur. Phys. J. C **78** 348 (2018).
[9] M. He. *et al.*, Phys. Rev. Lett. **124** 042301 (2020).
[10] Acharya, S. *et al.*, ALICE Collaboration, Phys. Rev. Lett. **128**, 012001 (2022).
[11] J.Altmann. *et al.*, arXiv:2405.19137.
[12] Acharya, S. *et al.*, ALICE Collaboration, arXiv:2405.14538.
[13] Acharya, S. *et al.*, ALICE Collaboration, Phys. Lett. B **846**, 137625 (2023).
[14] Acharya, S. *et al.*, ALICE Collaboration, Phys. Rev. D **110**, 032014 (2024).
[15] R. Ammar. *et al.*, CLEO Collaboration, Phys. Rev. Lett. **89**, 171803 (2002).
[16] Y. B. Li. *et al.*, Belle Collaboration, Phys. Rev. D **105**, L091101 (2022).
[17] T.M. Aliev. *et al.*, Phys. Rev. D **106**, 074022 (2022).
[18] Y.-K Hsiao. *et al.*, Eur. Phys. J. C **80**, 1066 (2020).
[19] C.-W. Liu, Phys. Rev. D **109**, 033004 (2024).
[20] Acharya, S. *et al.*, ALICE Collaboration, arXiv:2405.14571

© The Authors, published by EDP Sciences. This is an open access article distributed under the terms of the Creative Commons Attribution License 4.0 (https://creativecommons.org/licenses/by/4.0/).

Diffusion of heavy quarks in the early stage of high-energy nuclear collisions

Pooja[1,2,*], *Santosh Kumar* Das[2,**], *Vincenzo* Greco[3,4,***], and *Marco* Ruggieri[3,5,****]

[1]Department of Physics, University of Jyväskylä, P.O. Box 35, 40014 University of Jyväskylä, Finland
[2]School of Physical Sciences, Indian Institute of Technology Goa, Ponda-403401, Goa, India
[3]Department of Physics and Astronomy "Ettore Majorana", University of Catania, Via S. Sofia 64, I-95123 Catania, Italy
[4]INFN-Laboratori Nazionali del Sud, Via S. Sofia 62, I-95123 Catania, Italy
[5]INFN-Sezione di Catania, Via S. Sofia 64, I-95123 Catania, Italy

Abstract. We study the diffusion of heavy quarks in the early stage of high-energy nuclear collisions. The pre-equilibrium stage of relativistic heavy-ion collisions, commonly known as Glasma, evolves according to the classical Yang-Mills equations. Heavy quarks are coupled to the evolving Glasma fields via relativistic kinetic theory. We compute the momentum broadening, σ_p as well as the angular momentum fluctuations of heavy quarks in the early stage, which turn out to be anisotropic due to the anisotropy of the background gluon fields. We observe that $\sigma_p \propto t^2$ at very initial times. This non-Markovian diffusion of heavy quarks in the early stages is explained by the memory effect present in the gluon fields.

1 Introduction

Relativistic heavy-ion collisions (HICs) at the Large Hadron Collider (LHC) and the Relativistic Heavy Ion Collider (RHIC) offer unique opportunities to study hadronic matter under extreme conditions. Following the collision, a pre-equilibrium phase is formed, which transitions into a deconfined medium of quarks and gluons called the quark-gluon plasma (QGP), eventually leading to hadronization. The pre-equilibrium stage, the earliest phase of HICs, is characterized by high gluon occupation numbers and a highly non-linear regime referred to as the Glasma.

Heavy quarks (HQs) [1–15] are valuable probes of the early stages of HICs. Charm and beauty quarks, in particular, are produced with very short formation times that they can probe the entire evolution of the system, starting from the Glasma phase. Their formation time, estimated as $\tau_{\text{form}} \approx \frac{1}{2m}$,where m is the quark mass, is approximately 0.06 fm/c for charm quarks and 0.02 fm/c for beauty quarks. While the Glasma may only last for a fraction of a fm/c, the strong interactions between HQs and the dense gluon fields during this time can leave observable imprints on the final state. Therefore, studying HQs in the Glasma phase is

*e-mail: pooja19221102@iitgoa.ac.in
**e-mail: santosh@iitgoa.ac.in
***e-mail: greco@lns.infn.it
****e-mail: marco.ruggieri@dfa.unict.it

© The Authors, published by EDP Sciences. This is an open access article distributed under the terms of the Creative Commons Attribution License 4.0 (https://creativecommons.org/licenses/by/4.0/).

essential for a comprehensive understanding of relativistic HICs. In this work, we focus on the propagation of HQs in the very early phase of HICs, namely the Glasma phase.

2 Framework

In high-energy nuclear collisions, we utilize the effective theory of the color glass condensate (CGC) [16–18]. The CGC operates within the high-energy limit of Quantum Chromodynamics. The CGC framework conceptualizes the collision of two colliding objects as the interaction of two thin sheets of colored glass. During this collision, the interaction generates two sets of effective color charges, one for each sheet, connected by dense gluon field filaments. These gluon fields have a high occupation number, indicating a classical nature. These fields extend throughout the interaction region of the colliding objects and are collectively referred to as the Glasma [17]. Within the CGC framework, the Glasma represents the initial condition for the system produced in high-energy nuclear collisions.

In our work, we utilize the McLerran-Venugopalan (MV) model, where the static color charge densities ρ^a on the colliding nuclei L and R are postulated to be random variables. These variables follow a normal distribution such that

$$\langle \rho^a_{L,R}(\vec{x}_\perp) \rangle = 0, \tag{1}$$

$$\langle \rho^a_{L,R}(\vec{x}_\perp)\rho^b_{L,R}(\vec{y}_\perp) \rangle = (g\mu_{L,R})^2 \delta^{ab}\delta^{(2)}(\vec{x}_\perp - \vec{y}_\perp). \tag{2}$$

This study is constrained to SU(2) color group, hence $a, b = 1, 2, 3$. The parameter μ signifies the density of color charge carriers in the transverse plane, with $g\mu$ representing the color charge density. Importantly, $g^2\mu = O(Q_s)$, where Q_s stands for the saturation momentum scale.

The equations of motion governing the fields and their conjugate momenta, i.e., the classical Yang-Mills (CYM) equations, in the static box are given by

$$\frac{dA^a_i(x)}{dt} = E^a_i(x), \tag{3}$$

$$\frac{dE^a_i(x)}{dt} = \partial_j F^a_{ji}(x) + g f^{abc} A^b_j(x) F^c_{ji}(x). \tag{4}$$

The dynamics of the colored probes within the Glasma fields are governed by a set of equations known as the Wong equations [5–10, 19, 20]:

$$\frac{dx^i}{dt} = \frac{p^i}{E}, \tag{5}$$

$$\frac{dp^i}{dt} = gQ_a F^{i\nu}_a p_\nu, \tag{6}$$

$$\frac{dQ_a}{dt} = \frac{g}{E} f_{abc} A^i_b p^i Q_c, \tag{7}$$

where $i = x, y, z$ and $E = \sqrt{\vec{p}^2 + m^2}$ represent the relativistic energy of the HQ. Here, x^i, p^i, and Q_a denote the position, momentum, and effective color charge of the HQs, respectively.

3 Results

To determine Glasma parameters, we vary Q_s to obtain the QCD coupling g, see [7]. In Fig. 1, we show the evolution of transverse momentum broadening, $\sigma_p = \frac{1}{2}\langle (p_x(t) - p_{0x})^2 + (p_y(t) - p_{0y})^2 \rangle$, over proper time t for charm quarks in the Glasma fields for different values of

© The Authors, published by EDP Sciences. This is an open access article distributed under the terms of the Creative Commons Attribution License 4.0 (https://creativecommons.org/licenses/by/4.0/).

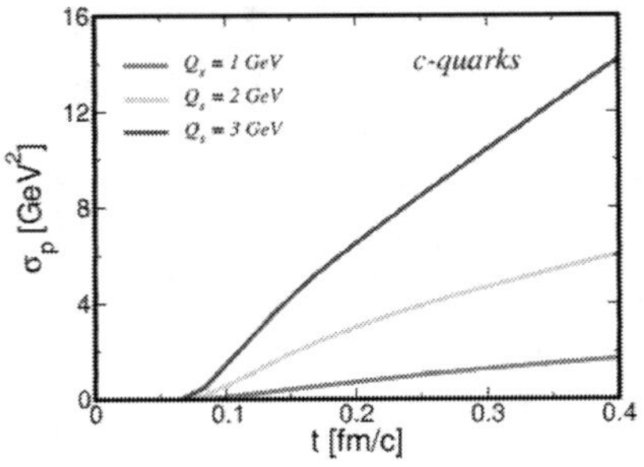

Figure 1. Transverse momentum broadening, σ_p w.r.t. proper time, t for charm quarks, considering initial $p_T = 0.5$ GeV. This calculation pertains to the evolving Glasma fields within a static box.

Q_s. Adjusting Q_s alters the energy density, thereby affecting the effective temperature of the Glasma medium where HQs diffuse. Consequently, a higher Q_s leads to increased diffusion of HQs in the Glasma fields.

An intriguing observation is that, in the very early time period, the relationship between σ_p and time is non-linear, in contrast to the standard Brownian motion (BM). This is followed by a linear growth of σ_p at a later stage. This discrepancy at initial times is attributed to the memory effects associated with the gluon fields within the bulk. These fields exert a correlated Lorentz force on the charm quarks, leading to the observed non-linearity of σ_p. As mentioned earlier, subsequent to the initial transient, σ_p exhibits a linear growth, resembling BM without drag, as observed in [6]. The transition between these two regimes occurs over a timescale of $t \approx \tau_{\rm mem}$ [6]. Hence, Fig. 1 highlights how the diffusion of the charm quarks during the very early stage of the evolution deviates from standard BM due to the memory of the background gluon fields in the evolving Glasma bulk.

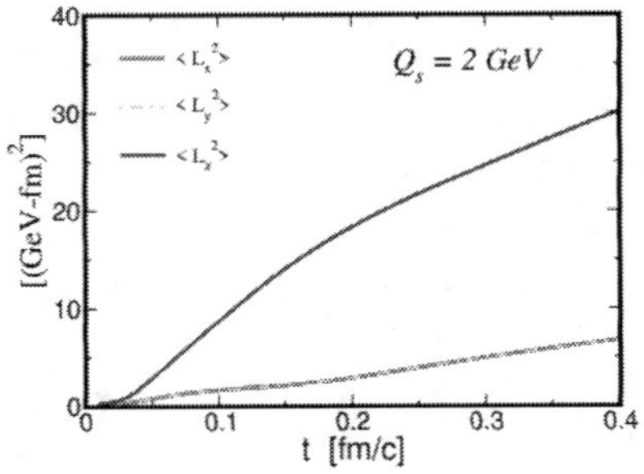

Figure 2. Evolution of the square of the components of orbital angular momentum with proper time for $m = 10$ GeV at $Q_s = 2$ GeV. The initial momentum is set with $p_T = 0.5$ GeV and $p_z = 0.5$.

Moreover, we study the fluctuations of angular momentum of HQs as well. To this end, we study the dynamics of the HQs in a classical color field within the non-relativistic limit, see [9, 20] for more a detailed framework. In Fig. 2, we present a plot depicting the evolution of squares of the components of angular momentum, namely $\langle L_x^2 \rangle$, $\langle L_y^2 \rangle$, and $\langle L_z^2 \rangle$ for $Q_s = 2$ GeV. Throughout the entire evolution, $\langle L_x \rangle = \langle L_y \rangle = \langle L_z \rangle = 0$, indicating that the quantities depicted in this figure represent the spreading of the components of $\vec{L}$.

© The Authors, published by EDP Sciences. This is an open access article distributed under the terms of the Creative Commons Attribution License 4.0 (https://creativecommons.org/licenses/by/4.0/).

Clearly, $\langle L_z^2 \rangle > \langle L_x^2 \rangle, \langle L_y^2 \rangle$; indicating that the interaction of HQs with the gluon fields leads to anisotropic fluctuations in orbital angular momentum. Futhermore, fluctuations of the spin angular momentum are much smaller in magnitude than orbital angular momentum fluctuations and remain isotropic, see [9]. Additionally, calculations show that spin and orbital angular momentum fluctuations are uncorrelated. Thus, we can approximate the fluctuations of total angular momentum, $\vec{J} = \vec{L} + \vec{S}$, as $\langle J_i^2 \rangle = \langle (L_i + S_i)^2 \rangle \approx \langle L_i^2 \rangle$.

4 Summary

We performed numerical simulations for the HQs in the background Glasma fields, focusing on linear momentum broadening and angular momentum fluctuations of HQs in the evolving Glasma fields. Our findings reveal that the diffusion of HQs in the pre-equlibrium phase of high-energy collisions is non-linear, influenced by strong, coherent gluon fields and significant memory effects. Additionally, we observed that the fluctuations of total angular momentum of HQs within the Glasma fields show pronounced anisotropy.

Acknowledgments

Pooja acknowledges IIT Goa and MHRD for funding this research.

References

[1] X. Dong and V. Greco, Prog. Part. Nucl. Phys. **104**, 97-141 (2019)
[2] S. K. Das, J. M. Torres-Rincon and R. Rapp, [arXiv:2406.13286 [hep-ph]]
[3] M. Ruggieri and S. K. Das, Phys. Rev. D **98**, no.9, 094024 (2018) [arXiv:1805.09617 [nucl-th]]
[4] Y. Sun, G. Coci, S. K. Das, S. Plumari, M. Ruggieri and V. Greco, Phys. Lett. B **798**, 134933 (2019)
[5] J. H. Liu, S. Plumari, S. K. Das, V. Greco and M. Ruggieri, Phys. Rev. C **102**, no.4, 044902 (2020)
[6] J. H. Liu, S. K. Das, V. Greco and M. Ruggieri, Phys. Rev. D **103**, no.3, 034029 (2021)
[7] P. Khowal, S. K. Das, L. Oliva and M. Ruggieri, Eur. Phys. J. Plus **137** (2022) no.3, 307
[8] M. Ruggieri, Pooja, J. Prakash and S. K. Das, Phys. Rev. D **106** (2022) no.3, 034032
[9] Pooja, S. K. Das, V. Greco and M. Ruggieri, Eur. Phys. J. Plus **138** (2023) no.4, 313
[10] Pooja, S. K. Das, V. Greco and M. Ruggieri, Phys. Rev. D **108** (2023) no.5, 054026
[11] Pooja, M. Y. Jamal, P. P. Bhaduri, M. Ruggieri and S. K. Das, [arXiv:2404.05315 [hep-ph]]
[12] D. Avramescu, V. Băran, V. Greco, A. Ipp, D. I. Müller and M. Ruggieri, Phys. Rev. D **107**, no.11, 114021 (2023)
[13] K. Boguslavski, A. Kurkela, T. Lappi and J. Peuron, JHEP **09** (2020), 077
[14] K. Boguslavski, A. Kurkela, T. Lappi, F. Lindenbauer and J. Peuron, Phys. Rev. D **109**, no.1, 014025 (2024)
[15] S. K. Das, PoS HardProbes2023 (2024), 011
[16] L. D. McLerran and R. Venugopalan, Phys. Rev. D **49**, 2233-2241 (1994)
[17] T. Lappi and L. McLerran, Nucl. Phys. A **772**, 200-212 (2006)
[18] F. Gelis, E. Iancu, J. Jalilian-Marian and R. Venugopalan, Ann. Rev. Nucl. Part. Sci. **60**, 463-489 (2010)
[19] S. K. Wong, Nuovo Cim. A **65**, 689-694 (1970)
[20] U. W. Heinz, Annals Phys. **161**, 48 (1985)

© The Authors, published by EDP Sciences. This is an open access article distributed under the terms of the Creative Commons Attribution License 4.0 (https://creativecommons.org/licenses/by/4.0/).

New insights into strange-quark hadronization measuring multiple (multi-)strange hadron production in small collision systems with ALICE

Sara Pucillo[1,*] for the ALICE Collaboration

[1]Dipartimento di Fisica, Università degli Studi di Torino, and Sezione INFN Torino

Abstract. Among the most important results from the Run-1 and Run-2 of the LHC is the observation of an enhanced production of (multi-)strange to non-strange hadron yields, gradually rising from low-multiplicity to high-multiplicity pp and p–Pb collisions, reaching values close to those measured in peripheral Pb–Pb collisions [1]. The observed behaviour cannot be quantitatively reproduced by any of the available QCD-inspired MC generators. In this contribution an extension of this study is presented: the measurement of the K^0_S, Λ, Ξ^- and Ω^- (together with antiparticles) multiplicity distributions in pp collisions at $\sqrt{s} = 5.02$ TeV as a function of the charged-particle multiplicity, together with the average probability for the multiplets production, extending the study of strangeness production beyond its average. This novel method, based on counting the number of strange particles event-by-event, represents a new test bench for production mechanisms, probing events with a large imbalance between strange and non-strange content.

1 Introduction

Among several probes of the quark–gluon plasma (QGP) formation, the so-called *Strangeness Enhancement* (SE) was one of the first proposed [2] and observed experimentally [3]. In addition to confirm the SE at the highest center-of-mass energy [4–6], the ALICE Collaboration carried out a comprehensive study of strange hadron production (relative to charged pions) as a function of the charged-particle multiplicity at midrapidity [1]. The main finding is that strangeness increases progressively with multiplicity in a compatible way across different collision systems and center-of-mass energies [1, 7, 8]. The invariance of the SE pattern on the colliding system suggests that the mechanisms at play in high-multiplicity pp interactions could be the same as those involved in particle formation in Pb–Pb collisions. Moreover the enhancement, previously unexpected in pp interactions, is proportional to the strangeness content in the hadron being the highest for the Ω^- baryon. However, the observed behaviour cannot be quantitatively reproduced by any of the available QCD-inspired models, suggesting that further developments are needed to obtain a complete microscopic understanding of strangeness production.

*e-mail: sara.pucillo@cern.ch

© The Authors, published by EDP Sciences. This is an open access article distributed under the terms of the Creative Commons Attribution License 4.0 (https://creativecommons.org/licenses/by/4.0/).

2 Measurement of (multi-)strange particle multiplicity distribution

To obtain more information about the strangeness production mechanism, the (multi-)strange particle multiplicity distribution, P(n_S), has been measured in pp collisions at $\sqrt{s} = 5.02$ TeV using a new technique based on counting the number of strange particles on an event-by-event basis. The signal extraction has been performed by developing a procedure based on signal/background weights obtained from the data after the application of selective cuts on topological features of the weak decay and performing particle identification of the daughters. As a first step, a fit to the 1-dimensional invariant mass spectrum given by the sum of a function for the signal (double-sided Crystal Ball [9]) and one for the background (polynomial of order 1 for all particles, except for Ω^- for which a polynomial of order 2 was chosen) was performed in different p_T and multiplicity bins. In this way, for each invariant mass and p_T range, it is possible to define a probability that the candidate is signal as the ratio between the value of the signal function and the total one, or background as the ratio between the value of the background function and the total one. After this first step, all events are re-spell-checked to apply the procedure to all the N candidates per event. Event by event, each candidate is considered, associating the probability for it to be signal or background (from the previous step). Then, all the products of the different probabilities associated to each candidate are computed and grouped according to the total signal yield. Consequently, it is possible to obtain that all candidates are signal, all are background, or all the intermediate situations, having for each event with N candidates a full raw probability spectrum spanning from 0 to N. Summing the result obtained for each event, it is possible to measure the spectra for each multiplicity class. In the Λ and $\overline{\Lambda}$ analysis, a feed-down correction is performed removing from the total candidates yield the fraction of Λ ($\overline{\Lambda}$) coming from the decay of Ξ^- ($\overline{\Xi}^+$) and Ξ^0. The correction for the detector response was obtained by performing a uni-dimensional Bayesian unfolding procedure [10] using a MC simulation that includes realistic strange particle p_T distributions and detector conditions. The probability of producing n K^0_S event is reported on the left in Fig. 1 for several multiplicity classes (including INEL > 0 [1]) showcasing that events with up to 7 K^0_S can be observed. The same procedure has been applied to obtain the (multi-)strange particle (antiparticle) multiplicity distribution for Λ ($\overline{\Lambda}$), Ξ^- ($\overline{\Xi}^+$) and Ω^- ($\overline{\Omega}^+$) measuring up to 5, 4 and 2 particles/event respectively and observing that, as expected at the LHC energies, the production of particles and antiparticles is the same from the highest to the lowest multiplicity class. It is worth noting that the measurement of the probability of producing n particles of a given species per event represents a unique opportunity to test the connection between charged and strange particle multiplicity production all the way to very extreme situations, e.g. spanning from events with 7 K^0_S at low average charged-particle multiplicity – where $\langle dN/d\eta \rangle_{|\eta|<0.5} \sim 3$, with potentially large fluctuations – to events with 0 K^0_S high multiplicity – where $\langle dN/d\eta \rangle_{|\eta|<0.5} \sim 20$.

3 Multiple strange hadron production yields and yield ratios

From the measurement of P(n_S) and using Eq. 1, it is possible to calculate the average probability $\langle Y_{n\text{-part}} \rangle$ for the production of n particles per event. On the right side of Fig. 1, the average production yield of 1, 2, 3, 4 and 5 K^0_S a function of the charged-particle multiplicity at midrapidity are reported respectively as black, red, blue, magenta and green markers. The increase with multiplicity is more than linear for the production of multiple strange hadrons

[1]INEL > 0 is an event class in which at least one reconstructed silicon pixel detectors tracklet is produced in the pseudorapidity interval $|\eta| < 1$ with respect to the beam, corresponding to about 75% of the total inelastic cross-section.

© The Authors, published by EDP Sciences. This is an open access article distributed under the terms of the Creative Commons Attribution License 4.0 (https://creativecommons.org/licenses/by/4.0/).

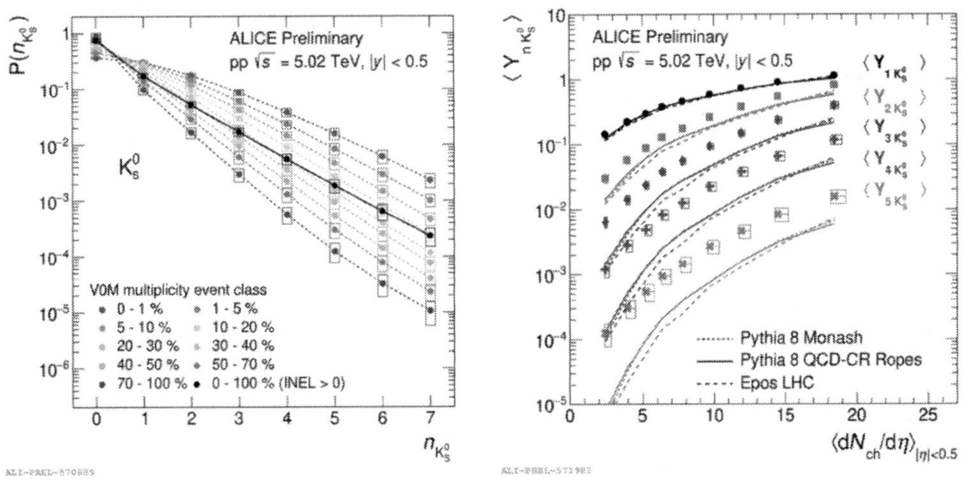

Figure 1: *Left*: K^0_S particle multiplicity distribution for several multiplicity classes. The distribution obtained for INEL > 0 has been reported in black and lines are drawn to guide the eye. *Right*: Multiple strange hadron production yields for K^0_S a function of the charged-particle multiplicity with model comparison. Model predictions obtained considering Pythia 8 Monash, Pythia 8 QCD-CR Ropes and Epos LHC are shown as dotted, continuous and dashed lines respectively.

and comparing the results with Pythia 8 Monash [11], Pythia 8 (QCD-CR) Ropes [12] and Epos LHC [13], one can see that the agreement worsens as the number of particles per event increases.

$$\langle Y_{n\text{-part}} \rangle = \sum_{i=0}^{\infty} \frac{i!}{n!(i-n)!} P_{i\text{-}part}. \tag{1}$$

Thanks to the measurement of multiple strange hadron production yields, one can explore several yield ratios to investigate the relative probability for a specific number of s-quarks to hadronize into different types of final hadrons. Firstly, yield ratios with a perfect strangeness balance ($\Delta S = 0$) as a function of the charged-particle multiplicity were considered, starting from the ratio between n Λ and n K^0_S. This is reported in the left side of Fig. 2, where an increasing pattern is observed when looking at multiple strange particle production, indicating that when the strangeness remains constant, the likelihood to obtain a baryon and not a meson rises with multiplicity. However, to better understand whether the observed increase is related to the production of the baryon itself or to the difference in mass, type or number of light quarks involved, other multiple strange yield ratios with perfect strangeness balance were considered. Figure 2 (right) shows the yield ratios between m baryons and n K^0_S, where a decreasing trend is observed when a larger number of light quarks are involved in the denominator. This is consistent with the idea that at high multiplicity it is simpler to pair s with u and d quarks, which are present in large abundance. Contrarily, in low multiplicity situations the surplus of s-quarks (fixed at 2 or 4 for the reported ratios) increases the probability of forming a multi-strange baryon. Looking at the model comparison, both Pythia 8 QCD-CR Ropes and Epos LHC can reproduce qualitatively the increasing or decreasing trends with multiplicity, but Epos LHC tends to underestimate the magnitude. Looking more closely, all trends are rather well reproduced by Pythia 8 QCD-CR Ropes suggesting that the strange quark production rate remains a puzzle, but once s

© The Authors, published by EDP Sciences. This is an open access article distributed under the terms of the Creative Commons Attribution License 4.0 (https://creativecommons.org/licenses/by/4.0/).

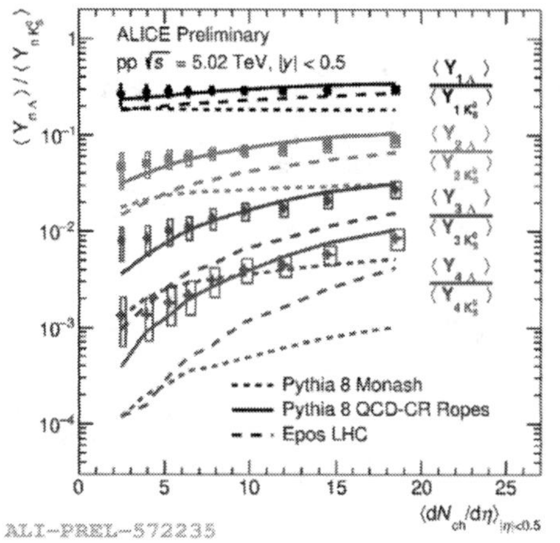

Figure 2: *Left.* Ratio $\langle Y_{n\Lambda}\rangle/\langle Y_{n\mathrm{K}^0_S}\rangle$ as a function of the charged-particle multiplicity. *Right.* Ratio $\langle Y_{m\mathrm{baryons}}\rangle/\langle Y_{n\mathrm{K}^0_S}\rangle$ as a function of the charged-particle multiplicity. In both plots there is a comparison to models: Pythia 8 Monash (dotted line), Pythia 8 QCD-CR Ropes (continuous line) and Epos LHC (dashed line).

is created the model of re-connection with light quarks catches the trends observed in the data.

In summary, the multiplicity distribution of (multi-)strange particles provides a valuable benchmark for testing production models in events with significant imbalances between charged and strange particle multiplicities. By analyzing $\mathrm{P}(n_S)$ distributions, strange hadron yields and their ratios, balanced by strangeness, can be calculated as a function of charged-particle multiplicity. Notably, the ratio $\langle Y_{n\Lambda}\rangle/\langle Y_{n\mathrm{K}^0_S}\rangle$ increases with multiplicity for$n > 1$, suggesting that factors beyond strangeness imbalance influence yield ratios. Conversely, ratios involving multi-strange baryons decline with multiplicity, implying that neither mass nor baryon number are directly tied to yield enhancement. These observed trends align with a basic quark coalescence model, and ongoing work with extreme strangeness differences (up to five) will allow to study the strangeness production at its extremes.

References

[1] ALICE Collaboration, Nat. Phys. **13**, 535-539 (2017).
[2] J.Rafelski and B.Müller, Phys. Rev. Lett. **56**, no 21 (1986).
[3] NA57 Collaboration, J. Phys. G: Nucl. Part. Phys **37**, no 4 (2010).
[4] ALICE Collaboration, Phys. Rev. B **734**, 409-410 (2014).
[5] ALICE Collaboration, Phys. Rev. B **728**, 25-38 (2014).
[6] ALICE Collaboration, Phys. Rev. B **758**, 389-401 (2016).
[7] ALICE Collaboration, Phys. Rev. C **99**, no 2, 024906 (2019).
[8] ALICE Collaboration, Eur. Phys.J.C **80**, no 2, 167 (2020).
[9] J.E. Gaiser, SLAC-0255, UMI-83-14449-MC, SLAC-R-0255, SLAC-R-255 (1982)
[10] G.D'Agostini, arXiv:1010.0632 (2010).
[11] C.Bierlich et al., JHEP 03, no 148 (2015).
[12] C.Bierlich, EPJ Web Conf. **171**, 14003 (2018).
[13] T.Pieroget al., Phys. Rev. C **92**, 034906 (2015).

© The Authors, published by EDP Sciences. This is an open access article distributed under the terms of the Creative Commons Attribution License 4.0 (https://creativecommons.org/licenses/by/4.0/).

Disoriented isospin condensates in heavy-ion collisions

Joseph Kapusta[1,*], *Scott* Pratt[2,**], and *Mayank* Singh[1,3,***]

[1]School of Physics and Astronomy, University of Minnesota, Minneapolis, MN 55455, USA
[2]Department of Physics and Astronomy and Facility for Rare Isotope Beams, Michigan State University, East Lansing, MI 48824, USA
[3]Department of Physics and Astronomy, Vanderbilt University, Nashville, TN 37240, USA

Abstract. Anomalous neutral to charged kaon correlations measured by the ALICE collaboration have defied usual explanations. We propose that the large fluctuations could arise because of a disoriented isospin condensate where there is an imbalance between up and down condensates at the time kaons hadronize. This could happen in heavy-ion collisions when the quark condensate re-forms as the system cools and the approximate chiral symmetry of QCD is broken. Within the linear sigma model, we show that the energy cost of forming a disoriented isospin condensate is small making it very plausible.

1 Introduction

The ALICE collaboration at the Large Hadron Collider (LHC) has reported anomalous fluctuations between charged and neutral kaons produced in Pb-Pb collisions at $\sqrt{s_{NN}} = 2.76$ TeV [1]. The fluctuations were significantly larger than what was predicted by conventional heavy-ion simulators like AMPT, HIJING and EPOS-LHC [2]. The measured quantity was ν−dynamic which is defined as [3]

$$\nu_{\rm dyn}(K_\pm, K_S) = \frac{\langle N_\pm(N_\pm - 1)\rangle}{\langle N_\pm\rangle^2} + \frac{\langle N_S(N_S - 1)\rangle}{\langle N_S\rangle^2} - 2\frac{\langle N_\pm N_S\rangle}{\langle N_\pm\rangle\langle N_S\rangle}. \quad (1)$$

Here, $N_\pm$ denotes the number of charged kaons while N_S is the number of neutral K-short kaons (K_S). The averaging is done over many events.

The ALICE collaboration found that the third term, the covariance between charged and neutral kaons was large and negative. This indicates that there were regions with an excess of either charged or neutral kaons relative to the other. There are a few processes like charge conservation, resonance decays or Bose symmetrization which can cause these local fluctuations [4]. But the measured $\nu_{\rm dyn}$ exceeded the expectations from these effects.

The above-mentioned effects tend to be local. Yet, the measured anomaly extended to over one unit in rapidity, indicating that these fluctuations are correlated over a long range. The fluctuations from known effects tend to be proportional to inverse multiplicity. When $\nu_{\rm dyn}$ was scaled with $\alpha = 1/\langle N_\pm\rangle + 1/\langle N_S\rangle \approx 6/N_K^{\rm tot}$, the anomaly was larger for events with more particles. Here $N_K^{\rm tot}$ is the total number of kaons and is expected to be proportional to the

*e-mail: kapusta@umn.edu
**e-mail: prattsc@msu.edu
***e-mail: mayank.singh@vanderbilt.edu

© The Authors, published by EDP Sciences. This is an open access article distributed under the terms of the Creative Commons Attribution License 4.0 (https://creativecommons.org/licenses/by/4.0/).

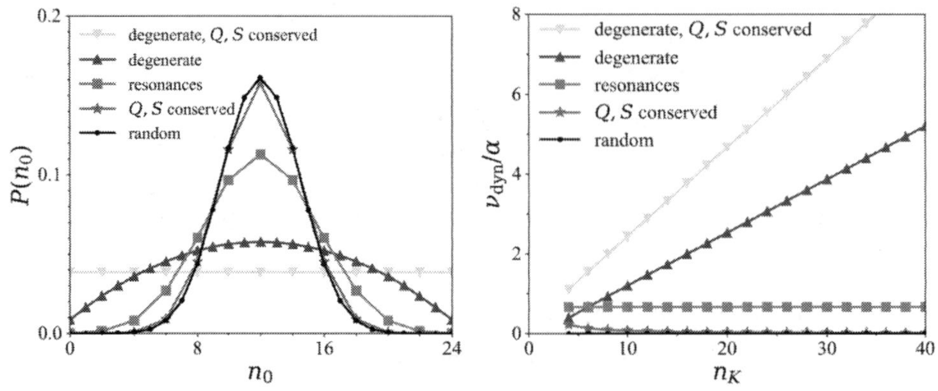

Figure 1. Probability density of n_0 kaons being neutral in a box of 24 kaons with different distributions (left). With inclusion of charge conservation, only even values of n_0 are allowed and $P(n_0)$ has been scaled by half. The ν_{dyn}/α for a single domain with n_K kaons (right).

total multiplicity. All this strongly suggests that the source of these correlations is unlikely to be local.

The disoriented isospin condensates (DIC) can explain the data. It will show a similar anomaly when ν_{dyn} is measured for a pair of particles one of which is rich in up quarks and the other in down quarks.

2 Domains of flat neutral kaon fraction

The ν_{dyn}/α can be calculated for simple kaon systems with different distributions as shown in figure 1 [5]. When the kaons are distributed randomly into its four types, then ν_{dyn} is identically zero. Charge and strangeness conservation gives it a non-zero value but that rapidly decreases to zero with increasing particle numbers. Resonance decays give a multiplicity independent ν_{dyn}/α. Within these simple systems, only a box of degenerate kaons can reproduce the experimental trend of ν_{dyn}/α increasing with multiplicity. While degenerate kaons are extremely unlikely in a heavy-ion collision it is important to note that a flat or near-flat neutral kaon fraction distribution has the right qualitative trend as compared to data. We show in the next section that DIC can produce a similar neutral kaon fraction distribution.

In a real experimental event, the kaon domains are not isolated. There can be causally disconnected volumes of kaons with different neutral kaon fractions. They have to be folded with each other and with thermal kaons which are produced with a nearly random distribution between different types. If the number of domains N_d is greater than 2, it can be shown that [3]

$$\nu_{\text{dyn}} = 4\beta_K \left(\frac{\beta_K}{3N_d} - \frac{1}{N_K^{\text{tot}}} \right). \tag{2}$$

Here N_d is the number of domains and β_K is the fraction of kaons coming from these domains. Assuming that the number of domains is proportional to the size of the system and that $\beta_K = \epsilon_\zeta V_d / m_K N_K^{\text{tot}}$ where ϵ_ζ is the energy density of domain formation, V_d is the total volume of all domains put together and m_K is the mass of a kaon, we can do a two-parameter fit to the ALICE data [5].

Figure 2 shows that a model with domains of flat neutral kaon fraction can explain the data. Only the five central points are fit as the number of domains fall below two for the last

© The Authors, published by EDP Sciences. This is an open access article distributed under the terms of the Creative Commons Attribution License 4.0 (https://creativecommons.org/licenses/by/4.0/).

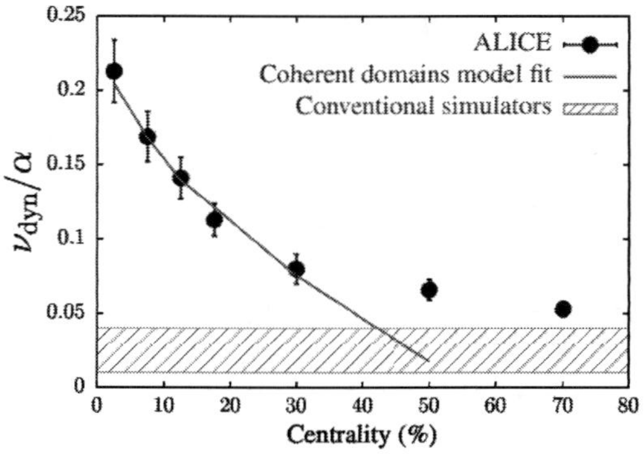

Figure 2. A model with domains of flat neutral kaon fraction can explain the data. The experimental data is from ref. [1].

two data points. In this regime eq. 2 is not valid. Also, note that the discrepancy between the data and the conventional models is small for larger centralities.

3 Disoriented isospin condensates

When the quark condensates re-form as the hot quark gluon plasma cools in heavy-ion collisions, they ultimately attain the vacuum equilibrium value where $\langle u\bar{u}\rangle = \langle d\bar{d}\rangle$. But they could have fluctuations over small volumes at finite temperatures as they re-form. These fluctuations could result in $\langle u\bar{u}\rangle \neq \langle d\bar{d}\rangle$ [6]. This can happen by combination between the isosinglet $\langle u\bar{u}\rangle + \langle d\bar{d}\rangle$ and isotriplet $\langle u\bar{u}\rangle - \langle d\bar{d}\rangle$ states. So the light quark condensate can range from all $\langle u\bar{u}\rangle$ to all $\langle d\bar{d}\rangle$. If the hadronization happens when DICs are present, they will leave their imprint on the final hadrons. If there is an excess of $\langle u\bar{u}\rangle$, then their combination with strange quarks and anti-quarks will lead to an excess of charged kaons over neutral kaons. The opposite will happen if $\langle d\bar{d}\rangle > \langle u\bar{u}\rangle$. So, if there are domains of DIC, they will lead to domains of flat neutral kaon fraction which will explain the ALICE data.

But how plausible is it to form DICs? We can calculate the statistical probability of condensate fluctuations using the 2+1 flavor linear sigma model [7]. The scalar fields can be related to the condensates as $\sigma_u = -\langle\bar{u}u\rangle/\sqrt{2}c'$, $\sigma_d = -\langle\bar{d}d\rangle/\sqrt{2}c'$, and $\zeta = -\langle\bar{s}s\rangle/\sqrt{2}c'$. Here c' is a constant. The up and down scalars can be related to σ as $\sigma_u = \sigma\cos\theta$ and $\sigma_d = \sigma\sin\theta$. Then θ ranges from 0 to $\pi/2$. The stable equilibrium value is obtained when $\theta = \pi/4$.

We can use lattice calculations to obtain the temperature dependence of quark condensates [8]. The energy cost for $\theta \neq \pi/4$ at temperature T is given as

$$\Delta U(\theta, T) = \frac{1}{2}\lambda\left[1 - \sin^2(2\theta)\right]\sigma^4 + c(T)\left[1 - \sin(2\theta)\right]\sigma^2\zeta + f_\pi m_\pi^2\left[1 - \frac{\cos\theta + \sin\theta}{\sqrt{2}}\right]\sigma. \quad (3)$$

The parameters are fixed to reproduce the known vacuum masses of mesons and are described in [5, 6]. The energy cost as a function of θ and temperature is plotted in figure 3 (left). At large temperatures, when the condensates have melted, there is negligible cost to θ fluctuations. At low temperatures, it becomes prohibitively expensive to form a DIC and θ is restricted to $\pi/4$. For intermediate temperatures, importantly around hadronization temperature, there is a finite chance of θ fluctuations.

© The Authors, published by EDP Sciences. This is an open access article distributed under the terms of the Creative Commons Attribution License 4.0 (https://creativecommons.org/licenses/by/4.0/).

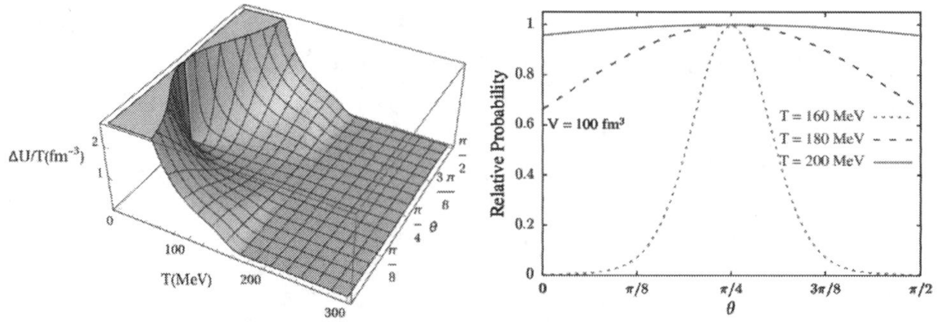

Figure 3. Energy cost of forming a DIC (left) and the relative probability for different values of θ for a domain of $V = 100$ fm^3.

We can also calculate the relative probability of having a particular θ value at different temperatures. For a domain of volume V, the relative probability is given by

$$P(\theta) = e^{-\Delta U(\theta,T)V/T}. \tag{4}$$

The relative probability is shown for $V = 100$ fm^3 in figure 3 (right). This will translate into similar plots for a neutral kaon fraction distribution relevant for the ALICE measurement.

4 Conclusions

The neutral to charged kaon fluctuations measured by ALICE collaboration cannot be accounted by known effects. They can be explained if domains of flat or near flat neutral kaon fractions exist in the system. Disoriented isospin condensates can form such domains. Thermodynamic analysis tells us that taking the system's temperature into account, the energy cost of forming DIC is really small and hence they are quite plausible.

Recently it was shown that DICs go to their equilibrium values on time scales much shorter than those relevant for heavy-ion collisions [9]. A full theory of DIC integrated with the hydrodynamic model of heavy-ion collisions can help us make predictions and constrain the physics of chiral symmetry restoration in these systems. This can be informed by lattice QCD calculations.

Acknowledgements This work was supported by the U.S. Department of Energy Grant Nos. DE-FG02-87ER40328 (JK and MS), DE-FG02-03ER41259 (SP) and DE-SC-0024347 (MS).

References

[1] S. Acharya *et al.* [ALICE], Phys. Lett. B **832**, 137242 (2022)
[2] R. Nayak, S. Dash, B. Nandi and C. Pruneau, Phys. Rev. C **101**, 054904 (2020)
[3] S. Gavin and J. I. Kapusta, Phys. Rev. C **65**, 054910 (2002)
[4] S. Cheng and S. Pratt, Phys. Rev. C **67**, 044904 (2003)
[5] J. I. Kapusta, S. Pratt and M. Singh, Phys. Rev. C **107**, 014913 (2023)
[6] J. I. Kapusta, S. Pratt and M. Singh, Phys. Rev. C **109**, L031902 (2024)
[7] J. Schaffner-Bielich and J. Randrup, Phys. Rev. C **59**, 3329-3342 (1999)
[8] A. Bazavov, *et al.* Phys. Rev. D **85**, 054503 (2012)
[9] O. Chabowski, J. I. Kapusta and M. Singh, [arXiv:2409.03711 [hep-ph]]

© The Authors, published by EDP Sciences. This is an open access article distributed under the terms of the Creative Commons Attribution License 4.0 (https://creativecommons.org/licenses/by/4.0/).

Isospin-symmetry breaking in kaon production: the role of charge-symmetry

Francesco Giacosa[1,2,*]

[1]Institute of Physics, Jan Kochanowski University, ul. Uniwersytecka 7, 25-406, Kielce, Poland
[2]Institute for Theoretical Physics, J. W. Goethe University, Max-von-Laue-Straße. 1, 60438 Frankfurt am Main, Germany

Abstract. An excess of charged versus neutral kaons has been recently reported by the NA61/SHINE collaboration. Similar excesses were also present in previous experiments, even if with larger errors. Models for hadron productions in heavy ion collisions systematically underestimate the measured charge-to-neutral kaon ratio. In this report, we comment on the role of charge-symmetry as a specific isospin transformation. In particular, we show that the ensemble of the initial colliding nuclei being charge-symmetric is sufficient for producing an equal number of charged and neutral kaons in the isospin-symmetry exact limit.

1 Introduction

A surprisingly large excess of charged over neutral kaons has been reported in Ar-Sc scattering by the NA61/SHINE experiment at CERN [1]. As argued in the companion theoretical paper in Ref. [2], two well established models for heavy ion collisions, the hadron resonance gas (HRG) and the Ultrarelativistic Quantum Molecular Dynamics (UrQMD) approaches [4, 5], are incapable of reproducing the experimental data. The HRG and UrQMD theoretical results are systematically smaller than the experimental ones, resulting in a theory-to-experiment mismatch of almost 5σ [2].

The charged-to-neutral kaon ratio is interesting since it quantifies isospin breaking. If an initial state with two nuclei containing an equal number of protons and neutrons is considered ($Q/B = 1/2$), isospin symmetry 'implies' an equal number of charged and neutral kaons. In this work we concentrate on this point. To this end, we make use of charge symmetry, which is a specific isospin symmetry that interchanges a member of a given isospin multiplet with (I, I_z) with the member $(I, -I_z)$, as for instance: $p \leftrightarrows n$, $K^+ \leftrightarrows K^0$, $\bar{K}^0 \leftrightarrows K^-$, $\pi^+ \leftrightarrows \pi^-$ etc. In particular, we show that the initial ensemble of scattering nuclei being charge-symmetry invariant leads to the prediction that the total charged and neutral kaon multiplicities are equal.

2 Charge-symmetry for the initial ensemble

The isospin operator $\hat{\mathbf{I}} = (\hat{I}_x, \hat{I}_y, \hat{I}_z)$ enters into the $SU(2)$ internal rotation described by the unitary operator $\hat{U}_I = e^{i\theta_k \hat{I}_k}$ (with $k = 1, 2, 3$): in the fundamental representation, $\hat{I}_k = \sigma_k/2$

*e-mail: fgiacosa@ujk.edu.pl

© The Authors, published by EDP Sciences. This is an open access article distributed under the terms of the Creative Commons Attribution License 4.0 (https://creativecommons.org/licenses/by/4.0/).

where σ_k are the Pauli matrices, according to which the light quark 'isodoublet' transforms as $(u,d)^t \to U_I(u,d)^t$. As a consequence, the quark u has eigenvalue $I_z = 1/2$, the quark d the eigenvalue $I_z = -1/2$ and both quarks carry the eigenvalue $I(I+1) = 3/2$ for the operator $\hat{\mathbf{I}}^2$, thus $I = 1/2$. The antiquark doublet $(-\bar{d}, \bar{u})^t$ transforms just as the quark one.

Kaons are grouped into the isodoublet ($I = 1/2$) states ($K^+ \equiv u\bar{s}, K^0 \equiv d\bar{s}$) and ($-\bar{K}^0 \equiv -s\bar{d}, K^- \equiv s\bar{u}$) that correspond naturally to the underlying isospin doublets (u,d) and $(-\bar{d}, \bar{u})$. Namely, they transform as (anti)quarks because the s-quark is left invariant by isospin.

Isospin is well conserved in low-energy QCD processes: the particle masses well reflect isospin symmetry [3], e.g. $(m_{\pi^+} - m_{\pi^0}) / (m_{\pi^+} + m_{\pi^0}) \simeq 0.017$ and $(m_{K^+} - m_{K^0}) / (m_{K^+} + m_{K^0}) \simeq -0.004$, and the treatment of pion-pion, pion-kaon, pion-nucleon, and nucleon-nucleon scattering closely follows the predictions of isospin symmetry [7, 8], see also the low-energy phenomenology within the extended linear sigma model [9] (for a recent review see Ref. [10]).

A specific isospin transformation, called charge symmetry $\hat{C}_I$ (not be confused with charge conjugation $\hat{C}$ that swaps particles in antiparticles) is realized by choosing $\hat{C}_I = e^{i\pi \hat{I}_y}$ (thus for $\theta_{1,3} = 0$ and $\theta_2 = \pi$). For the (anti)quarks and kaon doublets, $\hat{C}_I$ takes the matrix form

$$\hat{C}_I = \begin{pmatrix} 0 & 1 \\ -1 & 0 \end{pmatrix}, \tag{1}$$

thus interchanging $u \leftrightarrow d$ and $\bar{u} \leftrightarrow \bar{d}$, and conversely $K^+ \leftrightarrow K^0$ and $K^- \leftrightarrow \bar{K}^0$.

Next, we move to heavy-ion collisions and the related production of kaons. Let us consider, for simplicity, a certain reaction to the production of a single K^+: $A + A \to K^+X$, where X refers to other particles (but no kaons). Let us first assume that the initial state has total isospin $I = 0$. This case corresponds to both nuclei having $I = 0$. Indeed, nuclei usually appear in the lowest possible isospin state [11]. The corresponding matrix element is $A^{(0)}_{K^+X} = \left\langle K^+X|\hat{S}_{QCD}|AA, I = 0\right\rangle$, where $\hat{S}_{QCD}$ refers to the $\hat{S}$-matrix of QCD. Isospin symmetry implies that $\hat{U}_I^\dagger \hat{S}_{QCD} \hat{U}_I \simeq \hat{S}_{QCD}$. Considering the charge transformation $\hat{U}_I = \hat{C}_I$ we get:

$$A^{(0)}_{K^+X} = \left\langle K^+X|\hat{S}_{QCD}|AA, I = 0\right\rangle \simeq \left\langle K^0\tilde{X}|\hat{S}_{QCD}|AA, I = 0\right\rangle = A^{(0)}_{K^0\tilde{X}}, \tag{2}$$

where $\hat{C}_I |AA, I = 0\rangle = |AA, I = 0\rangle$ and $\hat{C}_I |X\rangle = |\tilde{X}\rangle$ have been used, the latter being the charge–symmetry rotated 'rest'. Above, $A^{(0)}_{K^0\tilde{X}}$ is the amplitude for the $K^0\tilde{X}$ formation. Thus, the (approximate) equality $A^{(0)}_{K^+X} \simeq A^{(0)}_{K^0\tilde{X}}$ implies that $A + A \to K^0\tilde{X}$ is equally probable as $A + A \to K^+X$. Extending the reasoning to any reaction involving kaons, one find that the number of produced K^+ equals the number of produced K^0 ($\langle K^+\rangle = \langle K^0\rangle$.). The same argument applied to the isodoublet pair K^- and $\bar{K}^0$ leads to $\langle K^-\rangle = \langle \bar{K}^0\rangle$.

Next, we consider a different initial state. For simplicity, for total isospin $I = 1$ one has

$$A^{(1)}_{K^+X} = \left\langle K^+X|\hat{S}_{QCD}|AA, I = 1\right\rangle \simeq -\left\langle K^0\tilde{X}|\hat{S}_{QCD}|AA, I = 1\right\rangle = -A^{(1)}_{K^0\tilde{X}}, \tag{3}$$

with $\hat{C}_I |AA, I = 1\rangle = -|AA, I = 1\rangle$. Clearly, the opposite sign does not change the final outcome for the cross-sections: $\langle K^+\rangle = \langle K^0\rangle$ and, analogously, $\langle K^-\rangle = \langle \bar{K}^0\rangle$ hold also in this case. This result can be easily generalized to any initial state with definite fixed isospin $I = I_0$ with $\hat{C}_I |AA, I = I_0\rangle = (-1)^{I_0} |AA, I = I_0\rangle$.

The arguments above do not, however, imply that $\langle K^+\rangle = \langle K^0\rangle$ and $\langle K^-\rangle = \langle \bar{K}^0\rangle$ hold in the most general case. As an illustrative example, we consider the following superposed initial state:

$$|AA\rangle = \alpha_0 |AA, I = 0\rangle + \alpha_1 |AA, I = 1\rangle \,. \tag{4}$$

© The Authors, published by EDP Sciences. This is an open access article distributed under the terms of the Creative Commons Attribution License 4.0 (https://creativecommons.org/licenses/by/4.0/).

The amplitude for K^+X production is given by

$$A_{|AA\rangle\to K^+X} = \alpha_0 A^{(0)}_{K^+X} + \alpha_1 A^{(1)}_{K^+X} \to \left|A_{|AA\rangle\to K^+X}\right|^2 = \left|\alpha_0 A^{(0)}_{K^+X} + \alpha_1 A^{(1)}_{K^+X}\right|^2 . \tag{5}$$

Using analogous steps as before:

$$A_{|AA\rangle\to K^0\tilde{X}} = \alpha_0 A^{(0)}_{K^0\tilde{X}} + \alpha_1 A^{(1)}_{K^0\tilde{X}} = \alpha_0 A^{(0)}_{K^+X} - \alpha_1 A^{(1)}_{K^+X} \to \left|A_{|AA\rangle\to K^0\tilde{X}}\right|^2 = \left|\alpha_0 A^{(0)}_{K^+X} - \alpha_1 A^{(1)}_{K^+X}\right|^2 . \tag{6}$$

Because of the different signs $\left|A_{|AA\rangle\to K^+X}\right|^2 \neq \left|A_{|AA\rangle\to K^0\tilde{X}}\right|^2$, implying that the production probabilities for K^+ and K^0 are different for the initial state $|AA\rangle$ of Eq. 4.

At this point, we assume that the initial ensemble is charge-symmetric. That means that the state

$$|AA\rangle_{C_I} = C_I\,|AA\rangle = \alpha_0\,|AA, I = 0\rangle - \alpha_1\,|AA, I = 1\rangle \tag{7}$$

occurs as often as the $|AA\rangle$ state of Eq. (4). By repeating the previous passages:

$$\left|A_{|AA\rangle_{C_I}\to K^+X}\right|^2 = \left|\alpha_0 A^{(0)}_{K^+X} - \alpha_1 A^{(1)}_{K^+X}\right|^2 , \quad \left|A_{|AA\rangle_{C_I}\to K^0\tilde{X}}\right|^2 = \left|\alpha_0 A^{(0)}_{K^+X} + \alpha_1 A^{(1)}_{K^+X}\right|^2 . \tag{8}$$

Since $|AA\rangle$ and $|AA\rangle_{C_I}$ are equally probable, the final K^+X rate is proportional to:

$$\frac{1}{2}\left|A_{|AA\rangle\to K^+X}\right|^2 + \frac{1}{2}\left|A_{|AA\rangle_{C_I}\to K^+X}\right|^2 = \left|\alpha_0 A^{(0)}_{K^+X}\right|^2 + \left|\alpha_1 A^{(1)}_{K^+X}\right|^2 , \tag{9}$$

where the interference terms disappear and equals the $K^0\tilde{X}$ production rate:

$$\frac{1}{2}\left|A_{|AA\rangle\to K^0\tilde{X}}\right|^2 + \frac{1}{2}\left|A_{|AA\rangle_{C_I}\to K^0\tilde{X}}\right|^2 = \left|\alpha_0 A^{(0)}_{K^+X}\right|^2 + \left|\alpha_1 A^{(1)}_{K^+X}\right|^2 . \tag{10}$$

Thus, upon averaging, the final state K^+X appears just as often as $K^0\tilde{X}$. Of course, this result applies if the initial ensemble is charge-symmetric. This point is well upheld because nuclei can be well described as clusters of α-particles [12].

These arguments can be repeated for any reaction involving kaons. Upon averaging over the initial states of a charge-symmetric ensemble guarantees that $\langle K^+\rangle = \langle K^0\rangle$ (and similarly that $\langle K^-\rangle = \langle \bar{K}^0\rangle$). Then, the charged and neutral multiplicities coincide:

$$\langle K^+\rangle + \langle K^-\rangle = \langle K^0\rangle + \langle \bar{K}^0\rangle . \tag{11}$$

The neutral kaons are not directly measured, since the physical states are the K_L and K_S which, by neglecting the (very) small CP violation, are given by: $K_{S(L)} = (K^0 \mp \bar{K}^0)/\sqrt{2}$. The emerging number of K_S and K_L is given by: $\langle K_S\rangle = \frac{1}{2}\langle K^0\rangle + \frac{1}{2}\langle \bar{K}^0\rangle = \langle K_L\rangle$. The final expected relation is $\langle K^+\rangle + \langle K^-\rangle = 2\langle K_S\rangle$, implying that the ratio

$$R_K = \frac{\langle K^+\rangle + \langle K^-\rangle}{2\langle K_S\rangle} = 1 \tag{12}$$

for colliding nuclei in the isospin-symmetric limit. This is why in Refs. [1, 2] the ratio R_K is studied both experimentally and theoretically.

3 Summary

In this work, we have shown that a charge-symmetric initial ensemble implies an equal number of charged and neutral kaons in the isospin-exact limit. As a consequence, the ratio R_K

© The Authors, published by EDP Sciences. This is an open access article distributed under the terms of the Creative Commons Attribution License 4.0 (https://creativecommons.org/licenses/by/4.0/).

of Eq. (12) is unity. Violations from this prediction appear because isospin symmetry is not exact. First of all, the mass of the charged kaons (493.677 MeV) is slightly smaller than the neutral ones (496.611 MeV), favoring charged kaons. This difference also affects resonance decays. A typical example is the decay of the $\phi(1020)$ meson into kaons, for which $\phi(1020) \to K^+K^-$ is 1.4 times larger than $\phi(1020) \to K^0\,\bar{K}^0$. Yet, the sum of all these effects leads only to a few percent modification of the relation $R_K = 1$. At high scattering energies ($\sqrt{s_{NN}} \gtrsim 15$ GeV), popular models predict that that $R_K \simeq 1.05$, which is sizably smaller than the central experimental values ($R_K \simeq 1.18$) of Ref. [1].

In the actual realized experiments the ratio $Q/B \simeq 0.4 < 1/2$. The surplus of neutrons favors the production of neutral kaons. This effect is small at high energies but becomes important at small ones ($\sqrt{s_{NN}} \lesssim 10$ GeV). The final mismatch of theory and experiment is quantified in Ref. [2] finding an overall $\sim 5\sigma$ effect. Future efforts to explain this 'kaon anomaly' are needed. On the theoretical side, one may explore models that may contain strong isospin breaking [13] (a possible role could be played by electromagnetic interaction). On the experimental side, the realization of experiments with nuclei with $Q/B = 1/2$ is highly desirable.

Acknowledgments The author thanks the co-authors of Ref. [2], A. Rybicki, L. Turko, and S. Mrówczyński for very useful discussions.

References

[1] H. Adhikary *et al.* [NA61/SHINE], [arXiv:2312.06572 [nucl-ex]].

[2] W. Brylinski, M. Gazdzicki, F. Giacosa, M. Gorenstein, R. Poberezhnyuk, S. Samanta and H. Stroebele, [arXiv:2312.07176 [nucl-th]].

[3] S. Navas *et al.* [Particle Data Group], Phys. Rev. D **110** (2024) no.3, 030001 doi:10.1103/PhysRevD.110.030001

[4] M. Bleicher and E. Bratkovskaya, Prog. Part. Nucl. Phys. **122** (2022), 103920 doi:10.1016/j.ppnp.2021.103920

[5] V. Vovchenko, M. I. Gorenstein and H. Stoecker, Phys. Rev. C **98** (2018) no.3, 034906 doi:10.1103/PhysRevC.98.034906

[6] M. Gazdzicki and O. Hansen, Nucl. Phys. A **528** (1991), 754-770 doi:10.1016/0375-9474(91)90261-4

[7] M. R. Pennington, J. Phys. Conf. Ser. **18** (2005), 1-73 doi:10.1088/1742-6596/18/1/001

[8] J. M. Alarcon, J. Martin Camalich and J. A. Oller, Phys. Rev. D **85** (2012), 051503 doi:10.1103/PhysRevD.85.051503

[9] D. Parganlija, P. Kovacs, G. Wolf, F. Giacosa and D. H. Rischke, Phys. Rev. D **87** (2013) no.1, 014011 doi:10.1103/PhysRevD.87.014011

[10] F. Giacosa, P. Kovács and S. Jafarzade, [arXiv:2407.18348 [hep-ph]].

[11] S. M. Lenzi and M. A. Bentley Springer, Berlin, Heidelberg (2009). https://doi.org/10.1007/978-3-540-85839-3 3

[12] T. Otsuka *et al*, Nature Commun. **13** (2022) no.1, 2234 doi:10.1038/s41467-022-29582-0

[13] J. Rafelski and M. Petran, Phys. Part. Nucl. **46** (2015) no.5, 748-755 doi:10.1134/S1063779615050238 [arXiv:2212.13302 [hep-ph]].

© The Authors, published by EDP Sciences. This is an open access article distributed under the terms of the Creative Commons Attribution License 4.0 (https://creativecommons.org/licenses/by/4.0/).

Production of Σ baryons as a function of multiplicity in pp collisions at the LHC with ALICE

Pavel Gordeev[1,*] for the ALICE Collaboration

[1]IUPAP

Abstract. In recent year, the production of charged Σ hyperons in pp and p–Pb collisions was measured by the ALICE collaboration using several techniques. Historically, the first one used the decay channel $\Sigma^+ \to p\pi^0$ with reconstruction of the proton with the tracking systems at mid-rapidity and the π^0 in the electromagnetic calorimeter EMCAL or via the photon conversion method. The second technique used the decay channel $\bar{\Sigma}^\pm \to \bar{n}\pi^\pm$ where charged pions are reconstructed with the tracking systems and the anti-neutron in the precise electromagnetic calorimeter PHOS. We present the transverse momentum spectra of charged Σ and its anti-particle measured in pp collisions at $\sqrt{s}$ = 13 and $\sqrt{s}$ = 5 TeV and compare them to predictions of several models. Finally, we discuss expectations from the Run 3 of LHC and in particular the possibility of Σ particle detection with the reconstruction of its decay after traversing several layers of the upgraded silicon tracker.

1 Introduction

The strangeness content of the final state in ultrarelativistic heavy-ion collisions has been studied so far through measurements of kaons, Λ, Ξ and Ω baryons in pp, pA and AA collisions, but not with Σ baryons yet. To date only Σ^0 in 7 TeV pp collisions have been measured by ALICE, while few other experiments have measured the charged states at lower pp ($p\bar{p}$) collision energies [1, 2].

Σ baryons contain a single strange quark and form a triplet, with the electric charge defined by its light quark content. In a thermal model, these states are abundant enough to carry a significant fraction of the strangeness produced in the collision. Therefore, the measurement of these states is interesting for the comparison with thermal model predictions. Measurement of the yield of Σ hyperons is also important to control feed-down to lighter states.

A reliable method of reconstructing and identifying Σ baryons is also important to the study of nucleon-Σ correlations, which will shed light on the presence of Σ-hyperons in neutron stars and constrain their Equation-of-State. So far, only p–Σ^0 interaction via femtoscopy was measured [3]. The measured correlation function is consistent with the p–(Λγ) baseline, which indicates the presence of an overall shallow strong potential, but with current uncertainties it is not possible to discriminate between different models. The experimental measurement is a challenging task because all decays of all Σ states involve neutral decay products, thus requiring high-resolution calorimeters or usage of photon conversion method (PCM).

*e-mail: pavel.gordeev@cern.ch

© The Authors, published by EDP Sciences. This is an open access article distributed under the terms of the Creative Commons Attribution License 4.0 (https://creativecommons.org/licenses/by/4.0/).

2 ALICE experiment in Run 2

The detailed description of the ALICE experiment can be found in [4]. The current analysis uses the Inner Tracking System (ITS) and Time Projection Chamber (TPC) for the reconstruction of tracks of charged particles. Charged particle identification is done via specific energy loss in TPC and with Time-Of-Flight (TOF) detector. The electromagnetic calorimeters, such as PHOton Spectrometer (PHOS), Electro Magnetic Calorimeter (EMCAL and DCAL), are used for reconstructing photons and anti-neutrons from decays. For multiplicity estimation, the V0 and Silicon Pixel Detectors (SPD) are used.

2.1 Σ^0 measurement in pp collisions at 7 TeV

Σ^0 is reconstructed via its decay into a photon and a Λ. Λ baryons are reconstructed by combining pairs of protons and pions, while photons are reconstructed via the PCM by electron-positron pair. Then, an invariant mass distribution is constructed using selected photon and Λ candidates (Fig. 1 (left panel)). This measurement provides a Σ^0 to Λ integrated yield ratio which complements world data from lower energies (Fig. 1 (right)).

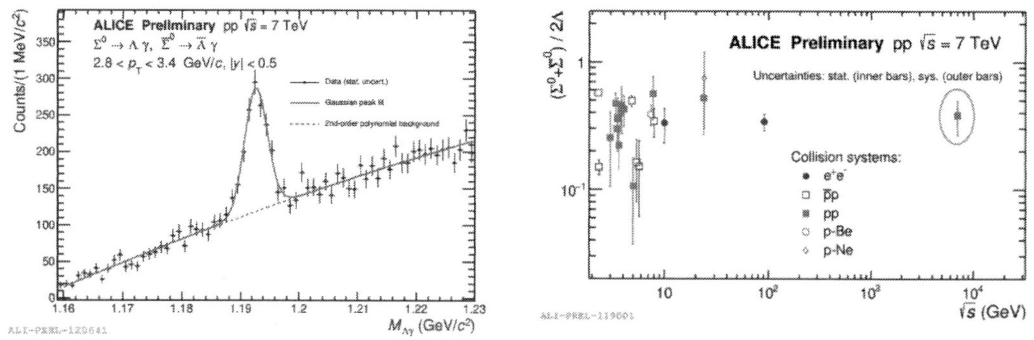

Figure 1. Left: invariant mass distribution of Λ and γ pairs in pp collisions at $\sqrt{s}$ = 7 TeV. Right: Σ^0 to Λ ratio as a function of $\sqrt{s}$ from different experiments

2.2 Σ^+ and $\bar{\Sigma}^-$ measurement in pp collisions at 13 TeV

The measurement of Σ^+ and $\bar{\Sigma}^-$ can be done via π^0 and p channel. The proton is detected directly in the tracking system, the π^0 is reconstructed through two-photon decay. Photons can be detected either in calorimeters or via PCM. Both PCM-Calo and PCM-PCM techniques give results that are in agreement with each other.

One can see comparison of resulting spectrum obtained in Minimum Bias pp collisions with different Monte-Carlo (MC) generators (Fig. 2). The spectrum is well reproduced by the EPOS LHC [5] generator, while PYTHIA8 [6] catches the shape, but underestimates the yield. PYTHIA6 does not describe the shape of the spectrum at low transverse momentum and underestimates the yield.

2.3 $\bar{\Sigma}^+$ and $\bar{\Sigma}^-$ measurement in pp and p–Pb collisions at 5 TeV

A more challenging experimental task is the measurement of $\bar{\Sigma}^\pm$ through the anti-neutron and charged pion channel. There is no hadron calorimeter to capture neutrons in the ALICE experiment, but one can try to use electromagnetic calorimeter to reconstruct anti-neutrons.

© The Authors, published by EDP Sciences. This is an open access article distributed under the terms of the Creative Commons Attribution License 4.0 (https://creativecommons.org/licenses/by/4.0/).

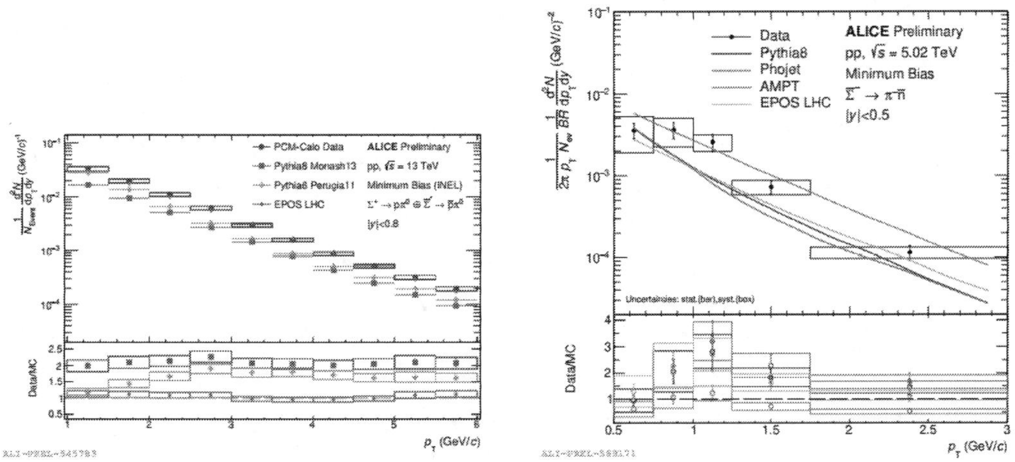

Figure 2. Left: Σ^+ ($\bar{\Sigma}^-$) spectrum in pp collisions at $\sqrt{s}$ = 13 TeV and comparison with different MC models. Right: $\bar{\Sigma}^-$ spectrum in pp collisions at $\sqrt{s}$ = 5.02 TeV and comparison with different MC models

Anti-neutron identification is done using the following parameters: the energy released during annihilation in the calorimeter, the neutrality of the cluster (clusters from charged particles are suppressed), and the dispersion of the cluster (shape of the cluster and number of cells).

However, it is not possible to measure an antineutron momentum directly. Therefore, time-of-flight information from PHOS is used to reconstruct the momentum. After that, $\bar{n}$ and $\pi^\pm$ candidates are combined to reconstruct secondary vertices to apply the topological selections that are used to increase the signal to background ratio.

$\bar{\Sigma}^\pm$ were measured both in pp and p–Pb collisions at 5.02 TeV. In Fig. 2 one can see $\bar{\Sigma}^-$ spectrum with different MC predictions in pp collisions at 5.02 TeV. EPOS LHC, PYTHIA8 and PHOJET [7] show good agreement with data points within large uncertainties. The AMPT [8] model have a good agreement at low p_T and overestimates the yield at high p_T.

3 Kink topology method in Run 3

The main changes from Run 2 to Run 3 are upgraded detectors to handle continuous readout and improved vertexing capabilities in the central barrel. In addition, an increase of statistics allows more precise measurements.

Σ-hyperons can be reconstructed making use of their kink-topology, reconstructed in the upgraded ITS (ITS2). The new ITS2 uses CMOS Monolithic Active Pixel Sensors, which significantly improve the impact parameter resolution and the tracking efficiency, especially for particles with low transverse momenta, as well as the readout-rate capability. This method is based on searching signals from particles in different detector layers, and requires more statistics to be effective, which Run 3 can provide. In Fig. 3 one can see invariant mass distributions of charged Σ baryons obtained for differently charged pions. However, two differently charged Σ cannot be distinguished from each other using this method. The ITS3 upgrade will help to reduce the width of the signal peaks.

© The Authors, published by EDP Sciences. This is an open access article distributed under the terms of the Creative Commons Attribution License 4.0 (https://creativecommons.org/licenses/by/4.0/).

4 Conclusion

For the first time at the LHC, the production of charged Σ-hyperons was measured. The method for anti-neutron reconstruction was proposed, which opens up a variety of new observables. Obtained results are consistent within uncertainties with EPOS LHC predictions and can be used to constrain other MC generators. More precise measurement of Σ-hyperons, Σ-hypernuclei search and hadron-Σ interactions measurement is foreseen at LHC with ALICE in Run 3 in 2022–2025.

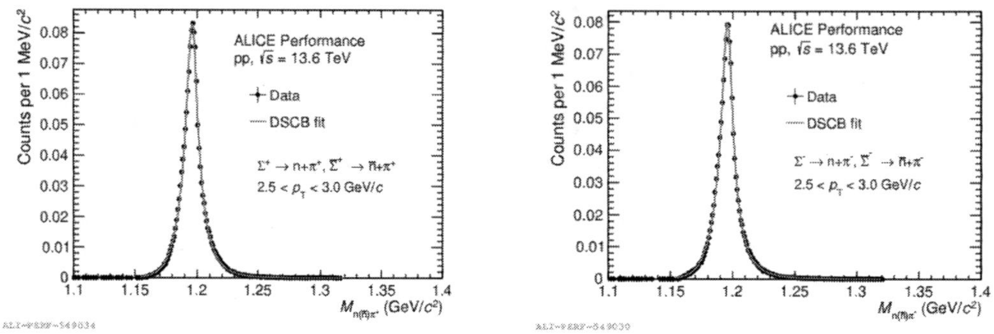

Figure 3. Invariant mass distribution of $n(\bar{n})$ and $\pi^{\pm}$ pairs in pp collisions at $\sqrt{s} = 13.6$ TeV

References

[1] L3 Collaboration, M. Acciarri et al., Inclusive Σ^{+} and Σ^{0} production in hadronic Z decays. Phys. Lett. B **479**, 79–88 (2000). https://doi.org/10.1016/S0370-2693(00)00369-5

[2] OPAL Collaboration, G. Alexander et al., Sigma+, Sigma0 and Sigma- hyperon production in hadronic Z0 decays. Z Phys C - Particles and Fields **73**, 587–599 (1997). https://doi.org/10.1007/s002880050350

[3] ALICE Collaboration, Acharya, Shreyasi et al, Investigation of the p–Σ0 interaction via femtoscopy in pp collisions. Phys. Lett. B **805**, 135419 (2020). https://doi.org/10.1016/j.physletb.2020.135419

[4] ALICE Collaboration, B. Abelev et al., Performance of the ALICE Experiment at the CERN LHC. Int. J. Mod. Phys. **A29**, 1430044 (2014). https://dx.doi.org/10.1142/S0217751X14300440

[5] T. Pierog, I. Karpenko, J. M. Katzy, E. Yatsenko, and K. Werner, EPOS LHC: Test of collective hadronization with data measured at the CERN Large Hadron Collider. Phys. Rev. C **92** no. 3, 034906 (2015). https://doi.org/10.1103/PhysRevC.92.034906

[6] C. Bierlich et al., A comprehensive guide to the physics and usage of PYTHIA 8.3. SciPost Phys. Codeb. **2022**, 8 (2022). https://doi.org/10.21468/SciPostPhysCodeb.8

[7] F. W. Bopp, R. Engel, and J. Ranft, Rapidity gaps and the PHOJET Monte Carlo. LAFEX International School on High-Energy Physics (LISHEP 98) Session A: Particle Physics for High School Teachers - Session B: Advanced School in HEP - Session C: Workshop on Diffractive Physics **3**, 729–741 (1998). https://doi.org/10.48550/arXiv.hep-ph/9803437

[8] Lin, Zi-Wei, et al., Multiphase transport model for relativistic heavy ion collisions. Physical Review C — Nuclear Physics **72(6)**, 064901 (2005). https://doi.org/10.1103/PhysRevC.72.064901

© The Authors, published by EDP Sciences. This is an open access article distributed under the terms of the Creative Commons Attribution License 4.0 (https://creativecommons.org/licenses/by/4.0/).

EPJ Web of Conferences **316**, 03004 (2025) https://doi.org/10.1051/epjconf/202531603004
SQM 2024

A Deep Learning Based Estimator for Light Flavour Elliptic Flow in Heavy Ion Collisions at LHC Energies

Gergely Gábor Barnaföldi[1,*], *Neelkamal* Mallick[1,2,3], *Suraj* Prasad[1,2], *Raghunath* Sahoo[2], and *Aditya* Nath Mishra[4]

[1]HUN-REN Wigner Research Center for Physics, P.O. Box 49, 1125 Budapest, Hungary
[2]Department of Physics, Indian Institute of Technology Indore, Simrol, Indore 453552, India
[3]University of Jyväskylä, Department of Physics, P.O. Box 35, FI-40014, Jyväskylä, Finland
[4]University Centre of Research and Development Department, Chandigarh University, Gharuan, Mohali-140413, Punjab, India

Abstract. We developed a deep learning feed-forward network for estimating elliptic flow (v_2) coefficients in heavy-ion collisions from RHIC to LHC energies. The success of our model is mainly the estimation of v_2 from final state particle kinematic information and learning the centrality and the transverse momentum (p_T) dependence of v_2 in wide p_T regime. The deep learning model is trained with AMPT-generated Pb-Pb collisions at $\sqrt{s_{NN}}$ = 5.02 TeV minimum bias events. We present v_2 estimates for $\pi^{\pm}$, $K^{\pm}$, and $p + \bar{p}$ in heavy-ion collisions at various LHC energies. These results are compared with the available experimental data wherever possible.

1 Introduction

Ultrarelativistic heavy-ion collisions have been studied extensively for decades in experiments at the Relativistic Heavy Ion Collider (RHIC BNL) and at the Large Hadron Collider (LHC CERN). In these collisions, a deconfined thermalized medium of quarks and gluons can be formed [1]. This medium of hot and dense state of the strongly interacting matter is called the quark-gluon plasma (QGP). Only signatures of the formation of QGP could be studied using various indirect effects such as jet quenching, strangeness enhancement, and quarkonia suppression since no direct observation is possible due to the short lifetime of the strongly interacting matter.

Transverse collective flow is another key observable, which is widely studied to investigate the properties of QGP in heavy-ion collisions [2]. This observable depends on the equation of state and transport coefficients of the system. Anisotropic flow signifies the formation of QGP medium in noncentral relativistic heavy-ion collisions. The pressure gradient formed in the hot and dense medium due to the initial spatial anisotropy can transform into final state momentum space azimuthal anisotropy. This momentum anisotropy could be expressed as the coefficients of the Fourier expansion of the azimuthal momentum distribution of the produced particles. The second-order flow coefficient is the so-called elliptic flow (v_2). Finite azimuthal anisotropy has been well observed in heavy-ion collision experiments so far

*e-mail: Barnafoldi.Gergely@wigner.hun-ren.hu

© The Authors, published by EDP Sciences. This is an open access article distributed under the terms of the Creative Commons Attribution License 4.0 (https://creativecommons.org/licenses/by/4.0/).

at RHIC and LHC energies up to higher-order cumulants with various analysis methods [3–6]. Here, we present our deep learning feed-forward network for estimating elliptic flow (v_2) coefficients, which we compare to heavy-ion collision data from RHIC to LHC energies.

2 The Model and the DNN architecture

Anisotropic flow can be measured and quantified by the coefficients of Fourier expansion of the azimuthal momentum distribution, given by [7]:

$$\frac{dN}{d\phi} = \frac{1}{2\pi}\left(1 + 2\sum_{n=1}^{\infty} v_n \cos\left[n(\phi - \psi_n)\right]\right) \text{ with } v_n = \langle\cos[n(\phi - \psi_n)]\rangle\,, \quad (1)$$

where, v_n denotes n^{th} order anisotropic flow coefficient, ϕ is the azimuthal angle, and ψ_n is the corresponding harmonic symmetry plane angle. In order to calculate the elliptic flow event-by-event, we have used the event plane method [8], and for simplicity, we have fixed the reaction plane angle, $\psi_R = 0$, which results in $v_2 = \langle\cos(2\phi)\rangle$.

A deep learning-based machine learning algorithm was developed to estimate the elliptic flow event-by-event. For training the deep neural network (DNN), we have used a multiphase transport (AMPT) model to simulate the dataset. AMPT is a Monte Carlo-based event simulator that is used to generate ultrarelativistic nucleus-nucleus collisions at RHIC and LHC energies [9]. AMPT has four components, namely, initialization of collisions by HIJING [10], parton transport by Zhang's Parton Cascade model [11], hadronization of the partons performed by spatial coalescence mechanism in string melting mode and Lund string fragmentation model in the default version of AMPT [12, 13], and finally, the hadron transport using a relativistic transport model [14, 15]. The DNN was trained with Pb-Pb collisions at $\sqrt{s_{NN}} = 5.02$ TeV minimum bias events with all charged particles having $p_T > 0.15$ GeV/c in pseudorapidity, $|\eta| < 0.8$.

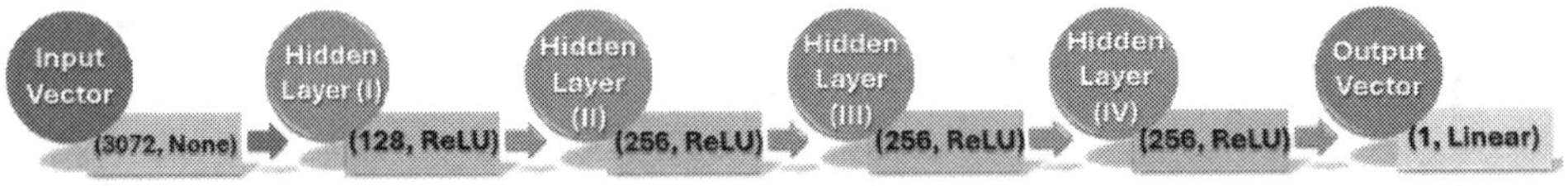

Figure 1. The structure of the DNN architecture used for the v_2 estimation. The number of nodes and the type of activation function used in each layer are denoted.

For this regression problem, the DNN consists of one input layer, four hidden layers, and one output layer, as depicted in Fig. 1. The inputs are given as "pictures" with pixels of normalized transverse momenta, mass and energy values on the pseudorapidity-azimuthal plane. This input with 3072 features are mapped to the first dense layer with 128 nodes, which is connected to the output layer via three hidden layers in succession, each having 256 nodes. The dense layers use the rectified linear unit as the activation function, and the output layer has a single node with a linear activation function. The DNN model uses the *adam* optimizer with *mean squared error* loss function. Details can be found in Refs. [16, 17].

3 Comparing DNN predictions to AMPT and to experimental data

The elliptic flow, $v_2(p_T)$ for identified hadrons: $\pi^{\pm}$, $K^{\pm}$, and $p+\bar{p}$ in nucleus-nucleus collisions at $\sqrt{s_{NN}} = 2.76$ TeV (Pb-Pb), 5.02 TeV (Pb-Pb), and 5.44 TeV (Xe-Xe) is plotted in Fig. 2

© The Authors, published by EDP Sciences. This is an open access article distributed under the terms of the Creative Commons Attribution License 4.0 (https://creativecommons.org/licenses/by/4.0/).

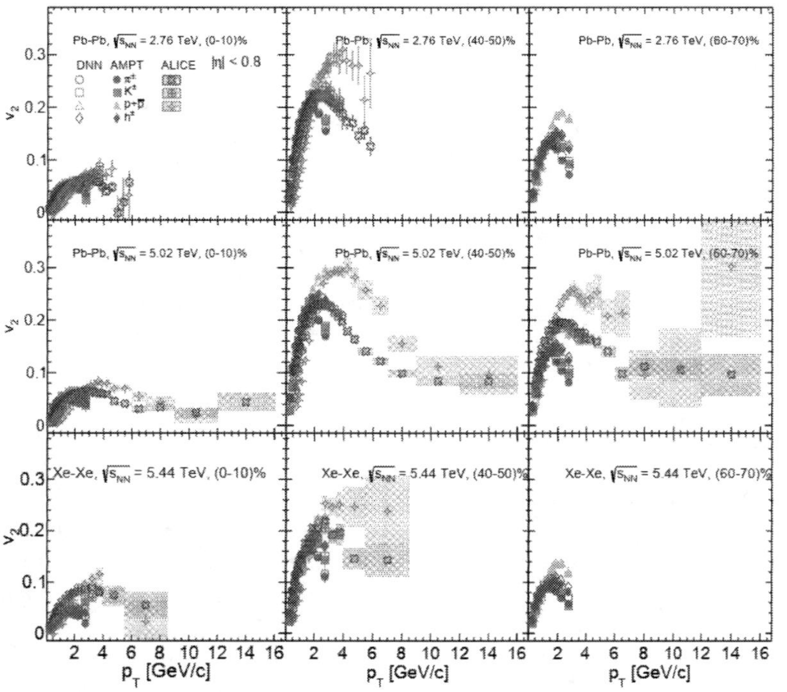

Figure 2. Centrality dependence of $v_2(p_T)$ for $\pi^{\pm}$, $K^{\pm}$, and $p + \bar{p}$, and all charged hadrons ($h^{\pm}$). The values from AMPT and the predictions from DNN are shown in comparison with ALICE data [18–20].

from top to bottom, respectively. To be comparable with ALICE data [18–20], the AMPT and DNN were simulated with tracks with $p_T > 0.5$ GeV/c and in midrapidity, $|y| < 0.5$. Three selected collision centrality ranges were used for the plots: 0-10%, 40-50% and 60-70%.

One can see from the results, that all particle species behave similarly: the magnitude of $v_2(p_T)$ increases with increasing p_T until it reaches a maximum value around $p_T \approx 2.0$ GeV/c, and then it starts to decrease beyond this point. The values of $v_2(p_T)$ from AMPT obtained in this region (*i.e.* $p_T \lesssim 2.0$–3.0 GeV/c) is comparable in magnitude with ALICE results for the individual particle cases. However, beyond this p_T value, AMPT fails to describe the data as $v_2(p_T)$ falls faster with increasing p_T since fragmentation takes over from coalescence at high p_T.

DNN predictions agree with AMPT values quite nicely up to $p_T \lesssim 4.0$–6.0 GeV/c. Beyond this p_T, the values from DNN start to differ from the AMPT calculated data points. The reason behind that is the statistics, which decreases for the higher $p_T \gtrsim 6.0$ GeV/c regions. The limited number of event provided less instances to the DNN model during the training process. For this reason, the mismatch between DNN and AMPT comes into the picture beyond $p_T \gtrsim 6.0$ GeV/c.

4 Discussion and Summary

In Refs. [16, 17], we demonstrated the applicability of a DNN-based machine learning model to evaluate the second-order anisotropic flow coefficient (v_2) event-by-event for identified

© The Authors, published by EDP Sciences. This is an open access article distributed under the terms of the Creative Commons Attribution License 4.0 (https://creativecommons.org/licenses/by/4.0/).

hadrons from final state particle kinematic information in heavy-ion collisions. The developed DNN model can well estimate v_2 for light-flavor identified particles such as $\pi^{\pm}$, $K^{\pm}$, and $p + \bar{p}$ in heavy-ion collisions at RHIC and LHC energies. Here, we compared the results to the data. The DNN was trained with AMPT data of minimum bias Pb-Pb collisions at $\sqrt{s_{NN}} = 5.02$ TeV and was able to learn and predict the centrality, hadron flavor, energy and transverse momentum dependence of elliptic flow for other collision systems at various energies. Results were presented for Pb-Pb collisions at $\sqrt{s_{NN}} = 2.76$ TeV, Xe-Xe collisions at $\sqrt{s_{NN}} = 5.44$ TeV in three centrality bins. We have seen DNN estimator and the AMPT data correlate well up to $p_T \lesssim 3$ GeV/c, where the training statistics issue vanishes. Comparison with available ALICE dataset [18–20] at LHC energies follows the trends of the AMPT with a high accuracy, which latter underestimates the data above $p_T \approx 2$ GeV/c, similarly as the DNN model. The obtained results suggest, that original data or Monte Carlo simulations with better agreement with data at high p_T, can train the DNN-estimator for more accurate predictions.

Acknowledgements

SP acknowledges the doctoral fellowship from UGC, Govt. of India. NM, SP and RS acknowledge the DAE-DST, Govt. of India funding under the mega-science project – "Indian participation in the ALICE experiment at CERN" bearing Project No. SR/MF/PS-02/2021-IITI (E-37123). GGB acknowledges the Hungarian National Research, Development and Innovation Office (NKFIH) under Contract No. OTKA K135515, 2021-4.1.2-NEMZ_KI-2024-00031, 2024-1.2.5-TÉT-2024-00022; Wigner Scientific Computing Laboratory (WSCLAB). The MoU between IIT Indore and HUN-REN Wigner RCP, Hungary, for the techno-scientific cooperation is highly appreciated. ANM would like to thank UCRD, Chandigarh University for their research facilities.

References

[1] S. A. Bass, M. Gyulassy, H. Stoecker and W. Greiner, J. Phys. G **25**, R1-R57 (1999).
[2] U. Heinz and R. Snellings, Ann. Rev. Nucl. Part. Sci. **63**, 123 (2013).
[3] J. Adams *et al.* [STAR Collaboration], Phys. Rev. Lett. **92**, 052302 (2004).
[4] K. Aamodt *et al.* [ALICE Collaboration], Phys. Rev. Lett. **105**, 252302 (2010).
[5] K. Aamodt *et al.* [ALICE Collaboration], Phys. Rev. Lett. **107**, 032301 (2011).
[6] B. B. Abelev *et al.* [ALICE Collaboration], Phys. Rev. C **90**, 054901 (2014).
[7] S. Voloshin and Y. Zhang, Z. Phys. C **70**, 665 (1996).
[8] M. Masera, G. Ortona, M. G. Poghosyan and F. Prino, Phys. Rev. C **79**, 064909 (2009).
[9] Z. W. Lin, C. M. Ko, B. A. Li, B. Zhang and S. Pal, Phys. Rev. C **72**, 064901 (2005).
[10] X. N. Wang and M. Gyulassy, Phys. Rev. D **44**, 3501 (1991).
[11] B. Zhang, Comput. Phys. Commun. **109**, 193 (1998).
[12] Z. w. Lin and C. M. Ko, Phys. Rev. C **65**, 034904 (2002).
[13] Y. He and Z. W. Lin, Phys. Rev. C **96**, 014910 (2017).
[14] B. Li, A. T. Sustich, B. Zhang and C. M. Ko, Int. J. Mod. Phys. E **10**, 267 (2001).
[15] B. A. Li and C. M. Ko, Phys. Rev. C **52**, 2037 (1995).
[16] N. Mallick *et al.*, Phys. Rev. D **105**, 114022 (2022).
[17] N. Mallick *et al.*, Phys. Rev. D **107**, 094001 (2023).
[18] S. Acharya *et al.* [ALICE Collaboration], JHEP **09**, 006 (2018).
[19] S. Acharya *et al.* [ALICE Collaboration], Phys. Rev. Lett., **123**, 14, 142301 (2019).
[20] B. B. Abelev *et al.* [ALICE], JHEP **06**, 190 (2015)

© The Authors, published by EDP Sciences. This is an open access article distributed under the terms of the Creative Commons Attribution License 4.0 (https://creativecommons.org/licenses/by/4.0/).

Testing CPT symmetry via precision mass measurements of multi-strange baryons in ALICE

Romain Schotter[1,*] on behalf of the ALICE Collaboration.

[1]Stefan Meyer Institute for Subatomic Physics, Austrian Academy of Sciences.

Abstract. In these proceedings, the measurements of the Ξ^-, $\overline{\Xi}^+$, Ω^-, $\overline{\Omega}^+$ masses and the mass differences between particle and anti-particle have been measured in pp collisions collected by the ALICE Collaboration during LHC Run 2. The results significantly improve the precision from previous experiments, thus allowing direct tests of CPT symmetry to an unprecedented level of precision in the multi-strange baryon sector.

1 Introduction

Fundamental symmetries stand as one of the most fruitful concepts in Modern Physics. They are of two kinds: continuous — such as the global translations in both space and time, or the Lorentz transformations — and discrete — for example, the space- (P) and time- (T) inversions, the charge conjugation (C), and their combined transformation given by CPT. In particular, the Lorentz and CPT symmetries are closely connected by the so-called CPT theorem which states that any local Lorentz-invariant quantum field theory must also be CPT invariant [1]. Consequently, CPT violation implies the breaking of the Lorentz symmetry, and vice versa [2]. Another implication of CPT symmetry is that matter and antimatter share the same properties (invariant mass, lifetime, etc). Most of the experimental checks of CPT invariance stem from these physical consequences.

The Particle Data Group (PDG) [3] compiles a large variety of CPT tests from many experiments and with a high degree of precision; so far, no CPT violation has been observed. However, for a certain number of them, there is some room for improvement. Most notably, we can mention the measurements of the mass difference between particle and anti-particle in the multi-strange baryon sector. The only test of this nature dates back to 2006 [4] for the Ξ^- and $\overline{\Xi}^+$, and from 1998 [5] for the Ω^- and $\overline{\Omega}^+$. As shown in Tab. 1, both studies suffer from limited statistics. A similar observation can be made about the mass values, and particularly for the Ω baryons where the measurements rely on no more than 100 Ω.

In these proceedings, we present a measurement of the mass difference of the Ξ^- and $\overline{\Xi}^+$, and of the Ω^- and $\overline{\Omega}^+$ baryons. The data samples are much larger than those exploited previously: ~2 400 000 ($\Xi^- + \overline{\Xi}^+$) and ~130 000 ($\Omega^- + \overline{\Omega}^+$) with little background. These direct measurements of the mass difference offer a test of CPT invariance to an unprecedented precision in the multi-strange baryon sector. The absolute masses are updated as well, with a precision substantially better than the previous measurements.

*e-mail: romain.schotter@cern.ch

© The Authors, published by EDP Sciences. This is an open access article distributed under the terms of the Creative Commons Attribution License 4.0 (https://creativecommons.org/licenses/by/4.0/).

Table 1. Particle properties with the last mass and mass difference measurements as of 2022, listed into [3–6]. Here, the mass difference refers to the normalized one, namely $(M_{\overline{\text{part}}} - M_{\text{part}})/M_{\text{average}}$.

Particle	Quark content	Mass measurement (MeV/c^2)	Sample	Relative mass difference measurement ($\times 10^{-5}$)	Sample
Ξ^-	dss	1321.70 ± 0.08 (stat.) ± 0.05 (syst.)	2500	2.5 ± 8.7 (tot.)	2500
$\overline{\Xi}^+$	$\bar{d}\bar{s}\bar{s}$	1321.73 ± 0.08 (stat.) ± 0.05 (syst.)	2300		2300
Ω^-	sss	1672 ± 1 (tot.)	100	1.44 ± 7.98 (tot.)	6323
$\overline{\Omega}^+$	$\bar{s}\bar{s}\bar{s}$	1673 ± 1 (tot.)	72		2607

2 Detector setup and data sample

All the aforementioned particles are detected at mid-rapidity ($|y| < 0.5$), using the central detectors of ALICE [7] at the LHC. The primary and secondary vertices are reconstructed using the Inner Tracking System (ITS), composed of six concentric layers of silicon detectors. The main tracking device is the Time Projection Chamber (TPC), which also provides particle identification of pions, kaons and protons based on their energy loss in the detector. The central part of the experiment is embedded in a large solenoid magnet (also called the L3 magnet), which offers three magnetic field configurations: +0.5, -0.5 and -0.2 T.

The mass of the particles of interest is measured in pp collisions at a centre-of-mass energy $\sqrt{s} = 13$ TeV, using approximately 2.2×10^9 minimum-bias events collected in 2016, 2017 and 2018. Only data taken with a magnetic field value of ± 0.5 T are considered.

3 Data analysis and study of the systematic uncertainties

In this measurement, the charged Ξ and Ω baryons are studied in their cascade decay channel: $\Xi^\pm \rightarrow \pi^\pm\Lambda \rightarrow \pi^\pm\pi^\pm p^\mp$ (with a branching ratio BR = 63.9 %) and $\Omega^\pm \rightarrow K^\pm\Lambda \rightarrow K^\pm\pi^\pm p^\mp$ (BR = 43.4 %) [3]. The reconstruction of these decay topologies is achieved by first reconstructing the Λ candidates, which are then matched with a pion or kaon track. To reduce the induced combinatorial background, various topological and kinematic cuts are used, similar to the methods described in [8].

The masses of the multi-strange baryons are measured through a fit of their invariant mass distributions, using the sum of two functions: a triple Gaussian for the peak and an exponential for the background. The measured mass corresponds to the center of the invariant mass peak, given by the position of the maximum of the triple Gaussian function, denoted as μ. The width (σ) provides an estimation of the mass resolution.

Figure 1 presents the invariant mass distributions of the Ξ^-, $\overline{\Xi}^+$, Ω^-, $\overline{\Omega}^+$ in pp collisions at $\sqrt{s} = 13$ TeV. Considering the reduced χ^2 values, all display a reasonably good fit. To ensure stable mass measurements, additional selections were applied. For instance, to deal with the residual distortions in the TPC, the analysis focuses solely on the positive z side of the detector where they are less pronounced. Following these selections, the combinatorial background is greatly reduced, and overall 15 281 ±128 Ξ^- (14 799 ±126 $\overline{\Xi}^+$) and 10 072 ±110 Ω^- (9 840 ±109 $\overline{\Omega}^+$) baryons were reconstructed. Although the final measurement relies only on a fraction of the initial data sample, the present results are still based on a statistics of strange baryons that is much larger than in previous measurements.

The dominant sources of systematic uncertainties originate from the candidate selections, the detector calibration, the finite precision on the magnetic field map, and the limited knowl-

© The Authors, published by EDP Sciences. This is an open access article distributed under the terms of the Creative Commons Attribution License 4.0 (https://creativecommons.org/licenses/by/4.0/).

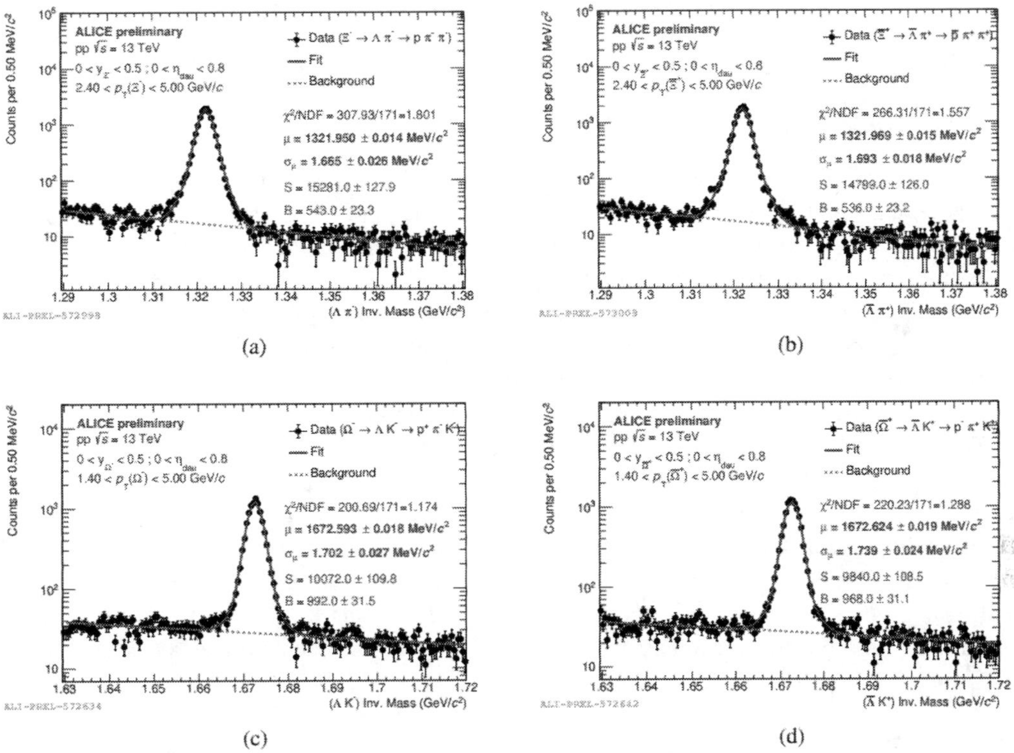

Figure 1. Invariant mass distributions of the Ξ^- (1(a)), $\overline{\Xi}^+$ (1(b)), Ω^- (1(c)) and $\overline{\Omega}^+$ (1(d)). The peak is modelled by a triple Gaussian function, and the background by an exponential function. The measured mass and mass resolution, with their associated statistical uncertainties, are displayed in bold font.

edge on the material distribution. Other systematic effects have also been studied, namely the mass extraction procedure, the contribution of pile-up, the precision on the tabulated masses of the decay daughters, and the correction on the mass offset in simulation. However, their contributions remain small and do not exceed 20 keV$/c^2$.

4 Results

The final values of the $\Xi^\pm$ and $\Omega^\pm$ masses are:

$$M(\Xi^-) = 1321.975 \pm 0.026\ (\text{stat.}) \pm 0.078\ (\text{syst.})\ \text{MeV}/c^2,$$

$$M(\overline{\Xi}^+) = 1321.964 \pm 0.024\ (\text{stat.}) \pm 0.083\ (\text{syst.})\ \text{MeV}/c^2,$$

$$M(\Omega^-) = 1672.511 \pm 0.033\ (\text{stat.}) \pm 0.102\ (\text{syst.})\ \text{MeV}/c^2,$$

$$M(\overline{\Omega}^+) = 1672.555 \pm 0.034\ (\text{stat.}) \pm 0.102\ (\text{syst.})\ \text{MeV}/c^2.$$

The final relative mass difference between particle and anti-particle are:

$$2 \times \frac{M(\overline{\Xi}^+) - M(\Xi^-)}{M(\overline{\Xi}^+) + M(\Xi^-)} = [-1.45 \pm 6.25\ (\text{tot.})] \times 10^{-5},$$

$$2 \times \frac{M(\overline{\Omega}^+) - M(\Omega^-)}{M(\overline{\Omega}^+) + M(\Omega^-)} = [3.28 \pm 4.47\ (\text{tot.})] \times 10^{-5},$$

© The Authors, published by EDP Sciences. This is an open access article distributed under the terms of the Creative Commons Attribution License 4.0 (https://creativecommons.org/licenses/by/4.0/).

where the total uncertainty is given by the quadratic sum of the statistical and systematic ones.

5 Discussion and conclusion

Our measurements can be compared to the previously measured values presented in Tab. 1. The uncertainty on the mass values has been reduced by approximately 15% for the Ξ and a factor 10 for the Ω baryons. On the mass difference, the precision has been improved by 40% and almost a factor 2 for Ξ and Ω baryons, respectively.

Considering our precision, the relative mass difference measurements of Ξ and Ω baryons are still compatible with 0 and further constrain the validity of CPT symmetry in the multi-strange baryon sector. While the Ω masses still agree with the PDG mass values in spite of the 10-fold improvement, the Ξ masses are 2.5σ larger than the tabulated values.

These may influence the study of QCD, particularly in the non-perturbative regime. For instance, lattice QCD relies on the mass of multi-strange baryons to convert its predictions in lattice units into physical units. We can think about the $(g_\mu - 2)$ prediction of the BMW Collaboration, using the mass of the Ω baryons [9], where our more precise mass values would lead all uncertainties from the physical input to be negligible; or the hadron mass spectroscopy that employs the mass of either Ξ or Ω baryons [10], which still needs to be consistent with the measured spectrum using our masses to set the physical scale.

References

[1] A. Kostelecky, The Status of CPT (1998), hep-ph/9810365, http://arxiv.org/abs/hep-ph/9810365

[2] M.S. Sozzi, Tests of discrete symmetries, J. Phys. G: Nucl. Part. Phys. **47**, 013001 (2019). https://doi.org/10.1088/1361-6471/ab4822

[3] Particle Data Group, Review of Particle Physics, PTEP **2022**, 083C01 (2022). https://doi.org/10.1093/ptep/ptac097

[4] J. Abdallah, et al., Masses, lifetimes and production rates of Ξ^- and $\overline{\Xi}^+$ at LEP 1, Phys. Lett. B **639**, 179 (2006). https://doi.org/10.1016/j.physletb.2006.06.029

[5] E756 Collaboration, Measurement of the properties of the $\overline{\Omega}^+$ and Ω^- hyperons, Phys. Rev. D **58**, 072002 (1998). https://doi.org/10.1103/PhysRevD.58.072002

[6] E.P. Hartouni, et al., Inclusive Production of Ω^- and $\overline{\Omega}^+$ by K_L^0-Carbon Interactions in the Energy Range 80-280 GeV/c, Phys. Rev. Lett. **54**, 628 (1985). https://doi.org/10.1103/PhysRevLett.54.628

[7] ALICE Collaboration, The ALICE experiment at the CERN LHC, J. Inst. **3**, S08002 (2008). https://doi.org/10.1088/1748-0221/3/08/S08002

[8] ALICE Collaboration, Multiplicity dependence of (multi-)strange hadron production in proton-proton collisions at $\sqrt{s} = 13$ TeV, Eur. Phys. J. C **80**, 167 (2020). https://doi.org/10.1140/epjc/s10052-020-7673-8

[9] S. Borsanyi, et al., Leading hadronic contribution to the muon magnetic moment from lattice QCD, Nature **593**, 51 (2021). https://doi.org/10.1038/s41586-021-03418-1

[10] Z. Fodor, C. Hoelbling, Light Hadron Masses from Lattice QCD, Rev. Mod. Phys. **84**, 449 (2012). https://doi.org/10.1103/RevModPhys.84.449

© The Authors, published by EDP Sciences. This is an open access article distributed under the terms of the Creative Commons Attribution License 4.0 (https://creativecommons.org/licenses/by/4.0/).

Scaling Properties of ϕ-Meson and Light Charged Hadron Production in Small and Large Systems at PHENIX

Rachid Nouicer[1,*] (for the PHENIX Collaboration)

[1]Physics Department, Brookhaven National Laboratory, Upton, New York 11973, United States

Abstract. Recent results on the identified charged-hadron ($\pi^{\pm}$, $K^{\pm}$, p, $\bar{p}$) production at midrapidity region ($|\eta| < 0.35$) have been measured by the PHENIX experiment in p + Al, ^{3}He + Au, Cu + Au collisions at $\sqrt{s_{NN}}$ = 200 GeV and U + U collisions at $\sqrt{s_{NN}}$ = 193 GeV. These measurements are presented through the invariant transverse-momentum (p_T) and transverse-mass (m_T) spectra for different collision centralities. The averaged freeze-out temperature value for different systems was found to be 166.1 ± 2.2 MeV, and do not exhibit any dependence on the collision centrality and $\langle N_{\text{part}} \rangle$ values. The particle ratios of K/π and p/π have been measured in different centrality ranges of large and small collision systems. The values of K/π ratios measured in all considered collision systems were found to be consistent with those measured in p+p collisions. Furthermore, the identified charged-hadron nuclear-modification factors (R_{AB}) are also presented. Enhancement of proton R_{AB} values over meson R_{AB} values was observed in central ^{3}He+Au, Cu+Au, and U+U collisions. The proton R_{AB} values measured in p+Al collision system were found to be consistent with R_{AB} values of ϕ, $\pi^{\pm}$, $K^{\pm}$, and π^0 mesons, suggesting that the size of the system produced in p+Al collisions is too small for recombination to cause a noticeable increase in proton production.

1 Introduction

Light hadrons are considerably produced in high-energy heavy-ion collisions and provide a wealth of information about properties of created QCD medium and reaction dynamics. These include, in particular, the implications of collective flow in small and large systems and the impact of recombination on baryon and strangeness enhancement. The system size dependence studies of different observable are crucial to investigate the properties of quark-gluon plasma (QGP) and hadronization based on initial conditions of the collisions [1].

This paper presents recently finalized measurements by PHENIX on the production of identified charged hadron ($\pi^{\pm}$, $K^{\pm}$, p, $\bar{p}$) in small and large system size, p+Al, p/d/^{3}He/Cu+Au collisions at $\sqrt{s_{NN}}$ = 200 GeV, and U+U collisions at $\sqrt{s_{NN}}$ = 193 GeV, and results were published recently [2]. The data sets used in the present analysis were collected by the PHENIX experiment [3]. The latter consisted of three components: global detectors, the central spectrometers, and the muon arm spectrometers. These measurements use the central spectrometers ($|\eta| < 0.35$), and provide precise tracking and particle identification for electrons, charged hadrons, and photons. Tracking and momentum determination are provided by drift chambers and pad chambers. Particle identification is provided by time of flight, ring-imaging Cherenkov detectors, and electromagnetic calorimeters.

*e-mail: nouicer@bnl.gov

© The Authors, published by EDP Sciences. This is an open access article distributed under the terms of the Creative Commons Attribution License 4.0 (https://creativecommons.org/licenses/by/4.0/).

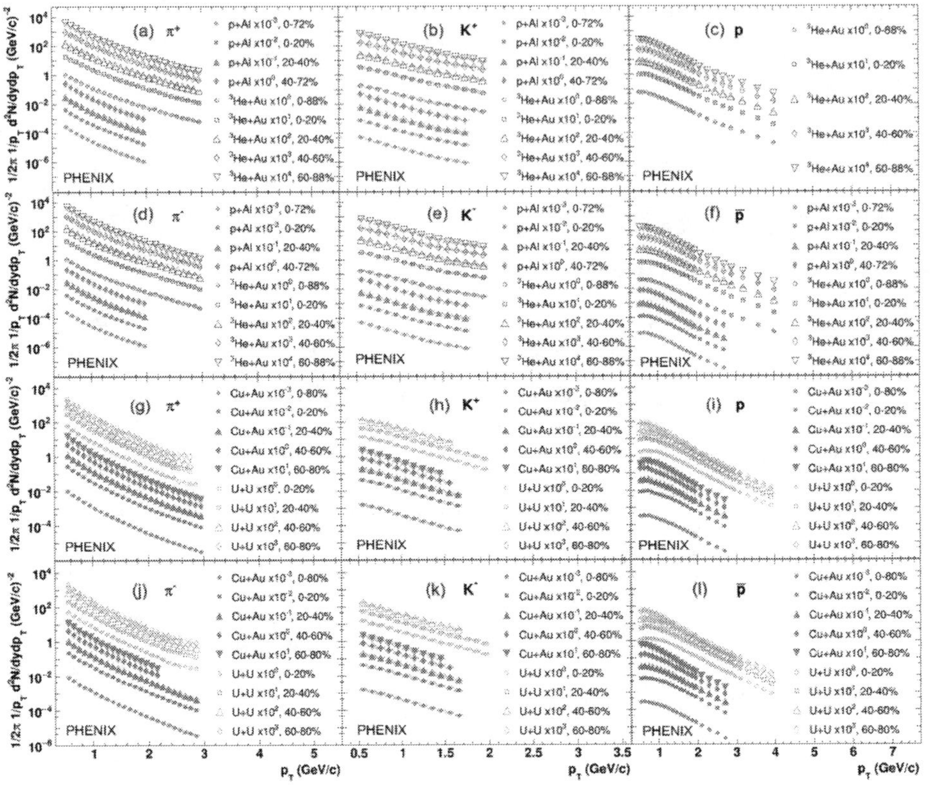

Figure 1. The π^+, π^-, K^+,K^-, p and $\bar{p}$ invariant p_T spectra measured in different centralities of p + Al, ^{3}He + Au, Cu + Au collisions at $\sqrt{s_{NN}}$ = 200 GeV and U + U collisions at $\sqrt{s_{NN}}$ = 193 GeV.

2 Identified Charged Hadron Results in Small and Large Systems

• **Invariant Spectra and Freeze-out Temperature**: figure 1 shows the $\pi^\pm$, $K^\pm$, p, and $\bar{p}$ invariant transverse momentum spectra measured in p+Al, ^{3}He+Au, Cu+Au collisions at $\sqrt{s_{NN}}$ = 200 GeV and in U + U collisions at $\sqrt{s_{NN}}$ = 193 GeV. We observe that the π, K, and p invariant spectra reveal different shapes as a function of p_T. To assess these differences, invariant-transverse-mass ($m_T = \sqrt{p_T^2 + m_0^2}$) spectra were calculated. We observe that the m_T invariant spectra of all identified charged hadrons have exponential form for $m_T < 1.5$ GeV and was approximated using following fit function:

$$\frac{1}{2\pi m_T}\frac{d^2N}{dm_T dy} = \frac{A}{2\pi T(T+m_0)}\exp\left(-\frac{m_T - m_0}{T}\right) \quad (1)$$

where T is the inverse-slope parameter, A is a normalization factor, and m_0 mass of the charged particle at rest. Figure 2(a) shows examples of T parameter vs. hadron mass (m_0) dependencies for different centralities of Cu + Au collisions. Ordering of pion, kaon, and proton inverse slope values $T_\pi < T_K < T_p$ can be seen in all centralities. By doing similar work in all systems, p+Al, ^{3}He+Au, and U + U, we obtained that the T values, calculated for pions in different centralities, are nearly of the same values in all collision systems. The T values, calculated for kaons, take intermediate values between pion and proton T-parameter values. The dotted lines on Fig. 2(a) represent linear fits of the $T(m_0)$ values from each centrality bin using thermal formula $T = T_0 + m\langle u_t\rangle^2$, where T_0 can be interpreted as a

© The Authors, published by EDP Sciences. This is an open access article distributed under the terms of the Creative Commons Attribution License 4.0 (https://creativecommons.org/licenses/by/4.0/).

freeze-out temperature and $\langle u_t \rangle$ as the average collective velocity for all particle species. The fit parameters for positively charged (T_0^+, $\langle u_t \rangle^+$) and negatively charged (T_0^-, $\langle u_t \rangle^-$) hadrons calculated in p+Al, ^{3}He+Au, Cu+Au, and U+U collision systems are presented in Fig. 2(b) as a function of $\langle N_{\rm part} \rangle$ values. The T_0 values calculated in collisions with different geometries and centralities were found to be coincident within uncertainties, indicating that the freeze-out temperature is approximately independent of $\langle N_{\rm part} \rangle$ values. The averaged T_0 value was found to be 166.1 ± 2.2 MeV and is shown in the Fig. 2(b) with red solid line.

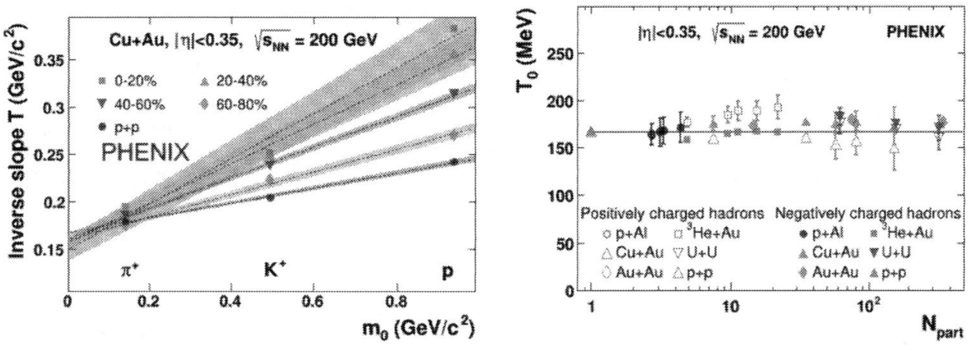

(a) Mass and centrality dependence of inverse slope (b) Freeze-out temperature as a function of $\langle N_{\rm part} \rangle$

Figure 2. (a) Mass and centrality dependence of inverse slope parameters T for π^+, K^+ and p in Cu+Au collisions. The dotted lines represent a linear fits (see text). (b) Freeze-out temperature (T_0) as a function of $\langle N_{\rm part} \rangle$ obtained for positively and negatively charged hadrons at different centralities of collision systems. The T_0 values measured in $p + p$ collisions are shown for comparison.

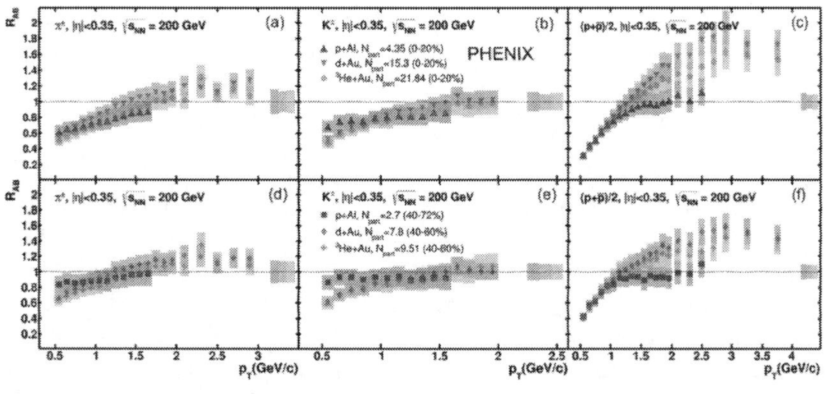

Figure 3. Identified charged hadron nuclear-modification factors as a function of p_T measured in central and peripheral p+Al, d + Au and ^{3}He+Au collisions. The dashed lines correspond to $R_{AB} = 1$ indicating absence of nuclear modification.

• **Particle Ratios:** the baryon production enhancement in nucleus-nucleus collisions is contemplated to be one of the signatures of QGP formation [4]. Based on this consideration, the ratios of p/π^+, $\bar{p}/\pi^-$ in different centralities of large (Cu+Au, U+U) and small ($p + p$, p+Al, ^{3}He+Au) collision systems have been measured and published recently by the PHENIX collaboration [2]. We observed in central collisions of large systems the p/π ratios reach the values of ≈ 0.6, but in peripheral collisions the values of p/π ratios are smaller than 0.4 in the whole p_T range. Furthermore, in small collision systems (p+Al, ^{3}He+Au), the values of p/π ratios are in good agreement to those measured in $p + p$ collisions [5].

© The Authors, published by EDP Sciences. This is an open access article distributed under the terms of the Creative Commons Attribution License 4.0 (https://creativecommons.org/licenses/by/4.0/).

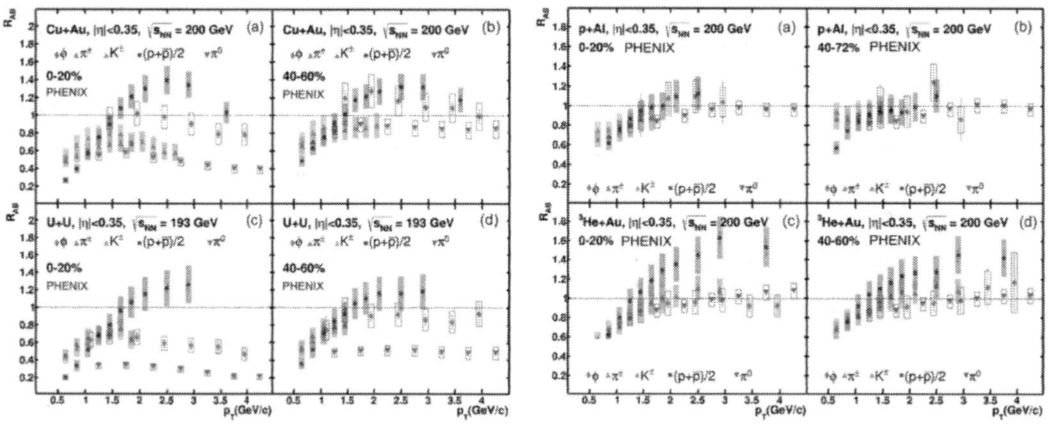

(a) R_{AA} for central and peripheral Cu + Au, and U + U (b) R_{AA} for central and peripheral p + Al, and ^{3}He+Au

Figure 4. R_{AB} for light hadron, ϕ, $\pi^{\pm}$, $K^{\pm}$, $(p+\bar{p})/2$ and π^0, as a function of p_T measured in central and peripheral: (a) in Cu+Au, and U+U collisions, and (b) p+Al and ^{3}He+Au collisions.

• **Nuclear Modification Factor:** figure 3 shows the identified charged-hadron R_{AB} values as a function of p_T in central and peripheral p+Al, d + Au, and ^{3}He+Au collisions. The R_{AB} values are found similar for collision systems with different geometries, but with the same $\langle N_{\text{part}}\rangle$ values, indicating that identified charged-hadron production depends only on system size and not geometry. In addition, we observe the following features: 1) the slope of $R_{AB}(p_T)$ in p+Al collisions is flatter than it is in ^{3}He+Au and d + Au collisions, and 2) proton R_{AB} values in p+Al collisions at the intermediate p_T range (1.0 GeV/c < p_T < 2.5 GeV/c) are equal to unity, while in ^{3}He+Au and d + Au collisions proton R_{AB} is above unity. These observations suggest that the differences between p+Al and d/^{3}He+Au might be caused by the size of the p+Al system being insufficient to observe an increase in proton production.

3 Comparison of Identified Charged Hadron with ϕ-Meson and π^0

Figures 4(a) and 4(b) show the comparison of R_{AB} values of identified charged-hadron with ϕ-meson and π^0 in small and large collision systems, respectively. All these measurements of R_{AB} were obtained by the PHENIX experiment [2, 6]. These results elucidate following features: 1) in large collision systems and in the ^{3}He+Au collision system, proton R_{AB} values are enhanced over all meson R_{AB} values. Knowing that the mass of the ϕ-meson m_{ϕ}=1019 MeV/c^2 is similar to the proton mass m_p=938 MeV/c^2, therefore the enhancement of proton R_{AB} values over ϕ-meson R_{AB} values suggests differences in baryon versus meson production instead of a simple mass dependence, 2) in p+Al collisions proton R_{AB} values and R_{AB} values of all measured mesons are in good agreement within uncertainties.

4 Summary

The PHENIX experiment has established a comprehensive program to study the light hadron (ϕ, $\pi^{\pm}$, $K^{\pm}$, $(p+\bar{p})/2$ and π^0) production in small and large collisions systems by measuring invariant transverse-momentum and transverse-mass spectra, particle ratios, and nuclear modification factors. These results were presented and discussed in the scope to study the properties of created QCD medium and reaction dynamics at RHIC energies.

© The Authors, published by EDP Sciences. This is an open access article distributed under the terms of the Creative Commons Attribution License 4.0 (https://creativecommons.org/licenses/by/4.0/).

References

[1] Rachid Nouicer, European Physical Journal Plus, **131** 3 (2016) 70

[2] N. J. Abdulameer et al. (PHENIX Collaboration), Phy. Rev. C **109**, 054910 (2024)

[3] K. Adcox et al. (PHENIX Collaboration), Nucl. Inst. Meth. Phys. Res., A **499**, 469 (2003)

[4] S. S. Adler et al. (PHENIX Collaboration), Phy. Rev. C **69**, 034909 (2004), K. Adcox et al. (PHENIX Collaboration), Phys. Rev. Lett. **88**, 242301 (2002)

[5] A. Adare et al. (PHENIX Collaboration), Phys. Rev. C **83**, 064903 (2011)

[6] A. Adare et al. (PHENIX Collaboration), Phys. Rev. C **83**, 064903 (2011), C. Aidala et al. (PHENIX Collaboration), Phys. Rev. C **98**, 054903 (2018), U. Acharya et al. (PHENIX Collaboration), Phys. Rev. C **105**, 064902 (2022)

© The Authors, published by EDP Sciences. This is an open access article distributed under the terms of the Creative Commons Attribution License 4.0 (https://creativecommons.org/licenses/by/4.0/).

Measuring the system size dependence of the strangeness production with ALICE

Roman Nepeivoda[1,*] for the ALICE Collaboration

[1]Lund University Department of Physics, Division of Particle and Nuclear Physics, Lund, Sweden

Abstract. The system size dependence of strangeness production in high-energy collisions was studied using Run 2 data from the ALICE experiment at the LHC. The analysis examines the ratios of strange hadron yields to pion yields as a function of mid-rapidity charged particle multiplicity $\langle \mathrm{d}N_{\mathrm{ch}}/\mathrm{d}\eta \rangle_{|\eta|<0.5}$, offering insights into the mechanisms driving strangeness enhancement across different collision systems (pp, p–Pb and Pb–Pb). The yield ratios exhibit a continuous rising trend from low-multiplicity pp collisions at $\sqrt{s}$ = 7 TeV to central Pb–Pb collisions at $\sqrt{s_{\mathrm{NN}}}$ = 5.02 TeV. Additionally, the evolution of the mean transverse momentum ($\langle p_{\mathrm{T}} \rangle$) for various strange hadrons ($\mathrm{K}^0_{\mathrm{S}}$, Λ ($\bar{\Lambda}$), $\Omega^{\pm}$, $\Xi^{\pm}$) is analyzed, revealing that $\langle p_{\mathrm{T}} \rangle$ does not connect smoothly between different collision systems, with harder spectra observed in high-multiplicity pp collisions compared to the peripheral Pb–Pb collisions. Finally, performance plots from LHC Run 3 are presented, showcasing the purity of Ξ^- and Ω^- invariant mass distributions and highlighting the enhanced detector capabilities in Run 3.

1 Introduction

Studying strangeness production in high-energy collisions is crucial for understanding the underlying dynamics of the quark-gluon plasma (QGP) and the hadronization process in quantum chromodynamics (QCD). Strangeness, a quantum number carried by strange quarks, is particularly sensitive to the conditions created in these collisions. Strangeness production has been widely studied across various systems and energies, from proton-proton (pp) [1, 2] and proton-lead (p–Pb) [3, 4] to lead-lead (Pb–Pb) [5, 6] collisions in the ALICE experiment at the Large Hadron Collider (LHC), providing key insights into the collective behavior of particles and the mechanisms driving hadronization.

The design of the ALICE experiment, optimized for the study of QGP in heavy-ion collisions, also allows for precise measurements in smaller systems, making it possible to compare the behavior of strange hadrons across different collision environments [7]. This comparative analysis is essential for probing the onset of collective effects and understanding the evolution of the QGP.

In heavy-ion collisions, where a QGP is expected to form, strangeness enhancement – defined as an increase in the production of strange hadrons relative to non-strange hadrons – has been considered a signature of QGP formation. However, recent observations in smaller systems, such as pp [1, 2] and p–Pb [3, 4] collisions, have challenged this traditional view.

*e-mail: roman.nepeivoda@cern.ch

© The Authors, published by EDP Sciences. This is an open access article distributed under the terms of the Creative Commons Attribution License 4.0 (https://creativecommons.org/licenses/by/4.0/).

These findings suggest that strangeness enhancement may not be exclusive to heavy-ion collisions. Rather, mechanisms other than QGP formation – such as increased parton density or the presence of strong color fields – could also play a significant role in driving strangeness enhancement in smaller collision systems, calling for a deeper understanding of the underlying physics.

2 ALICE detector

The ALICE detector is specifically designed to study the properties of the QGP formed in heavy-ion collisions. It comprises several subsystems used for tracking, vertexing, and particle identification. The Time Projection Chamber (TPC) [8] serves as the primary tracking device, offering excellent spatial resolution and particle identification capabilities through specific ionization energy loss ($\mathrm{d}E/\mathrm{d}x$) measurements. The Inner Tracking System (ITS) [9] provides precise vertex reconstruction and complements the TPC in tracking low-momentum particles. The Time-of-Flight (TOF) [10] detector further aids in particle identification by measuring the flight time of particles with high precision.

3 Strangeness scaling with charged-particle multiplicity

One of the key observables in the studies of strange hadron production is the ratio of p_{T}-integrated strange hadron yields to pion yields as a function of the charged-particle multiplicity at mid-rapidity ($\langle \mathrm{d}N_{\mathrm{ch}}/\mathrm{d}\eta\rangle_{|\eta|<0.5}$). As multiplicity increases, this ratio rises, particularly for hadrons with higher strangeness content, offering a direct comparison of strangeness production across different collision systems and energies.

Previous measurements have demonstrated that these ratios follow a smooth, continuous trend from low-multiplicity pp collisions to central Pb–Pb collisions [1–7]. The left panel of Fig. 1 shows the preliminary ALICE results for the ratios of integrated yields between strange hadrons and pions in Pb–Pb collisions at $\sqrt{s_{\mathrm{NN}}} = 5.02$ TeV (squares). These are compared with results from pp collisions at $\sqrt{s} = 7$ TeV [1] and $\sqrt{s} = 13$ TeV [2] (open and closed circles, respectively), p–Pb collisions at $\sqrt{s_{\mathrm{NN}}} = 5.02$ TeV [3] (diamonds), and Pb–Pb collisions at $\sqrt{s_{\mathrm{NN}}} = 2.76$ TeV [5, 6] (stars). The new data reconfirm this smooth scaling, suggesting that the increase in strange hadron production relative to pions is largely driven by the charged-particle multiplicity rather than being system- or energy-specific. Notably, a previously observed hint of a decreasing Ξ/π ratio observed in central Pb–Pb collisions at $\sqrt{s_{\mathrm{NN}}} = 2.76$ TeV is not supported by the new data at $\sqrt{s_{\mathrm{NN}}} = 5.02$ TeV, reinforcing the consistent scaling of strangeness production across different collision systems and energies.

Another critical observable is the mean transverse momentum ($\langle p_{\mathrm{T}}\rangle$) of strange hadrons, which provides insights into collective motion and kinetic freeze-out conditions. Interestingly, rather than showing a smooth connection across different collision systems, the evolution of $\langle p_{\mathrm{T}}\rangle$ with increasing multiplicity exhibits system-dependent behaviour, as seen in the right panel of Fig. 1. Harder spectra are observed in high-multiplicity pp collisions compared to peripheral Pb–Pb collisions. It is important to note that the $\langle p_{\mathrm{T}}\rangle$ data are gathered at the same center-of-mass energy of $\sqrt{s_{\mathrm{NN}}} = 5.02$ TeV for different collision systems, ensuring consistent comparisons. However, the same multiplicities across different systems do not necessarily reflect the same underlying particle production mechanisms. In high-multiplicity pp collisions, there is a bias toward events dominated by hard scatterings. This contrasts with peripheral Pb–Pb collisions where similar multiplicities could arise from several softer pp interactions. As discussed in [11], this difference can be explained using the core-corona formalism, where higher p_{T} values are associated with particles originating from the "core", a

© The Authors, published by EDP Sciences. This is an open access article distributed under the terms of the Creative Commons Attribution License 4.0 (https://creativecommons.org/licenses/by/4.0/).

radially expanding medium, rather than from a "corona", which behaves as a decaying static droplet.

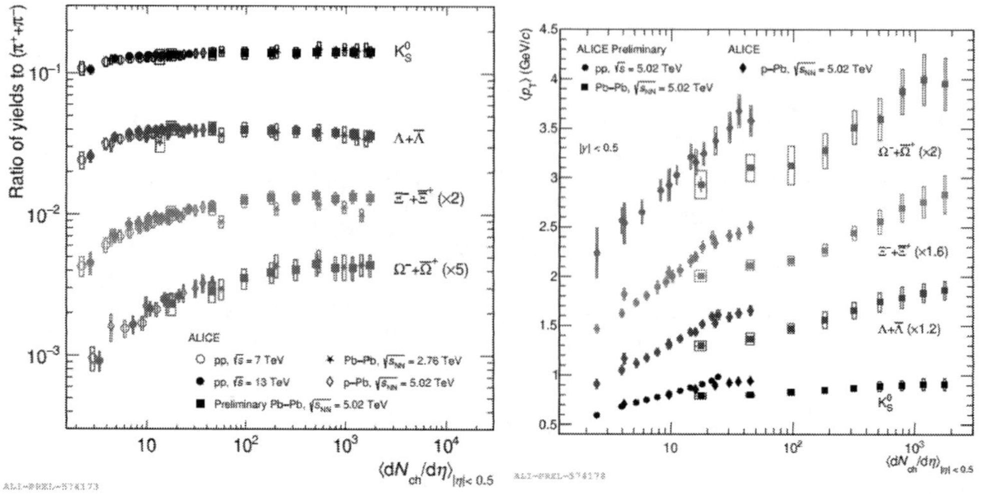

Figure 1. (left) The ratio of p_T-integrated K^0_S, Λ ($\bar{\Lambda}$), Ω, and Ξ strange particle yields to the corresponding pion yields as a function of $\langle dN_{ch}/d\eta\rangle_{|\eta|<0.5}$ in Pb–Pb collisions at $\sqrt{s_{NN}}$ = 2.76 TeV [5, 6] (stars) and $\sqrt{s_{NN}}$ = 5.02 TeV (squares) compared to pp collisions at $\sqrt{s}$ = 7 TeV [1] and $\sqrt{s}$ = 13 TeV [2] (open and closed circles, respectively), and p–Pb collisions at $\sqrt{s_{NN}}$ = 5.02 TeV [3] (diamonds). (right) The average transverse momentum $\langle p_T\rangle$ dependence on the particle density in Pb–Pb collisions at $\sqrt{s_{NN}}$ = 5.02 TeV (square markers). Results for pp and p–Pb [3, 4] collisions at the same center-of-mass energy are shown as dots and diamonds, respectively. Error bars indicate statistical uncertainties, while boxes represent total systematic uncertainties.

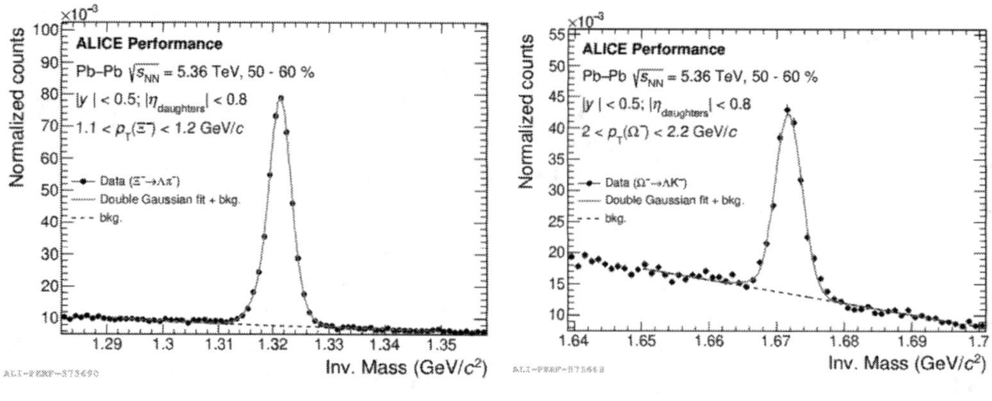

Figure 2. Invariant mass distributions of Ξ^- (left) and Ω^- (right) baryons measured in Pb–Pb collisions at $\sqrt{s_{NN}}$ = 5.36 TeV in the 50–60% centrality class. The left panel shows the invariant mass distribution of $\Xi^- \rightarrow \Lambda\pi^-$ in $1.1 < p_T(\Xi^-) < 1.2$ GeV/c, while the right panel shows the invariant mass distribution of $\Omega^- \rightarrow \Lambda K^-$ in $2.0 < p_T(\Omega^-) < 2.2$ GeV/c. The data are fitted with a double Gaussian function for the signal plus a linear background component. The clear peaks around 1.322 GeV/c^2 and 1.672 GeV/c^2 correspond to the reconstructed masses of the Ξ^- and Ω^- baryons, respectively.

© The Authors, published by EDP Sciences. This is an open access article distributed under the terms of the Creative Commons Attribution License 4.0 (https://creativecommons.org/licenses/by/4.0/).

4 Performance of cascade reconstruction in Run 3

As ALICE transitioned into LHC Run 3, a substantial upgrade of its components has been undertaken, also allowing to enhance its performance in the reconstruction of strange particles. The new ITS2 [12] provides improved spatial resolution, while the TPC's GEM-based readout system ensures better handling of higher interaction rates [13]. These upgrades allow for data collection at higher rates with greater precision. This is evidenced by the ALICE minimum bias Pb–Pb dataset collected during Run 3, which is in one year of data taking already 40x larger than that from Run 2. The reconstructed invariant mass distributions of Ξ^- and Ω^- baryons, shown in Fig. 2, demonstrate the resolution and background suppression achieved with the upgraded detectors.

These advancements are not merely incremental; they mark a significant leap forward in the experiment's capability to probe the QGP and explore the properties of QCD in heavy-ion collisions. As Run 3 progresses, the enhanced precision in strange hadron measurements will pave the way for deeper insights into the behavior of matter under extreme conditions, contributing to a broader understanding of the strong interaction.

References

[1] ALICE Collaboration, S. Acharya et al., "Multiplicity dependence of light-flavor hadron production in pp collisions at $\sqrt{s}$ = 7 TeV". Phys. Rev. C **99(2)**, 024906 (2019).
[2] ALICE Collaboration, S. Acharya et al., "Multiplicity dependence of (multi-)strange hadron production in proton-proton collisions at $\sqrt{s}$ = 13 TeV". Eur. Phys. J. C **80**, 1-26 (2020).
[3] ALICE Collaboration, J. Adam et al., "Multi-strange baryon production in p–Pb collisions at $\sqrt{s_{\rm NN}}$ = 5.02 TeV". Phys. Lett. B **758**, 389-401 (2016).
[4] ALICE Collaboration, B. Abelev et al., "Multiplicity dependence of pion, kaon, proton and Λ production in p–Pb collisions at $\sqrt{s_{\rm NN}}$ = 5.02 TeV". Phys. Lett. B **728**, 25–38 (2014).
[5] ALICE Collaboration, B. Abelev et al., "Multi-strange baryon production at mid-rapidity in Pb–Pb collisions at $\sqrt{s_{\rm NN}}$ = 2.76 TeV". Phys. Lett. B **728**, 216-227 (2014).
[6] ALICE Collaboration, B. Abelev et al., "K^0_S and Λ production in Pb–Pb collisions at $\sqrt{s_{\rm NN}}$ = 2.76 TeV". Phys. Rev. Lett. **111**, 222301 (2013).
[7] ALICE Collaboration, J. Adam et al., "Enhanced production of multi-strange hadrons in high-multiplicity proton–proton collisions". Nat. Phys. **13**, 535-539 (2017).
[8] ALICE Collaboration, J.Alme et al., "The ALICE TPC, a large 3-dimensional tracking device with fast readout for ultra-high multiplicity events". Nucl. Instrum. Meth. A **622**, 316-367 (2010).
[9] G. Contin and ALICE Collaboration, "Performance of the present ALICE Inner Tracking System and studies for the upgrade". JINST **7**, C06007 (2012).
[10] A. Akindinov et al., "Performance of the ALICE Time-Of-Flight detector at the LHC". Eur. Phys. J. Plus **128**, 44 (2013).
[11] K. Werner, "Core-corona procedure and microcanonical hadronization to understand strangeness enhancement in proton-proton and heavy ion collisions in the EPOS4 framework". Phys. Rev. C **109(1)**, 014910 (2024).
[12] F. Reidt and ALICE Collaboration, "Upgrade of the ALICE ITS detector". Nucl. Instrum. Meth. A **1032**, 166632 (2022).
[13] ALICE Collaboration, J. Adolfsson et al., "The upgrade of the ALICE TPC with GEMs and continuous readout". JINST **16**, P03022 (2021).

© The Authors, published by EDP Sciences. This is an open access article distributed under the terms of the Creative Commons Attribution License 4.0 (https://creativecommons.org/licenses/by/4.0/).

Strangeness production in pp and pPb collisions at LHCb

Clara Landesa Gómez[1,*] on behalf of the LHCb Collaboration

[1]IGFAE (Universidade de Santiago de Compostela),
Rúa de Xoaquín Díaz de Rábago, Santiago de Compostela, Spain

Abstract. Strange hadron production provides information about the hadronisation process in high-energy hadron collisions. Strangeness enhancement has been interpreted as a signature of quark-gluon plasma formation in heavy-ion collisions, and recent observations of strangeness enhancement in small collision systems have challenged conventional hadronisation models. With its forward geometry and excellent particle identification capabilities, the LHCb detector is well-suited to study strangeness production in a unique kinematic region. Recent studies of strangeness production with the LHCb detector are presented, including measurements of strangeness enhancement in the charm- and beauty-hadron systems.

1 Introduction

The Quark-Gluon Plasma (QGP) is a phase of the QCD phase diagram where quarks and gluons appear as free particles in thermal equilibrium. The QGP is predicted to have been formed microseconds after the Big Bang and to constitute the inner core of neutron stars. This state of matter can be recreated in a laboratory colliding heavy nuclei at relativistic energies.

Due to the abundance of freely interacting quarks and gluons, the thermalised medium produced in a heavy-ion collision is expected to achieve nearly flavour-chemical equilibrium, yielding an enhanced proportion of strange quarks [1] with respect to *cold* (non-thermalised) hadronic matter. After the collision, the temperature of the system drops, and it undergoes the phase transition, quarks form into hadrons (hadronisation). Shortly after the phase transition, the relative abundance of strange baryons and mesons are fixed at chemical freeze-out according to the Bose-Einstein or Fermi distribution [2].

Therefore, it is expected that the yield of strange hadrons will increase with the size of the QGP formed. This phenomenon, known as *strangeness enhancement* is one of the first probes postulated as a signature of QGP formation. In a heavy-ion collision, the size of the QGP is proportional to the number of nucleons that participate in the collision (participants): higher at lower impact parameter. The number of participants can be related to final-state quantities such as the charged particle multiplicity [3].

Smaller systems, such as pPb and pp collisions, are not expected to achieve enough energy density to form a thermalised medium. However, over recent years there has been an increase in interest in the search for QGP traits in these systems. The ALICE Collaboration has measured a strangeness enhancement in high multiplicity pp collisions [4]. In this paper, we summarise recent LHCb results for heavy flavour strange hadrons in small systems.

*e-mail: clara.landesa.gomez@cern.ch

© The Authors, published by EDP Sciences. This is an open access article distributed under the terms of the Creative Commons Attribution License 4.0 (https://creativecommons.org/licenses/by/4.0/).

The LHCb detector, at the LHC, is a single-arm forward spectrometer covering the pseudorapidity range $2 \lesssim \eta \lesssim 5$ [5] in the laboratory frame. It is composed of several subdetectors. The tracking system consists of the VErtex LOcator system (VELO) and several planar tracking stations. The VELO is a silicon strip detector whose main purpose is to reconstruct the beams interaction point (primary vertex). Particle identification in LHCb is provided by electromagnetic and hadronic calorimeters, two Cherenkov detectors and muon trackers.

The design of the LHCb detector is optimised to study heavy flavour physics and therefore provides the opportunity to study strangeness enhancement in the relatively unexplored heavy flavour sector. Furthermore, heavy hadrons constitute exceptional probes for hadronisation. The usual model to describe hadronisation in hadronic collisions is fragmentation, where showers of partons produced by outgoing quarks form into hadrons. However, in the case of a very dense medium, readily available quarks with overlapping wave functions could combine into colour singlets, hadronising through coalescence. Coalescence is thought to be the leading mechanism in high-energy heavy-ion collisions where the QGP is formed. In pp and pPb collisions, if coalescence emerges as hadronisation mechanism, the production of strange to non-strange heavy hadrons could increase with particle multiplicity, provided there is strangeness enhancement. This effect would be more pronounced at low transverse momentum (p_{T}) where there is a higher density of light particles in the underlying event.

2 Recent LHCb results of strangeness production in small systems

The following sections constitute a collection of recent LHCb measurements of the ratios of strange to non-strange heavy hadrons in small collision systems: pp and pPb.

2.1 B_s^0 over B^0 production ratio in pp collisions at $\sqrt{s}$ = 13 TeV

The pp collisions data used for this measurement [6] were collected at a centre-of-mass energy $\sqrt{s} = 13$ TeV, corresponding to a total integrated luminosity of $5.4\,\mathrm{fb}^{-1}$. The B_s^0/B^0 ratio was studied for several normalised multiplicity intervals. The chosen multiplicity proxies are the number of tracks reconstructed by the VELO ($\mathrm{N}_{\mathrm{tracks}}^{\mathrm{VELO}}$), and a subset of these that point backwards with respect to LHCb ($\mathrm{N}_{\mathrm{tracks}}^{\mathrm{back}}$), in the pseudorapidity interval $-3.5 < \eta < -1.5$. The B_s^0/B^0 ratio consistently increases with $\mathrm{N}_{\mathrm{tracks}}^{\mathrm{VELO}}$, which is not the case for $\mathrm{N}_{\mathrm{tracks}}^{\mathrm{back}}$. The lack of dependence on $\mathrm{N}_{\mathrm{tracks}}^{\mathrm{back}}$ could indicate that the mechanism responsible for the increase in the ratio is related to the local particle density. The ratio as a function of multiplicity is studied for several p_{T} intervals. Figure 1 shows that there is only a significant increase of the ratio at the lowest p_{T} interval, where the B mesons have p_{T} smaller or similar to their mass.

2.2 Strange over non-strange charm hadron production ratios in pPb collisions

In pPb collisions, the kinematic reach of LHCb is extended by reversing the direction of the proton and the lead beams, accessing forward ($1.5 \lesssim y^* \lesssim 4$) and backward ($-5 \lesssim y^* \lesssim -2.5$) rapidities[1]. The pPb collision data were acquired during two LHC runs, the first in 2013, at $\sqrt{s_{\mathrm{NN}}} = 5.02$ TeV, with an integrated luminosity of $1.1\,\mathrm{nb}^{-1}$ ($0.4\,\mathrm{nb}^{-1}$) for forward (backward) collisions, and the second in 2016 at $\sqrt{s_{\mathrm{NN}}} = 8.16$ TeV, with $12.5\ \mathrm{nb}^{-1}$ ($19.3\,\mathrm{nb}^{-1}$).

In the charm sector, we first look at the results of prompt D-meson production ratios D_s^+/D^+ and D_s^+/D^0 measured in pPb collisions at 5 TeV [7]. The ratios do not show significant dependence with p_{T} or rapidity.

[1] y^* is the nucleon-nucleon centre-of-mass rapidity. Since the proton beam had a higher energy per nucleon than the Pb beam, the nucleon-nucleon centre-of-mass was moving in the laboratory frame with a rapidity of -0.465 in the direction of the proton beam, causing the pPb and Pbp configurations to have different $|y^*|$ acceptances.

© The Authors, published by EDP Sciences. This is an open access article distributed under the terms of the Creative Commons Attribution License 4.0 (https://creativecommons.org/licenses/by/4.0/).

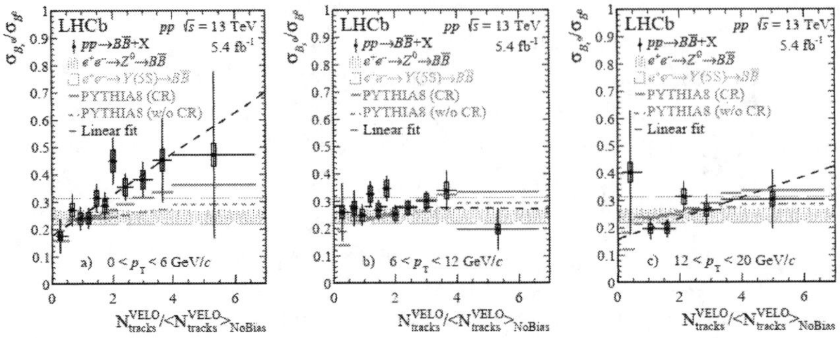

Figure 1. Ratio of cross sections $\sigma_{B_s^0}/\sigma_{B^0}$ measured against normalised $\mathrm{N}^{\mathrm{VELO}}_{\mathrm{tracks}}$, for three p_{T} ranges. The normalisation factor, $\langle \mathrm{N}^{\mathrm{VELO}}_{\mathrm{tracks}} \rangle_{\mathrm{NoBias}}$, is the average number of VELO tracks for unbiased events, which are selected based on the Large Hadron Collider beam clock, without any other trigger requirements. The vertical error bars (boxes) represent point-to-point uncorrelated (fully correlated) uncertainties. The horizontal bands show the values measured in e^+e^- collisions. From Ref. [6].

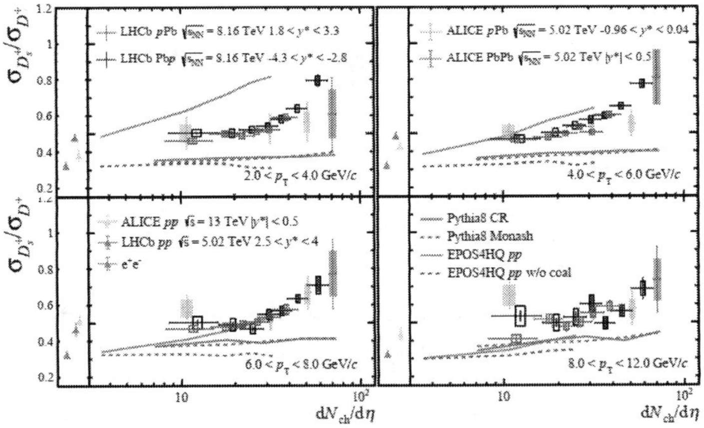

Figure 2. Cross-section ratio of D_s^+ over D^+ in different p_{T} intervals. The red (black) points show the results measured by LHCb for forward (backward) pPb collisions at $\sqrt{s_{\mathrm{NN}}} = 8.16\,\mathrm{TeV}$. The measurements from the ALICE Collaboration for pPb and PbPb collisions at $\sqrt{s_{\mathrm{NN}}} = 5.02\,\mathrm{TeV}$ are included as well as the results from e^+e^- collisions and pp collisions. The vertical error bars show the statistical uncertainties and the boxes show the systematic uncertainties. Calculations from Pythia 8 and EPOS4HQ with and without coalescence mechanism are also shown. From Ref. [8].

The higher luminosity 8.16 TeV data sample allows for the study of the D_s^+ over D^+ production ratio with more precision [8]. The charged particle multiplicity in the forward region ($dN_{ch}/d\eta$) is obtained from a subset of $\mathrm{N}^{\mathrm{VELO}}_{\mathrm{tracks}}$ that is used to reconstruct the primary vertex, applying corrections based on simulation. Figure 2 shows the dependence of the ratio with $dN_{ch}/d\eta$ for different p_{T} intervals. At low multiplicity, the results resemble those of pp collisions, while at high multiplicities the ratio increases and is compatible with measurements from PbPb collisions. This trend is observed both in forward and backward rapidities. The data is compared to predictions from Pythia8 and EPOS4HQ. All models show some discrepancies with data, although EPOS4HQ depicts the increasing trend when accounting for coalescence as hadronisation mechanism.

© The Authors, published by EDP Sciences. This is an open access article distributed under the terms of the Creative Commons Attribution License 4.0 (https://creativecommons.org/licenses/by/4.0/).

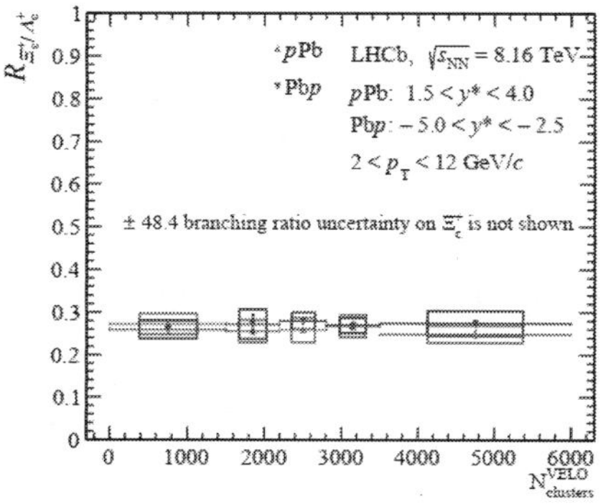

Figure 3. The LHCb measurements of the Ξ_c^+ over Λ_c^+ multiplicity ratio as a function of $\rm N_{clusters}^{VELO}$ for forward (backward) pPb collisions are shown as red (upside down blue) triangles. The error bars represent the statistical uncertainties, while the squares indicate the systematic uncertainty. The 48.4% branching ratio uncertainty of the Ξ_c^+ is not accounted for. From Ref. [9].

The baryon ratio of Ξ_c^+ over Λ_c^+ production is measured in pPb collisions at $\sqrt{s_{\rm NN}} = 8.16\,\text{TeV}$ [9]. The sample was divided into multiplicity classes using the number of energy clusters deposited in the VELO stations ($\rm N_{clusters}^{VELO}$). As shown in Figure 3, the ratio does not show a significant dependence on multiplicity.

3 Summary

The relative production of several strange to non-strange hadrons was studied in small collision systems at LHCb. In pp collisions at $\sqrt{s} = 13$ TeV, there is an observed enhancement of the B_s^0/B^0 ratio at high multiplicity and low $p_{\rm T}$, which is qualitatively compatible with a coalescence hadronisation mechanism. This enhancement is not observed when using backward tracks as multiplicity estimator. In pPb collisions at $\sqrt{s_{\rm NN}} = 5.02\,\text{TeV}$ the D_s^+/D^+ and D_s^+/D^0 ratios do not show a clear trend with $p_{\rm T}$ or η. In pPb collisions at $\sqrt{s_{\rm NN}} = 8.16\,\text{TeV}$, a significant enhancement of the D_s^+/D^+ ratio is observed in events with high multiplicity. However, the baryon ratio Ξ_c^+/Λ_c^+ appears to be constant when studied as a function of $\rm N_{clusters}^{VELO}$.

References

[1] J. Rafelski, *et al.*, Strangeness Production in the Quark-Gluon Plasma, Phys. Rev. Lett. **48**, 1066 (1982), [Erratum: Phys.Rev.Lett. 56, 2334 (1986)].

[2] P. Braun-Munzinger, *et al.*, Chemical freezeout and the QCD phase transition temperature, Phys. Lett. B **596**, 61 (2004), `nucl-th/0311005`.

[3] M.L. Miller, *et al.*, Glauber modeling in high energy nuclear collisions, Ann. Rev. Nucl. Part. Sci. **57**, 205 (2007), `nucl-ex/0701025`.

[4] J. Adam et al. (ALICE), Enhanced production of multi-strange hadrons in high-multiplicity proton-proton collisions, Nature Phys. **13**, 535 (2017), `1606.07424`.

[5] A.A. Alves, Jr., *et al.* (LHCb), The LHCb Detector at the LHC, JINST **3**, S08005 (2008).

[6] R. Aaij et al. (LHCb), Evidence for modification of b quark hadronization in high-multiplicity pp collisions at $\sqrt{s} = 13$ TeV, Phys. Rev. Lett. **131**, 061901 (2023), `2204.13042`.

[7] R. Aaij, *et al.* (LHCb), Measurement of prompt D^+ and D_s^+ production in pPb collisions at $\sqrt{s_{\rm NN}} = 5.02$ TeV, JHEP **01**, 070 (2024), `2309.14206`.

[8] R. Aaij, *et al.* (LHCb), Observation of strangeness enhancement with charmed mesons in high-multiplicity pPb collisions at $\sqrt{s_{\rm NN}}$ =8.16 TeV, Phys. Rev. D **110**, L031105 (2024), `2311.08490`.

[9] R. Aaij et al. (LHCb), Measurement of Ξ_c^+ production in pPb collisions at $\sqrt{s_{\rm NN}}$ =8.16 TeV at LHCb, Phys. Rev. C **109**, 044901 (2024), `2305.06711`.

© The Authors, published by EDP Sciences. This is an open access article distributed under the terms of the Creative Commons Attribution License 4.0 (https://creativecommons.org/licenses/by/4.0/).

p–ϕ femtoscopic correlation analysis using a dynamical model

Kenshi Kuroki[1,*] and *Tetsufumi* Hirano[1,**]

[1]Department of Physics, Sophia University, Tokyo 102–8554, Japan

Abstract. We analyse the p–ϕ correlation functions in high-multiplicity proton+proton collisions at the LHC using a dynamical model, DCCI2, and discuss the effects of collision dynamics on the p–ϕ femtoscopic study. Collision dynamics, such as collective expansion and hadronic rescatterings, leads to the relative momentum-dependent non-Gaussian source functions. This results in deviations in the correlation functions compared to those using the Gaussian source function adopted in existing studies. Our analysis shows the importance of using the source functions that reflect more realistic collision dynamics for future precision analysis of hadron interactions via femtoscopy.

1 Introduction

Femtoscopic analysis using two-particle momentum correlations in high-energy nuclear collisions has gained significant attention as a novel phenomenological approach to understanding low-energy hadron interactions. According to the Koonin-Pratt formula [1, 2], the correlation function is interpreted as a convolution of the source function, which represents the space-time structure of the generated matter reflecting the dynamics of the nuclear collisions, and the square of the relative wave function, which reflects the final-state interaction between the pair of interest. Thus, inputting the source function allows us to extract information about hadron interaction from the measured hadron correlation. Recent active studies have demonstrated the effectiveness of femtoscopy in studying hadron interactions (see, e.g., Ref. [3] for a review). Femtoscopy is now advancing to explore less understood interactions, such as the baryon-vector meson interaction.

Recently, the ALICE Collaboration measured the proton–ϕ meson correlation function in high-multiplicity (0–0.17%) proton+proton collisions at $\sqrt{s}$ = 13 TeV [4], revealing a spin-averaged attractive p–ϕ interaction with an unexpectedly negligible inelastic contribution. A subsequent study [5] reanalysed the correlation data to disentangle the two spin components of the interaction, $^2S_{1/2}$ and $^4S_{3/2}$. By adopting the $N\phi(^4S_{3/2})$ potential from the (2+1)-flavour lattice QCD calculation near the physical point [6], the counterpart $N\phi(^2S_{1/2})$ potential was constrained from the comparison with the experimental correlation function. The result indicated the controversial existence of a p–ϕ bound state with a binding energy in the range $[12.8, 56.1]$ MeV.

Although simple Gaussian source functions have been assumed in existing studies, the actual hadron emission should reflect the complex dynamics of high-energy nuclear collisions.

*e-mail: k-kuroki-e23@eagle.sophia.ac.jp
**e-mail: hirano@sophia.ac.jp

© The Authors, published by EDP Sciences. This is an open access article distributed under the terms of the Creative Commons Attribution License 4.0 (https://creativecommons.org/licenses/by/4.0/).

Thus, we analyse the correlation functions using the source functions from a state-of-the-art dynamical model and discuss the effects of collision dynamics on the p–ϕ femtoscopy.

2 Model and Analysis

As significant femtoscopic correlations appear only in the region of low relative momentum q, we focus exclusively on the s-wave scattering. Assuming the spherical source function, the p–ϕ correlation function $C(q)$ at the pair rest frame is expressed as

$$C(q) = 1 + \int_0^\infty \mathrm{d}r\, 4\pi r^2 S(q;r)\Big\{|\varphi_0(q;r)|^2 - [j_0(qr)]^2\Big\}, \tag{1}$$

where r is the relative separation of the pair at their emission, and the source function $S(q;r)$ corresponds to its probability density distribution.[1] While φ_0 and the spherical Bessel function j_0 correspond to the s-wave parts of the radial wave functions with and without interaction, respectively, the function inside the curly bracket represents the increase or decrease of the wave function squared at each r due to the interaction. The wave function φ_0 can be obtained by solving the Schrödinger equation with the p–ϕ interaction.

For the ${}^4S_{3/2}$ channel, we employ the lattice QCD potential [6], which is overall attractive but does not support bound states. For the ${}^2S_{1/2}$ channel, we follow the same parameterisation as in Ref. [5] motivated by the lattice QCD $V^{(3/2)}$, while neglecting channel-couplings for simplicity:

$$V^{(1/2)}(r) = \beta\Big[a_1 e^{-(r/b_1)^2} + a_2 e^{-(r/b_2)^2}\Big] + a_3 m_\pi^4\Big[1 - e^{-(r/b_3)^2}\Big]^2 \frac{e^{-2m_\pi r}}{r^2}, \tag{2}$$

where $a_{1,2,3}$ and $b_{1,2,3}$ are fixed parameters listed in Table 1 fitted to the lattice QCD data in the ${}^4S_{3/2}$ channel [6], and they are assumed to be common to both $V^{(3/2)}$ and $V^{(1/2)}$. The adjustable parameter β is later constrained by comparison with the experimental correlation function. Note that for $\beta = 1$, the potential (2) is equivalent to $V^{(3/2)}$.

Table 1. Parameter values in $V^{(1/2)}$.

Parameter	a_1 [MeV]	a_2 [MeV]	a_3 [fm^5]	b_1 [fm]	b_2 [fm]	b_3 [fm]
Value	−371	−119	−1.62	0.13	0.3	0.63

To generate the source function, we employ the dynamical core-corona initialisation model (DCCI2) [7], which is a state-of-the-art hydrodynamics-based model. Since DCCI2 treats the QGP fluids (core) and the non-equilibrium partons (corona) simultaneously, it is well-suited for describing high-multiplicity proton+proton collisions. From the phase space distribution of p and ϕ from DCCI2 at their last interacting points with the surrounding hadron gas, we obtain the p–ϕ source function $S(q;r)$ which reflects the collision dynamics. Note that we turn off the decay of ϕ to avoid the complicated reconstruction.

3 Results

We impose almost the same event selection and kinematic cuts as in the ALICE measurement [4]. From 2.3×10^6 minimum bias DCCI2 simulation events for proton+proton collisions at $\sqrt{s} = 13$ TeV, we select the top 0.17% events of the charged hadron multiplicity in

[1] Although the relative momentum-dependence of the source function is usually neglected, the source function in principle depends on q.

© The Authors, published by EDP Sciences. This is an open access article distributed under the terms of the Creative Commons Attribution License 4.0 (https://creativecommons.org/licenses/by/4.0/).

the pseudorapidity range $-3.7 < \eta_p < -1.7$ and $2.8 < \eta_p < 5.1$ with at least one charged hadron in the range $|\eta_p| < 1$. Then, we impose the transverse sphericity cut, $0.7 < S_{\mathrm{T}} < 1.0$, and finally obtain 3 702 events. In each event, protons are selected from the pseudorapidity range $|\eta_p| < 0.8$ and the transverse momentum range $0.5 < p_{\mathrm{T}} < 4.05$ GeV, while ϕ mesons are selected from the pseudorapidity range $|\eta_p| < 0.8$. Note that we employ the event-mixing method to generate the source function in the present study in order to increase statistics.

3.1 Source Function

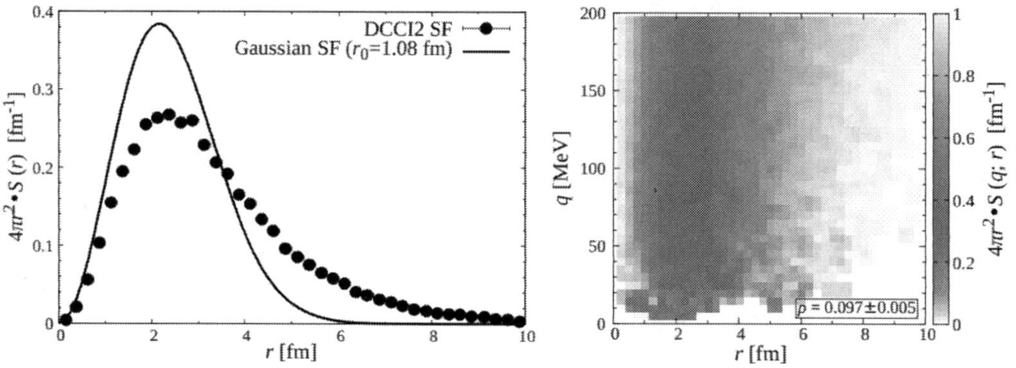

Figure 1. Relative momentum q-integrated (left panel) and q-differential (right panel) p–ϕ source function multiplied by the Jacobian. The Gaussian source function adopted in the previous studies (solid line in the left panel) is also shown for reference.

Figure 1 (left) shows the q-integrated source function from DCCI2 compared to the previously adopted Gaussian source function with the source size $r_0 = 1.08$ fm [4, 5]. Due to the non-Gaussian long tail, which mainly comes from the proton rescatterings with the surrounding pion gas, the source size from DCCI2 becomes larger.

Figure 1 (right) shows the q-differential source function. In general, the collision dynamics, such as the collective expansion of the generated matter, makes a correlation between the momentum space and the coordinate space, which leads to the relative momentum-dependent source function. In fact, the source function from DCCI2 within a current kinematic setup slightly depends on q with a small positive Pearson correlation coefficient $\rho(q, r) \approx 0.1$. The interesting point here is that the source size in the $q \lesssim 20$ MeV region is significantly smaller than in the other regions.

3.2 Correlation Function

Figure 2 shows the resulting correlation functions of each spin channel and their weighted average. Compared to the results using the Gaussian source function, the correlation from DCCI2 is slightly weaker due to the larger source size. In addition, one can see the intriguing behaviours in the small q region due to a significantly smaller source size in this region.

Finally, we vary the adjustable parameter β in $V^{(1/2)}$ and compare the correlation function with the ALICE data [4]. Since $C^{(3/2)}$ is anchored by the lattice QCD potential, the spin-averaged correlation function C^{tot}, which can be compared with the experimental data, changes only through the change in $C^{(1/2)}$. In the case of $\beta = 6$, $C^{(1/2)}$ increases from unity, leading to an overestimation of C^{tot}. On the other hand, for the cases $\beta = 7$ and 8, the overall

© The Authors, published by EDP Sciences. This is an open access article distributed under the terms of the Creative Commons Attribution License 4.0 (https://creativecommons.org/licenses/by/4.0/).

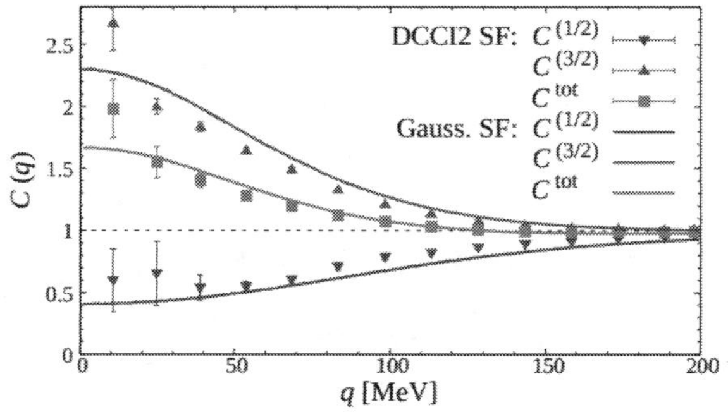

Figure 2. p–ϕ correlation functions using the DCCI2 source function (plots) compared with those using the Gaussian source function (lines) for the $^2S_{1/2}$ channel (blue), $^4S_{3/2}$ channel (green), and spin-average: $C^{\text{tot}} = \frac{1}{3}C^{(1/2)} + \frac{2}{3}C^{(3/2)}$ (red). The parameter $\beta = 7$ is chosen in $V^{(1/2)}$.

decrease of $C^{(1/2)}$ makes C^{tot} consistent with the ALICE data. If we increase β further, the suppression of $C^{(1/2)}$ becomes weaker, and thus C^{tot} starts to overestimate the data. From these results, β should be around 7 to 8 within the current setup to reproduce the experimental data, which means that our femtoscopic analysis using DCCI2 also suggests the existence of a p–ϕ bound state in the $^2S_{1/2}$ channel with a binding energy of around 10 to 70 MeV, in agreement with a previous study [5].

4 Summary

We analysed the effects of the collision dynamics on the p–ϕ femtoscopy in high-multiplicity proton+proton collisions at the LHC by utilising a dynamical model, DCCI2. A non-Gaussian long tail appears in the source function mainly due to the hadronic rescatterings, leading to a slightly weaker correlation compared to that using the Gaussian source function adopted in previous studies. In addition, we revealed that the source function slightly depends on the relative momentum due to the collision dynamics, which affects the correlation function, especially, in the small relative momentum regions. These findings strongly suggest the importance of using the source function that reflects more realistic collision dynamics for future precision hadron interaction studies via femtoscopy. Finally, from the comparison with the ALICE data, our analysis also supports the existence of a p–ϕ bound state.

Acknowledgements

The authors thank Y. Kanakubo for generously providing the DCCI2 data essential for this study. The work by T. H. was partly supported by JSPS KAKENHI Grant No. JP23K03395.

References

[1] S.E. Koonin, Phys. Lett. B **70**, 43 (1977). 10.1016/0370-2693(77)90340-9
[2] S. Pratt, Phys. Rev. Lett. **53**, 1219 (1984). 10.1103/PhysRevLett.53.1219
[3] L. Fabbietti, V. Mantovani Sarti, O. Vazquez Doce, Ann. Rev. Nucl. Part. Sci. **71**, 377 (2021). 10.1146/annurev-nucl-102419-034438
[4] S. Acharya et al. (ALICE), Phys. Rev. Lett. **127**, 172301 (2021). 10.1103/PhysRevLett.127.172301
[5] E. Chizzali, Y. Kamiya, R. Del Grande, T. Doi, L. Fabbietti, T. Hatsuda, Y. Lyu, Phys. Lett. B **848**, 138358 (2024). 10.1016/j.physletb.2023.138358
[6] Y. Lyu, T. Doi, T. Hatsuda, Y. Ikeda, J. Meng, K. Sasaki, T. Sugiura, Phys. Rev. D **106**, 074507 (2022). 10.1103/PhysRevD.106.074507
[7] Y. Kanakubo, Y. Tachibana, T. Hirano, Phys. Rev. C **105**, 024905 (2022). 10.1103/PhysRevC.105.024905

© The Authors, published by EDP Sciences. This is an open access article distributed under the terms of the Creative Commons Attribution License 4.0 (https://creativecommons.org/licenses/by/4.0/).

Measurement of p–Ξ^- ($\bar{p}$–$\bar{\Xi}^+$) Correlation Function in Isobar and Au+Au Collisions at $\sqrt{s_{\mathrm{NN}}}$ = 200 GeV with the STAR Detector

Boyang Fu for the STAR Collaboration[1,*]

[1]Institute of Particle Physics and Key Laboratory of Quark & Lepton Physics (MOE), Central China Normal University, Wuhan, 430079, China.

Abstract. Understanding the strong interactions between baryons, especially hyperon-nucleon (Y-N) interactions, is crucial for comprehending the equation-of-state (EoS) of the nuclear matter and inner structure of neutron star. In these proceedings, we present the measurements of p–Ξ^- ($\bar{p}$–$\bar{\Xi}^+$) correlation functions with high statistics in Isobar (Ru+Ru, Zr+Zr) and Au+Au collisions at $\sqrt{s_{\mathrm{NN}}}$ = 200 GeV by the STAR experiment. With the Lednický-Lyuboshitz approach, the source size and strong interaction parameters of p–Ξ^- ($\bar{p}$–$\bar{\Xi}^+$) pairs are extracted.

1 Introduction

Baryon interactions are of fundamental interest in nuclear physics and astrophysics. Further, the EoS of the nuclear matter at high baryon density that governs the properties inside neutron stars depends on baryon interactions [1]. Among all the baryon interactions, interactions between Y-N and hyperon-hyperon (Y-Y) are of utmost importance to solve the hyperon puzzle [2]. p–Ξ^- ($\bar{p}$–$\bar{\Xi}^+$) interactions with strangeness $S = -2$ sector have long been of interest in theory and experiments as they are proposed to be a potential decay channel for the H-dibaryon (consisting of six quarks: $uuddss$) particle [3].
In heavy-ion collisions, two-particle femtoscopy is a powerful and unique method for extracting information about the spatio-temporal properties of the source, characterising the final state interactions (FSI), and searching for the possible bound states. In these proceedings, we present new results of p–Ξ^- ($\bar{p}$–$\bar{\Xi}^+$) correlation using high statistics data from Isobar and Au+Au collisions at $\sqrt{s_{\mathrm{NN}}}$ = 200 GeV by the STAR experiment. In the following, we will only use p–Ξ^- to denote the p–Ξ^- and $\bar{p}$–$\bar{\Xi}^+$ pairs.

2 Correlation Function

The observable of interest in femtoscopy is the two-particle correlation function. The correlation function can be expressed theoretically as $C(k^*) = \int d^3r^* S(r^*)|\Psi(r^*, k^*)|^2$, where k^* and r^* are the relative momentum and relative distance of the pair of interest in the Pair rest Frame (PRF) reference. $S(r^*)$ is the source function and $\Psi(r^*, k^*)$ represents the wave function of the pair of interest. Experimentally, this correlation function is computed as

*e-mail: boyangfu@mails.ccnu.edu.cn

© The Authors, published by EDP Sciences. This is an open access article distributed under the terms of the Creative Commons Attribution License 4.0 (https://creativecommons.org/licenses/by/4.0/).

$C(k^*) = \mathcal{N}[A(k^*)/B(k^*)]$, where $A(k^*)$ are the correlated pairs in same-event, and $B(k^*)$ is the uncorrelated background pairs obtained from mixed-events. The normalization parameter $\mathcal{N}$ is chosen such that the mean value of the correlation function equals unity for k^* [200, 400] MeV/c.

The correlation function is parameterized using the Lednický-Lyuboshitz (LL) model [4], which considers a static spherical Gaussian source under a smoothness approximation [5], convoluted with an S-wave function. In LL model, the complex scattering amplitude $f^S(k^*)$ with Coulomb interaction is evaluated via the effective range approximation,

$$f^S(k^*) = \left[\frac{1}{f_0^S} + \frac{1}{2}d_0^S k^{*2} - \frac{2}{a_c}h(\eta) - ik^* A_c(\eta)\right]^{-1} \quad (1)$$

where f_0^S is scattering length, d_0^S is effective range of the interaction, a_c is the Bohr radius, $h(\eta)$ is complex function, η and A_c are Coulomb related factor, and S denotes the total spin of the pair. A spin-averaged fit is considered in this work.

3 Results

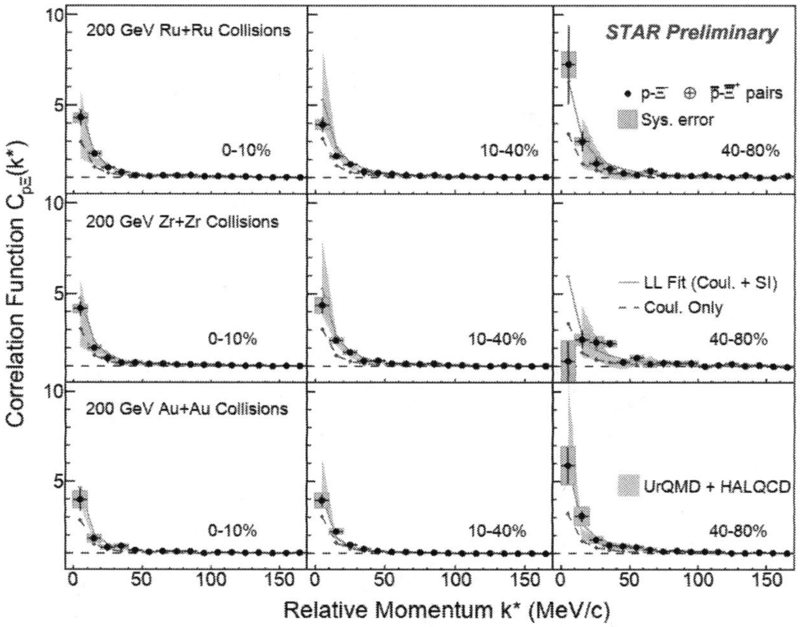

Figure 1. Measured $p{-}\Xi^-$ correlation functions in 0-10% (left), 10-40% (middle) and 40-80% (right) centrality in Ru+Ru (top), Zr+Zr (middle) and Au+Au (bottom) collisions at $\sqrt{s_{\mathrm{NN}}} = 200$ GeV. The black vertical bars and boxes represent the statistical and systematic uncertainties, respectively. In each colliding system, the measured correlation function of three centralities are fitted simultaneously with Lednický-Lyuboshitz function, shown as magenta lines, including effects of Coulomb and strong interactions. Correlation functions with pure Coulomb are shown as blue dashed lines. The orange bands represent the results from UrQMD + HALQCD simulation [6-7].

Figure 1 shows the measured $p{-}\Xi^-$ correlation function as a function of k^* in three centrality bins in Isobar and Au+Au collisions at $\sqrt{s_{\mathrm{NN}}} = 200$ GeV. In this analysis, the residual correlation from $p{-}\Xi(1530)$ has been removed and we neglect other pairs contribution. The

© The Authors, published by EDP Sciences. This is an open access article distributed under the terms of the Creative Commons Attribution License 4.0 (https://creativecommons.org/licenses/by/4.0/).

measured correlation functions show clear enhancement at low k^* in all centrality classes and become more pronounced in peripheral (40%-80%) collisions. The correlation functions are simultaneously fitted with LL model with five free parameters (Source size (R_G) for three centrality bins, a common f_0 and d_0) in each collision system using Bayesian method [8]. It is clear that only Coulomb interaction cannot account for the full measured correlation functions. Results of the UrQMD + HALQCD calculations [6-7] are shown as orange band in the Fig. 1. The transverse momentum (p_T) spectra of single particles from the UrQMD model is employed to generate phase space, the interaction potential is provided by Lattice QCD [6-7]. The simulated model results describe the data reasonably well in all centrality classes.

Figure 2 shows the probability density distribution of the strong interaction parameters f_0 and d_0 for p–Ξ^- pairs for each collision systems obtained from the LL fit. It is found that the f_0 and d_0 are consistent with Isobar and Au+Au collisions within 1σ. Using same fitting procedure, the results from UrQMD + HALQCD simulation are consistent with data within 1σ. The f_0 is determined in the range [0.57, 0.90] fm, which is relatively small, indicating that the attractive interaction between p–Ξ^- pairs is shallow. Figure 3 panel (b) shows the comparison with HALQCD theory, it is observed that the HALQCD prediction is within good agreement with experimental data. This is the first time, the strong interaction parameters for p–Ξ^- pairs are extracted.

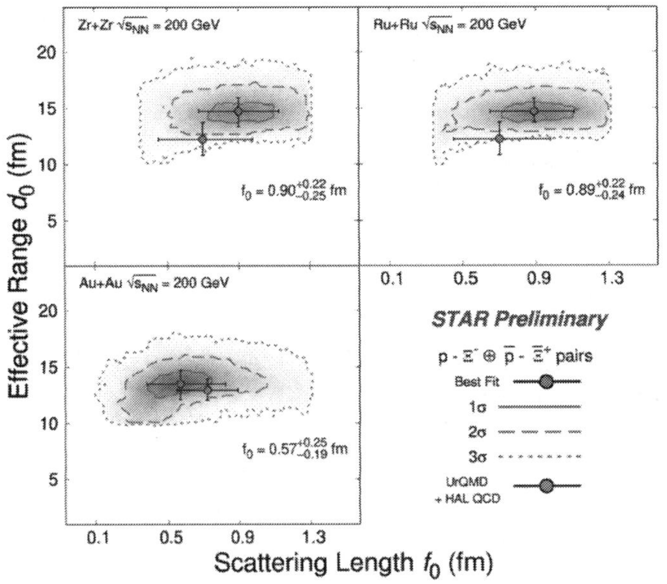

Figure 2. The p–Ξ^- spin-averaged scattering parameters (f_0) and effective range (d_0) obtained from simultaneous fit in Isobar and Au+Au collisions by Bayesian method. The best fit results are shown in red solid points. The purple bands and dashed lines represent the 1σ to 3σ standard deviations of the fit. The blue solid points represent the f_0 and d_0 extracted from UrQMD + HALQCD simulation with the same method.

The obtained source size R_G for p–Ξ^- pairs in different centrality classes from the LL fit are presented in Fig. 3 panel (a). It is observed that the R_G exhibit a monotonic decrease from central to peripheral collisions. Among different collision systems, the extracted R_G tends to follow a linear distribution as a function of $dN_{ch}/d\eta$ (charged particles rapidity density). The same fitting were performed to UrQMD simulations and the resulting radius from model are close to data.

© The Authors, published by EDP Sciences. This is an open access article distributed under the terms of the Creative Commons Attribution License 4.0 (https://creativecommons.org/licenses/by/4.0/).

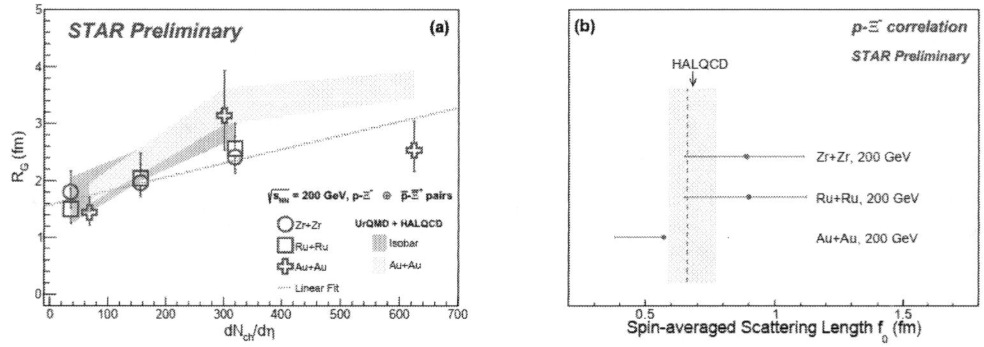

Figure 3. (a) p–Ξ^- source size R_G extracted from LL function as a function of $dN_{ch}/d\eta$ in Isobar and Au+Au collisions. The gray dotted line shows the linear fit to all data points. The pink and blue bands represent the R_G extracted from UrQMD + HALQCD simulation. (b) Comparison of p–Ξ^- scattering length f_0 between experimental data and HALQCD theory predictions. The black points represent extracted f_0 from Ru+Ru, Zr+Zr and Au+Au collisions at $\sqrt{s_{NN}}$ = 200 GeV. The blue band represent theory prediction from HALQCD Collaboration, the width of band represents the statistical uncertainties [6].

4 Summary

In these proceedings, we present the systematic measurements of p–Ξ^- correlation functions in Isobar and Au+Au collisions at $\sqrt{s_{NN}}$ = 200 GeV. For the first time, the source size and strong interaction parameters are extracted for p–Ξ^- pairs in each collision systems with Lednický-Lyuboshitz approach. It is observed that, from central to peripheral collisions, the R_G decreases. Calculations from the transport model UrQMD coupled with Lattice QCD potential provide a good description of the data. The extracted strong interaction parameters (f_0, d_0) are consistent between Isobar and Au+Au collisions. The f_0 has a rather small value indicating there is a shallow (Compared to typical *N*-*N* interaction) attractive interaction in p–Ξ^- pairs. Our measurements provide strong constraints towards the understanding of *Y*-*N* interactions.

5 Acknowledgements

This work was supported in part by National Key Research and Development Program of China (No.2022YFA1604900, No.2020YFE0202002), National Natural Science Foundation of China (No.12122505) and the Fundamental Research Funds for the Central Universities (CCNU220N003).

References

[1] M. Lisa et al., Ann. Rev. Nucl. Part. Sci. **55**, 357–402 (2005)
[2] J. M. Lattimer et al., Science **304**, 536 (2004)
[3] R. L. Jaffe, Phys. Rev. Lett. **38**, 195 (1977)
[4] R. Lednicky et al., Sov. J. Nucl. Phys. **35**, 770 (1982)
[5] S. Pratt et al., Phys. Rev. C **56**, 1095 (1997)
[6] Y. Kamiya et al., Phys. Rev. C **105**, 014915 (2022)
[7] K. Sasaki et al., Nucl. Phys. A **998** 121737 (2020)
[8] H. Mäntysaari et al., Phys. Lett. B **833**, 137348 (2022)

© The Authors, published by EDP Sciences. This is an open access article distributed under the terms of the Creative Commons Attribution License 4.0 (https://creativecommons.org/licenses/by/4.0/).

 https://doi.org/10.1051/epjconf/202531603011

Measurement of strange baryon production in charged-particle jets in pp and p-Pb collisions with ALICE

Gijs van Weelden[1,2,*], on behalf of the ALICE Collaboration

[1]Nikhef, Science Park 105, 1098 XG Amsterdam
[2]Utrecht University, Heidelberglaan 8, 3584 CS Utrecht

Abstract. The enhancement of the ratio of strange to non-strange hadrons with increasing charged-particle multiplicity density is often interpreted as a signature of the medium formed in heavy-ion collisions. However, this effect has also been observed in smaller systems, such as pp and p–Pb collisions. Understanding this effect requires a detailed description of the production mechanisms of strange hadrons in and out of jets. In this manuscript published results on the production of K^0_S, $\Lambda(\bar{\Lambda})$, $\Xi^{\pm}$ and $\Omega^{\pm}$ hadrons in charged-particle jets and the underlying event in pp and p-Pb collisions measured by ALICE are presented. In addition, a new analysis of in-jet fragmentation into Λ and K^0_S hadrons in pp collisions at $\sqrt{s}$ = 13.6 TeV from LHC Run 3 data is shown.

1 Introduction

In heavy-ion collisions, a medium of deconfined quarks and gluons is formed. One signature of this medium is the enhancement of the ratio of strange to non-strange hadrons with increasing charged-particle multiplicity density. However, this effect has also been observed in smaller systems, such as pp and p–Pb collisions, where such a medium is not expected.

The yields of K^0_S, Λ ($\bar{\Lambda}$), $\Xi^{\pm}$, and $\Omega^{\pm}$ hadrons were measured as a function of charged-particle multiplicity in pp, p–Pb and Pb–Pb collisions by the ALICE Collaboration [1]. The yields of these strange hadrons were found to increase with multiplicity at a rate faster than that of the pion yields. Questions remain on the mechanisms driving the production of these hadrons and their increase with increasing charged-particle multiplicity. In these proceedings, possible ways to disentangle the strangeness production mechanisms in jets and in the underlying event (UE) are investigated. A published analysis [2] on the production of K^0_S, Λ ($\bar{\Lambda}$), $\Xi^{\pm}$, and $\Omega^{\pm}$ in charged-particle jets and in the UE in pp and p–Pb collisions is presented. In particular, this is the first measurement of $\Omega^{\pm}$ production in jets. Additionally, a new analysis is discussed on in-jet fragmentation into K^0_S and Λ ($\bar{\Lambda}$) using Run 3 ALICE data.

2 Strange hadron yields in charged-particle jets

Jets are reconstructed using the anti-k_T algorithm [3, 4], with a jet resolution parameter R = 0.4, using tracks with transverse momentum $p_{T,track} > 0.15$ GeV/c and pseudorapidity $|\eta_{track}| < 0.8$, across the full acceptance in azimuth φ. The pseudorapidity of reconstructed

*e-mail: g.van.weelden@cern.ch

© The Authors, published by EDP Sciences. This is an open access article distributed under the terms of the Creative Commons Attribution License 4.0 (https://creativecommons.org/licenses/by/4.0/).

jets is required to be in the range $|\eta_{\rm jet}| < 0.35$, ensuring the jets are fully contained within the acceptance region of reconstructed weak decays. In addition, a transverse momentum cut of $p_{\rm T,jet} > 10$ GeV/c is applied to ensure the jet originates from a hard scattering process.

This analysis considers the V0 particles ($\rm K^0_S$, Λ, and $\bar{\Lambda}$), as well as the cascades (here meaning $\Xi^{\pm}$ and $\Omega^{\pm}$). The following decay channels were searched for, with their corresponding branching ratios (*BR*):

$$
\begin{aligned}
\mathrm{K^0_S} &\rightarrow \pi^+ + \pi^- & BR &= (69.20 \pm 0.05)\%, \\
\Lambda(\bar{\Lambda}) &\rightarrow \mathrm{p}(\bar{\mathrm{p}}) + \pi^-(\pi^+) & BR &= (63.9 \pm 0.5)\%, \\
\Xi^-(\bar{\Xi}^+) &\rightarrow \Lambda(\bar{\Lambda}) + \pi^-(\pi^+) & BR &= (99.887 \pm 0.035)\%, \\
\Omega^-(\bar{\Omega}^+) &\rightarrow \Lambda(\bar{\Lambda}) + \mathrm{K}^-(\mathrm{K}^+) & BR &= (67.8 \pm 0.7)\%.
\end{aligned}
$$

The V0 and cascade candidates are reconstructed based on the topology of their weak decays, within a pseudorapidity range of $|\eta| < 0.75$. The purity is improved by identifying charged tracks based on their energy deposition in the Time Projection Chamber subdetector. The invariant mass is used to extract the signal and combinatorial background. In addition, candidates that satisfy the reconstruction criteria for multiple particles, are rejected.

Particles are defined as within the jet cone (JC) if their distance in the $\eta - \varphi$ plane with respect to the jet axis is less than the jet resolution parameter. This includes both particles associated with the hard scattering and particles from the UE. The UE contribution is estimated with the perpendicular cone method (PC) [2], where the yields of strange hadrons are measured within a cone of size R, placed at the same η as the jet and at an angle $\Delta\varphi = \pi/2$.

The $p_{\rm T}$-differential yields $\mathrm{d}\rho/\mathrm{d}p_{\rm T}$ of the particles are calculated by

$$\frac{\mathrm{d}\rho}{\mathrm{d}p_{\rm T}} = \frac{1}{N_{\rm ev}} \times \frac{1}{A_{\rm acc}} \times \frac{\mathrm{d}N}{\mathrm{d}p_{\rm T}}, \tag{1}$$

where $\mathrm{d}N/\mathrm{d}p_{\rm T}$ is the $p_{\rm T}$-differential particle production yields, $N_{\rm ev}$ is the number of events, $A_{\rm acc}$ is the area of the $\eta - \varphi$ acceptance for a given selection. The $p_{\rm T}$-differential yields of in-jet particles can then be extracted by subtracting the PC selected yields from the JC selected yields.

Figure 1 shows the baryon-to-meson and baryon-to-baryon ratios for the measured strange particles in pp collisions. For each of these ratios, the UE yields closely follow the inclusive yields. For the baryon-to-meson ratios and Ξ/Λ, an enhancement is seen in the inclusive and UE yields with respect to the in-jet yields. For the Ω/Λ and Ω/Ξ ratios, this enhancement is not seen, but the uncertainties are too large to make any definitive statements.

Figure 2 shows the $\Xi/\mathrm{K^0_S}$ yield ratio in p–Pb collisions for different multiplicity bins: high (0-10%), intermediate (10-40%) and low (40-100%). The UE yields show an enhancement with increasing multiplicity, whereas the in-jet particle yields are consistent within uncertainties across multiplicity. A similar enhancement is found for the ratios of $\Lambda/\mathrm{K^0_S}$ and Ξ/Λ. Due to the low number of Ω candidates in the studied data sample, the ratios of $\Omega/\mathrm{K^0_S}$, Ω/Λ, and Ω/Ξ were only measured for Minimum Bias (MB) p–Pb collisions. The results are found to be consistent within uncertainties for MB pp and MB p–Pb collisions.

3 Jet fragmentation

A different approach to studying the production of strange hadrons in jets is to measure the jet fragmentation functions. The jet fragmentation function for a hadron is defined as the number density of hadrons in a jet, that carry a fraction z of the longitudinal momentum of the jet. The

© The Authors, published by EDP Sciences. This is an open access article distributed under the terms of the Creative Commons Attribution License 4.0 (https://creativecommons.org/licenses/by/4.0/).

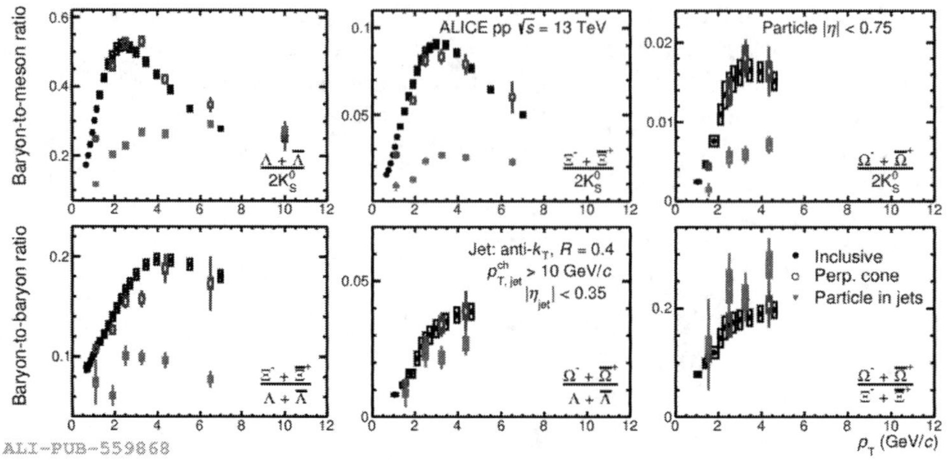

Figure 1: p_T-dependent baryon-to-meson (top) and baryon-to-baryon (bottom) yield ratios for inclusive, in-jet, and PC selected particles in pp collisions.

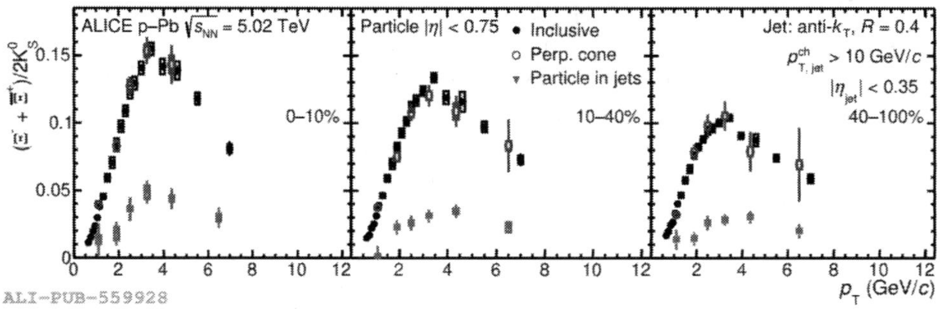

Figure 2: Ξ/K^0_S yield ratios for inclusive, in-jet, and PC selected particles in p–Pb collisions in different multiplicity classes.

longitudinal momentum fraction is calculated by projecting the particle 3-momentum vector $\mathbf{p}_i$ onto that of the jet ($\mathbf{p}_{jet}$) and normalising to the jet momentum

$$z_i = \frac{\mathbf{p}_i \cdot \mathbf{p}_{jet}}{\mathbf{p}^2_{jet}}. \tag{2}$$

Due to the relatively long lifetimes of K^0_S (2.6844 cm/c) and Λ (7.89 cm/c) [5], their decay daughters are removed from the jet clustering input by removing tracks with a large distance of closest approach to the primary vertex. Therefore, the V0 candidates are reconstructed separately and added to the jet clustering input, yielding a collection of charged-particle + V0 jets (ch+V0 jets).

To illustrate the performance of the ch+V0 jet approach in Run 3 ALICE data, figure 3 shows the invariant mass spectra of V0 particles in ch+V0 jets around the K^0_S invariant mass. V0 particles with $0.5 < z < 0.7$ were selected in jets with $10 < p_{T,jet} < 20$ GeV/c (left) and $20 < p_{T,jet} < 40$ GeV/c (right). The clearly visible invariant mass peaks with large numbers of candidates in the Run 3 data enables precise measurements of V0 fragmentation distributions in jets and study of the dependence on the jet momentum.

© The Authors, published by EDP Sciences. This is an open access article distributed under the terms of the Creative Commons Attribution License 4.0 (https://creativecommons.org/licenses/by/4.0/).

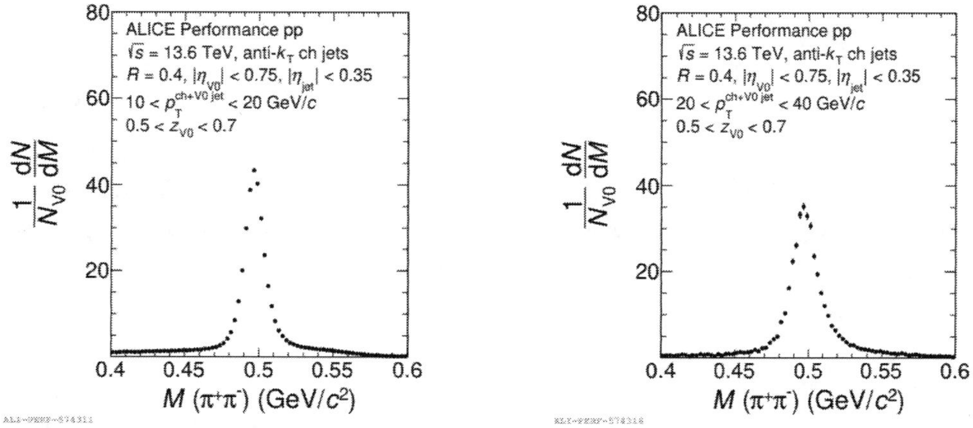

Figure 3: Invariant mass spectrum of V0 candidates in ch+V0 jets around the K^0_S mass, with $0.5 < z_{V0} < 0.7$. Left: $10 < p_{T,jet} < 20$ GeV/c, right: $20 < p_{T,jet} < 40$ GeV/c.

4 Conclusions

Studying the production of strange hadrons in jets can provide insight into the mechanisms that drive the increase of the strange to non-strange hadron ratio with charged-particle multiplicity density. In these proceedings, the first measurement of $\Omega^{\pm}$ baryons in jets was presented, as well as results on K^0_S, Λ ($\bar{\Lambda}$), and $\Xi^{\pm}$ yields in jets, in pp and p–Pb collisions. It is observed that the baryon-to-meson and the 2-strange-to-1-strange yield ratios show an enhancement with multiplicity in the UE, but not in jets. Also, an approach to measuring in-jet fragmentation into K^0_S and Λ ($\bar{\Lambda}$) in pp collisions using jets constructed from both charged particles and reconstructed V0 candidates was shown. This approach showcases the possibility to identify V0 particles at high z in Run 3 ALICE data.

References

[1] J. Adam *et al.* [ALICE Collaboration], Nature Phys. **13** (2017), 535-539 doi:10.1038/nphys4111 [arXiv:1606.07424 [nucl-ex]].
[2] S. Acharya *et al.* [ALICE Collaboration], JHEP **07** (2023), 136 doi:10.1007/JHEP07(2023)136 [arXiv:2211.08936 [nucl-ex]].
[3] M. Cacciari, G. P. Salam and G. Soyez, Eur. Phys. J. C **72** (2012), 1896 doi:10.1140/epjc/s10052-012-1896-2 [arXiv:1111.6097 [hep-ph]].
[4] M. Cacciari and G. P. Salam, Phys. Lett. B **641** (2006), 57-61 doi:10.1016/j.physletb.2006.08.037 [arXiv:hep-ph/0512210 [hep-ph]].
[5] S. Navas *et al.* [Particle Data Group], Phys. Rev. D **110** (2024) no.3, 030001 doi:10.1103/PhysRevD.110.030001

© The Authors, published by EDP Sciences. This is an open access article distributed under the terms of the Creative Commons Attribution License 4.0 (https://creativecommons.org/licenses/by/4.0/).

EPJ Web of Conferences **316**, 03012 (2025) https://doi.org/10.1051/epjconf/202531603012
SQM 2024

Preliminary results on the transverse flow of charged K mesons emitted from Ag+Ag collisions at beam energy of 1.58 GeV/nucleon measured with HADES

Jan Orlinski[1,*] for the HADES Collaboration[**]

[1]Faculty of Physics, University of Warsaw

Abstract. In this contribution, preliminary results on the flow of charged K mesons emitted from semicentral Ag+Ag collisions at a beam kinetic energy of 1.58 GeV/nucleon measured with HADES are presented as $v_{1,2}$ coefficients mapped in the scaled rapidity – transverse momentum phase space. The coefficients, extracted by fitting a Fourier series to the azimuthal angle distributions, are corrected for the finite resolution of the event plane reconstruction, but not yet for inefficiencies related to the local track density in the detector. The $K^{\pm}$ flow is compared to the corresponding distributions for protons emitted from the same system. These comparisons suggest a repulsive potential between the K^+ and nuclear matter, while conclusions for the K^- are not possible to be drawn at this point, due to high statistical uncertainties.

1 Introduction and motivation

Heavy-ion collisions at energies of few *A*GeV allow the study of nuclear matter at extreme temperatures, pressures and densities [1, 2] by analysing the kinematic distributions of particles emitted from collisions and deducing the nature of the hot and dense nuclear matter via model comparisons. In this work, transverse flow is understood as anisotropies of emission in the azimuthal angle (with respect to the beam axis), sensitive to multiple properties of nuclear matter, like the Equation of State or partial restoration of chiral symmetry [3, 4].

For K mesons, an important open question is their interaction with nucleons, as well as with the $q\bar{q}$ condensate, in the expanding fireball. It was predicted in Ref. [5] that the potential of the kaon-nucleon (KN) interaction should be repulsive for the K^+ and attractive for the K^-. The consequence of this interaction should be the changing of kaon mass and decay constant in dense nuclear matter [3]. Results on charged kaon flow in Ni+Ni collisions at a beam kinetic energy of 1.91*A* GeV published by the FOPI Collaboration [6] appear to support this hypothesis, although no final statement can be made as of now, in part due to limited statistics in the samples analysed therein.

The Ag+Ag collisions at beam energy of 1.58 GeV/nucleon, measured by the HADES spectrometer, may be essential in answering these open questions, due to the acceptance and precision of HADES, as well as the large statistics collected during the Ag+Ag run. By studying the transverse flow of $K^{\pm}$ mesons with unprecedentedly low uncertainties and a large coverage of the phase space, we hope to significantly further our understanding of this effect.

*e-mail: Jan.Orlinski@fuw.edu.pl
**www.hades.gsi.de

© The Authors, published by EDP Sciences. This is an open access article distributed under the terms of the Creative Commons Attribution License 4.0 (https://creativecommons.org/licenses/by/4.0/).

2 The HADES spectrometer

HADES [7], standing for High-Acceptance Di-Electron Spectrometer, is a fixed-target experiment measuring mainly products of relativistic heavy-ion collisions, although some experiments with elementary beams were also conducted. A sketch of the setup is presented in Figure 1 and a short overview is provided in this section.

An essential component of the spectrometer are the four Multiwire Drift Chamber (MDC) planes separated by the superconducting magnet. This allows a momentum measurement based on the deflection of tracks in the known magnetic field. The relative precision of momentum reconstruction is 4 % [7], when using the Runge-Kutta method to solve the differential equations of motion of a charged particle in the known magnetic field.

By extrapolating the tracks to a common vertex in the target and to the detectors of the META system (Time Of Flight and Resistive Plate Chamber) one can obtain a path length for each track. This extrapolation can be matched to the nearest signal in TOF/RPC, and the event time from the START detector. Thus, the path length and the travel time within the detector can be obtained: two components of the velocity ($\vec{v}$). In combination with the reconstructed momentum ($\vec{p}$), this can be used to calculate the mass (m) for each track, based on the relativistic formula $\vec{p} = \gamma m\vec{v}$. The HADES detectors and track reconstruction methods are described in detail in Ref. [7].

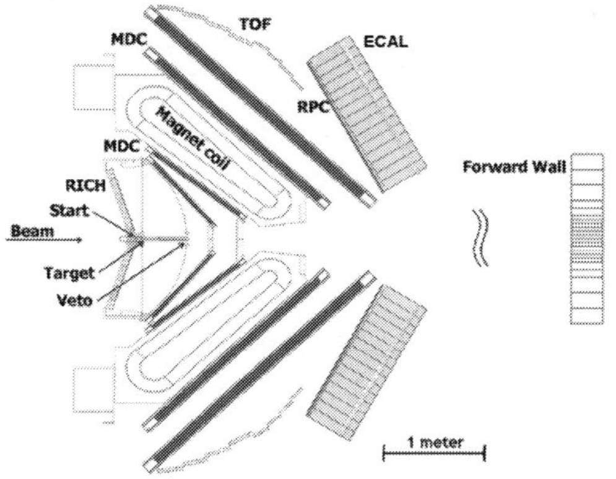

Figure 1. A schematic view of the HADES spectrometer as it was used for the Ag+Ag beamtime in March 2019. The distance between the main body and the Forward Wall detector is not to scale [7].

3 The flow extraction method

The K$^{\pm}$ mesons signal is extracted by modelling the signal and background in experimental mass distributions within cells of the momentum space in the scaled rapidity y_0[1], transverse momentum p_t and the azimuthal angle with respect to the event plane $\Delta\phi$. From these, a $\Delta\phi$ distribution can be obtained for every point in (y_0, p_t) and modelled with a cosine Fourier series:

$$\frac{dN}{d\Delta\phi} = N\left(1 + 2\sum_n v_n \cos(n\Delta\phi)\right), \tag{1}$$

[1]Scaled rapidity is defined as $y_0 = \frac{y - y_{CM}}{y_{CM}}$, where y is the rapidity of a track and y_{CM} is the center-of-mass rapidity, both calculated in the Laboratory frame.

© The Authors, published by EDP Sciences. This is an open access article distributed under the terms of the Creative Commons Attribution License 4.0 (https://creativecommons.org/licenses/by/4.0/).

yielding maps of v_1 and v_2, describing the directed and elliptic flow, respectively, as functions of transverse momentum and scaled rapidity.

The distributions are then corrected for the finite resolution of the event-plane reconstruction. The event plane in HADES is reconstructed based on charged hits from the collision spectators in the Forward Wall detector located a few meters downstream from the main setup (see Fig. 1). This is done only with some limited resolution and the random smearing in this reconstruction dampens the measured anisotropies. Correction factors for the strength of each harmonic are calculated according to the method described by J.-Y. Ollitrault in Ref. [8].

The presented distributions are not yet corrected for relative efficiency losses in areas of large local track density, which causes a lack of symmetry with respect to mid-rapidity, which would be expected from the system's geometry. Other efficiency effects are negligible, since they are dependent on p_t and y_0, and no integrated distributions are shown herein.

4 Results and discussion

In Figure 2 the experimentally found v_1 coefficient is shown for the K^+ (left panel) and the K^- (right panel). The distributions are shown as functions of scaled rapidity, for consecutive bins of p_t. The distributions are superimposed onto the proton flow measured in the p_t interval of 500-600 MeV/c. As the distributions are not corrected for the aforementioned occupancy effects (expected to affect more the y_0>0 region), and thus not symmetrical, it is currently hard to draw any physical conclusions. This is complicated further by the lack of systematic uncertainties in the analysis. However, the behaviour of K^+ mesons in comparison to that of the protons, could be seen as an indication of the so-called *antiflow* effect, where the studied particles have an inverse slope of $v_1(y_0)$ than that of the bulk matter. The appearance of this effect, within the statistical uncertainties, seems to be p_t-dependent. For the K^-, current statistical uncertainties themselves make any speculation difficult.

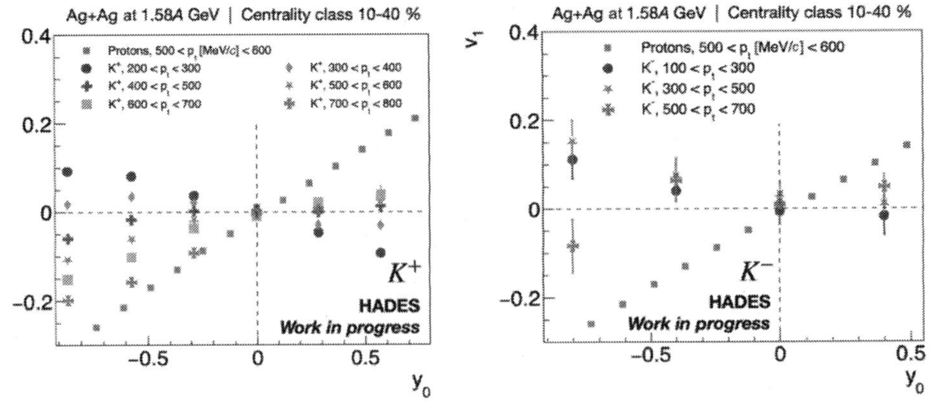

Figure 2. Distribution of v_1 coefficients describing the directed flow of K^+ mesons (left panel) and K^- (right) as function of scaled rapidity y_0 and in different p_t ranges. The flow of protons with $500 < p_t$ [MeV/c] < 600 emitted from the same system is shown in grey. For many points the statistical uncertainties are smaller than the markers; systematic uncertainties are not evaluated as of yet.

In Figure 3, the v_2 profiles are shown for both charges (K^+ on the left, K^- – right) as a function of transverse momentum, for different y_0 bins. The mid-rapidity proton flow is superimposed onto the plot. Here, the occupancy corrections were also not applied and the systematics were not evaluated. Moreover, the statistical uncertainties are higher for the v_2 coefficient, than for v_1. Thus, it is impossible to draw firm physical conclusions as of yet.

© The Authors, published by EDP Sciences. This is an open access article distributed under the terms of the Creative Commons Attribution License 4.0 (https://creativecommons.org/licenses/by/4.0/).

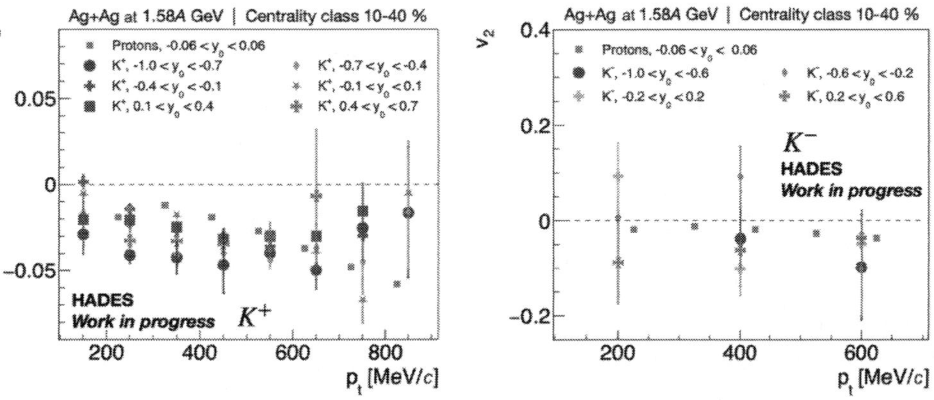

Figure 3. Distribution of v_2 coefficients describing the elliptic flow of K^+ mesons (left panel) and K^- (right) as function of transverse momentum p_t and in different y_0 ranges. The flow of protons emitted from the same system at midrapidity is also shown in grey. Systematic uncertainties are not evaulated as of yet.

5 Summary and outlook

The unprecedented precision of the HADES experiment for Ag+Ag collisions at a beam kinetic energy of 1.58 GeV/nucleon may significantly advance our understanding of the KN interaction in hot and dense nuclear matter, by measuring the transverse flow of charged K mesons. In this contribution, the work-in-progress results of this measurement for Ag+Ag collisions at beam energy of 1.58 GeV/nucleon are shown. The results still require efficiency corrections and an evaluation of systematic uncertainties, however the achieved precision and coverage are already promising. The results allow us to speculate, that – once they are finalized – they will provide valuable insight into the interaction of K mesons with nuclear matter.

References

[1] C. Hartnack *et al.*, Phys. Rept. **510**, 119-200 (2012) https://doi.org/10.1016/j.physrep.2011.08.004
[2] P. Danielewicz *et al.*, Science **298**, 1592-1596 (2002) https://doi.org/10.1126/science.1078070
[3] T. Song *et al.*, Phys. Rev. C **103**, 044901 (2021) https://doi.org/10.1103/PhysRevC.103.044901
[4] Y. Nara, A. Onishi, Phys. Rev. C **105**, 014911 (2022) https://doi.org/10.1103/PhysRevC.105.014911
[5] J. Schaffner-Bielich, I. Mishustin, J. Bondorf, Nucl. Phys. A **625**, 325 (1997) https://doi.org/10.1016/S0375-9474(97)81464-9
[6] V. Zinyuk *et al.* (FOPI Collaboration), Phys. Rev. C **90**, 025210 (2014) https://doi.org/10.1103/PhysRevC.90.025210
[7] G. Agakichiev *et al.* (HADES Collaboration), Eur. Phys. J. A **41**, 243-277 (2009) https://doi.org/10.1140/epja/i2009-10807-5
[8] J.-Y. Ollitrault, arXiv:9711003v2 [nucl-ex] (2018) https://doi.org/10.48550/arXiv.nucl-ex/9711003

© The Authors, published by EDP Sciences. This is an open access article distributed under the terms of the Creative Commons Attribution License 4.0 (https://creativecommons.org/licenses/by/4.0/).

Strangeness production in fixed-target collisions at LHCb

Chiara Lucarelli[1,*] on behalf of the LHCb collaboration

[1]INFN Firenze, Via Sansone, 1, 50019 Sesto Fiorentino (FI), Italy

Abstract. Studies of strangeness production in high-energy fixed-target collisions provide information on hadronization and serve as important inputs to models of particle production in cosmic rays. Thanks to its forward geometry and the possibility to inject gas into the beam pipe, the LHCb detector with the SMOG system is ideally suited to study strangeness production in fixed-target collisions at the LHC. We present recent LHCb results on strangeness production in fixed-target proton-nucleus collisions, including studies of hyperon production and polarization.

1 Introduction

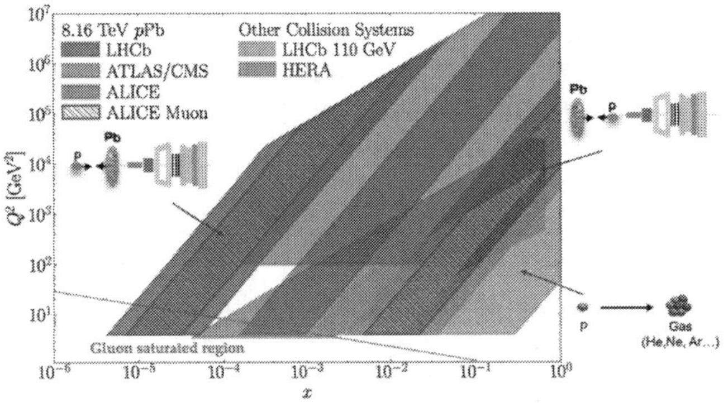

Figure 1. The LHCb acceptance in the Q^2/ Bjorken-x plane compared to other experiments.

The LHCb experiment is a single-arm forward spectrometer covering the pseudorapidity range between $2 < \eta < 5$ [1]. Originally designed to study heavy hadrons produced in *pp* collisions at the LHC, it has proven during the 2011-2018 data-taking years to be a general purpose experiment in the forward direction. Among the experiments operating at the LHC, LHCb has the unique ability to inject gases in the accelerator beam-pipe, a feature originally conceived to exploit the proton-gas collisions in order to reconstruct the LHC beam transverse profiles and measure the luminosity. Since 2015, leveraging on the forward geometry of the detector and its excellent particle reconstruction and identification performance, the gas injection system SMOG has also been used as a gaseous target, turning LHCb into the highest-energy fixed-target experiment ever. During Run 2, samples with both proton and lead ion beams were collected injecting noble gases (He, Ne, Ar),

*e-mail: chiara.lucarelli@cern.ch

© The Authors, published by EDP Sciences. This is an open access article distributed under the terms of the Creative Commons Attribution License 4.0 (https://creativecommons.org/licenses/by/4.0/).

covering a rapidity region in the centre of mass system ($-3 < y^* < 0$) complementary to the ones covered by the collider mode samples. This configuration provides unique physics opportunities at the LHC. The high Bjorken-x and moderate Q^2 region, mostly unexplored by previous experiments, can be accessed with SMOG and collision systems with atomic numbers between the proton and lead can be studied, providing inputs to many research fields such as heavy flavour production, nuclear structure and cosmic rays astrophysics (Fig. 1) [2].

2 $\bar{p}$ production from antihyperon decays in *p*He collisions at $\sqrt{s_{\rm NN}} = 110$ GeV

The antimatter abundance in cosmic rays is a sensitive probe for indirect dark matter searches. Space-based experiments, such as AMS-02, have been measuring the antimatter flux and have highlighted possible discrepancies with the spectra expected from the production in cosmic rays spallation with the interstellar medium, mainly composed of hydrogen and helium. However, the interpretation of these results is still limited by our knowledge of the antimatter production in known processes [3]. Exploiting the injection of helium through SMOG, LHCb performed the first-ever measurements of prompt antiproton ($\bar{p}$) production in pHe collisions at $\sqrt{s_{\rm NN}} = 110$ GeV, contributing to improvements of the modelling of secondary $\bar{p}$ production in space [4]. The non-prompt $\bar{p}$ contribution originating from anti-hyperon decays, called "detached", has been addressed in a separate measurement [5]. The detached-to-prompt $\bar{p}$ production ratio has been measured using two complementary approaches, exploiting the prompt result in the ratio. In the exclusive approach, the dominant contribution coming from the $\bar{\Lambda} \to \bar{p}\pi^+$ is reconstructed by exploiting the secondary vertex topology without the use of any particle identification information. A second inclusive approach considers the detached $\bar{p}$ from all anti-hyperon species. The $\bar{p}$ candidates are identified using the Ring Imaging Cherenkov detectors, and the prompt and detached contributions are determined based on each candidates consistency with originating from the pHe collision point. The dominant uncertainty due to the luminosity cancels out in the ratio in both cases. Both approaches highlight a larger contribution in the ratio with respect to the predictions of the most commonly used theoretical models. The results will contribute to reducing the uncertainty on the secondary $\bar{p}$ production cross-section in the collisions between the cosmic rays and the interstellar medium (Fig. 2).

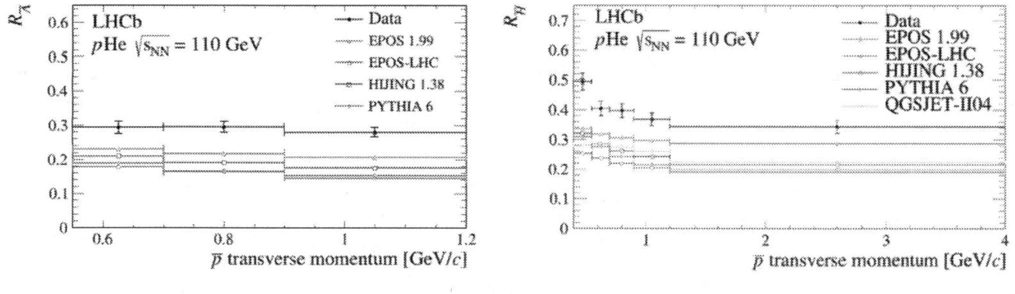

Figure 2. Results for the detached-to-prompt antiproton ratio measurement as a function of the transverse momentum. Both the exclusive (left) and inclusive (right) approaches show an excess with respect to the predictions.

© The Authors, published by EDP Sciences. This is an open access article distributed under the terms of the Creative Commons Attribution License 4.0 (https://creativecommons.org/licenses/by/4.0/).

3 Λ transverse polarization in *p*Ne collisions at $\sqrt{s_{NN}} = 68.4$ GeV

The spontaneous transverse polarization of Λ hyperons was first observed in 1976 in inclusive production by an unpolarized 300 GeV proton beam on a beryllium target [6]. While several mechanisms are under study to explain the polarized production of the *Lambda* in collisions involving nuclei, as of today, the polarization in light systems has been generally explained through the introduction of transverse-momentum-dependent (TMD) fragmentation functions. In particular, the polarizing fragmentation function $D_{1T}^{\perp}$ describes the fragmentation of an unpolarized quark into a transversely polarized hadron accounting for spin and momentum correlations [7]. Given that the determination of TMD fragmentations is based on experimental data, several experiments have been performed to try to describe the Λ polarization. LHCb performed the Λ transverse polarization measurement considering the *p*Ne sample at $\sqrt{s_{NN}} = 68.4$ GeV [8]. The decays of the $\Lambda \to p\pi^-$ and $\bar{\Lambda} \to \bar{p}\pi^+$ are exploited to measure the polarization. Given the strong parity violation of these decays, the angular distribution of the products show a large asymmetry. Considering the Λ rest frame, the proton is preferentially emitted along the direction of the Λ spin and the magnitude of the polarization is obtained from the linear fit to the angular distribution of the proton. The final polarization obtained are $P_\Lambda = 0.029 \pm 0.019$ (stat) $\pm$ 0.012 (syst) and $P_{\bar{\Lambda}} = 0.003 \pm 0.023$ (stat) $\pm$ 0.014 (syst) for the Λ and $\bar{\Lambda}$, respectively. The results as a function of the x-Feynman variable x_F have also been compared with previous experiments (Fig. 3). Even though the results cover a wide range in the kinematics and collisions systems, the polarization obtained is in agreement with previous measurements, in particular with the HERA-B results, which cover a similar x_F region.

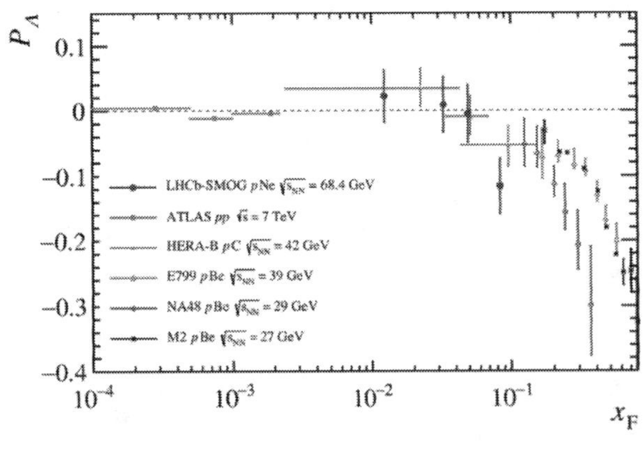

Figure 3. Comparison of the magnitude of the Λ hyperson polarization as a function of x_F obtained in the experiments performed up to the time of writing.

4 LHCb fixed-target Run3 upgrade

In preparation for the LHC Run 3, which began in 2022, the LHCb detector went through an upgrade, and a new gas injection system, SMOG2, was installed [9]. It consists of a 20 cm-long storage cell installed 44 cm upstream of the nominal LHCb interaction point and a new gas feed system that allows a fine and automatic control of the gas flow, pressure, and injected species. The confined interaction region allows for a two-orders-of-magnitude increases in instantaneous luminosity, and new non-noble gases, such as hydrogen (H_2), deuterium (D_2) and oxygen (O_2), can be injected. Moreover, the separation of the beam-beam and beam-gas interaction regions makes it possible to acquire data from both types of collisions simultaneously, making LHCb the first experiment running at the same time with two different interaction systems and collision energies. The first physics data have been collected during may 2024 with all available gases, recorded simultaneously with *pp* collisions [10]. Figure 4

© The Authors, published by EDP Sciences. This is an open access article distributed under the terms of the Creative Commons Attribution License 4.0 (https://creativecommons.org/licenses/by/4.0/).

shows the recorded integrated luminosity during May 2024, as well as an example of the invariant mass distribution of reconstructed $\Lambda \to p\pi^-$ candidates in the data-taking periods with simultaneous $pp + p$Ne collisions. While the data acquisition lasted 2 hours, more than twenty millions Λ candidates were reconstructed. A very large sample of charm hadrons can therefore be expected from the SMOG2 system. The upgraded fixed-target system will therefore contribute to a "significant extension of the LHC complex physics reach at a limited cost", as highlighted in the 2020 European Strategy for Particle Physics Update [11].

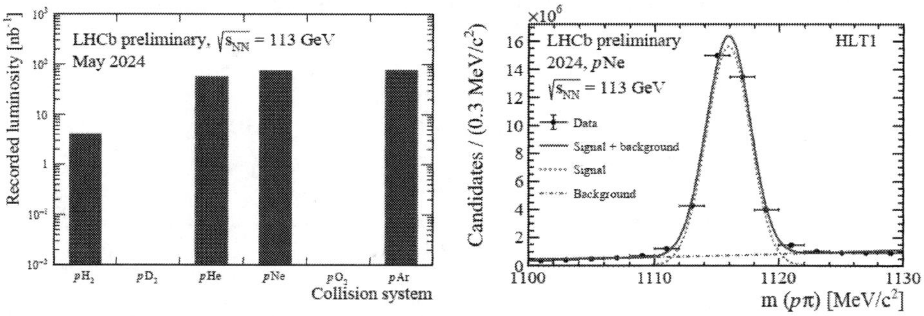

Figure 4. The left plot shows the integrated luminosity recorded during May 2024, corresponding to around two hours of injections per gas. The right plot shows the invariant mass distribution for $\Lambda \to p\pi^-$ decay candidates.

References

[1] Aaij, R. et al., The LHCb Detector at the LHC. Journal of Instrumentation **3**, 08 S08005 (2008). 10.1088/1748-0221/3/08/S08005

[2] Bursche, A. et al., Physics opportunities with the fixed-target program of the LHCb experiment using an unpolarized gas target. CERN-LHCb-PUB-2018-015 (2018).

[3] Giesen, G. et al., AMS-02 antiprotons, at last! Secondary astrophysical component and immediate implications for Dark Matter. JCAP **09**, 023 (2015). 10.1088/1475-7516/2015/09/023

[4] Aaij, R. et al., Measurement of Antiproton Production in pHe Collisions at $\sqrt{s_{NN}}$ = 110 GeV. Phys. Rev. Lett. **121**, 22 222001 (2018). 10.1103/PhysRevLett.121.222001

[5] Aaij, R. et al., Measurement of antiproton production from antihyperon decays in pHe collisions at $\sqrt{s_{NN}}$ = 110 GeV. Eur. Phys. J. C **83**, 6 543 (2023). 10.1140/epjc/s10052-023-11673-x

[6] Bunce, G. et al., Λ^0 Hyperon Polarization in Inclusive Production by 300-GeV Protons on Beryllium. Phys. Rev. Lett. **36**, 19 1113–1116 (1976). 10.1103/PhysRevLett.36.1113

[7] Boer, D. et al., Spin asymmetries in jet-hyperon production at LHC. Physics Letters B **659**, 1 127-136 (2008). 10.1016/j.physletb.2007.10.059

[8] Aaij, R. et al., Transverse polarization measurement of Λ hyperons in pNe collisions at $\sqrt{s_{NN}}$ = 68.4 GeV with the LHCb detector. Submitted to JHEP (2024). arXiv:2405.11324

[9] Boente Garcia, O., A high-density gas target at the LHCb experiment. Submitted to Physical Review Accelerators and Beams (2024). arXiv:2407.14200

[10] Aaij, R. et al., First results from SMOG2 2024. LHCB-FIGURE-2024-005 (2024).

[11] Ellis, R. K. et al., Physics Briefing Book: Input for the European Strategy for Particle Physics Update 2020. CERN-ESU-004 (2019).

© The Authors, published by EDP Sciences. This is an open access article distributed under the terms of the Creative Commons Attribution License 4.0 (https://creativecommons.org/licenses/by/4.0/).

Dependence of net-hyperon production at mid-rapidity on beam energy and its implication on baryon number carrier

Chun Yuen Tsang[1,2]

[1]Brookhaven National Laboratory, Upton, NY, USA
[2]Kent State University, 800 E Summit St, Kent, OH, USA

Abstract. The conventional picture of baryon number is that each valence quark inside a baryon carries 1/3 unit of baryon number. However, an alternative picture exists where the center of a Y-shaped topology of gluon fields, called the baryon junction, carries 1 unit of baryon number. Previous analysis suggests that the nature of the baryon number carrier is sensitive to net-proton yields at mid-rapidity in heavy-ion collisions, and experimental measurements potentially challenge the conventional picture. In this study, we analyzed published data from the BES-I program at RHIC, which performed Au+Au collisions at center-of-mass energies ranging from 7.7 to 200 GeV. We investigated the rapidity dependence of net-hyperon yields after correcting for strangeness production suppression, and found that the net-hyperon yields at mid-rapidity follow an exponential dependence on the beam rapidity. This behavior is consistent with Regge model predictions. Moreover, the exponential slopes for net-Λ, net-Ξ, and net-Ω are consistent with each other, suggesting that the baryon transport is flavor blind. Conventional models like `PYTHIA`, which use valence quarks as the baryon number carrier, have difficulty reproducing this transport behavior.

1 Introduction

The baryon number (B) is a conserved quantity in quantum chromodynamics (QCD). While mesons, like pions and kaons, carry zero baryon number ($B = 0$), baryons such as protons and neutrons carry one unit ($B = 1$). The conventional quark model assigns 1/3 unit of the baryon number to each quark. However, an alternative theory proposes that the entire unit of the baryon number is carried by the baryon junction, a non-perturbative Y-shaped topology of gluons connected to all three valence quarks. Neither scenario has been verified experimentally. Recent proposals have outlined experimental approaches to distinguish these scenarios in high-energy hadronic and photon-induced collisions [1].

Here, we report the dependence of net-hyperon yields, after correcting for the strange quark production suppression, on δy by compiling published data on hyperon production at mid-rapidity in heavy-ion collisions. Here, $\delta y = y_{beam} - y_{CM}$, where y_{beam} is the beam rapidity and $y_{CM} \sim 0$ denotes the rapidity where the net-hyperon yields are measured. Such a dependence is quantified by fitting the net-hyperon yields with $f(\delta y) \sim e^{-\alpha_B \delta y}$. Since baryon junction is flavor blind [1], one would expect the transport behavior of hyperons to resemble that of inclusive baryons. Specifically, we will inspect if the fitted α_B for net-Λ, net-Ξ and net-Ω are consistent with each other and with that of net-p at various centralities. Furthermore,

© The Authors, published by EDP Sciences. This is an open access article distributed under the terms of the Creative Commons Attribution License 4.0 (https://creativecommons.org/licenses/by/4.0/).

predictions of α_B for net-hyperons from various versions and tunes of the PYTHIA event generator will be compared to our results.

2 Beam energy dependence of net-hyperon yields

Using published data from RHIC BES-I program, we compute net-Λ, net-Ξ and net-Ω yields, $(dN/dy)_{\Lambda-\bar{\Lambda}}/\langle N_{\text{part}}\rangle$, $(dN/dy)_{\Xi-\bar{\Xi}}/\langle N_{\text{part}}\rangle$, and $(dN/dy)_{\Omega-\bar{\Omega}}/\langle N_{\text{part}}\rangle$, within the rapidity range of $|y| < 0.5$ in Au+Au collisions at center-of-mass energies ($\sqrt{s_{\text{NN}}}$) of 7.7, 11.5, 19.6, 27, 39, 62.4 and 200 GeV [2–5]. Here $\langle N_{\text{part}}\rangle$ is the average number of participating nucleons in a given centrality class of Au+Au collisions.

Since strange quarks must be pair-produced, we need to account for the effects associated with the strange quark production to isolate the effect of baryon transport. We gauge the difficulty of strange quark production using the $K^-(s\bar{u})/\pi^-(d\bar{u})$ yield ratio in Au+Au collisions. The correction for the strangeness production suppression is performed by dividing the net-hyperon yields by $(K^-/\pi^-)^n$, where n is the number of valence strange quarks in the hyperon. Figure 1 shows that net-hyperon yields follow exponential dependence on δy as the Regge theory predicted. The fitted α_B values are shown in Fig. 2a, and they agree with each other within 2σ for all hyperons and net-protons. There is a mild centrality dependence for peripheral collisions, but it disappears if we use K^-/π^- from 0-10% centrality for all centralities instead as Fig. 2b shows. It indicates that this centrality dependence is a result of changes in K^-/π^- ratio in peripheral collisions, rather than a change in baryon transport mechanism.

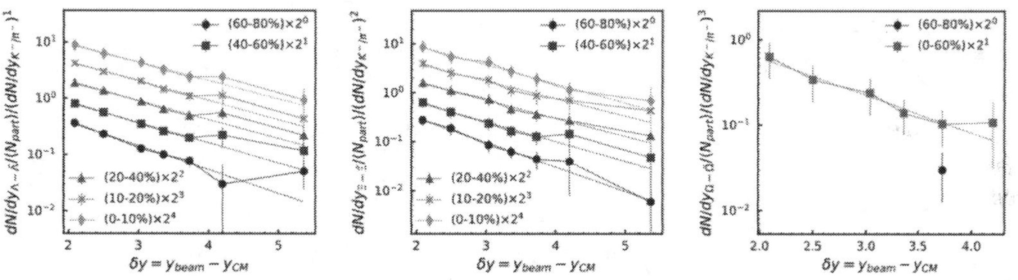

Figure 1: Net-Λ (left), net-Ξ (central) and net-Ω (right) yields within $|y| < 0.5$ as a function of δy in Au+Au collisions. These yields are scaled by $(K^-/\pi^-)^n$ in Au+Au collisions, where n is the number of valence strange quarks in the hyperon.

The independence of α_B on centrality shows that α_B is not affected by superfluous processes such as multiple scattering and reflects the inherent properties of baryon stopping mechanism, and therefore Au+Au results should be comparable to results from systems of other sizes. When other published data for Pb+Pb collisions at CERN SPS [6, 7] and by ALICE [8, 9], for 200 GeV Cu+Cu collisions by STAR [3], and for 200 GeV and 900 GeV $p+p$ collisions by STAR [10] and ALICE [11] are included in the fit for central collisions, α_B values remain unchanged with combined fits show $\alpha_B(\Lambda) = 0.72 \pm 0.06$, $\alpha_B(\Xi) = 0.93 \pm 0.12$ and $\alpha_B(\Xi) = 1.02 \pm 0.32$. Values for Au+Au only fit can be found in Table 1.

3 Comparison between data and PYTHIA

Here, we compare slope parameters in $p+p$ collisions predicted by the PYTHIA event generator, which employs valence quarks as the baryon number carriers, to those in Au+Au

© The Authors, published by EDP Sciences. This is an open access article distributed under the terms of the Creative Commons Attribution License 4.0 (https://creativecommons.org/licenses/by/4.0/).

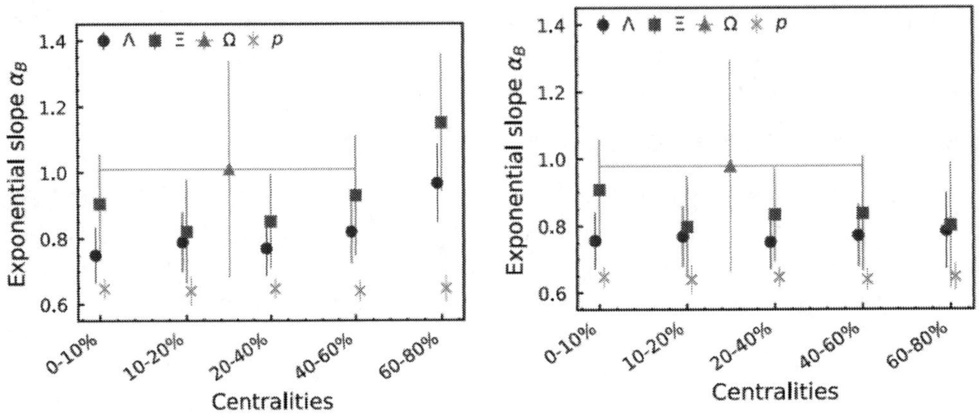

(a) K^-/π^- varies with centralities. (b) Always use K^-/π^- at 0-10% centrality.

Figure 2: Comparison of slope parameters for net-Λ, net-Ξ, net-Ω and net-p as a function of centrality from Au+Au collisions at $\sqrt{s_{\rm NN}}$ = 7.7 - 200 GeV. (a) used K^-/π^- at varying centralities, while (b) only used K^-/π^- at 0-10% for net-hyperon of different centralities.

Table 1: α_B values from measurements and PYTHIA.

	Data (0-60%)	Ver. 6.4	Ver. 6.4 (P0)	Ver. 6.4 (P12)	Ver. 8.3	Ver. 8.3 CR
Λ	0.72 ± 0.07	2.58 ± 0.03	1.15 ± 0.01	0.81 ± 0.01	1.18 ± 0.01	0.88 ± 0.01
Ξ	0.85 ± 0.13	N.A.	0.73 ± 0.05	0.52 ± 0.05	0.62 ± 0.08	0.55 ± 0.06
Ω	0.98 ± 0.32	N.A.	0.25 ± 0.10	0.04 ± 0.15	N.A.	N.A.
p	0.65 ± 0.07	0.74 ± 0.05	0.72 ± 0.02	0.35 ± 0.01	0.98 ± 0.02	0.69 ± 0.02

collisions. The fitted α_B values from PYTHIA are compared to measurements in Table 1. Values for some configurations of PYTHIA are marked as "N.A.", corresponding to cases where the PYTHIA distribution was not fitted due to presence of negative net-hyperon yield at any beam energy.

Not only does PYTHIA fail to reproduce the measured α_B, it also shows strong variations across hyperon species. PYTHIA 8.3 CR, which includes a partial implementation of the baryon junction picture, performs better than its non-CR counterpart as its predicted α_B values are closer to the measured values for both net-proton and net-Λ. This underscores the potential of implementing the baryon junction mechanism. However, it still fails to reproduce the slope parameter for net-Ξ and net-Ω, and does not eliminate the flavor dependence, as α_B values change significantly between net-Λ, and net-Ξ. This calls for a genuine inclusion of the baryon junction mechanism in event generators for a more rigorous test.

4 Summary

We investigated the validity of the baryon junction picture by analyzing net-hyperon yields in Au+Au collisions from RHIC BES-I program. After accounting for the difficulty associated with the strange quark production by dividing net-hyperon yield with $(K^-/\pi^-)^n$, net-hyperon yields at mid-rapidity ($|y| < 0.5$) exhibit the expected exponential dependence on beam rapidity. The extracted exponential slopes (α_B) for net-Λ, net-Ξ, and net-Ω are consistent with

© The Authors, published by EDP Sciences. This is an open access article distributed under the terms of the Creative Commons Attribution License 4.0 (https://creativecommons.org/licenses/by/4.0/).

each other within uncertainty, which suggests a common transport mechanism for different hyperons and supports a flavor-blind baryon junction scenario. Predictions from different versions and tunes of PYTHIA event generator, which assigns baryon number to valence quarks in the incoming protons, are unable to replicate the transport behaviors for all the baryons examined.

References

[1] N. Lewis, W. Lv, M.A. Ross, C.Y. Tsang et al., Search for baryon junctions in photonuclear processes and isobar collisions at RHIC, Eur. Phys. J. C **84**, 590 (2024), 2205.05685. 10.1140/epjc/s10052-024-12834-2

[2] J. Adams et al. (STAR), Scaling Properties of Hyperon Production in Au+Au Collisions at $\sqrt{s}$ = 200-GeV, Phys. Rev. Lett. **98**, 062301 (2007), nucl-ex/0606014. 10.1103/PhysRevLett.98.062301

[3] G. Agakishiev et al. (STAR), Strangeness Enhancement in Cu+Cu and Au+Au Collisions at $\sqrt{s_{NN}}$ = 200 GeV, Phys. Rev. Lett. **108**, 072301 (2012), 1107.2955. 10.1103/PhysRevLett.108.072301

[4] J. Adam et al. (STAR), Strange hadron production in Au+Au collisions at $\sqrt{s_{\rm NN}}$ = 7.7, 11.5, 19.6, 27, and 39 GeV, Phys. Rev. C **102**, 034909 (2020), 1906.03732. 10.1103/PhysRevC.102.034909

[5] M.M. Aggarwal et al. (STAR), Strange and Multi-strange Particle Production in Au+Au Collisions at $\sqrt{s_{NN}}$ = 62.4 GeV, Phys. Rev. C **83**, 024901 (2011), [Erratum: Phys.Rev.C 107, 049903 (2023)], 1010.0142. 10.1103/PhysRevC.83.024901

[6] C. Alt et al. (NA49), Energy dependence of Lambda and Xi production in central Pb+Pb collisions at A-20, A-30, A-40, A-80, and A-158 GeV measured at the CERN Super Proton Synchrotron, Phys. Rev. C **78**, 034918 (2008), 0804.3770. 10.1103/PhysRevC.78.034918

[7] F. Antinori et al. (NA57), Enhancement of hyperon production at central rapidity in 158-A-GeV/c Pb-Pb collisions, J. Phys. G **32**, 427 (2006), nucl-ex/0601021. 10.1088/0954-3899/32/4/003

[8] S. Schuchmann, Ph.D. thesis, Frankfurt U. (2015)

[9] B.B. Abelev et al. (ALICE), Multi-strange baryon production at mid-rapidity in Pb-Pb collisions at $\sqrt{s_{NN}}$ = 2.76 TeV, Phys. Lett. B **728**, 216 (2014), [Erratum: Phys.Lett.B 734, 409–410 (2014)], 1307.5543. 10.1016/j.physletb.2014.05.052

[10] B.I. Abelev et al. (STAR), Strange particle production in p+p collisions at $\sqrt{s}$ = 200-GeV, Phys. Rev. C **75**, 064901 (2007), nucl-ex/0607033. 10.1103/PhysRevC.75.064901

[11] K. Aamodt et al. (ALICE), Strange particle production in proton-proton collisions at sqrt(s) = 0.9 TeV with ALICE at the LHC, Eur. Phys. J. C **71**, 1594 (2011), 1012.3257. 10.1140/epjc/s10052-011-1594-5

© The Authors, published by EDP Sciences. This is an open access article distributed under the terms of the Creative Commons Attribution License 4.0 (https://creativecommons.org/licenses/by/4.0/).

Strange hadron production at high baryon density

Hongcan Li[1,2,*] on the behalf of the STAR Collaboration

[1]Central China Normal University, Wuhan 430079, China
[2]University of Chinese Academy of Sciences, Beijing 101408, China

Abstract. Strange hadrons have been suggested as sensitive probes of the properties of the nuclear matter created in heavy-ion collisions. At few-GeV collision energies, the formed medium is baryon-rich due to baryon stopping effect. In these proceedings, the recent results on strange hadron production in Au+Au collisions at $\sqrt{s_{NN}}$ = 3.2, 3.5, 3.9 and 4.5 GeV with the fixed-target mode from the STAR Beam Energy Scan phase-II program are presented. The transverse momentum spectra, rapidity density distributions, excitation function and centrality dependence of strange hadrons (K^0_S, Λ, Ξ^-) are shown. These results are compared with those from higher collision energies and physics implications are discussed by comparing to the transport model calculations.

1 Introduction

Relativistic heavy-ion collisions provide an excellent opportunity to study the quark-gluon plasma (QGP). Searching for quantum chromodynamics (QCD) critical point, studying properties of QGP and exploring QCD phase diagram are major physical goals of the STAR experiment at the Relativistic Heavy-Ion Collider (RHIC) where the Beam Energy Scan (BES) program was developed. In BES-II program, by fixed-target (FXT) mode, the collision energy of per nucleon pair in Au+Au collision reach down to 3 GeV, where the baryon chemical potential of the created nuclear matter reaches 750 MeV. Properties of the collision system at the high baryon density may be different compared to the QGP where partonic interactions dominate.

Hadrons containing s and/or $\bar{s}$ quarks are called strange hadrons. There is initially no strange hadron colliding so all the final state ones are producted by the collision process, which indicates that the production mechanism of strange hadrons is highly correlated to the reaction mechanism governing the hadronic collision [1–4]. At the STAR FXT energies, strange hadrons are produced near or below the threshold energy, then their yields, especially the excitation function of multi-strange (anti-)hyperons, may provide strong constraints on the equation-of-state (EoS) of the medium created in heavy-ion collisions.

2 Experimental and data analysis

For the BES-II program, STAR upgraded its inner Time Projection Chamber (iTPC) and its end-cap Time of Flight (eTOF)[5, 6]. This improved the detector acceptance and particle identification capabilities. STAR then collected approximately 10 more times collision events than BES-I, which provided more accurate measurements.

In this analysis, we used the dataset of Au+Au collisions at $\sqrt{s_{NN}}$ = 3.2, 3.5, 3.9 and 4.5 GeV. TPC and TOF detectors are used for particle identification and for short-lived particle

*e-mail: lihc@mails.ccnu.edu.cn

© The Authors, published by EDP Sciences. This is an open access article distributed under the terms of the Creative Commons Attribution License 4.0 (https://creativecommons.org/licenses/by/4.0/).

reconstruction. The strange hadrons K^0_S, Λ and Ξ^- are reconstructed by using the hadronic decay channels: $K^0_S \rightarrow \pi^+ + \pi^-$, $\Lambda \rightarrow p + \pi^-$ and $\Xi^- \rightarrow \Lambda + \pi^-$. The KFParticle Finder package was used for the strange hadron reconstruction process [7].

3 Results and discussions

3.1 Transverse momentum spectra and rapidity density distribution

Figure.1 shows the transverse momentum (p_T) spectra of K^0_S, Λ and Ξ^- in central (0-10%) Au+Au collisions. Thanks to the large acceptance of the STAR detector and FXT mode setup, we can measure almost the full rapidity range from beam backward rapidity to middle rapidity. Because the measured p_T range cannot reach down to 0, we use different fit functions for extrapolating the data to unmeasured regions. The blast-wave function is used to fit K^0_S and Λ spectra, while the m_T-exponential function is used to fit Ξ^- spectra.

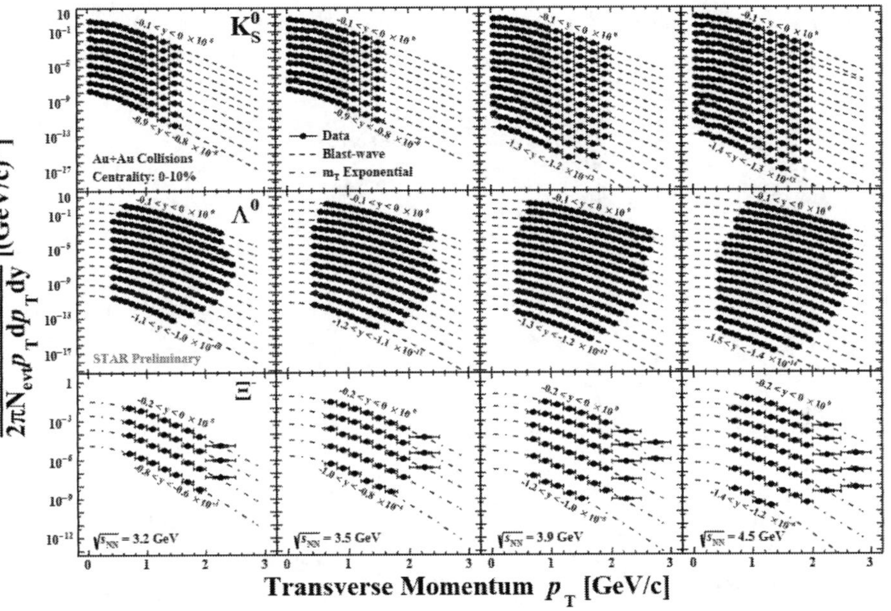

Figure 1. Transverse momentum spectra of K^0_S, Λ and Ξ^- in Au+Au central collision (0-10%) at $\sqrt{s_{NN}}$ = 3.2, 3.5, 3.9 and 4.5 GeV. The black solid circles is the measured data points. Dash lines correspond to blast-wave or m_T-exponential fits.

Figure.2 shows the rapidity distributions (dN/dy) of K^0_S, Λ and Ξ^- in central (0-10%) Au+Au collisions. They are obtained by integrating the p_T spectra with measured data points and fitting extrapolation function. We observe that the strange hadron yields increase with collision energy. The dN/dy shape plateaus at mid-rapidity and is gaussian-like at backward rapidity. In order to describe the dN/dy shape, dN/$dy \propto \frac{1}{Cosh(y^2/\sigma^2)}$ is chosen and used for fitting data points

3.2 Strangeness excitation function

Excitation functions are obtained from the aforementioned measurements, which is middle rapidity yield of per average participating nucleon number $\langle N_{part} \rangle$ as a function of collision energy, and $\langle N_{part} \rangle$ is estimated by Glauber model. Figure.3 shows the excitation function in Au+Au central collisions.

The STAR-FXT energies crossed NN collision threshold energy of Ξ^- production, so we can see the Ξ^- yield increases rapidly near the threshold. The rate of increase decreases as

© The Authors, published by EDP Sciences. This is an open access article distributed under the terms of the Creative Commons Attribution License 4.0 (https://creativecommons.org/licenses/by/4.0/).

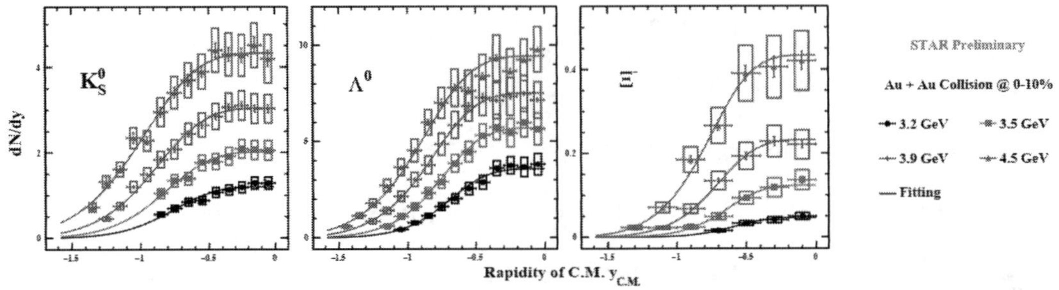

Figure 2. Rapidity distribution of K^0_S, Λ and Ξ^- in Au+Au central collisions (0-10%) at $\sqrt{s_{NN}}$ = 3.2, 3.5, 3.9 and 4.5 GeV. The vertical line is the statistical uncertainties, and the box is the systematic uncertainties.

the collision energy moves away from the threshold, and approximately remains constant at $\sqrt{s_{NN}}$ ~ 20 GeV or higher. At low energy K^0_S yield are blow the Λ ones but that cross at ~ 8 GeV indicating a transition from the baryon-dominated to the meson-dominated matter. The enhancement in baryon yield relative to meson at low energies is likely driven by stronger baryon stopping in the few-GeV energy region.

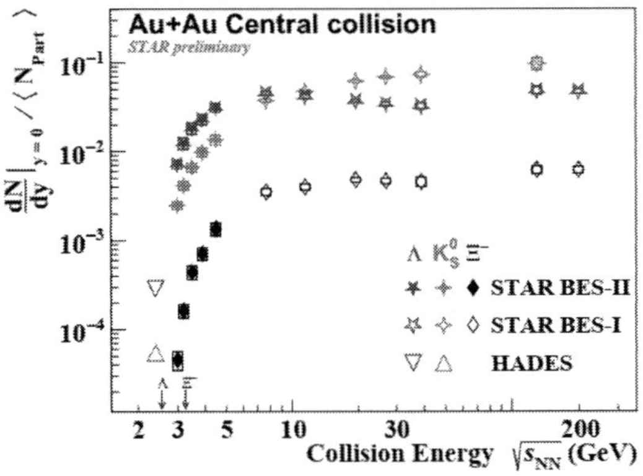

Figure 3. Excitation function of K^0_S, Λ and Ξ^- in Au+Au the most central collision. The solid points is STAR BES-II FXT-mode results and the open points is STAR BES-I and HADES results[8–11]. The NN collision threshold energies of Λ and Ξ^- are marked by two arrows.

3.3 Scaling property of centrality dependence

In order to quantitatively describe the centrality of heavy-ion collisions, we select $\langle N_{part} \rangle$ to represent centrality of collision. For determining the centrality dependence of the yields, we fit a power law function and extract a scaling parameter α_S.

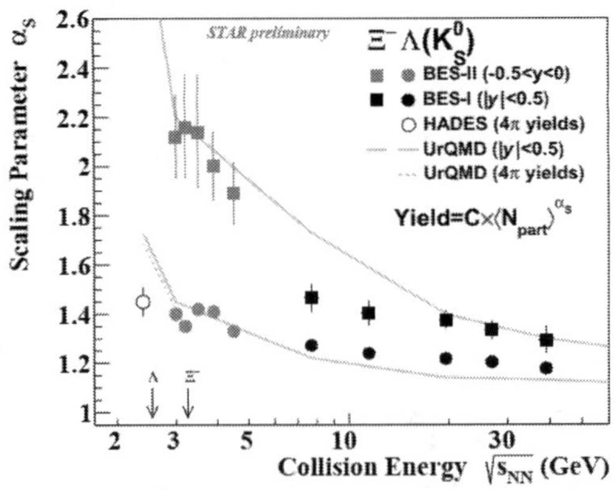

Figure 4. Scaling parameter α_S of K^0_S, Λ and Ξ^- as a function of collision energy. Results from the transport model UrQMD is show as gray lines.

© The Authors, published by EDP Sciences. This is an open access article distributed under the terms of the Creative Commons Attribution License 4.0 (https://creativecommons.org/licenses/by/4.0/).

Figure.4 shows the scaling parameter of K^0_S, Λ and Ξ^- as a function of collision energy. Single strange hadron production K^0_S and Λ is associated production via NN $\rightarrow$ NΛK, therefore their scaling parameter is extracted simultaneously. The scaling parameter is greater than unity, which indicates that their yield increase more rapidly than the increase of the number of participating nucleons. Double strange hadron production Ξ^- has larger scaling parameter than K^0_S and Λ, which indicates that Ξ^- yield increase more rapidly. This may be because Ξ^- has different production channel than NN $\rightarrow$ NΞKK. The scaling parameter decreases with increasing collision energy which shows the centrality dependence of strange hadron yield weakening as collision energy increases.

We also compare this result with transport model UrQMD[12] which qualitatively reproduces the energy dependence, but it cannot quantitatively describe in all energies, especially $\sqrt{s_{\rm NN}}$ = 7.7 to 11.5 GeV for Ξ^-, which may be due to missing medium effects.

4 Summary

In these proceeding, we report yield measurements of K^0_S, Λ and Ξ^- in Au+Au collisions at $\sqrt{s_{\rm NN}}$ = 3.2, 3.5, 3.9 and 4.5 GeV. Their dN/dy and the corresponding strangeness excitation function are presented. The decreasing α_S represents centrality dependence of yield as energy increase is shown.

Acknowledgement: This work was supported in part by the National Natural Science Foundation of China (Grant No. 12375134 and No. 12305146), the National Key Research and Development Program of China under Grant No. 2020YFE0202002, and the Fundamental Research Funds for the Central Universities (Grant No. CCNU22QN005).

References

[1] P. Koch, et al. Physics Reports **142**, 167 (1986) 4. https://www.sciencedirect.com/science/article/abs/pii/0370157386900967
[2] C. Blume and C. Markert. Progress in Particle and Nuclear Physics **66**, 834 (2011) 4. https://www.sciencedirect.com/science/article/pii/S0146641011000834
[3] J.H. Chen, et al. arXiv:2407.02935 **[nucl-ex]**. https://arxiv.org/abs/2407.02935
[4] F. Liu, et al. SCIENTIA SINICA Physica, Mechanica & Astronomica **53**, 290003 (2023) 9. https://www.sciengine.com/SSPMA/doi/10.1360/SSPMA-2023-0167
[5] STAR Collaboration. Journal of Physics: Conference Series **779**, 012082 (2017) 1. https://iopscience.iop.org/article/10.1088/1742-6596/779/1/012082
[6] STAR Collaboration, CBM Collaboration eTOF Group. arXiv:1609.05102 **[nucl-ex]**. https://arxiv.org/abs/1609.05102
[7] I. Kisel. Journal of Physics: Conference Series **1602**, 012006 (2020) 1. https://dx.doi.org/10.1088/1742-6596/1602/1/012006
[8] STAR Collaboration. Journal of High Energy Physics **2024**, 139 (2024). https://link.springer.com/article/10.1007/JHEP10(2024)139
[9] STAR Collaboration. Physics Letters B **831**, 137152 (2022). https://www.sciencedirect.com/science/article/pii/S0370269322002866
[10] STAR Collaboration. Physics Review C **102**, 034909 (2020). https://journals.aps.org/prc/abstract/10.1103/PhysRevC.102.034909
[11] HADES Collaboration. Physics Letters B **793**, 457 (2019). https://www.sciencedirect.com/science/article/pii/S0370269319303272
[12] S.A. Bass, et al. Progress in Particle and Nuclear Physics **41**, 255 (1998). https://www.sciencedirect.com/science/article/pii/S0146641098000581

© The Authors, published by EDP Sciences. This is an open access article distributed under the terms of the Creative Commons Attribution License 4.0 (https://creativecommons.org/licenses/by/4.0/).

Measurement of charged and neutral kaons in Ar+Sc collisions at the NA61/SHINE experiment

Tatjana Susa[1,*]
for the NA61/SHINE Collaboration

[1]Ruđer Bošković Institute

Abstract. NA61/SHINE is a large-acceptance fixed-target experiment conducted at the CERN SPS. The experiment has measured charged kaon production in 0-10% central Ar+Sc collisions at beam momenta of 13*A*, 19*A*, 30*A*, 40*A*, 75*A*, and 150*A* GeV/*c*. In this report, we present the measurements of K^0_S production and charged-to-neutral kaon ratio in 0-10% central Ar+Sc collisions at 75*A* GeV/*c*. The results are compared to the charged and neutral kaon production in other collision systems measured by NA61/SHINE, as well as to model predictions and data from other experiments.

1 Introduction

NA61/SHINE [1] is a fixed-target experiment located on the H2 beam line of the CERN SPS accelerator complex. Its primary detection system comprises four Time Projection Chambers (TPCs), which enable precise measurement of particle momenta and provide particle identification through specific energy loss ($\mathrm{d}E/\mathrm{d}x$). Additionally, two Time-of-Flight (ToF) walls improve kaon identification capabilities (*tof* measurement). The TPC setup enables particle detection at transverse momenta as low as $p_T = 0$ GeV/c across a wide range of forward rapidities. A high-resolution hadron calorimeter, the Projectile Spectator Detector, measures forward-going energy and is used to determine the centrality of nuclear collisions.

The primary goal of the strong interaction program of the NA61/SHINE experiment is to discover the critical point of strongly interacting matter, study the properties of the onsets of deconfinement and fireball, and measure the open charm production. To accomplish this goal, a two-dimensional scan of the phase diagram of strongly interacting matter is conducted by varying the beam momentum (13*A*-150/158*A* GeV/*c*) and the size of the colliding nuclei (*p*+*p*, *p*+Pb, Be+Be, Ar+Sc, Xe+La, Pb+Pb).

2 Kaon production

NA61/SHINE has measured the inclusive spectra and mean multiplicities of $K^\pm$ mesons in the 10% most central Ar+Sc collisions at beam momenta of 13*A*, 19*A*, 30*A*, 40*A*, 75*A*, and 150*A* GeV/*c* (corresponding to $\sqrt{s_{NN}}$ = 5.12, 6.12, 7.62, 8.77, 11.9 and 16.8 GeV) [2]. Charged kaons were identified by $\mathrm{d}E/\mathrm{d}x$ and combined *tof*-$\mathrm{d}E/\mathrm{d}x$ method. At high momenta, in the region of the relativistic rise, the $\mathrm{d}E/\mathrm{d}x$ measurement alone was sufficient to identify

*e-mail: tatjana.susa@irb.hr

© The Authors, published by EDP Sciences. This is an open access article distributed under the terms of the Creative Commons Attribution License 4.0 (https://creativecommons.org/licenses/by/4.0/).

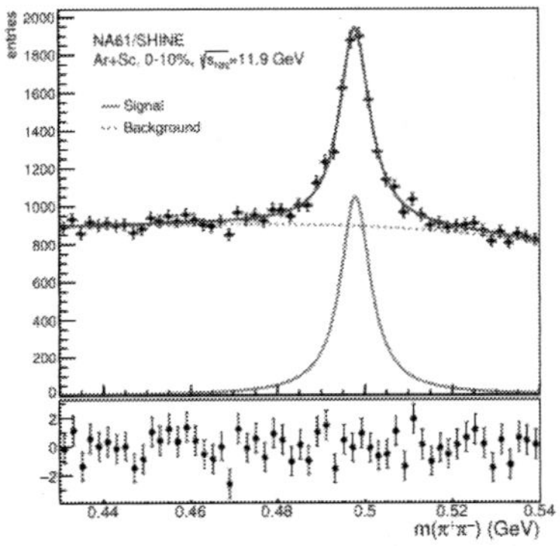

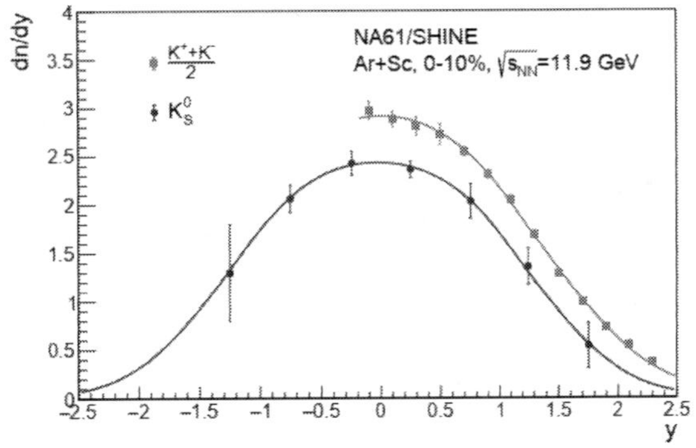

Figure 1. Fitted invariant mass distribution of K^0_S candidates in rapidity - transverse momentum bin $y \in (-1.0, -0.5)$, $p_T \in (1.2, 1.5)$ GeV/c. The bottom panel shows the difference between the experimental data and the fitted (signal+background) function, divided by the experimental uncertainty (left). The rapidity spectrum of K^0_S mesons in 0-10% central Ar+Sc collisions at 75A GeV/c is shown by blue circles [5]. Green squares represent the averaged spectrum of charged kaons [2]. Total uncertainties are drawn. Blue and green curves show the result of double Gaussian fit (Eq. 1) to the K^0_S and averaged fit for charged kaons spectra, respectively (right).

charged kaons. At lower momenta, where the bands for different particle species overlap, the additional measurement of *tof* was used for unambiguous kaon identification. These two methods cover most of the relevant rapidity and transverse momentum space, especially the mid-rapidity region of K^+ and K^- spectra. The yields of charged K mesons at mid-rapidity in the 10% most central Ar+Sc collisions at beam momentum of 75A GeV/c were determined in the interval of collision center-of-mass rapidity $0.0 < y < 0.2$ and amount to $(\mathrm{d}n/\mathrm{d}y)(K^+)_{y\approx 0} = 3.732 \pm 0.016$ (*stat*) $\pm$ 0.148 (*sys*) and $(\mathrm{d}n/\mathrm{d}y)(K^-)_{y\approx 0} = 2.029 \pm 0.012$ (*stat*) $\pm$ 0.069 (*sys*).

K^0_S mesons were identified by the characteristic topology of their decay into charged pions ($K^0_S \to \pi^+ + \pi^-$, with a branching ratio BR = 69.2%). The identification process involved pairing all positively charged pion candidates with negatively charged ones. These pairs were then tracked backward through the NA61/SHINE magnetic field, and the minimum distance of closest approach (DCA) between the particles was determined. A pair was considered a K^0_S candidate if the DCA was less than 1 cm. The invariant mass of the resulting decay particle pairs forms a distribution that contains both the K^0_S signal and background contributions, such as those from Λ decays, photon conversions, and random particle crossing. A K^0_S signal appears as a peak on top of a smooth background. The raw number of K^0_S in different rapidity - transverse momentum bins was determined by fitting the invariant mass spectrum. The fit function consisted of a Lorentzian to describe the K^0_S signal and a third-order Chebyshev polynomial to account for the background. An example of the invariant mass distribution, together with the signal and background fits, is shown in Fig. 1 (left). The raw number of K^0_S was corrected for losses due to the geometrical acceptance, reconstruction efficiency, and selection criteria applied in the analysis. The corrections were based on simulated Ar+Sc events produced by the EPOS [3] event generator. The simulated particles were tracked through the NA61/SHINE detector using the GEANT framework [4]. The simulated events were reconstructed with the same software as the experimental events and the same selection criteria were applied. The reliability of the K^0_S reconstruction and correction procedure was validated by studying the K^0_S lifetime distributions in seven rapidity bins, for details see Ref. [5]. The

© The Authors, published by EDP Sciences. This is an open access article distributed under the terms of the Creative Commons Attribution License 4.0 (https://creativecommons.org/licenses/by/4.0/).

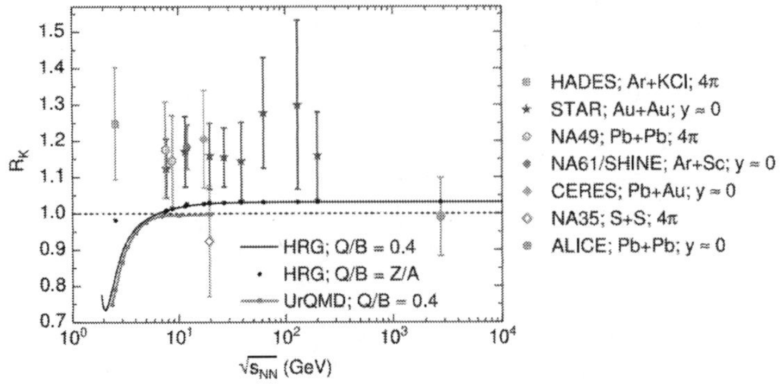

Figure 2. Charged-to-neutral kaon ratio as a function of collision energy. Experimental data are shown by symbols with total uncertainties. HRG baseline for electric-to-baryon charge $Q/B = 0.4$ is shown by a black line. HRG baseline for Q/B values specified accordingly to given types of colliding nuclei is represented by black dots. UrQMD model results are shown by grey squares. Figure from Ref. [7] with later changes.

extracted mean K^0_S lifetimes agree within uncertainties with the PDG [6] value in all rapidity bins indicating a good quality of the analysis.

The double-differential yield of K^0_S meson in 0-10% central Ar+Sc collisions at 75A GeV/c was calculated in seven rapidity bins in the range $-1.5 < y < 2.0$ and nine transverse momentum bins spanning $0.0 < p_T < 2.7$ GeV/c. The transverse momentum spectra were fitted using an exponential function $f(p_T) = A \cdot p_T \cdot \exp(-\sqrt{p_T^2 + m_0^2}/T)$, where A is a normalisation factor, m_0 is the mass of K^0_S, and T represents the inverse slope parameter. The yield of K^0_S in each rapidity bin was determined by integrating the fitted exponential function. The resulting rapidity spectrum, shown in Fig. 1 (right), was fitted with a double-Gaussian function:

$$f(y) = A_1 \cdot \exp\left(-\frac{1}{2}\frac{(y-\mu)^2}{\sigma^2}\right) + A_2 \cdot \exp\left(-\frac{1}{2}\frac{(y+\mu)^2}{\sigma^2}\right), \tag{1}$$

where A_1 and A_2 are the normalisation factors, μ represents the displacement of each Gaussian from $y = 0$, and σ is the standard deviation. The mean multiplicity of K^0_S meson, calculated as the integral of fitted double-Gaussian function, is $\langle K^0_S \rangle = 6.49 \pm 0.10\ (stat) \pm 0.83\ (sys)$.

The yield of K^0_S meson at mid-rapidity, determined from the fit at $y = 0$, was found to be $(dn/dy)_{y\approx 0} = 2.433 \pm 0.027\ (stat) \pm 0.102\ (sys)$.

3 Charged-to-neutral kaon ratio

In Ar+Sc collisions, an approximately equal abundance of charged and neutral kaons is expected as Ar and Sc nuclei are nearly isospin symmetric (valence quarks u = d within 6%), for details see Ref. [5] and references therein. However, experimental data from NA61/SHINE [5] show a significant difference. The rapidity distribution of K^0_S mesons, compared with the averaged distribution of K^+ and K^- mesons, Fig. 1 (right), reveals a much higher yield of charged kaons across the entire rapidity range. At mid-rapidity, the charged-to-neutral kaon ratio defined as $R_K = (K^+ + K^-)/(2K^0_S)$ was found to be 1.184 ± 0.061.

In Fig. 2, this charged-to-neutral kaon ratio was compared with a compilation of results on K^+, K^- and K^0_S production from other experiments and predictions from the Ultrarelativistic

© The Authors, published by EDP Sciences. This is an open access article distributed under the terms of the Creative Commons Attribution License 4.0 (https://creativecommons.org/licenses/by/4.0/).

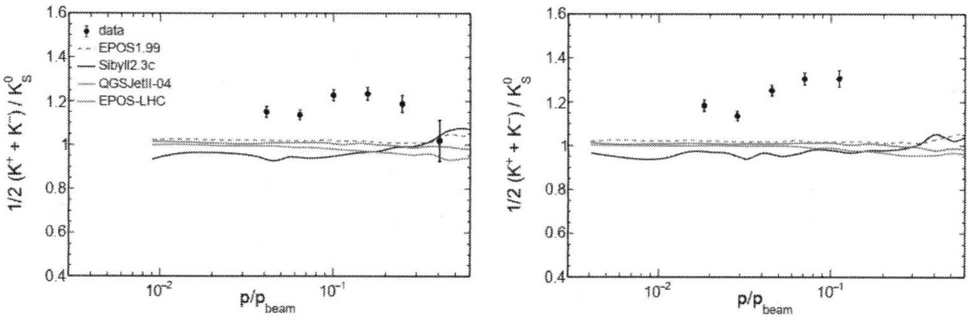

Figure 3. Charged-to-neutral kaon ratio in π^-+C reactions at 158 GeV/c (left) and 350 GeV/c (right) plotted as a function of scaled laboratory momentum p/p_{beam}, adopted from Ref. [10]. Only statistical uncertainties are drawn, while systematic uncertainties are estimated to be 5-10%.

Quantum Molecular Dynamics (UrQMD) [8] and Hadron Resonance Gas (HRG) [9] models. While data from other experiments, despite their large uncertainties, support NA61/SHINE results, both models fail to describe the observed charged-to-neutral kaon ratio.

Ratios $R_K > 1$ were also observed in π^-+C collisions at beam momenta of 158 and 350 GeV/c. These measurements, shown in Fig. 3, were compared with the predictions of several microscopic models. None of the models can reproduce the observed charged-to-neutral kaon ratio.

4 Summary

NA61/SHINE has measured charged kaon production in 0-10% central Ar+Sc collisions at several beam momenta: 13A, 19A, 30A, 40A, 75A, and 150A GeV/c. K^0_S production has been measured in 0-10% central Ar+Sc collisions at 75A GeV/c. At this beam momentum, the ratio of charged-to-neutral kaons, defined as $R_K = (K^+ + K^-)/(2K^0_S)$, was found to be 1.184 ± 0.061. The $R_K > 1$ has been measured also in π^-+C collisions at 150 and 350 GeV/c. The UrQMD and Hadron Resonance Gas models do not reproduce the experimental results on charged-to-neutral kaon ratio in A+A collisions. Models fail to describe charged-to-neutral kaon ratio also for small asymmetric π^-+C systems.

References

[1] N. Abgrall *et al.* [NA61/SHINE Collab.], JINST **9**, P06005 (2014).
[2] H. Adhikary *et al.* [NA61/SHINE Collab.], Eur. Phys. J. C **84**, 416 (2024).
[3] K. Werner, Nucl. Phys. Proc. Suppl. **175-176**, 81 (2008).
[4] Brun, R. Hagelberg, M. Hansroul, and J. C. Lassalle, Tech. Rep. CERN-DD-78-2-REV, CERN-DD-78-2, CERN, Geneva (1978).
[5] H. Adhikary *et al.* [NA61/SHINE Collab.], arXiv:2312.06572 [nucl-ex].
[6] S. Navas *et al.* [Particle Data Group Collab.], Phys. Rev. D **110**, 030001 (2024).
[7] W. Brylinski, M. Gazdzicki, F. Giacosa, M. Gorenstein, R. Poberezhnyuk, S. Samanta, and H. Stroebele, arXiv:2312.07176 [nucl-th].
[8] M. Bleicher *et al.*, J. Phys. G **25**, 1859 (1999).
[9] V. Vovchenko and H. Stoecker, Comput. Phys. Commun. **244**, 295 (2019).
[10] H. Adhikary *et al.* [NA61/SHINE Collab.], Phys. Rev. D **107**, 062004 (2023).

© The Authors, published by EDP Sciences. This is an open access article distributed under the terms of the Creative Commons Attribution License 4.0 (https://creativecommons.org/licenses/by/4.0/).

EPJ Web of Conferences **316**, 04001 (2025) https://doi.org/10.1051/epjconf/202531604001
SQM 2024

Heavy Flavour Energy Loss in Small and Large Systems

Coleridge Faraday[1,*] and *W. A.* Horowitz[1,2,]

[1]Department of Physics, University of Cape Town, Private Bag X3, Rondebosch 7701, South Africa
[2]Department of Physics, New Mexico State University, Las Cruces, New Mexico, 88003, USA

Abstract. We present suppression results for high-p_T D and π mesons produced in $p/d + A$ and $A + A$ collisions at RHIC and LHC. These results are computed using a convolved elastic and radiative energy loss model, which receives small system size corrections to both the elastic and radiative energy loss. We observe that suppression in small systems is almost entirely due to elastic energy loss; furthermore, we find that our model is acutely sensitive to the transition between hard thermal loop and vacuum propagators in the elastic energy loss. Finally, we consider the central limit theorem approximation, which is commonly used to model the elastic energy loss distribution as Gaussian.

1 Introduction

Over the past decade, the observation of quark-gluon plasma (QGP) signatures in small collision systems has created considerable interest within the heavy-ion physics community. Traditional QGP signatures such as quarkonium suppression [1], strangeness enhancement [2], and elliptic flow [3]—typically associated with large heavy-ion collisions—have also been detected in small systems. These findings raise questions about whether QGP forms in even the smallest collision systems.

One of the most robust probes of QGP formation is jet quenching, often measured in terms of the nuclear modification factor R_{AB} for the collision $A + B$. This factor quantifies the degree of energy loss experienced by high-p_T partons as they traverse the medium. A measured $R_{AB} \sim 0.2$ for leading pions in central Au + Au collisions at RHIC [4] has been attributed to significant medium-induced partonic energy loss. By contrast, photons, which do not undergo strong interactions in the medium, exhibit $R_{AB} \sim 1$ [5].

In small systems, determining R_{AB} is more challenging due to centrality bias, leading to an ambiguous measured suppression pattern. Centrality bias refers to a non-trivial correlation between the hard and soft modes of the QGP, potentially leading to an inaccurate normalization for the R_{AB}. This issue is especially pronounced in small systems. LHC data from central p + Pb collisions finds $R_{AB} \simeq 1$–1.2 for pions and D mesons [6, 7] at low- to moderate-p_T, suggesting no QGP formation. Conversely, PHENIX observes $R_{AB} \simeq 0.75$ in central d + Au collisions, qualitatively consistent with QGP formation [8].

The situation on the theoretical front, however, has its own set of challenges. Theoretical energy loss models generally rely on approximations valid only for large systems. For instance, the Djordjevic-Gyulassy-Levai-Vitev (DGLV) framework for radiative energy loss [9] assumes a large pathlength L, neglecting terms proportional to $e^{-\mu L}$, where μ is the Debye

*e-mail: frdcol002@myuct.ac.za

© The Authors, published by EDP Sciences. This is an open access article distributed under the terms of the Creative Commons Attribution License 4.0 (https://creativecommons.org/licenses/by/4.0/).

mass. A *short pathlength correction* (SPL) has been derived that restores these terms [10]. Similar large system size assumptions appear in elastic energy loss models, which often assume a Gaussian energy loss distribution based on the central limit theorem, which is inapplicable in small systems [11]. In this work, we examine a more realistic Poisson distribution for elastic energy loss [11]—which is valid for all system sizes—and compare it to the Gaussian distribution at the level of R_{AB}. Furthermore we interrogate a particular uncertainty in the elastic energy loss, which stems from the transition between hard thermal loop (HTL) and vacuum propagators, by computing results with Braaten and Thoma (BT) elastic energy loss model [12] as well as with HTL-only elastic energy loss [11].

2 Results

We present a model for perturbative QCD energy loss which includes both radiative and elastic contributions. The radiative component is modelled using the DGLV framework, with and without short pathlength corrections (DGLV and DGLV+SPL) [9, 10]. For the elastic component, we compare results using BT [12], Gaussian HTL-only (Gauss. HTL), and Poisson HTL-only (Poiss. HTL) models [11]. Thermodynamic and geometrical quantities are calculated from IP-Glasma initial conditions [13] coupled with subsequent Bjorken expansion. A detailed description of the model may be found in our other works [14, 15].

In Fig. 1, we compare R_{AB} results for pions and D mesons produced in central Pb + Pb, Au + Au, p + Pb, and d + Au collisions at RHIC and LHC. Details are found in the caption of Fig. 1. There are six model results presented which are calculated by varying the radiative energy loss between the DGLV and the DGLV + SPL models, and separately the elastic energy loss between the BT, Poiss. HTL and Gauss. HTL models.

From Fig. 1, we first compare the R_{AB} calculated with the DGLV and DGLV + SPL radiative energy loss models to understand the impact of the SPL correction. Including the SPL correction greatly reduces the π^0 and $h^\pm$ suppression, but only mildly reduces D meson suppression. This flavor-dependence is due to the SPL correction being significantly larger for gluons compared to quarks, over and above the usual $C_A/C_F = 9/4$ colour scaling [10, 14]. We also observe that the magnitude of the SPL correction grows in p_T, which is due to the different asymptotics of DGLV and DGLV + SPL: $\Delta E_{\text{DGLV}} \sim \ln E$ and $\Delta E_{\text{DGLV + SPL}} \sim E$ respectively [10]. The $\sqrt{s}$ dependence of the SPL correction is due primarily to the larger proportion of gluons which fragment to pions at LHC compared to RHIC [15] and the larger experimentally accessible p_T range. Finally, we observe that the SPL correction is fractionally larger in small systems compared to larger systems due to the $e^{-\mu L}$ scaling of the SPL correction [10, 14].

In Fig. 1, we additionally compare the R_{AB} calculated with the BT and Gauss. HTL elastic energy loss models, which is sensitive to the uncertainty in the transition between HTL and vacuum propagators. We observe that this uncertainty leads to an $O(30–80)\%$ effect in large systems and an $O(20–40)\%$ effect in small systems. Finally, we compare the Poiss. HTL and Gauss. HTL results to understand the effect of the commonly used central limit theorem approximation. We observe that this effect is $O(5–10)\%$ in large systems and $< 2\%$ in small systems. This agreement between Poisson and Gaussian results cannot be attributed to convergence under the central limit theorem, as one would expect the results to align more closely in larger systems than in smaller ones. In actuality, the agreement is because the R_{AB} in small systems depends mostly on the zeroth and first moments of the elastic energy loss distributions—which are constrained to be the same for Gaussian and Poisson distributions—while in large systems the radiative energy loss dominates over the elastic [15]. The Poisson and Gaussian results are more similar at LHC energies because the steeper production spectrum at RHIC makes higher-order moments more important than at LHC.

© The Authors, published by EDP Sciences. This is an open access article distributed under the terms of the Creative Commons Attribution License 4.0 (https://creativecommons.org/licenses/by/4.0/).

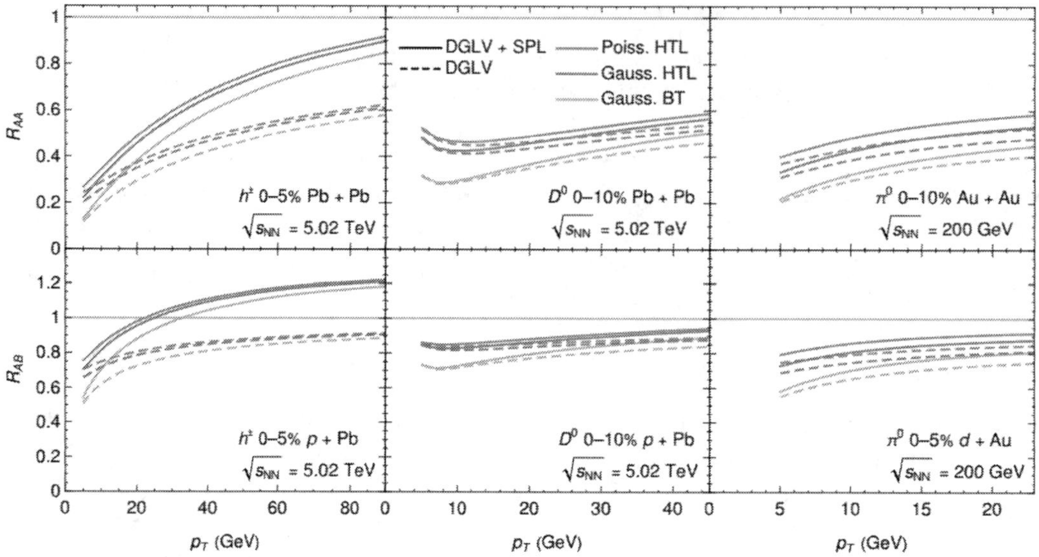

Figure 1. The top row displays R_{AA} results from our model for charged hadrons produced in 0–5% centrality Pb + Pb collisions at $\sqrt{s_{NN}}$ = 5.02 TeV; D^0 mesons produced in 0–10% centrality Pb + Pb collisions at $\sqrt{s_{NN}}$ = 5.02 TeV; and neutral pions produced in 0–10% centrality Au + Au collisions at $\sqrt{s_{NN}}$ = 200 GeV. The bottom row shows R_{AB} for charged hadrons produced in 0–5% centrality p + Pb collisions at $\sqrt{s_{NN}}$ = 5.02 TeV; D mesons produced in 0–10% centrality p + Pb collisions at $\sqrt{s_{NN}}$ = 5.02 TeV; and neutral pions produced in 0–5% d + Au collisions at $\sqrt{s_{NN}}$ = 200 GeV. Model results are produced by varying the radiative energy loss between the DGLV [9] and DGLV + SPL models [10], and separately the elastic energy loss between the BT [12], Gauss. HTL, and Poiss. HTL models [11].

3 Conclusions

We have presented suppression results for high-p_T D and π mesons produced in $p/d + A$ and $A + A$ collisions at RHIC and LHC, from a model which receives small system size corrections to both the radiative and elastic energy loss. We found that an uncertainty in the transition between HTL and vacuum propagators in the elastic energy loss imparts a O(40–80)% uncertainty on the R_{AB} for all systems and final states at low- to moderate-p_T. Furthermore, we found that the application of the central limit theorem to model the elastic distribution as Gaussian produced a negligible effect at the level of the R_{AB}, particularly so in small systems. We attributed this to the fact that for small suppression the R_{AB} depends mostly on the zeroth and first moments of the elastic energy loss, which are constrained to be identical for Gaussian and Poisson distributions, while for larger momenta radiative energy loss dominates.

Future work may involve fitting the strong coupling α_s to data to determine whether observed small and large system suppression can be simultaneously described by a single model. Additionally, one may further analyse the sensitivity to various uncertainties in elastic and radiative energy loss, including the large formation time assumption, which we previously found was not satisfied self-consistently in the DGLV model [14, 16]. Moreover, placing energy loss models on a more rigorous theoretical foundation could involve carefully treating the kinematics [17], computing the scales at which the coupling runs at finite temperature and system size, and understanding the validity of various approximations in the calculations [18]. Lastly, investigating QGP formation in small systems may also include examining the impact

© The Authors, published by EDP Sciences. This is an open access article distributed under the terms of the Creative Commons Attribution License 4.0 (https://creativecommons.org/licenses/by/4.0/).

of small system size on the coupling [19], thermodynamics [20], the equation of state [21], and jet substructure observables [22].

Acknowledgements

CF and WAH thank the South African National Research Foundation and SA-CERN Collaboration for financial support.

References

[1] J. Adam et al. (ALICE), JHEP **06**, 050 (2016), 1603.02816. 10.1007/JHEP06(2016)050
[2] B.B. Abelev et al. (ALICE), Phys. Lett. B **728**, 25 (2014), 1307.6796. 10.1016/j.physletb.2013.11.020
[3] S. Chatrchyan et al. (CMS), Phys. Lett. B **718**, 795 (2013), 1210.5482. 10.1016/j.physletb.2012.11.025
[4] K. Adcox et al. (PHENIX), Phys. Rev. Lett. **88**, 022301 (2002), nucl-ex/0109003. 10.1103/PhysRevLett.88.022301
[5] S.S. Adler et al. (PHENIX), Phys. Rev. Lett. **94**, 232301 (2005), nucl-ex/0503003. 10.1103/PhysRevLett.94.232301
[6] J. Adam et al. (ALICE), Phys. Rev. C **94**, 054908 (2016), 1605.07569. 10.1103/PhysRevC.94.054908
[7] G. Aad et al. (ATLAS), JHEP **07**, 074 (2023), 2211.15257. 10.1007/JHEP07(2023)074
[8] N.J. Abdulameer et al. (PHENIX) (2023), 2303.12899.
[9] M. Djordjevic, M. Gyulassy, Nucl. Phys. A **733**, 265 (2004), nucl-th/0310076. 10.1016/j.nuclphysa.2003.12.020
[10] I. Kolbe, W.A. Horowitz, Phys. Rev. C **100**, 024913 (2019), 1511.09313. 10.1103/PhysRevC.100.024913
[11] S. Wicks, PhD thesis (2008)
[12] E. Braaten, M.H. Thoma, Phys. Rev. D **44**, R2625 (1991). 10.1103/PhysRevD.44.R2625
[13] B. Schenke, C. Shen, P. Tribedy, Phys. Rev. C **102**, 044905 (2020), 2005.14682. 10.1103/PhysRevC.102.044905
[14] C. Faraday, A. Grindrod, W.A. Horowitz, Eur. Phys. J. C **83**, 1060 (2023), 2305.13182. 10.1140/epjc/s10052-023-12234-y
[15] C. Faraday, W.A. Horowitz (2024), 2408.14426.
[16] C. Faraday, W.A. Horowitz, Assumption Breakdown in Radiative Energy Loss, in *67th Annual Conference of the South African Institute of Physics* (2023), 2309.06246
[17] H. Clayton, M.D. Sievert, W.A. Horowitz, Eur. Phys. J. C **82**, 437 (2022), 2110.14737. 10.1140/epjc/s10052-022-10386-x
[18] W.A. Horowitz, B.A. Cole, Phys. Rev. C **81**, 024909 (2010), 0910.1823. 10.1103/PhysRevC.81.024909
[19] W.A. Horowitz, J.F.D. Plessis, Phys. Rev. D **105**, L091901 (2022), 2203.01259. 10.1103/PhysRevD.105.L091901
[20] S. Mogliacci, I. Kolbé, W.A. Horowitz, Phys. Rev. D **102**, 116017 (2020), 1807.07871. 10.1103/PhysRevD.102.116017
[21] W. Horowitz, A. Rothkopf, SciPost Phys. Proc. **10**, 025 (2022), 2109.01422. 10.21468/SciPostPhysProc.10.025
[22] I. Kolbé (2023), 2303.14166.

© The Authors, published by EDP Sciences. This is an open access article distributed under the terms of the Creative Commons Attribution License 4.0 (https://creativecommons.org/licenses/by/4.0/).

Charm and Bottom hadron production with a coalescence plus fragmentation hadronization approach: AA system size scan down to pp collisions

Salvatore Plumari[1,2,*], *Vincenzo* Minissale[1,3,**], and *Vincenzo* Greco[1,2,***]

[1]Dipartimento di Fisica e Astronomia "E. Majorana", Università degli Studi di Catania, Via S. Sofia 64, I-95125 Catania, Italy
[2]Laboratori Nazionali del Sud, INFN-LNS, Via S. Sofia 62, I-95123 Catania, Italy
[3]INFN Sezione di Catania, Via Santa Sofia,64 - 95123 Catania, Italy

Abstract. Heavy baryon production in *pp*, *pA*, and *AA* collisions at RHIC and LHC energies provides new challenges to the theoretical understanding of heavy-quark hadronization. In recent years, a hybrid coalescence-plus-fragmentation model has successfully predicted the large baryon/meson ratio observed in AA and pp collisions. This approach also predicts an enhancement of Ξ_c and Ω_c production in pp collisions that is in agreement with the recent ALICE data. In this paper, we provide predictions for bottom hadrons in PbPb and pp collisions at LHC energies. The comparison between charm and bottom hadron production offers new insights into both the hadronization mechanisms. Finally, the extension of the model to multi-charmed baryons like Ξ_{cc} and Ω_{ccc} has shown that multi-charmed baryon yields, especially Ω_{ccc}, are sensitive to the thermalization of charm quarks, offering insights into hadronization and quark equilibration dynamics across different systems.

1 Introduction

Data from RHIC and LHC experiments have provided evidence for the creation of a quark-gluon plasma (QGP). Heavy hadrons, such as D and B mesons and Λ_c or Λ_b baryons, which contain charm or bottom quarks, can be considered excellent probes for studying the properties of the QGP. This is because heavy quarks (HQs), due to their large masses and well-understood production mechanisms via perturbative QCD (pQCD), can probe the entire evolution of the QGP. Key observables in the study of the QGP are the transverse momentum spectra and the nuclear modification factor (R_{AA}) of heavy hadrons, which provide insights into the medium created and the interactions of heavy quarks with it. Studies on $R_{AA}(p_T)$ of D mesons, along with elliptic flow measurements ($v_2(p_T)$), suggest that charm quarks not only interact with the QGP, losing their energy, but may also participate in the collective flow of the medium [1–12]. On the other hand, recent experimental results at LHC energies have shown an enhancement of Λ_c/D^0, first observed in *PbPb* collisions at top LHC energies and later, in high-multiplicity proton-proton (pp) collisions. The Λ_c production is much

*e-mail: salvatore.plumari@dfa.unict.it
**e-mail: vincenzo.minissale@dfa.unict.it
***e-mail: greco@lns.infn.it

© The Authors, published by EDP Sciences. This is an open access article distributed under the terms of the Creative Commons Attribution License 4.0 (https://creativecommons.org/licenses/by/4.0/).

higher than predicted by standard fragmentation models based on data from e^+e^- or ep collisions. These findings in small systems like pp collisions suggest that hadronization may involve mechanisms typically associated with medium creation in heavy-ion collisions, as generally assumed in hybrid approaches to hadronization, such as coalescence and fragmentation. In the literature, several models, such as coalescence and fragmentation mechanisms, have been proposed to explain these phenomena (see ref. [13] for a recent systematic comparison between different models). Finally, recent studies have shown qualitative agreement with experimental data, reinforcing the hypothesis that, in high-multiplicity proton-proton (pp) collisions, the production of heavy hadrons may follow mechanisms similar to those observed in heavy-ion collisions [14–16].

2 Hadronization model

The coalescence approach was initially introduced as a hadronization mechanism for Heavy Ion collisions at RHIC, to explain the p_T spectra, baryon-to-meson ratio, and elliptic flow splitting between mesons and baryons [17, 18]. Applied at LHC energies was shown to be able to get a successfull description of light hadron spectra and baryon-to-meson ratios [19]. In recent years, this approach has been extended to heavy-flavor hadronization in AA collisions [20–23] and recently also including multi-charmed hadrons like Ξ_{cc} and Ω_{ccc} [24]. The model is based on the Wigner formalism for quark coalescence.

In this approach the momentum spectrum of hadrons formed by coalescence of quarks can be written as:

$$\frac{dN_H}{dyd^2P_T} = g_H \int \prod_{i=1}^{N_q} \frac{d^3p_i}{(2\pi)^3 E_i} p_i \cdot d\sigma_i \; f_{q_i}(x_i, p_i) C_H(x_1 ... x_{N_q}, p_1 ... p_{N_q}) \, \delta^{(2)}\left(P_T - \sum_{i=1}^{n} p_{T,i}\right)$$

The statistical factor g_H represents the probability of forming a colorless hadron from quarks and antiquarks with spin 1/2. $d\sigma_i$ refers to an element of a space-like hypersurface, while f_{q_i} denotes the quark (anti-quark) phase-space distribution functions for the i-th quark (anti-quark). The function $C_H(x_1 ... x_{N_q}) = C^{N_q-1} f_H(x_1 ... x_{N_q}, p_1 ... p_{N_q})$ Where the Wigner function $f_H(x_1 ... x_{N_q}, p_1 ... p_{N_q})$ describes the spatial and momentum distribution of quarks in a hadron, and the factor C^{N_q-1} is fixed so that the total coalescence probability gives $\lim_{p\to 0} P^{tot}_{coal} = 1$. The Wigner function used assumes a Gaussian form in both position and momentum space as $f_H(...) = \prod_{i=1}^{N_q-1} 8 \exp\left(-\frac{xr_i^2}{\sigma_{ri}^2} - p_{ri}^2 \sigma_{ri}^2\right)$, where x_{ri} and p_{ri} represent the 4-vectors for relative coordinates. The widths σ_{ri} are connected to the the root mean square charge radius of the hadron produced. For details about the widths and statistical factor employed in the calculations see [14, 21, 24, 25]. As established in previous coalescence studies [4, 19–21, 26], the probability of coalescence is a decreasing function with the transverse momentum (p_T), leading to independent fragmentation to become the dominant mechanism for hadron production at high p_T. In our model, a smooth transition between the low and high p_T regimes is achieved by introducing a fragmentation probability $P_{frag}(p_T)$. Therefore, HQ with non-zero p_T undergoes hadronization via either coalescence or fragmentation. Given the coalescence probability $P_{coal}(p_T)$, the complementary fragmentation probability is defined as $P_{frag}(p_T) = 1 - P_{coal}(p_T)$. For the results presented, we apply the Peterson fragmentation for charmed hadrons [27] and the Kartvelishvili fragmentation for bottom hadrons [28].

3 Single and multi-charmed hadrons from pp to AA collisions

In $Au + Au$ collisions at RHIC energies or $Pb + Pb$ collisions at LHC energies, the hadronization via coalescence plus fragmentation predicts an enhancement of baryon production, notably increasing Λ_c yields for $p_T < 3 - 4,$ GeV at both RHIC and LHC resulting in a rise and

© The Authors, published by EDP Sciences. This is an open access article distributed under the terms of the Creative Commons Attribution License 4.0 (https://creativecommons.org/licenses/by/4.0/).

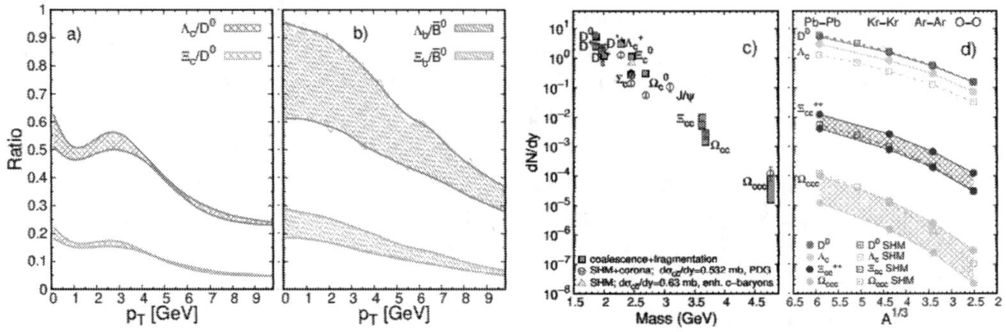

Figure 1. (Left plots) Comparison between Λ_c/D^0,Ξ_c/D^0 (a) and Λ_b/B^0,Ξ_b/B^0 (b) as a function of p_T for pp collision at $5.02\,TeV$. The bands correspond to the sensitivity of the ratio with a variation of the Wigner widths. (Right plots) panel c), Charmed hadrons yields in PbPb collision at 5.02TeV. Coalescence plus fragmentation results shown by orange band while SHM results shown by blue open circles [29]. The band corresponds to the yields obtained with realistic distribution for the lower limit and the one obtained with thermal distribution for the upper limit. In panel d) dN/dy at mid-rapidity of single and multi-charmed hadrons as a function of the size of the systems. Open symbols are for SHM calculation and full symbols are results obtained within coalescence plus fragmentation.

fall heavy flavour baryon-to-meson ratio. On the other hand, in pp collisions, assuming the formation of a QGP medium [30, 31], the model's predictions for D^0, D^+, D^*, and Λ_c are similar to those from the Statistical Hadronization Model (SHM), but significant differences arise for Ξ_c and Ω_c when comparing models with the same number of resonances predicting larger values [14, 32]. In the left panel of Fig.1 a), it is shown that coalescence plus fragmentation approach predict in a natural way a charmed baryon/meson ratio consistent with the recent experimental data available for pp collisions at LHC energies [33]. In Fig.1 b) it is shown the baryon-to-meson ratios for $\Lambda_b/\bar{B}^0$, $\Xi_b^0/\bar{B}^0$ in pp collisions at $\sqrt{s}$ = 5.02 TeV. The hybrid coalescence-plus-fragmentation model predicts higher baryon ratios in the bottom sector than in the charm sector. For Λ_b, coalescence enhances the Λ_b/B^0 ratio to ≈ 0.9 at low p_T, about 50% larger than the Λ_c/D^0 ratio, with a decreasing trend at higher p_T, consistent with charmed baryons. The $\Xi_b^0/\bar{B}^0$ ratio is ≈ 0.3 at low momenta, smaller than $\Lambda_b/\bar{B}^0$ but similar to experimental data and theoretical predictions for Ξ_c/D^0 [14].

In Fig.1 c), the yields for single and multi-charmed hadrons in *PbPb* collisions at mid-rapidity ($0-10\%$ centrality) are compared between the coalescence-plus-fragmentation approach and the Statistical Hadronization Model (SHM), which includes an extended set of charmed baryons as suggested in [23, 29, 32]. The red band represents the yield range, where the lower limit corresponding to a realistic charm distribution and the upper limit to a thermal distribution. For multi-charmed baryons (Ξ_{cc}, Ω_{cc}, Ω_{ccc}), the model shows high sensitivity to the charm distribution. A thermalized charm distribution leads to higher yields, which is consistent with the SHM results. This behaviour is related to the fact that more charms clustering at low momentum, facilitating recombination. This shows that the yields for multi-charmed baryons are more sensitive to the charm distribution compared to single-charmed hadrons. The Fig.1 d) illustrates the production of single and multi-charmed hadrons in different collision systems, including $Pb+Pb$, $Kr+Kr$, $Ar+Ar$, and $O+O$. The initial p_T distribution for these systems is based on FONLL calculations, evolving within a QGP medium modeled by a relativistic Boltzmann approach, as discussed in [24]. As the collision system size decreases, the fireball's lifetime and the charm quark's time in the QGP are reduced; therefore, smaller systems exhibit flatter final p_T distributions compared to larger systems like $Pb+Pb$ [24].

© The Authors, published by EDP Sciences. This is an open access article distributed under the terms of the Creative Commons Attribution License 4.0 (https://creativecommons.org/licenses/by/4.0/).

The dN/dy yield, derived from coalescence plus fragmentation for each species (shown as different bands), is plotted against the system size $A^{1/3}$ and compared to SHM results (open symbols). Similarly to Fig.1 c) the upper band limit is for a full charm thermalization, while the lower limit the distribution from the Boltzmann transport approach it has been used. The comparison with SHM reveals that, although similar, the scaling of dN/dy differs as system size decreases.

References

[1] P.B. Gossiaux, J. Aichelin, Phys. Rev. C **78**, 014904 (2008).
[2] M. He, R.J. Fries, R. Rapp, Phys. Rev. Lett. **110**, 112301 (2013).
[3] J. Uphoff, O. Fochler, Z. Xu, C. Greiner, Phys. Lett. **B717**, 430 (2012).
[4] S. Cao, G.Y. Qin, S.A. Bass, Phys. Rev. **C92**, 024907 (2015).
[5] A. Beraudo, A. De Pace, M. Monteno, M. Nardi, F. Prino, JHEP **05**, 279 (2021).
[6] M.L. Sambataro, Y. Sun, V. Minissale, S. Plumari, V. Greco, Eur. Phys. J. C **82**, 833 (2022).
[7] M.L. Sambataro, V. Minissale, S. Plumari, V. Greco, Phys. Lett. B **849**, 138480 (2024).
[8] S. Plumari, G.L. Guardo, F. Scardina, V. Greco, Phys. Rev. **C92**, 054902 (2015).
[9] F. Scardina, S.K. Das, V. Minissale, S. Plumari, V. Greco, Phys. Rev. **C96**, 044905 (2017).
[10] S. Plumari, G. Coci, V. Minissale, S.K. Das, Y. Sun, V. Greco, Phys. Lett. B **805**, 135460 (2020).
[11] Y. Sun, S. Plumari, V. Greco, Eur. Phys. J. C **80**, 16 (2020).
[12] L. Oliva, S. Plumari, V. Greco, JHEP **05**, 034 (2021).
[13] J. Zhao et al., (2023) 2311.10621.
[14] V. Minissale, S. Plumari, V. Greco, Phys. Lett. B **821**, 136622 (2021).
[15] A. Beraudo, A. De Pace, D. Pablos, F. Prino, M. Monteno, M. Nardi, Phys. Rev. D **109**, L011501 (2024).
[16] J. Zhao, J. Aichelin, P.B. Gossiaux, K. Werner, Phys. Rev. D **109**, 054011 (2024).
[17] V. Greco, C. Ko, P. Levai, Phys.Rev.Lett. **90**, 202302 (2003).
[18] R. Fries, B. Muller, C. Nonaka, S. Bass, Phys. Rev. Lett. **90**, 202303 (2003).
[19] V. Minissale, F. Scardina, V. Greco, Phys. Rev. C **92**, 054904 (2015), 1502.06213.
[20] Y. Oh, C.M. Ko, S.H. Lee, S. Yasui, Phys. Rev. C **79**, 044905 (2009), 0901.1382.
[21] S. Plumari, V. Minissale, S.K. Das, G. Coci, V. Greco, Eur. Phys. J. **C78**, 348 (2018).
[22] S. Cho, K.J. Sun, C.M. Ko, S.H. Lee, Y. Oh, Phys. Rev. C **101**, 024909 (2020).
[23] M. He, R. Rapp, Phys. Rev. Lett. **124**, 042301 (2020), 1905.09216.
[24] V. Minissale, S. Plumari, Y. Sun, V. Greco, Eur. Phys. J. C **84**, 228 (2024), 2305.03687.
[25] V. Minissale, V. Greco, S. Plumari, (2024), 2405.19244.
[26] P.B. Gossiaux, R. Bierkandt, J. Aichelin, Phys. Rev. C **79**, 044906 (2009), 0901.0946.
[27] C. Peterson, D. Schlatter, I. Schmitt, P.M. Zerwas, Phys. Rev. **D27**, 105 (1983).
[28] V.G. Kartvelishvili, A.K. Likhoded, V.A. Petrov, Phys. Lett. B **78**, 615 (1978).
[29] A. Andronic, et al. JHEP **07**, 035 (2021).
[30] R.D. Weller, P. Romatschke, Phys. Lett. B **774**, 351 (2017), 1701.07145.
[31] M. Greif, C. Greiner, B. Schenke, S. Schlichting, Z. Xu, Phys. Rev. D **96**, 091504 (2017).
[32] M. He, R. Rapp, Phys. Lett. B **795**, 117 (2019).
[33] S. Acharya et al. (ALICE), Phys. Rev. C **104**, 054905 (2021), 2011.06079.

© The Authors, published by EDP Sciences. This is an open access article distributed under the terms of the Creative Commons Attribution License 4.0 (https://creativecommons.org/licenses/by/4.0/).

EPJ Web of Conferences **316**, 04003 (2025) https://doi.org/10.1051/epjconf/202531604003
SQM 2024

Investigating Bottom Quark Energy Loss, Hadronization, and B Meson Nuclear Modification Factors in B^+ and B_s^0 Decays: Insights from CMS in pp, pPb, and PbPb Collisions

Jhovanny Mejia on behalf of the CMS Collaboration[1,*]

[1]Centro de Investigación y de Estudios Avanzados del Instituto Politécnico Nacional Mexico City, Mexico

Abstract. In this paper, using the exclusive decay channels $B_s^0 \rightarrow J/\psi\phi$ and $B^+ \rightarrow J/\psi K^+$, we present two studies made by the CMS Collaboration. The first is the p_T-differential cross sections of B_s^0 and B^+ mesons in pp collisions at 5.02 TeV. By utilizing previous PbPb collision data at the same nucleon-nucleon (NN) center of mass energy, R_{AA} factors for the B mesons are determined. The second one corresponds to the measurement of the B^+ meson production cross section with respect to meson transverse momentum inclusively and, for the first time, in different charged particle multiplicity ranges. This study is conducted in proton-lead collisions at a per nucleon center-of-mass energy of 8.16 TeV.

1 Introduction

Quantum chromodynamics (QCD) matter at high density under extreme conditions can form an exotic medium known as the quark-gluon plasma (QGP). In this state, quarks and gluons are no longer confined within hadrons but form a strongly interacting collective medium. Extensive research on the QGP has been conducted with heavy ion collisions. Recent observations of QGP-like phenomena in small collision systems, such as proton-proton (PP) and proton-lead (pPb) collisions [1, 2], challenge our understanding of high-energy heavy ion physics [3]. In particular for beauty quarks, there are still significant theoretical uncertainties in the hadronization process in the presence of the QGP because of a lack of understanding of the phenomena that can affect the production of hadrons. In particular, the strangeness content of the medium formed in heavy ion collisions is enhanced since the QGP temperature is larger than the strange quark mass. Therefore, through a quark-recombination mechanism, production of hadrons containing strangeness is expected to be enhanced relative to hadrons that do not contain strange quarks [4]. Recent studies of open-charm [5] and open-beauty [6] mesons in PbPb collisions at a nucleon-nucleon center-of-mass energy of 5.02 TeV at the LHC hinted an enhanced production of strange-charm and strange-beauty hadrons. Recently, this line of studies has been extended to heavy quarkonia B_c [7] and exotic hadron $X(3872)$ [8] production in PbPb collisions. To further investigate the hadronization of heavy quarks, the measurements of various flavors of beauty hadrons are essential, though they are notoriously difficult because of relatively small production cross sections compared to charm hadrons, large and varied background sources, and detector resolution limitations. In this paper, two recent results in this field are reviewed.

*e-mail: jhovanny.andres.mejia.guisao@cern.ch

© The Authors, published by EDP Sciences. This is an open access article distributed under the terms of the Creative Commons Attribution License 4.0 (https://creativecommons.org/licenses/by/4.0/).

EPJ Web of Conferences **316**, 04003 (2025) https://doi.org/10.1051/epjconf/202531604003
SQM 2024

2 Constraining bottom quark energy loss and hadronization with B^+ and B_s^0 nuclear modification factors in pp and PbPb collisions at 5.02 TeV with CMS

The CMS Collaboration reports on the production yields of B_s^0 and B^+ mesons in proton-proton (pp) collisions recorded in 2017 at the CERN LHC with a center-of-mass energy of 5.02 TeV. Utilizing an integrated luminosity of 302.3 pb^{-1}, the analysis focuses on the exclusive decay channels $B_s^0 \rightarrow J/\psi\, \phi$ and $B^+ \rightarrow J/\psi\, K^+$ in a transverse momentum (p_T) range of $7 < p_T(B) < 50$ GeV and a rapidity interval $|y| < 2.4$. Combining with the previous measurements in lead-lead collisions at the same center of mass, we measure the nuclear modification factor, R_{AA}, of the B mesons [4].

The p_T-differential production cross sections of B^+ (left) and B_s^0 mesons are presented in Fig. 1. The result is compared to the cross sections obtained from FONLL calculations [9], which are obtained by scaling the FONLL total b quark production by the world-average production fractions of B^+ (B_s^0) of 40.2 (10.5)%. The calculated FONLL reference spectra are consistent with both mesons' measured pp spectra within the quoted uncertainties.

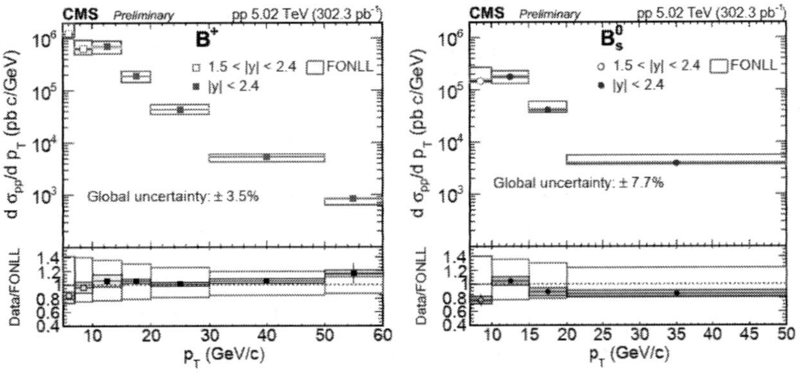

Figure 1. The p_T-differential cross section of B^+ (left) and B_s^0 (right) mesons in pp collisions at 5.02 TeV. The vertical bars (shaded boxes) correspond to statistical (systematic) uncertainties. The global systematic uncertainty includes uncertainties in the integrated luminosity and the external branching fractions. The comparison with the calculation from FONLL (open boxes) is also shown. The bottom panels show that the measured cross sections deviate from the FONLL calculations by less than 20%, and they are consistent within uncertainty [4].

The nuclear modification factors of B^+ (left) and B_s^0, are shown in Fig. 2 as a function of p_T. For B^+, a strong suppression is observed, and R_{AA} is around 0.3 to 0.6 in $7 < p_T < 50$ GeV. In $10 < p_T < 50$ GeV, R_{AA} of B_s^0 is also smaller than unity but compatible within uncertainties than that of B^+. Higher statistics in PbPb collisions are needed to verify this effect.

The B^+ and B_s^0 R_{AA} are compared in Fig. 3 to the CMS measurements of the R_{AA} of charged particles [10], D^0 mesons [11] performed at the same energy and in a similar centrality range. For B^+ mesons, the R_{AA} values are consistent with those of charged particles and D^0 mesons for $p_T > 10$ GeV. At lower p_T a hint of a reduced level of suppression is present for B^+ mesons, which is in line with expectations based on the quark mass dependence of parton energy loss.

© The Authors, published by EDP Sciences. This is an open access article distributed under the terms of the Creative Commons Attribution License 4.0 (https://creativecommons.org/licenses/by/4.0/).

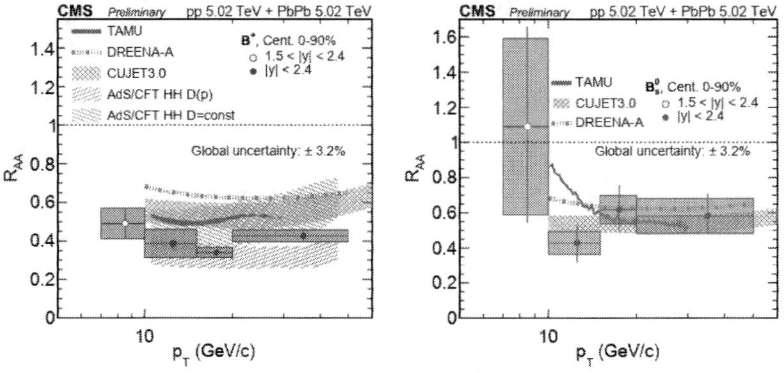

Figure 2. The p_T dependence of the nuclear modification factor of B^+ (left) and B_s^0 mesons in PbPb collisions at 5.02 TeV. The vertical bars (boxes) correspond to statistical (systematic) uncertainties. The global systematic uncertainty comprises the uncertainties in luminosity, T_{AA} and the number of minimum-bias events. Theoretical calculations are also shown for comparison [4].

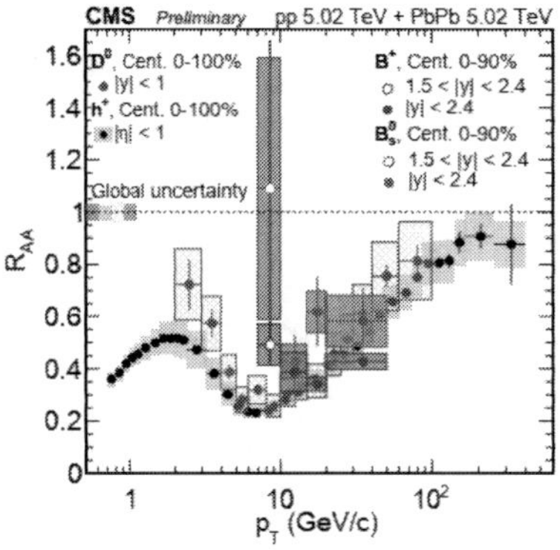

Figure 3. Comparison of the nuclear modification factor of B^+ and B_s^0 mesons with D^0 mesons [11] and charged hadrons [10] in PbPb collisions at 5.02 TeV. The global systematic uncertainties, depicted in shaded boxes around R_{AA} = 1, comprise the uncertainties in T_{AA}, the number of minimum-bias events, and the luminosity in pp collisions [3].

3 Measurement of the B^+ differential cross section as a function of transverse momentum and multiplicity in pPb collisions at $\sqrt{s_{NN}}$ = 8.16 TeV

The measurement of the B^+ meson production cross section with respect to the transverse momentum inclusively and, for the first time, in different charged particle multiplicities, in proton-lead collisions at a per nucleon center-of-mass energy of 8.16 TeV was performed. The data were collected by the CMS during 2016 and correspond to an integrated luminosity of 175 nb^{-1}. The measurement uses the exclusive decay channel $B^+ \rightarrow J/\psi K^+$ [3].

The results of the measurement of the differential cross section, for the phase space region $3 < p_T(B) < 50$ GeV and $|y(B)| < 1.8$, are displayed in Fig. 4 (left). In the same plot, the pp FONLL predictions [9] at the same energy are superimposed. The FONLL reference cross sections are obtained by scaling the FONLL total b quark production by the world-average B^+ production fraction of 40.2%. The FONLL predictions are scaled by the atomic mass of the Pb nucleus (A = 208), to consider the number of binary NN collisions.

The theoretical predictions of FONLL are in good agreement with the measurements. The event-by-event charged-particle multiplicity is defined using tracks originating from the PV, which must satisfy the following criteria: the impact parameter significance $(d/\sigma(d))$

© The Authors, published by EDP Sciences. This is an open access article distributed under the terms of the Creative Commons Attribution License 4.0 (https://creativecommons.org/licenses/by/4.0/).

measured with respect to the PV must be less than 3, both along $(d_z/\sigma(d_z))$, and transverse $(d_T/\sigma(d_T))$ to the beam direction, and the relative p_T uncertainty, $\sigma(p_T)/p_T$, must be less than 10%. To ensure high tracking efficiency and to reduce the rate of misreconstructed tracks, the requirements $|\eta| < 2.4$ and $p_T > 0.4$ GeV must be satisfied. The quantity N_{trk}^{corr} is the corresponding multiplicity corrected for detector and algorithm inefficiencies in the same kinematic region. Figure 4 (right) shows the differential cross section in p_T bins divided into classes of charged particle multiplicity.

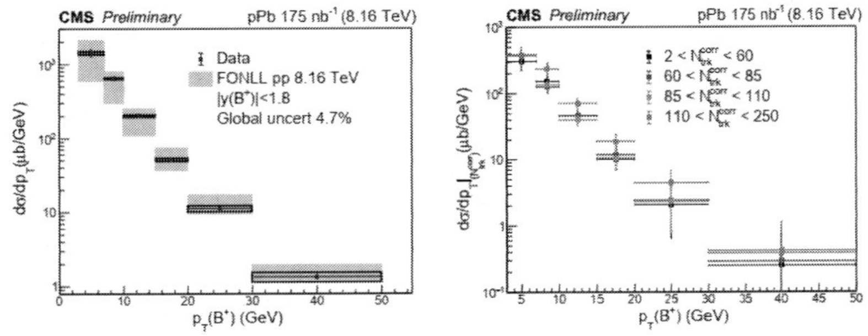

Figure 4. (Left) B^+ differential cross section in p_T bins. The vertical bars (boxes) represent the statistical (systematic) uncertainties. The global systematic uncertainty, not included in the data points, comprises the uncertainties in the integrated luminosity measurement and the B meson branching fractions. Results are compared to FONLL calculations, scaled by the number of binary NN collisions, represented by squares. The yellow boxes represent the theoretical uncertainties from FONLL [9]. (Right) B^+ differential cross section shown in p_T bins divided into classes of reconstructed track multiplicity. The vertical bars represent the total uncertainties [3].

4 Summary

The exclusive decay channels $B_s^0 \rightarrow J/\psi\phi$ and $B^+ \rightarrow J/\psi K^+$ are investigated. The differential cross sections are well-described by fixed-order plus next-to-leading logarithm calculations. Utilizing pp and PbPb collision data at 5.02 TeV R_{AA} factors for the B mesons are determined. Additionally, conducted in proton-lead collisions at 8.16 TeV, the measurement of the B^+ meson production cross section is presented with respect to meson p_T inclusively and, for the first time, in different charged particle multiplicity ranges.

References

[1] CMS collaboration, Phys. Rev. Lett. **115**, 012301 1 2015.
[2] CMS collaboration, Eur. Phys. J. C **77**, 269 4 2017.
[3] CMS collaboration, CMS-PAS-HIN-22-001, 2023.
[4] CMS collaboration, CMS-PAS-HIN-21-014, 2023.
[5] ALICE collaboration, J. High Energ. Phys. **10**, 174 2018.
[6] CMS collaboration, Phys. Lett. B **796**, 168 2019.
[7] CMS collaboration, Phys. Rev. Lett. **128**, 252301 25 2022.
[8] CMS collaboration, Phys. Rev. Lett. **128**, 032001 3 2022.
[9] M. Cacciari and others, J. High Energ. Phys. 137 2012.
[10] CMS collaboration, J. High Energ. Phys. **04**, 039 2017.
[11] CMS collaboration, Phys. Lett. B **782**, 474 2018.

© The Authors, published by EDP Sciences. This is an open access article distributed under the terms of the Creative Commons Attribution License 4.0 (https://creativecommons.org/licenses/by/4.0/).

First $D^0 + \overline{D}^0$ measurement in nucleus-nucleus collisions at SPS energies with NA61/SHINE

Anastasia Merzlaya[1,*] for the NA61/SHINE Collaboration

[1]University of Oslo

Abstract. The measurement of open charm meson production provides a tool for the investigation of the properties of the hot and dense matter created in nucleus-nucleus collisions at relativistic energies. In particular, charm mesons are of vivid interest in the context of the study of the nature of the phase-transition between confined hadronic matter and the quark-gluon plasma. Recently, the NA61/SHINE experimental setup was upgraded with a high spatial resolution Vertex Detector which enables the reconstruction of secondary vertices from open charm meson decays.

The first direct neutral D meson yields at the SPS energy regime will be shown. The analysis used the 20% most central Xe+La collisions at 150A GeV/c from the data set collected in 2017. It allowed the estimation of the mid-rapidity yields ($\mathrm{d}N/\mathrm{d}y$) of $D^0 + \overline{D}^0$ using their $\pi^{\pm} + K^{\mp}$ decay channels and phase space corrections derived from the three models AMPT, PHSD, and PYTHIA/Angantyr. The results will be compared and discussed in the context of several model calculations including statistical and dynamical approaches.

1 Introduction

The study of open charm production is a sensitive tool for detailed investigations of the properties of hot and dense matter formed in nucleus-nucleus collisions. Since heavy quarks are produced in hard scattering processes in the early stage of the collision, by studying them one can get insight to the properties of the created medium. In particular, such measurements are of vivid interest at the SPS energies, which are close to the quark-gluon plasma creation threshold.

Also such study gives a unique opportunity to test the validity of theoretical models based on perturbative Quantum Chromodynamics and Statistical model approaches for nucleus-nucleus collisions at SPS energies, which provide very different predictions for charm yields [1]. However, up to now there were no open charm measurements in this energy regime.

2 Open charm measurements by NA61/SHINE

The SPS Heavy Ion and Neutrino Experiment (NA61/SHINE) [2] at CERN was designed for studies of the properties of the onset of deconfinement and search for the critical point of strongly interacting matter by investigating p+p, p+A and A+A collisions at different

*e-mail: anastasia.merzlaya@cern.ch

© The Authors, published by EDP Sciences. This is an open access article distributed under the terms of the Creative Commons Attribution License 4.0 (https://creativecommons.org/licenses/by/4.0/).

beam momenta from 13A to 150/158A GeV/c for ions and up to 400 GeV/c for protons. The main tracking detectors are four Time Projection Chambers (TPC) which provide momentum, charge and particle identification (PID) for the particles produced in the collision.

The heavy-ion programme of the NA61/SHINE experiment at CERN SPS has been expanded to allow precise measurements of particles with short lifetime. To meet the challenges of open charm measurements, NA61/SHINE was upgraded with the Small Acceptance Vertex Detector (SAVD). SAVD was composed of four detector planes located 5, 10, 15 and 20 cm downstream of the target. The planes were constructed using position-sensitive MIMOSA-26AHR pixel sensors [3]. SAVD provided precise vertex reconstruction with excellent spatial resolution. The obtained transverse and longitudinal primary vertex resolution was $\sigma_{x,y} \approx 1$ μm and $\sigma_z = 15$ μm, respectively. The SAVD acceptance for D^0, $\overline{D}^0$ was $-0.5 < y < 1.0$, $0.2 < p_T < 2.0$ GeV/c, and the analysis presented here was performed in this kinematic range. More details about SAVD, track and vertex reconstruction can be found in Refs. [4, 5].

The $D^0 + \overline{D}^0$ reconstruction is done via $\pi^\pm + K^\mp$ decay channel. The SAVD tracks matched to TPC tracks are used to search for the $D^0 + \overline{D}^0$ signal. In the analysis presented here, the PID information was not used. Each track is paired with all the opposite-sign tracks from the same event and is assumed to be either a kaon or a pion, thus each pair contributes twice in the combinatorial invariant mass distribution. In order to suppress the combinatorial background of the invariant mass distribution, the following cuts are applied:

- cut on the daughter track impact parameter, $d > 36$ μm;
- cut on the distance of closest approach between the daughter tracks, $DCA < 42$ μm;
- cut on the longitudinal distance between the D^0 decay vertex candidate and the primary vertex scaled with the gamma factor, $V_z/\gamma > 0.15$ μm;
- cut on the impact parameter of the back-extrapolated D^0 candidate momentum vector, $D < 20$ μm;
- cut on the D^0 candidate momentum in laboratory frame, $13 < p < 38$ GeV/c.

Figure 1 shows the $K^\pm\pi^\mp$ invariant mass distribution after the applied cuts. The peak corresponding to $D^0 + \overline{D}^0$ has a statistical significance, $S/\sqrt{S+B}$, of approximately 6.

3 Analysis results

In order to determine the necessary corrections and calculate the absolute value of $D^0 + \overline{D}^0$ yield, simulations with GEANT4 were performed. The input to GEANT4 consists of events generated by the EPOS model [6], while the embedded D^0, $\overline{D}^0$ mesons were generated with phase space distributions obtained from the three models: AMPT [7], PHSD [8] and PYTHIA/Angantyr [9], which predict quite different (y, p_T) distribution of open charm.

After applying corrections from Monte Carlo, the corrected yield for each phase space assumption was calculated. The results are presented in Table 1. The visible yield $N(D^0 + \overline{D}^0)_{visible}$ corresponds to the yield in the $-0.5 < y < 1.0$, $0.2 < p_T < 2.0$ GeV/c rapidity-transverse momentum range covered by the SAVD acceptance. In order to obtain the rapidity density, $\mathrm{d}N(D^0 + \overline{D}^0)/\mathrm{d}y\ _{-0.5<y<1.0}$, and the 4π integrated yield, $\langle D^0 + \overline{D}^0 \rangle$, the extrapolation factors for each of the models were determined. The AMPT model predicts that about 84% of $D^0 + \overline{D}^0$ are produced in the SAVD acceptance, while PHSD and PYTHIA/Angantyr predict this fraction to be approximately 67%. Thus, $\langle D^0 + \overline{D}^0 \rangle$ differ significantly between the model assumptions.

© The Authors, published by EDP Sciences. This is an open access article distributed under the terms of the Creative Commons Attribution License 4.0 (https://creativecommons.org/licenses/by/4.0/).

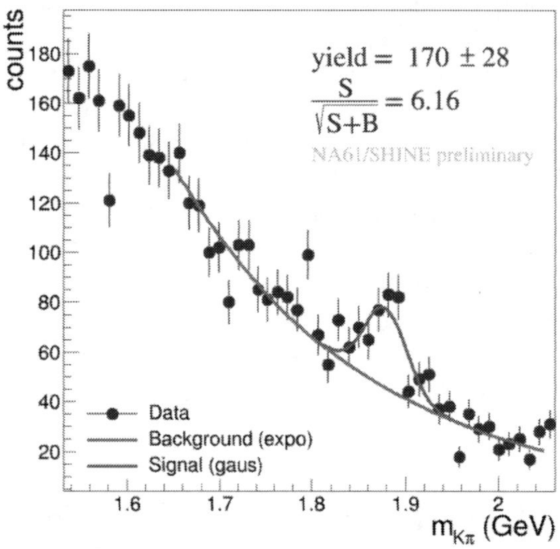

Figure 1. The $K^{\pm}\pi^{\mp}$ invariant mass distribution after the applied cuts for Xe+La collisions at 150*A* GeV/*c* taken in 2017. The invariant mass distribution was fitted (red line) using an exponential function to describe the background and a Gaussian to describe the $D^0 + \overline{D}^0$ signal contribution. The indicated uncertainties are statistical only.

Table 1. Results for the visible yield, dN/dy at mid-rapidity and yield in 4π of $D^0 + \overline{D}^0$ obtained assuming the phase space distribution for AMPT, PHSD and PYTHIA/Angantyr models. The first indicated uncertainty corresponds to the statistical uncertainty and the second – to the systematic uncertainty. The systematic uncertainty does not include the model-dependent phase space correction.

correction with:	$N(D^0 + \overline{D}^0)_{visible}$	$\frac{dN(D^0+\overline{D}^0)}{dy}\Big\vert_{-0.5<y<1.0}$	$\langle D^0 + \overline{D}^0 \rangle$
AMPT	0.184 ± 0.032	0.129 ± 0.023 ± 0.035	0.218 ± 0.039 ± 0.060
PHSD	0.204 ± 0.036	0.148 ± 0.026 ± 0.036	0.303 ± 0.054 ± 0.074
PYTHIA/Angantyr	0.201 ± 0.035	0.147 ± 0.026 ± 0.037	0.300 ± 0.052 ± 0.075

Figure 2 shows the comparison of the obtained 4π yield $\langle D^0 + \overline{D}^0 \rangle$ with the results of various model predictions. The precision of the data is sufficient to discriminate between the current model predictions. While microscopic models (AMPT, PYTHIA/Angantyr, PHSD, HSD [10]) tend to significantly underestimate $D^0 + \overline{D}^0$ yield, ALCOR [11] and SMES [12] models are overestimating it.

An estimation based on measurements of the $D^0 + \overline{D}^0$ cross-section in different p+A collision systems and in a range of collision energies, $\sqrt{s_{NN}}$, spanning between 20 and 40 GeV [13], scaled by the number of binary collisions in order to correspond to central Xe+La reactions, yields a value that is below the NA61/SHINE measurement with a statistical significance of 2-3σ.

4 Summary and outlook

The presented analysis shows the first-ever direct measurement of open charm in nucleus-nucleus collisions at SPS energies. The precision of this result is sufficient to disentangle between the different theoretical predictions.

During CERN's long-shutdown 2 the TPC readout and DAQ of NA61/SHINE were upgraded increasing the data taking rate from 80 Hz to 1.2 kHz [14]. Within the upgrade also a new Vertex Detector based on ALPIDE sensors was installed. In 2022–2023 NA61/SHINE collected at the level of 180 M events of Pb+Pb collisions at 150*A* GeV/*c*. These data will

© The Authors, published by EDP Sciences. This is an open access article distributed under the terms of the Creative Commons Attribution License 4.0 (https://creativecommons.org/licenses/by/4.0/).

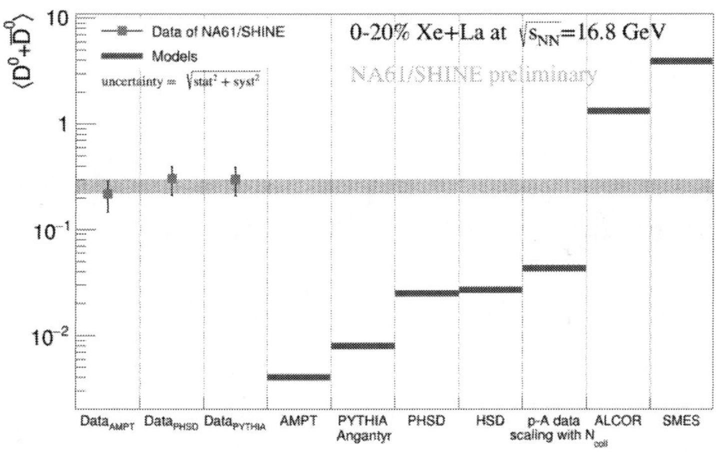

Figure 2. Comparison of the obtained result to theoretical model predictions. The red band indicates the theoretical uncertainty of the result due to the unknown phase space distribution of D^0, $\overline{D}^0$.

allow double differential measurements of D^0 and $\overline{D}^0$ in rapidity and transverse momentum, as well as measurements of other charmed hadrons. This study will provide a further, better insight into charm production mechanisms at energies close to the QGP creation threshold.

This work was supported by the Norwegian Financial Mechanism 2014–2021 (grant 2019/34/H/ST2/00585).

References

[1] A. Aduszkiewicz *et al.* (NA61/SHINE Collaboration), CERN-SPSC-2018-008 / SPSC-P-330-ADD-10 (2018)
[2] N. Abgrall *et al.* (NA61/SHINE Collaboration), JINST **9**, P06005 (2014)
[3] M. Deveaux *et al.*, J. Instrum. **6**, C02004 (2011)
[4] A. Merzlaya, PhD thesis, https://cds.cern.ch/record/2771816 (2021)
[5] A. Aduszkiewicz *et al.*, Eur. Phys. J. C **83** no.6, 471 (2023)
[6] T. Pierog *et al.*, Phys. Rev. C **92** no.3, 034906 (2015)
[7] Z. W. Lin *et al.*, Phys. Rev. C **72**, 064901 (2005)
[8] W. Cassing and E. L. Bratkovskaya, Nucl. Phys. A **831**, 215-242 (2009)
[9] C. Bierlich *et al.*, JHEP **10**, 134 (2018)
[10] O. Linnyk *et al.*, Int. J. Mod. Phys. E **17**, 1367-1439 (2008)
[11] P. Levai *et al.*, J. Phys. G **27**, 703-706 (2001)
[12] M. Gazdzicki and C. Markert, Acta Phys. Polon. B **31** no.4, 965-983 (2000)
[13] P. Braun-Munzinger *et al.*, Eur. Phys. J. C **1**, 123-130 (1998)
[14] H. Adhikary *et al.* (NA61/SHINE Collaboration), CERN-SPSC-2023-030 / SPSC-SR-336 (2023)

© The Authors, published by EDP Sciences. This is an open access article distributed under the terms of the Creative Commons Attribution License 4.0 (https://creativecommons.org/licenses/by/4.0/).

The role of strangeness in heavy-quark hadronisation from small to large collision systems with ALICE

Mattia Faggin[1,*], on behalf of the ALICE Collaboration

[1]CERN, Espl. des Particules 1, 1211 Meyrin (CH)

Abstract. Production measurements of strange hadrons originating from the hadronisation of charm quarks (prompt) and from beauty-hadron decays (non-prompt) offer a unique tool to study the heavy-quark hadronisation across different collision systems. This contribution discusses the final results of the ALICE Collaboration obtained by measuring strange D mesons in pp, p–Pb, and Pb–Pb collisions collected during the LHC Run 2. The first measurement of the production of orbitally excited charm-strange mesons in pp collisions is also reported. Additionally, the production measurements of prompt D_s^+ mesons are compared to those of non-strange mesons across the different collision systems, along with the measurement of non-prompt D_s^+ mesons in heavy-ion collisions. Lastly, the first studies of strange and non-strange D mesons using the large data sample of pp collisions at $\sqrt{s}$ =13.6 TeV harvested from the start of LHC Run 3 are presented.

1 Introduction

Due to their large mass ($m_c \sim 1.3$ GeV$/c^2$, $m_b \sim 4.2$ GeV$/c^2$), charm and beauty quarks are mainly produced in hard scattering processes among partons in the early stages of a hadronic collision at the LHC. In ultrarelativistic heavy-ion collisions, they are produced before the quark gluon plasma (QGP) formation, and since their number is preserved during the QGP phase they experience its full evolution. Therefore, charm and beauty quarks are unique probes of the colour-rich environment produced in hadronic collisions at the LHC, and their dynamics can be studied by reconstructing charm and beauty hadrons. Their measurement in pp collisions are important to test the predictions of perturbative QCD (pQCD)-based calculations and to put constraints on heavy-quark production, parton distribution functions (PDFs) and hadronisation processes. They also provide a fundamental reference to understand the results in larger collision systems. In fact, such measurements in p–Pb collisions are sensitive to cold-nuclear-matter effects, such as PDF modification in bound nucleons, while those in heavy-ion collisions are influenced by the in-medium parton energy loss, collective motion and modified hadronisation. In particular, the formation of heavy-flavour (HF) hadrons in Pb–Pb collisions can occur via the coalescence of a charm or beauty quark with lighter ones thermalised in the QGP. Interestingly, a smooth strangeness enhancement as function of event multiplicity has been observed at the LHC, without a clear collision system dependence [1]. One can speculate whether a similar interplay between an enhanced formation of $s\bar{s}$ pairs and hadronisation via coalescence occur in the three collision systems. This can be effectively studied via the measurement of strange heavy-flavour hadron production.

*e-mail: mattia.faggin@cern.ch

© The Authors, published by EDP Sciences. This is an open access article distributed under the terms of the Creative Commons Attribution License 4.0 (https://creativecommons.org/licenses/by/4.0/).

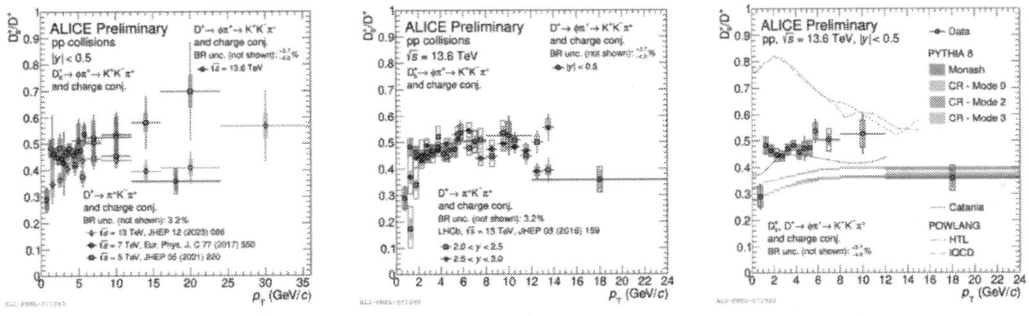

Figure 1. p_{T}-differential $\mathrm{D_s}^+/\mathrm{D}^+$ ratio at midrapidity in pp collisions at $\sqrt{s}$ = 13.6 TeV by ALICE compared with the same result at lower collision energies at midrapidity (left) and at forward rapidity by LHCb (middle), as well as with model predictions (right).

2 The ALICE experiment

The measurements described above are possible with the data collected during the LHC Run 1 and 2 by the ALICE experiment. The innermost subsystem in the central barrel ($|\eta| < 0.9$) is the Inner Tracking System (ITS), a silicon tracker fundamental to reconstruct and resolve the beam collision point (primary vertex, PV) and the heavy-flavour hadron decay point. The Time Projection Chamber (TPC), a gaseous detector, is the main tracker of the experiment, which also provides particle identification (PID) via the measurement of ionization energy loss per unit length $\mathrm{d}E/\mathrm{d}x$. Finally, the Time-Of-Flight (detector) provides PID via the measurement of the time-of-flight of charged particles. During the long shutdown 2 (LS2) of the LHC the experiment underwent a major upgrade program. The most significant upgrades involved the ITS, the readout of the TPC, now utilising Gas Electron Multipliers (GEMs), and the new Fast Interaction Trigger (FIT) detector. Thanks to these upgrades, during the LHC Run 3 ALICE can collect significantly larger data samples in a continuous readout mode and can exploit a $\sim \times 2$ better pointing resolution of tracks to the PV compared to Run 2.

3 Measurements of charm-strange meson production

The prompt and non-prompt $\mathrm{D_s}^+/(\mathrm{D}^+ + \mathrm{D}^0)$ p_{T}-differential ratios have been measured in pp collisions at $\sqrt{s}$ = 5.02 TeV and $\sqrt{s}$ = 13 TeV [2, 3]. The results do not show any significant dependence neither on p_{T} nor on collision energy. The same holds also for the charm and beauty p_{T}-integrated strange-to-non-strange fragmentation-fraction ratio $f_{\mathrm{s}}/(f_{\mathrm{u}} + f_{\mathrm{d}})$ [3, 4]. They are compatible within each other, and with the results from $\mathrm{e^+e^-}$, $\mathrm{e^-p}$ and $\mathrm{p\bar{p}}$ collisions at lower energy.

ALICE provided the first measurement of the p_{T}-differential production-yield ratio of prompt $\mathrm{D_s}^+/\mathrm{D}^+$ mesons at midrapidity in pp collisions at $\sqrt{s}$ = 13.6 TeV (Fig. 1) using a sample of $\mathcal{L} \sim 1\ \mathrm{pb}^{-1}$ collected in 2022. This measurement extends down to $p_{\mathrm{T}} = 0.5$ GeV/c and it is $\times 2$ more granular in the range $1 < p_{\mathrm{T}} < 6$ GeV/c than the result obtained with the Run 2 sample at $\sqrt{s}$ = 13 TeV. The results do not show any significant collision-energy and rapidity dependence, being compatible with the same results at midrapidity at $\sqrt{s}$ = 5.02 TeV, $\sqrt{s}$ = 7 TeV and $\sqrt{s}$ = 13 TeV [2, 4, 5], as well as with those at forward rapidity in pp collisions at $\sqrt{s}$ = 13 TeV [6]. In the right panel of Fig. 1, the measurement is compared with predictions from model calculations assuming different mechanisms for the heavy-quark hadronisation. The measured $\mathrm{D_s}^+/\mathrm{D}^+$ ratio is underestimated by the predictions of PYTHIA

© The Authors, published by EDP Sciences. This is an open access article distributed under the terms of the Creative Commons Attribution License 4.0 (https://creativecommons.org/licenses/by/4.0/).

8 [7], where colour reconnections among multi-parton interactions (MPI) beyond leading-colour topologies are assumed to explain the observed baryon enhancement at midrapidity. The measurement is also compared with calculations from POWLANG [8] and Catania [9] models, in which the formation of a thermalised system of light partons is foreseen, and the heavy quarks can hadronise via coalescence. The measurement is significantly overestimated by POWLANG, while the prediction from the Catania model better describes the data.

The production of heavy-strange hadrons in pp collisions has been studied also as a function of event multiplicity [10]. The prompt p_{T}-differential $\mathrm{D_s}^+/\mathrm{D}^0$ ratio does not show any significant dependence on p_{T} or multiplicity. The measurement shows some tension with the predictions from the Canonical-Ensemble Statistical Hadronisation Model (CE-SHM [11]), where the quantum-number conservation volume is assumed to increase with the event multiplicity. In fact, the model calculations overestimate the measurement in high-multiplicity pp collisions. The $\mathrm{D_s}^+/\mathrm{D}^0$ ratio is well described by the predictions of PYTHIA 8. On the contrary, PYTHIA 8 predictions significantly underestimate the prompt Ξ_c^0 production and the p_{T}-differential Ξ_c^0/D^0 ratio in pp collisions at the LHC [12]. The charm-quark fragmentation fraction into Ξ_c^0 has been observed to be compatible with that of prompt $\mathrm{D_s}^+$ in both pp and p–Pb collisions. The charm-quark fragmentation fractions $f(\mathrm{c} \to \mathrm{h_c})$ in these two collision systems are compatible within each other and show a significant enhancement of baryon relative production compared to collision systems involving leptons (e.g. $\mathrm{e^+e^-}$, $\mathrm{e^-p}$) [13]. Further information about the charm-quark hadronisation and its dependence on collision system and multiplicity are provided by the measurement of $\mathrm{D}_{s1}^{+}(2536)$ and $\mathrm{D}_{s2}^{*+}(2573)$ resonance production. The measured $\mathrm{D}_{s1}^{+}(2536)/\mathrm{D_s}^+$ and $\mathrm{D}_{s2}^{*+}(2573)/\mathrm{D_s}^+$ ratios do not show a significant dependence on the event multiplicity within the current uncertainties.

The production of prompt $\mathrm{D_s}^+$ mesons at midrapidity has been measured by ALICE also in Pb–Pb collisions at $\sqrt{s_{\mathrm{NN}}} = 5.02$ TeV. The prompt-$\mathrm{D_s}^+$ nuclear modification factor (R_{AA}) has been observed to be significantly lower than unity for $p_{\mathrm{T}} \gtrsim 4$ GeV/c, as a consequence of the in-medium energy loss of the charm quark. In addition, the $R_{\mathrm{AA}}(\mathrm{D_s}^+)$ has been observed to be larger than that of non-strange D mesons for $p_{\mathrm{T}} < 6$ GeV/c. This is in agreement with the scenario of a relative enhancement of $\mathrm{D_s}^+$ production in Pb–Pb collisions compared to pp due to the interplay of the charm-quark hadronisation via coalescence and the enhanced strangeness production in the QGP [14]. A hint of similar ordering between strange and non-strange D-meson R_{AA} has been observed also for non-prompt D mesons [15]. In addition, the measurements show a hint of higher R_{AA} for non-prompt $\mathrm{D_s}^+$ compared to that of prompt $\mathrm{D_s}^+$ for $p_{\mathrm{T}} < 6$ GeV/c, showing sensitivity to the different diffusion and collisional energy loss for charm and beauty quarks in the QGP, given that the spatial diffusion coefficient is proportional to the inverse of the mass. The current experimental uncertainties are still too large to be sensitive to the different radiative energy loss for charm and beauty at high p_{T} due to the dead-cone effect. However, the comparison with the predictions from TAMU transport model [16, 17], which overestimates the measurement of prompt $R_{\mathrm{AA}}(\mathrm{D_s}^+)$ at high p_{T}, indicates that the heavy-quark radiative energy loss in the QGP, ignored in these model calculations, is important to describe the measurement. Finally, ALICE measured the elliptic flow coefficient (v_2) of prompt $\mathrm{D_s}^+$ and observed a deviation from 0 of about 6.4σ. With the current uncertainties, no significant differences are observed with respect to the non-strange D-meson v_2. The measured $v_2(\mathrm{D_s}^+)$ is well described for $p_{\mathrm{T}} < 10$ GeV/c by transport models implementing charm-quark hadronisation via coalescence with a flowing strange-quark.

In conclusion, the ALICE Collaboration measured the prompt and non-prompt strange D-meson production in pp and Pb–Pb collisions. Such measurements are useful to probe the charm- and beauty-quark hadronisation in hadronic collisions and their dynamics in the QGP produced in heavy-ion collisions at the LHC. More insights will be provided by the analysis of the Run 3 data.

© The Authors, published by EDP Sciences. This is an open access article distributed under the terms of the Creative Commons Attribution License 4.0 (https://creativecommons.org/licenses/by/4.0/).

References

[1] S. Acharya et al. (ALICE), Multiplicity dependence of π, K, and p production in pp collisions at $\sqrt{s}$ = 13 TeV, Eur. Phys. J. C **80**, 693 (2020), 2003.02394. 10.1140/epjc/s10052-020-8125-1

[2] S. Acharya et al. (ALICE), Measurement of beauty and charm production in pp collisions at $\sqrt{s} = 5.02$ TeV via non-prompt and prompt D mesons, JHEP **05**, 220 (2021), 2102.13601. 10.1007/JHEP05(2021)220

[3] S. Acharya et al. (ALICE), Measurement of beauty-quark production in pp collisions at $\sqrt{s} = 13$ TeV via non-prompt D mesons (2024), 2402.16417.

[4] S. Acharya et al. (ALICE), Charm production and fragmentation fractions at midrapidity in pp collisions at $\sqrt{s}$ = 13 TeV, JHEP **12**, 086 (2023), 2308.04877. 10.1007/JHEP12(2023)086

[5] S. Acharya et al. (ALICE), Measurement of D-meson production at mid-rapidity in pp collisions at $\sqrt{s} = 7$ TeV, Eur. Phys. J. C **77**, 550 (2017), 1702.00766. 10.1140/epjc/s10052-017-5090-4

[6] R. Aaij et al. (LHCb), Measurements of prompt charm production cross-sections in *pp* collisions at $\sqrt{s}$ = 13 TeV, JHEP **03**, 159 (2016), [Erratum: JHEP 09, 013 (2016), Erratum: JHEP 05, 074 (2017)], 1510.01707. 10.1007/JHEP03(2016)159

[7] J.R. Christiansen, P.Z. Skands, String Formation Beyond Leading Colour, JHEP **08**, 003 (2015), 1505.01681. 10.1007/JHEP08(2015)003

[8] A. Beraudo, A. De Pace, D. Pablos, F. Prino, M. Monteno, M. Nardi, Heavy-flavor transport and hadronization in pp collisions, Phys. Rev. D **109**, L011501 (2024), 2306.02152. 10.1103/PhysRevD.109.L011501

[9] V. Minissale, S. Plumari, V. Greco, Charm hadrons in pp collisions at LHC energy within a coalescence plus fragmentation approach, Phys. Lett. B **821**, 136622 (2021), 2012.12001. 10.1016/j.physletb.2021.136622

[10] S. Acharya et al. (ALICE), Observation of a multiplicity dependence in the pT-differential charm baryon-to-meson ratios in proton–proton collisions at s=13 TeV, Phys. Lett. B **829**, 137065 (2022), 2111.11948. 10.1016/j.physletb.2022.137065

[11] Y. Chen, M. He, Charged-particle multiplicity dependence of charm-baryon-to-meson ratio in high-energy proton-proton collisions, Phys. Lett. B **815**, 136144 (2021), 2011.14328. 10.1016/j.physletb.2021.136144

[12] J. Zhu (ALICE), Latest ALICE results on charm and beauty hadronisation in hadronic collisions, EPJ Web Conf. **296**, 15008 (2024), 2312.09705. 10.1051/epjconf/202429615008

[13] S. Acharya et al. (ALICE), Charm fragmentation fractions and $c\bar{c}$ cross section in p–Pb collisions at $\sqrt{s_{\mathrm{NN}}}$ = 5.02 TeV (2024), 2405.14571.

[14] S. Acharya et al. (ALICE), Measurement of prompt D_s^+-meson production and azimuthal anisotropy in Pb–Pb collisions at $\sqrt{s_{NN}}$=5.02TeV, Phys. Lett. B **827**, 136986 (2022), 2110.10006. 10.1016/j.physletb.2022.136986

[15] S. Acharya et al. (ALICE), Measurement of beauty-strange meson production in Pb–Pb collisions at sNN=5.02TeV via non-prompt Ds+ mesons, Phys. Lett. B **846**, 137561 (2023), 2204.10386. 10.1016/j.physletb.2022.137561

[16] M. He, R.J. Fries, R. Rapp, Heavy Flavor at the Large Hadron Collider in a Strong Coupling Approach, Phys. Lett. B **735**, 445 (2014), 1401.3817. 10.1016/j.physletb.2014.05.050

[17] M. He, R. Rapp, Hadronization and Charm-Hadron Ratios in Heavy-Ion Collisions, Phys. Rev. Lett. **124**, 042301 (2020), 1905.09216. 10.1103/PhysRevLett.124.042301

© The Authors, published by EDP Sciences. This is an open access article distributed under the terms of the Creative Commons Attribution License 4.0 (https://creativecommons.org/licenses/by/4.0/).

Open heavy flavor production at LHCb

Jianqiao Wang[1,*], on behalf of the LHCb Collaboration

[1]Center for High Energy Physics, Tsinghua University, Beijing, CHINA

Abstract. The LHCb experiment is a dedicated heavy-flavor experiment at the LHC and is uniquely well-suited to studying heavy-flavor production in heavy-ion collisions. Open heavy-flavor production studies at LHCb provide strong constraints on nuclear parton densities and probe the hadronization process in the hot, dense, nuclear media produced in heavy-ion collisions. Recent measurements of heavy-flavor production in heavy-ion collisions with the LHCb detector are presented, including studies of charm and beauty baryon production.

1 Introduction

Heavy quarks are created at the early stage of ultra-relativistic heavy-ion collisions. Their production can be strongly affected by interactions with deconfined hot nuclear matter, know as quark gluon plasma (QGP). Cold nuclear matter (CNM) effects are also present in the collisions and can be investigated in small systems such as proton-nucleus (pA) collisions, where QGP effects are expected to be negligible. One of the dominant CNM effects is the modification of the initial parton densities, compared to that in free nucleons. These effects can be described by nuclear parton distribution functions (nPDFs) [1, 2] and color glass condensate (CGC) [3] models. Heavy flavor particles also provide unique insights into hadronisation mechanism in heavy-ion collisions. Parton fragmentation is considered as a universal hadronisation process across different collisions [4]. In heavy-ion collisions, quark coalescence can arise as an alternative mechanism [5, 6]. Coalescence will be more important in large systems as the particle density becomes higher. Enhanced baryon-to-meson ratio compared to e^-e^+ is an important signature of quark coalescence. As the enhancement has also been observed in small systems, there have been more discussions on whether hot nuclear matter is created in these collisions.

The LHCb detector is a single-armed detector designed for studying heavy flavor particles. It has become a general-purpose detector for studying pp, pPb, PbPb and fixed-target collisions [7, 8]. These proceedings present the LHCb result of Λ_b^0/D^0 production ratio in high-multiplicity pp collisions at $\sqrt{s} = 13$ TeV [9]. The measurements of charm hadron production in pPb [13–16] and fixed-target pNe and PbNe collisions [17, 18] are also discussed.

2 Λ_b^0/B^0 production ratio in pp collisions at $\sqrt{s} = 13$ TeV

The Λ_b^0/B^0 production ratio is studied as a function of particle multiplicity in pp collisions [9]. The multiplicity is characterised with the total track multiplicity measured in the Vertex Locator (VELO) $N_{\text{VELO}}^{\text{tracks}}$. The result is shown in the left panel of Fig. 1. The points are consistent

*e-mail: jianqiao.wang@cern.ch

© The Authors, published by EDP Sciences. This is an open access article distributed under the terms of the Creative Commons Attribution License 4.0 (https://creativecommons.org/licenses/by/4.0/).

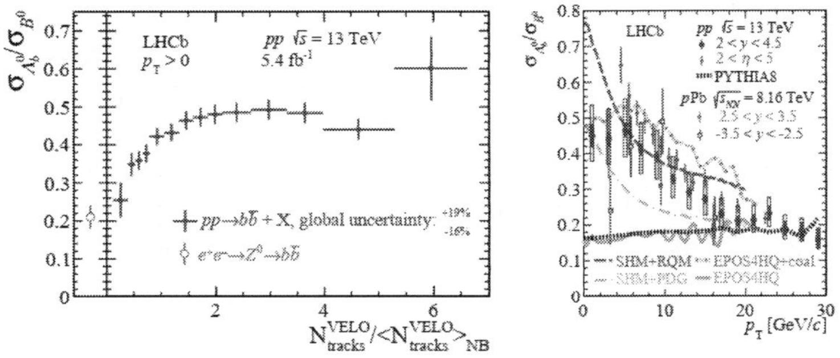

Figure 1. Λ_b^0/B^0 production ratio as a function of (left) number of VELO tracks $N_{\text{VELO}}^{\text{tracks}}$ and (right) p_{T}. $N_{\text{VELO}}^{\text{tracks}}$ is normalised by its averaged value in no-biased data.

with e^+e^- results [10] at very low multiplicity, but increase to about 0.5 as multiplicity increases. These results could indicate that coalescence emerges as an additional production mechanism for baryons at high multiplicity, where multiple quark wave functions overlap. The production ratio is also shown as a function of p_{T} in the right panel of Fig. 2. The ratio is significantly enhanced at low p_{T} and converges to e^+e^- data with increasing p_{T}, which is consistent with the expectation that coalescence is more dominant at low p_{T}. Predictions from statistical hadronisation model (SHM) [11] and EPOS4 [12] also give better descriptions of the ratio by considering coalesence contribution.

3 Open charm production in pPb collisions

LHCb pPb data correspond to center-of-mass energies of $\sqrt{s_{\text{NN}}} = 5.02\,\text{TeV}$ and 8.16 TeV and include both forward and backward kinematic regions, which cover both shadowing and anti-shadowing regions. Various open-charm hadrons are measured in pPb collisions, including D^+ and D_s^+ production at $\sqrt{s_{\text{NN}}} = 5.02\,\text{TeV}$ [13], and D^0 [14], Ξ_c^+ [16], D^+ and D_s^+ [15] at $\sqrt{s_{\text{NN}}} = 8.16\,\text{TeV}$. The nuclear modification factor $R_{p\text{Pb}}$ can be derived by comparing the cross-sections in pPb with that in pp collisions. Nuclear modification factors of D^+, D_s^+ and D^0 mesons at $\sqrt{s_{\text{NN}}} = 5.02\,\text{TeV}$ are shown in Fig. 2. Nuclear modification factors for all

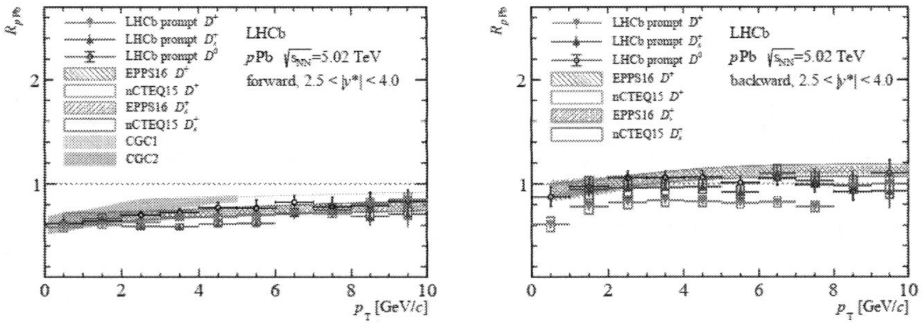

Figure 2. Nuclear modification factor $R_{p\text{Pb}}$ of D mesons in pPb collisions at $\sqrt{s_{\text{NN}}} = 5.02\,\text{TeV}$ at (left) forward and (right) backward rapidities.

mesons show significant suppression at forward rapidity, suggesting the existence of nuclear shadowing. While for backward rapidity $R_{p\text{Pb}}$ is less suppressed, which is consistent with

© The Authors, published by EDP Sciences. This is an open access article distributed under the terms of the Creative Commons Attribution License 4.0 (https://creativecommons.org/licenses/by/4.0/).

the expectation of anti-shadowing. The nuclear modification factor for D^+ mesons is slightly smaller than that for D_s^+ mesons. It may hint at possible strangeness enhancement.

The forward-backward produciton ratio R_{FB} can be measured when a pp reference is not available. R_{FB} of D^+, D_s^+ and Ξ_c^+ particles are shown in Fig. 3. Nuclear shadowing is also verified from the suppression at low p_{T}. At high p_{T}, R_{FB} of D^+, D_s^+ mesons show a rising trend toward unity, similar to that of D^0 mesons at $\sqrt{s_{\mathrm{NN}}} = 8.16\,\mathrm{TeV}$ [14]. However, $R_{\mathrm{FB}}(\Xi_c^+)$ shows no significant dependence on p_{T}.

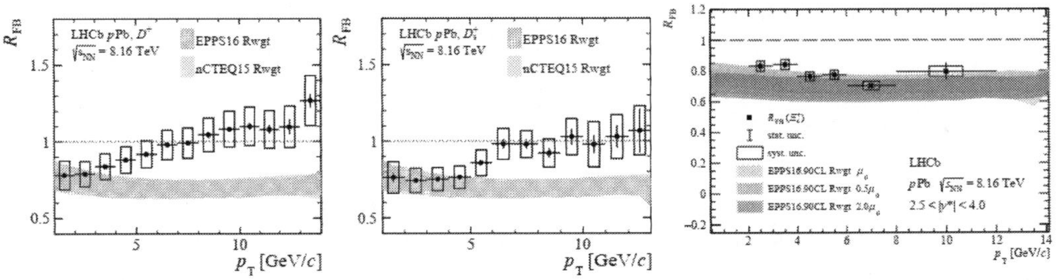

Figure 3. Forward-backward production ratio R_{FB} of (left) D^+, (middle) D_s^+, and (right) Ξ_c^+ particles as a function p_{T} in pPb collisions at $\sqrt{s_{\mathrm{NN}}} = 8.16\,\mathrm{TeV}$.

4 Open charm production in fixed-target collisions

The LHCb detector provides unique fixed-target mode at the LHC, with various species of noble gas target (He, Ne, Ar). The $D^0 - \overline{D}^0$ production asymmetry is measured to study the intrinsic charm in nucleons [17]. The result is presented in the left panel of Fig. 4. An asymmetry as large as 15% is observed at the most backward rapidity, and MS model with 1% intrisic charm and 10% recombination gives the best description of the data [19]. The J/ψ and D^0 production cross-sections are also measured in PbNe collisions [18]. The $J/\psi/D^0$ ratios as a function of N_{coll} in both pNe and PbNe collisions are shown together in Fig. 4. A power-law dependence of $R_{J/\psi/D^0} \propto (N_{\mathrm{coll}})^{\alpha'-1}$ is observed across different collisions, and no evidence of anomalous suppression or QGP formation is seen.

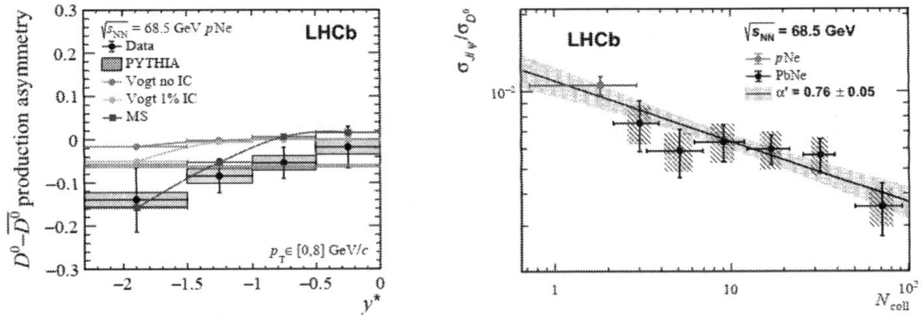

Figure 4. (left) $D^0 - \overline{D}^0$ production asymmetry as a function of y^* in pNe collisions and (right) $J/\psi/D^0$ ratio as a function of N_{coll} at $\sqrt{s_{\mathrm{NN}}} = 68.5\,\mathrm{GeV}$.

© The Authors, published by EDP Sciences. This is an open access article distributed under the terms of the Creative Commons Attribution License 4.0 (https://creativecommons.org/licenses/by/4.0/).

5 Conclusion

The LHCb experiment has strong capabilities to study heavy flavor particles. The Λ_b^0/B^0 ratio shows an enhancement in high multiplicity pp collisions compared to e^+e^- results, indicating the coalescence contribution in b-quark hadronization. The nuclear modification factor and forward backward production ratios are measured in pPb collisions. The significant suppression at forward rapidity agrees with the expectation from nuclear shadowing. D^0 production and $J/\psi/D^0$ ratio are measured in fixed-target pNe and PbNe collisions. No evidence for anomalous suppression or existence of hot nuclear medium is found in these systems.

References

[1] K. Kovařík and A. Kusina et al. Phys. Rev., D93, (2016) 8, 085037. https://doi.org/10.1103/PhysRevD.93.085037.
[2] K. J. Eskola and P. Paakkinen and H. Paukkunen and C. A. Salgado. Eur. Phys. J. C77, (2017), 3. https://doi.org/10.1140/epjc/s10052-017-4725-9.
[3] F. Gelos and E. Iancu and J. Jalilian-Marian and R. Venugopalan. Ann. Rev. Nucl. Part. Sci. 60 (2010) 463-489. https://doi.org/10.1146/annurev.nucl.010909.083629.
[4] B. Andersson and G. Gustafson and G. Ingelman and T. Sjostrand. Phys. Rept. 97 (1983) 31–145. https://doi.org/10.1016/0370-1573(83)90080-7
[5] R. J. Fries and B. Müller and C. Nonaka and S. A. Bass. Phys. Rev. Lett. 90 (2003) 202303. https://doi.org/10.1103/PhysRevLett.90.202303
[6] V. Greco and C. M. Ko and P. Levai. Phys. Rev. Lett. 90 (2003) 202302. https://doi.org/10.1103/PhysRevLett.90.202302
[7] LHCb Collaboration. JINST 3 (2008) S08005. https://doi.org/10.1088/1748-0221/3/08/S08005
[8] LHCb Collaboration. Int. J. Mod. Phys. A. 30 (2015) 1530022. https://doi.org/10.1142/S0217751X15300227
[9] LHCb Collaboration. Phys. Rev. Lett. 132 (2024) 081901. https://doi.org/10.1103/PhysRevLett.132.081901
[10] HFLAV Collaboration. Eur. Phys. J. C81 (2021) 226 https://doi.org/10.1140/epjc/s10052-020-8156-7
[11] M. He and R. Rapp. Phys. Rev. Lett. 131 (2023) 012301. https://doi.org/10.1103/PhysRevLett.131.012301
[12] J. Zhao and J. Aichelin and P. B. Gossiaux and K. Werner. Phys.Rev.D 109 (2024) 5, 054011. https://doi.org/10.1103/PhysRevD.109.054011
[13] LHCb Collaboration. JHEP 01 (2024) 070 https://doi.org/10.1007/JHEP01(2024)070
[14] LHCb Collaboration. Phys.Rev.Lett. 131 (2023) 102301 https://doi.org/10.1103/PhysRevLett.131.102301
[15] LHCb Collaboration. Phys.Rev.D 110 (2024) L031105 https://doi.org/10.1103/PhysRevLett.131.102301
[16] LHCb Collaboration. Phys. Rev. C 109 (2024) 044901 https://doi.org/10.1103/PhysRevC.109.044901
[17] LHCb Collaboration. Eur. Phys. J. C83 (2023) 541 https://doi.org/10.1140/epjc/s10052-023-11641-5
[18] LHCb Collaboration. Eur. Phys. J. C83 (2023) 658 https://doi.org/10.1140/epjc/s10052-023-11674-w
[19] R. Maciula and A. Szczurek. Phys. Lett. B835 (2022) 137530 https://doi.org/10.1016/j.physletb.2022.137530

© The Authors, published by EDP Sciences. This is an open access article distributed under the terms of the Creative Commons Attribution License 4.0 (https://creativecommons.org/licenses/by/4.0/).

A potential approach to the X(3872) thermal behaviour

Miguel A. Escobedo[1,*], *Néstor* Armesto[2], *Elena G.* Ferreiro[2], and *Víctor* López-Pardo[2]

[1]Departament de Física Quàntica i Astrofísica and Institut de Ciències del Cosmos, Universitat de Barcelona (IEEC-UB), Martí i Franquès 1, 08028 Barcelona, Catalonia, Spain
[2]Instituto Galego de Física de Altas Enerxías (IGFAE), Universidade de Santiago de Compostela, E-15782, Galicia, Spain.

Abstract. We study the potential of X(3872) at finite temperature in the Born-Oppenheimer approximation under the assumption that it is a tetraquark. We argue that, at large number of colors, it is a good approximation to assume that the potential consists in a real part plus a constant imaginary term. The real part is then computed adapting an approach by Rothkopf and Lafferty and using as input lattice QCD determinations of the potential for hybrids. This model allows us to qualitatively estimate at which temperature range the formation of a heavy tetraquark is possible, and to propose a qualitative picture for the dissociation of the state in a medium. Our approach can be applied to other suggested internal structures for the X(3872) and to other exotic states. This work summarizes the results of [1].

1 Introduction

In nature, there exist quarkonium-like particles whose quantum numbers and properties cannot be explained by the simple quark-antiquark model. Among them, we focus on the $X(3872)$, whose internal structure is still a matter of debate. There are two competing models: the tetraquark and the hadronic molecule. On one hand, the tetraquark is a compact bound state of four quarks, in our case two heavy and two light. On the other hand, a hadronic molecule is formed by two heavy-light mesons joint by the strong force analogue of van der Waals interaction.

In order to understand the internal structure of the $X(3872)$, a common approach is to formulate a theoretical model and then check if it is compatible with the observed properties of the state. These properties include, for example, its quantum numbers, spectroscopy, and decay channels. Recently, the $X(3872)$ has been observed in heavy-ion collisions [2]. This opens the possibility to gain information about structure of the bound state in a different way. That is, studying how the presence of the quark-gluon plasma modifies its behavior.

In this manuscript, we are going to discuss the modification to the potential of the $X(3872)$ at finite temperature. First, let us review briefly the state-of-art of conventional quarkonium in heavy-ion collisions. In recent years, it has been understood that the potential has both a real and an imaginary part [3]. The origin of this imaginary part is the inelastic collision of quarkonium with medium particles. In order to understand the role of the imaginary part in the evolution of the state of quarkonium the formalism of open quantum system was

*e-mail: miguel.a.escobedo@fqa.ub.edu

© The Authors, published by EDP Sciences. This is an open access article distributed under the terms of the Creative Commons Attribution License 4.0 (https://creativecommons.org/licenses/by/4.0/).

applied (see [4] for a review). The approach has been used to obtain phenomenological predictions that agree with observations [5]. Using this framework, it was found that solving the Schrödinger equation with a complex potential is a good approximation when the binding energy is much smaller than the temperature and when regeneration effects are not considered. This motivates us to find the potential of the $X(3872)$ at finite temperature.

2 Theoretical framework

We are going to use the Born-Oppenheimer approximation. We assume that heavy quarks move non-relativistically around the center-of-mass, with velocity v. The fact that $\Lambda_{QCD} \gg E \sim m_Q v^2$ (where E is the binding energy) implies that the dynamics of light quarks and gluons is much faster than that of the evolution of the bound state. This involves a two-step approximation. First, we compute the potential taking the heavy quarks as static color sources. After this, we solve the Schrödinger equation with that potential.

First, let us discuss the potential at $T = 0$. Ideally, we would like to use a lattice QCD potential for the tetraquark as a starting point. However, this computation is not yet available. Instead, we use hybrid data and make the approximation that the tetraquark potential would behave qualitatively similar to the hybrid potential. Morevover, we assume a single channel approximation, where the heavy quarks do not contribute to the spin of the tetraquark. We do not expect that any of this approximations will change the qualitative picture and a quantitative study is out of the scope of this manuscript. The hybrid lattice potential that we are going to use is taken from [6]. It happens that this potential is well fitted by the following formulate

$$V(r,0) = \frac{A_{-1}}{r} + A_0 + A_2 r^2 . \tag{1}$$

Note that the previous formula is accurate for the distances probed in the lattice computation and, therefore, useful to study whether or not bound state formation is possible. However, eq. (1) is not valid at large distances in which we know, thanks to effective string theory, that the potential rises linearly.

Next, we must discuss how to extend this potential to finite temperature. We are going to study separately the real and the imaginary part of the potential. For the real part of the potential, we use the approach developed in [7]. There, the authors were able to reproduce the static potential of quarkonium at finite temperature using as imput the lattice potential at $T = 0$. The real part of the potential is obtained from the following convolution

$$\mathrm{Re}\, V(\mathbf{p}) = \mathrm{Re}\left(\frac{V_{vac}(\mathbf{p})}{\epsilon(\mathbf{p}, m_D)}\right), \tag{2}$$

where ϵ is the medium permittivity in the HTL approximation, m_D is the Debye mass and V_{vac} is the potential at $T = 0$. In our case, starting from eq. (1), we obtain the following real part of the potential at finite temperature

$$\Re[V(r, m_D)] = A_{-1}\left(m_D + \frac{e^{-m_D r}}{r}\right) + A_0 + A_2\left[\frac{6}{m_D^2}(1 - e^{-m_D r}) - \left(2r^2 + \frac{6r}{m_D}\right)e^{-m_D r}\right]. \tag{3}$$

Now, let us discuss the imaginary part of the potential. In the case of conventional quarkonium, it has the following properties. At short distances, the imaginary potential goes like r^2 because the medium sees quarkonium as a small dipole. At long distances, the heavy quarks are not correlated, so the imaginary part of the potential is equal to $-i$ times the decay width of a single heavy quark. Between these two limits, we expect that the imaginary part of the

© The Authors, published by EDP Sciences. This is an open access article distributed under the terms of the Creative Commons Attribution License 4.0 (https://creativecommons.org/licenses/by/4.0/).

potential is a smoothly increasing function. Indeed, the imaginary potential of quarkonium computed in the HTL approximation fulfills these properties. Knowing this, it is easy to infer how the imaginary part of the potential of a tetraquark would behave. In this case, the heavy quarks are in an octet state. When $r \to 0$, the medium sees the heavy quark pair as a non-relativistic heavy gluon, and the imaginary part of the potential will be $-i/2$ times the decay width of a heavy gluon. At large distances, the two quarks are uncorrelated, similar to the color singlet case. At intermediate distances, we expect that the imaginary part of the potential is a smooth function that interpolates between these two extreme regimes. However, in the large N_c limit, the decay width of a heavy gluon is equal to that of two heavy quarks. Therefore, we can take the imaginary part of the potential to be a constant. Finally, we have to make an educated guest of the size of this constant. We choose the following

$$\Gamma = A_{-1}T + A_2\frac{T}{m_D^2}\,. \tag{4}$$

The rationale for this formula is the following. We expect that the same parameters that are involved in the real part of the potential also appear here. The rest is fixed by dimensional analysis.

3 Results

The dissociation temperature is the temperature above which the bound state no longer exists in the medium. It is obtained by solving the Schrödinger equation using the complex potential. Since in our case the imaginary part is a constant, it factors out and does not affect the solution. The result that we found is that the dissociation temperature is around $T_d \sim 250\,\text{MeV}$.

The survival probability can be computed using the following formula

$$S(t) = e^{-\int_{t_0}^{t}\Gamma(T(\tau),\tau)d\tau}\,. \tag{5}$$

Note that in our case the survival probability takes a particularly simple form because the

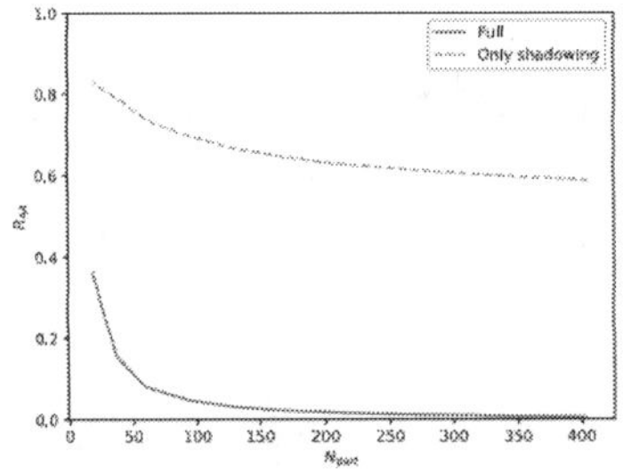

Figure 1. Prediction for R_{AA} of $X(3872)$ at LHCb conditions. The dashed line considers only cold nuclear matter effects, following the model discussed in [8]

decay width is a constant that does not depend on the form of the wave function. Using this formula, we can compute the nuclear modification factor, R_{AA}, once we know how the temperature seen by the bound state changes with time and position. To do this we follow the lines of [8]. We consider a Bjorken expansion starting at $t = 0.6\,\text{fm}$ and finishing at the time at which the temperature goes below $175\,\text{MeV}$, close to the phase transition. The initial

© The Authors, published by EDP Sciences. This is an open access article distributed under the terms of the Creative Commons Attribution License 4.0 (https://creativecommons.org/licenses/by/4.0/).

temperature at a given point is computed using a model that takes into account shadowing. If this temperature is larger than T_d then the state is not form. Otherwise, we compute the survival probability. The results obtained following this procedure can be seen in fig. 1, where we have considered LHCb conditions. We observe a strong suppression and only a mild influence of cold nuclear matter effects. We note that, at the moment, we have not yet considered recombination effects.

4 Conclusions

In this manuscript, we have studied exotic quarkonia in heavy-ion collisions with the aim of shedding light on their internal structure. We have developed a qualitative model for the potential of the $X(3872)$. Our results indicate that, for this state, suppression is dominated by screening while only a mild contribution from the imaginary part of the potential is observed.

We have received financial support from Xunta de Galicia (Centro singular de investigación de Galicia accreditation 2019-2022, ref. ED421G-2019/05), by European Union ERDF, by the "María de Maeztu" Units of Excellence program MDM2016-0692, and by the Spanish Research State Agency under project PID2020-119632GBI00. This work has received funding from the European Research Council project ERC-2018-ADG-835105 YoctoLHC and from the European Union's Horizon 2020 research and innovation programme under grant agreement No. 824093. The work of MAE has been supported by the Maria de Maetzu excellence program under project CEX2019-000918-M, by projects PID2019-105614GB-C21 and PID2022-136224NB-C21 funded by MCIN/AEI/10.13039/501100011033, and by grant 2021-SGR-249 of Generalitat de Catalunya. VLP has been supported by Xunta de Galicia under project ED481A 2022/286.

References

[1] N. Armesto, E.G. Ferreiro, M.A. Escobedo, V. López-Pardo, A potential approach to the X(3872) thermal behavior, Phys. Lett. B **854**, 138760 (2024), 2401.10125. 10.1016/j.physletb.2024.138760

[2] A.M. Sirunyan et al. (CMS), Evidence for X(3872) in Pb-Pb Collisions and Studies of its Prompt Production at $\sqrt{s_{NN}}$=5.02 TeV, Phys. Rev. Lett. **128**, 032001 (2022), 2102.13048. 10.1103/PhysRevLett.128.032001

[3] M. Laine, O. Philipsen, P. Romatschke, M. Tassler, Real-time static potential in hot QCD, JHEP **03**, 054 (2007), hep-ph/0611300. 10.1088/1126-6708/2007/03/054

[4] Y. Akamatsu, Quarkonium in quark–gluon plasma: Open quantum system approaches re-examined, Prog. Part. Nucl. Phys. **123**, 103932 (2022), 2009.10559. 10.1016/j.ppnp.2021.103932

[5] N. Brambilla, M.A. Escobedo, A. Islam, M. Strickland, A. Tiwari, A. Vairo, P. Vander Griend, Regeneration of bottomonia in an open quantum systems approach, Phys. Rev. D **108**, L011502 (2023), 2302.11826. 10.1103/PhysRevD.108.L011502

[6] S. Capitani, O. Philipsen, C. Reisinger, C. Riehl, M. Wagner, Precision computation of hybrid static potentials in SU(3) lattice gauge theory, Phys. Rev. D **99**, 034502 (2019), 1811.11046. 10.1103/PhysRevD.99.034502

[7] D. Lafferty, A. Rothkopf, Improved Gauss law model and in-medium heavy quarkonium at finite density and velocity, Phys. Rev. D **101**, 056010 (2020), 1906.00035. 10.1103/PhysRevD.101.056010

[8] M.A. Escobedo, E.G. Ferreiro, Simple model to include initial-state and hot-medium effects in the computation of quarkonium nuclear modification factor, Phys. Rev. D **105**, 014019 (2022), 2110.12295. 10.1103/PhysRevD.105.014019

© The Authors, published by EDP Sciences. This is an open access article distributed under the terms of the Creative Commons Attribution License 4.0 (https://creativecommons.org/licenses/by/4.0/).

EPJ Web of Conferences **316**, 04008 (2025) https://doi.org/10.1051/epjconf/202531604008
SQM 2024

Quarkonium production in pp and heavy-ion collisions

Taesoo Song[1,*], *Joerg* Aichelin[2,3], *Jiaxing* Zhao[4,5], *Pol B.* Gossiaux[2], and *Elena* Bratkovskaya[1,4,5]

[1]GSI Helmholtzzentrum für Schwerionenforschung GmbH, Planckstrasse 1, 64291 Darmstadt, Germany
[2]SUBATECH UMR 6457 (IMT Atlantique, Université de Nantes, IN2P3/CNRS), 4 Rue Alfred Kastler, F-44307 Nantes, France
[3]Frankfurt Institute for Advanced Studies, Ruth-Moufang-Strasse 1, 60438 Frankfurt am Main, Germany
[4]Institute for Theoretical Physics, Johann Wolfgang Goethe Universität, Frankfurt am Main, Germany
[5]Helmholtz Research Academy Hessen for FAIR (HFHF),GSI Helmholtz Center for Heavy Ion Research. Campus Frankfurt, 60438 Frankfurt, Germany

Abstract. We describe quarkonium production in pp and heavy-ion collisions by using the Remler's formalism where quarkonium density operator is applied to all possible combination of heavy quark and heavy antiquark pairs. In pp collisions heavy (anti)quark momentum is provided by the PYTHIA event generator after rescaling p_T and rapidity to imitate the FONLL calculations. Then spatial separation between heavy quark and heavy antiquark is introduced based on the uncertainty principle. In heavy-ion collisions quarkonium wavefunction changes with temperature assuming heavy quark potential equals the free energy of heavy quark pair in heat bath. The density operator is updated whenever heavy quark or heavy antiquark scatters in quark-gluon plasma (QGP). Our results are consistent with the experimental data on bottomonia from ALICE and CMS Collaborations assuming that the interaction rate of bottom (anti)quark in bottomonium is suppressed to 10 % that of unbound bottom (anti)quark. We also find that off-diagonal recombination of bottomonium from two different initial pairs barely happens even in Pb+Pb collisions at $\sqrt{s_{NN}} = 5.02$ TeV.

1 Introduction

Ultra-relativistic heavy-ion collisions produce an extremely hot and dense matter which is supposed to cross over the phase boundary between hadron gas and quark-gluon plasma. Quarkonium is a flavorless bound state of heavy quark pair, whose suppression in heavy-ion collisions was suggested as a signature for QGP formation, because the binding of heavy quark pair changes through the color screening and scattering with thermal partons [1]. However, it was discovered that J/ψ is less suppressed in mid-rapidity as well as in higher energy collisions at the top energies of RHIC and LHC, which indicates the regeneration of quarkonium from the uncorrelated heavy quark pairs [2]. So the complete description of quarkonium production in heavy-ion collisions requires suppression as well as regeneration.

*e-mail: t.song@gsi.de

© The Authors, published by EDP Sciences. This is an open access article distributed under the terms of the Creative Commons Attribution License 4.0 (https://creativecommons.org/licenses/by/4.0/).

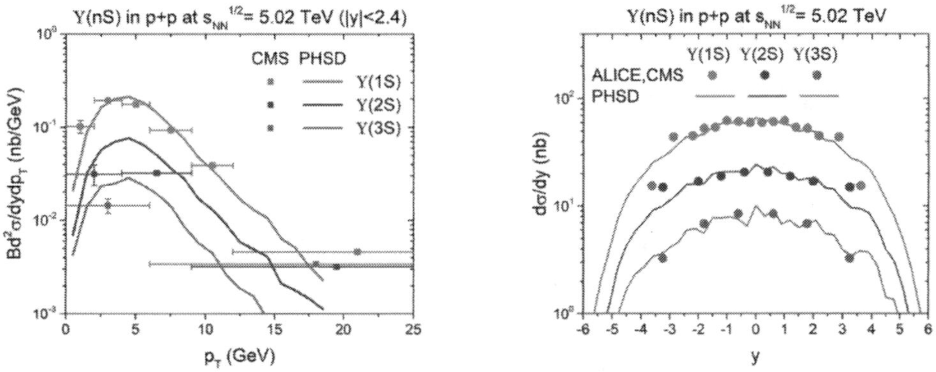

Figure 1. (Left) p_T spectra and (right) rapidity distributions of $\Upsilon(1S)$, $\Upsilon(2S)$ and $\Upsilon(3S)$ including the feed-out from excited states in pp collisions at $\sqrt{s} = 5.02$ TeV, which are compared with experimental data from the CMS and ALICE Collaborations [6, 7].

Quarkonium production takes two steps. First, heavy quark pair is produced and then the produced heavy quark pair forms a bound state by emitting soft gluons to be a color singlet. The first step is a hard process which can be described by pQCD, while the second step is a soft process which requires a model. In heavy-ion collisions, the first step is modified by so-called cold nuclear matter effects such as shadowing effects and the second step by hot nuclear matter effects. In this study we try to describe quarkonium production not only in pp collisions but also in heavy-ion collisions by using the Remler's formalism.

2 Remler's formalism for quarkonium production

The probability that a quarkonium eigenstate i with momentum $\mathbf{P}$ is produced at $\mathbf{R}$ is given by [3, 4]

$$(2\pi)^3 \frac{dN_i}{d^3R d^3P} = \sum \int \frac{d^3r d^3p}{(2\pi)^3} \Phi_i^W(\mathbf{r},\mathbf{p}) \prod_{j>2} \int \frac{d^3r_j d^3p_j}{(2\pi)^{3(N-2)}} W^{(N)}(\mathbf{r_1},\mathbf{p_1},...,\mathbf{r_N},\mathbf{p_N}), \quad (1)$$

where $\mathbf{R} = (\mathbf{r_1}+\mathbf{r_2})/2$, $\mathbf{r} = \mathbf{r_1}-\mathbf{r_2}$, $\mathbf{P} = \mathbf{p_1}-\mathbf{p_2}$, $\mathbf{p} = (\mathbf{p_1}-\mathbf{p_2})/2$ with $\mathbf{r_k}$ and $\mathbf{p_k}$ being respectively the position and momentum of k'th (anti)heavy quark; $\Phi_i^W(\mathbf{r},\mathbf{p})$ is the Wigner density of quarkonium wavefunction which is nonvanishing only for flavorless and colorless combination; $W^{(N)}(\mathbf{r_1},\mathbf{p_1},\mathbf{r_2},\mathbf{p_2},...,\mathbf{r_N},\mathbf{p_N})$ is density matrix of the N (anti)heavy quarks produced in a proton-proton or a heavy-ion collision, which is simply taken for the classical phase space density distribution [4]. The summation in the right hand side implies that all possible combinations of heavy quark pairs are considered to form quarkonium state i.

In pp collisions the momentum of heavy quark and that of heavy antiquark are generated by the PYTHIA event generator followed by rescaling the transverse momentum and rapidity to imitate the FONLL calculations [5]. The cross section for charm production in pp collisions is parameterized from the experimental data [5]. The spatial separation between heavy quark and heavy antiquark is introduced based on the uncertainty principle [4].

Fig. 1 shows the p_T spectra and rapidity distributions of $\Upsilon(1S)$, $\Upsilon(2S)$, and $\Upsilon(3S)$ including the feed-out from excited states in pp collisions at $\sqrt{s} = 5.02$ TeV. One can see that the results are consistent with the experimental data from CMS and ALICE Collaborations [6, 7].

© The Authors, published by EDP Sciences. This is an open access article distributed under the terms of the Creative Commons Attribution License 4.0 (https://creativecommons.org/licenses/by/4.0/).

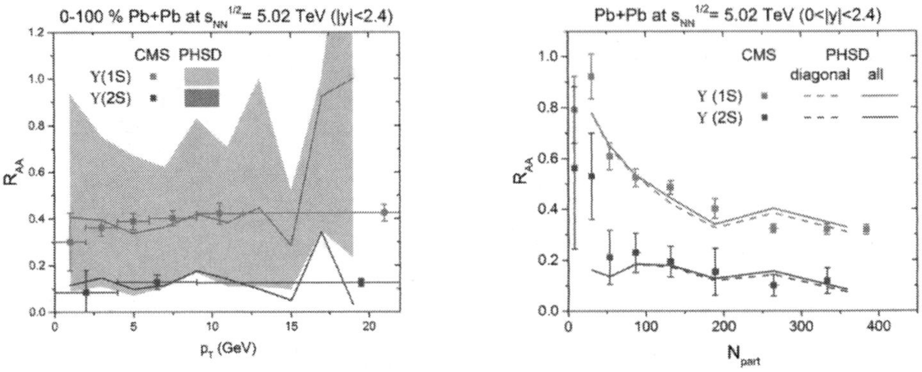

Figure 2. R_{AA} of $\Upsilon(1S)$ and $\Upsilon(2S)$ as a function of (left) transverse momentum and (right) number of participants in Pb+Pb collisions at $\sqrt{s_{NN}}$ = 5.02 TeV in comparison with the experimental data from the CMS Collaboration [6].

A hot dense matter produced in relativistic heavy-ion collisions delays the formation of quarkonium, because quarkonium cannot be formed above its dissociation temperature. We assume that the free energy of heavy quark pair is equivalent to the heavy quark potential [8]. In this case the dissociation temperature of $\Upsilon(1S)$ is roughly 3 T_c, while those of excited states are near T_c [4]. The first Wigner projection of Eq. (1) is carried out when the local temperature of heavy quark pair becomes lower than the dissociation temperature of quarkonium state i. After that, whenever scattering happens to heavy quark or heavy antiquark, the Wigner production is updated [9], considering temperature-dependent quarkonium radius, which is obtained by solving Schrödinger equation with the heavy quark potential.

The left panel of Fig. 2 displays R_{AA} of $\Upsilon(1S)$ and $\Upsilon(2S)$ as a function of transverse momentum in Pb+Pb collisions at $\sqrt{s_{NN}}$ = 5.02 TeV. The upper limit of the magenta band indicates R_{AA} of $\Upsilon(1S)$ at the dissociation temperature of $\Upsilon(1S)$ and the lower limit of the band R_{AA} of $\Upsilon(1S)$ at T_c. One can see that Wigner density of $\Upsilon(1S)$ decreases with time, because bottom and antibottom quarks are more and more separated from each other. The red solid line indicates R_{AA} of $\Upsilon(1S)$ at T_c, assuming only 10 % of (anti)bottom quark scatterings update(change) the Wigner density of $\Upsilon(1S)$, which is consistent with the experimental data from the CMS Collaboration [6]. In this case the interaction rate of $\Upsilon(1S)$ is roughly estimated to be 40-100 MeV at T=0.2-0.4 GeV [4, 10]. This reduction is reasonable, because quarkonium is a color singlet and part of heavy quark interaction will be canceled with that of heavy antiquark interaction. We also note that the feed-down from excited state bottomonia to $\Upsilon(1S)$ in heavy-ion collisions is less than 10 %. As for $\Upsilon(2S)$ its dissociation temperature is close to T_c. That is why the blue band is so narrow that it looks like a line and the interaction rate of $\Upsilon(2S)$ cannot be constrained by R_{AA} of $\Upsilon(2S)$.

The right panel shows R_{AA} as a function of the number of participants. The solid lines include all contributions, while the dashed lines only diagonal contribution which originates from initial bottom quark pairs. One can see that the off-diagonal contribution which originate from two different initial bottom quark pairs is little even in central Pb+Pb collisions at $\sqrt{s_{NN}}$ = 5.02 TeV, because the number of produced bottom quark pairs is not so large.

© The Authors, published by EDP Sciences. This is an open access article distributed under the terms of the Creative Commons Attribution License 4.0 (https://creativecommons.org/licenses/by/4.0/).

3 Conclusion

The description of quarkonium production in heavy-ion collisions is often separated from the quarkonium production in pp collisions, though the nuclear modification factor (R_{AA}) is defined as their ratio. In this study we have used an unified method, the Remler's formalism, for quarkonium production in pp as well as in heavy-ion collisions. Assuming that the interaction rate of bottom (anti)quark bound in quarkonium is 10 % that of unbound bottom (anti)quark, our results are consistent with the experimental data in Pb+Pb collisions at $\sqrt{s_{NN}}$ = 5.02 TeV. Furthemore, we have found that the off-diagonal contribution from two different initial pairs is little even at LHC.

Acknowledgements

We acknowledge support by the Deutsche Forschungsgemeinschaft (DFG, German Research Foundation) through the grant CRC-TR 211 'Strong-interaction matter under extreme conditions' - Project number 315477589 - TRR 211. This work is also supported by the European Union's Horizon 2020 research and innovation program under grant agreement No 824093 (STRONG-2020). The computational resources have been provided by the LOEWE-Center for Scientific Computing and the "Green Cube" at GSI, Darmstadt, and by the Center for Scientific Computing (CSC) of the Goethe University, Frankfurt.

References

[1] T. Matsui and H. Satz, J/ψ Suppression by Quark-Gluon Plasma Formation. Phys. Lett. B **178**, 416 (2019). https://doi.org/10.1016/0370-2693(86)91404-8

[2] L. Manceau et al. (ALICE), J/\psi production in pp collisions and in Pb-Pb collisions at the LHC with the ALICE experiment. Nucl. Phys. B Proc. Suppl. **234**, 321 (2013). https://doi.org/10.1016/j.nuclphysbps.2012.12.041

[3] E. A. Remler, Use of the Wigner Representation in Scattering Problems. Annals Phys. **95**, 455 (1975). https://doi.org/10.1016/0003-4916(75)90065-2

[4] T. Song, J. Aichelin, J. Zhao, P. B. Gossiaux and E. Bratkovskaya, Bottomonium production in pp and heavy-ion collisions. Phys. Rev. C **108**, 054908 (2023). https://doi.org/10.1103/PhysRevC.108.054908

[5] T. Song, H. Berrehrah, D. Cabrera, J. M. Torres-Rincon, et al., Tomography of the Quark-Gluon-Plasma by Charm Quarks. Phys. Rev. C **92**, 014910 (2015). https://doi.org/10.1103/PhysRevC.92.014910

[6] A. M. Sirunyan et al. (CMS), Measurement of nuclear modification factors of Υ(1S), Υ(2S), and Υ(3S) mesons in PbPb collisions at $\sqrt{s_{NN}}$ = 5.02 TeV. Phys. Lett. B **790**, 270 (2019). https://doi.org/10.1016/j.physletb.2019.01.006

[7] S. Acharya et al. (ALICE), Inclusive quarkonium production in pp collisions at $\sqrt{s}$ = 5.02 TeV. Eur. Phys. J. C **83**, 61 (2023). https://doi.org/10.1140/epjc/s10052-022-10896-8

[8] O. Kaczmarek et al., Heavy quark free energies and the renormalized Polyakov loop in full QCD. Prog. Theor. Phys. Suppl., **153**, 287 (2004). https://doi.org/10.1143/PTPS.153.287

[9] T. Song, J. Aichelin and E. Bratkovskaya, Charmonium production in a thermalizing heat bath. Phys. Rev. C **107**, 054906 (2023). https://doi.org/10.1103/PhysRevC.108.054908

[10] A. Andronic at el., Comparative study of quarkonium transport in hot QCD matter. Eur. Phys. J. A **60**, 88 (2024). https://doi.org/10.1140/epja/s10050-024-01306-6

© The Authors, published by EDP Sciences. This is an open access article distributed under the terms of the Creative Commons Attribution License 4.0 (https://creativecommons.org/licenses/by/4.0/).

Investigating the interplay between initial hard processes and final-state effects measuring prompt and non-prompt J/ψ

Maurice Coquet[1,*], on behalf of the ALICE Collaboration

[1]SUBATECH, Nantes University, IMT Atlantique, IN2P3/CNRS, 4 rue Alfred Kastler, 44307 Nantes, France

Abstract. Charmonium production in high-energy collisions can be split into a prompt and a non-prompt component. Both components can be distinguished experimentally by studying displaced topology, and represent valuable tools to investigate the properties of the strongly interacting medium produced in ultra-relativistic heavy-ion collisions. In particular, the study of non-prompt charmonium production can give access to the beauty-hadron production cross section and can be used to investigate the in-medium energy loss of beauty quarks. In these proceedings, the recent measurement of prompt and non-prompt J/ψ carried out by the ALICE Collaboration in Pb–Pb collisions at midrapidity ($|y| < 0.8$) at $\sqrt{s_{\rm NN}} = 5.02$ TeV, will be presented. Moreover, thanks to the installation of the new muon forward tracker (MFT), the prompt/non-prompt charmonium separation will be possible in LHC Run 3 also at forward rapidity ($2.5 < y < 4$).

1 Introduction

Heavy quarks, charm and beauty quarks, are excellent probes to study the strongly interacting medium produced in ultra-relativistic heavy-ion collisions, the quark-gluon plasma (QGP). Indeed, due to their large masses, these quarks are produced by the initial hard scatterings at the beginning of the collision and experience the full evolution of the collision. In particular, charmonia, bound states of charm and anti-charm quarks, are extensively studied in heavy-ion collisions to access properties of the QGP and its evolution. In such collisions, the production of charmonium is affected by mechanisms which modify their production with respect to smaller collision systems. First, charmonium states are expected to experience dissociation in the QGP by color-charge screening and dynamical dissociation through interactions with medium constituents [1]. In addition, at LHC energies, a large number of $c\bar{c}$ pairs is produced, sufficient so that charmonium production is enhanced via the recombination of uncorrelated $c\bar{c}$ pairs [2, 3]. This mechanism is known as regeneration. Finally, gluon energy loss in the QGP is expected to affect the production of heavy quarks, and thus of charmonium, at high $p_{\rm T}$ [4], and cold nuclear matter effects such as nuclear shadowing of the parton distribution functions [5] must also be taken into account in charmonium production. A commonly studied charmonium state is the vector state J/ψ . Its production in high-energy collisions includes prompt and non-prompt contributions, which can be distinguished based on the displacement

*e-mail: maurice.louis.coquet@cern.ch

© The Authors, published by EDP Sciences. This is an open access article distributed under the terms of the Creative Commons Attribution License 4.0 (https://creativecommons.org/licenses/by/4.0/).

of the production vertex of the state. On the one hand, prompt J/ψ are produced directly at the collision vertex between the two colliding hadrons, or from the feed-down of excited charmonium states. Hence, prompt J/ψ can probe the suppression and regeneration mechanisms inside the QGP. On the other hand, non-prompt J/ψ are produced through the decay of hadrons containing beauty quarks. Their study can thus provide information on the partonic energy loss of the beauty quark, as well as its transport coefficients inside the QGP.

With the ALICE apparatus, charmonium can be measured either at midrapidity ($|y| < 0.8$) in the dielectron decay channel, or at forward rapidity ($2.5 < y < 4$) in the dimuon decay channel. In Run 2, the separation between prompt and non-prompt contribution to J/ψ production was only achievable at midrapidity, using the Inner Tracking System (ITS) for vertexing, while the forward muon arm was able to perform only inclusive measurements. A detailed description of the ALICE apparatus can be found in Ref.[6].

2 Results

In Run 2, the fraction of non-prompt J/ψ was measured by ALICE at midrapidity in Pb–Pb collisions at 5.02 TeV, down to $p_T = 1.5$ GeV/c [7]. The extraction of this fraction is based on a two-dimensional maximum-log-likelihood fit using two variables: the invariant mass of the dielectron candidates, and the pseudo-proper decay length x defined as $x = c \times L_{xy} \times m_{J/\psi}/p_T$, where $L_{xy} = (\vec{L} \times \vec{p_T})/p_T$ is the projection in the transverse plane of the vector pointing from the primary vertex to the J/ψ decay vertex, and $m_{J/\psi}$ is the J/ψ mass.

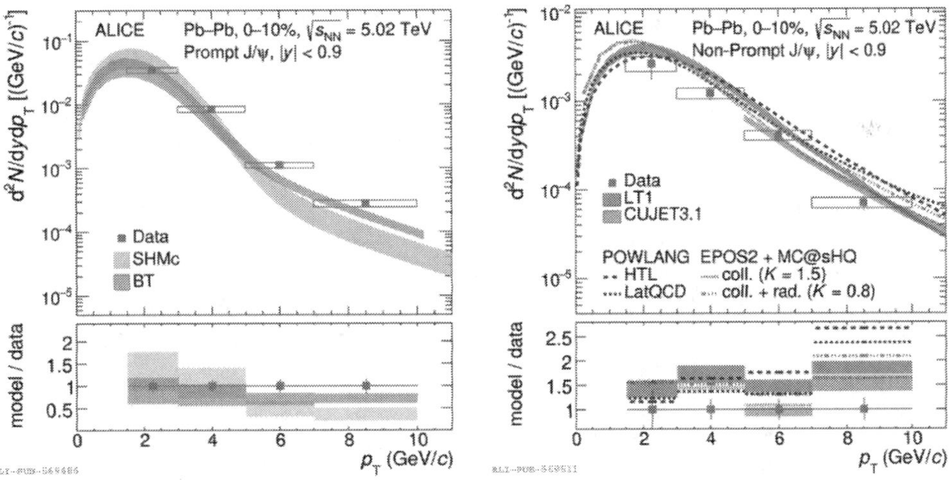

Figure 1. Prompt (left) and non-prompt (right) J/ψ yields as a function of p_T in the 0 – 10% centrality class compared with models [7]. Vertical error bars and boxes represent statistical and systematic uncertainties, respectively. Bottom panels show the ratios between models and data.

The prompt and non-prompt J/ψ p_T spectra in Pb–Pb are computed by combining the non-prompt fraction with the measured inclusive J/ψ p_T-dependent cross section [8]. The obtained cross sections are compared with available models. The statistical hadronization model (SHMc) [9] assumes that all bound states are produced at hadronization and that their yields follow thermal weights. The Boltzmann transport model (BT) is based on a detailed balance equation, involving continuous dissociation and regeneration of J/ψ states during the QGP phase [10]. Figure 1 (left) shows a comparison of these two models with the ALICE Pb–Pb data in the 0-10% centrality class. For $p_T < 5$ GeV/c, both models are in agreement

© The Authors, published by EDP Sciences. This is an open access article distributed under the terms of the Creative Commons Attribution License 4.0 (https://creativecommons.org/licenses/by/4.0/).

with the data, while for $p_T > 5$ GeV/c, the statistical model tends to underpredict the data. In SHMc, two contributions to J/ψ production are considered: core, which corresponds to thermal production at phase boundary, and corona, corresponding to the cooler edges of the fireball where no QGP is expected to form, and where J/ψ production is assumed to be captured by scaled pp distributions. The uncertainty associated with SHMc plotted on Fig.1 (left) corresponds to variation of the fraction of corona J/ψ . For the BT model, the uncertainty corresponds to variation of the initial $c\bar{c}$ cross section.

The non-prompt cross section is compared on Fig.1 (right) with models based on Langevin equations, such as LT1 [12] or POWLANG [13], as well as jet energy-loss framework (CUJET [14]) and EPOS2+MC@sHQ [15] which employs the EPOS2 event generator coupled with a Monte Carlo treatment of Boltzmann transport for the heavy quarks. As shown on Fig.1 (right), these models overpredict the data for $p_T > 7$ GeV/c. This is particularly the case for the POWLANG model which includes only collisional energy loss for the b-quark, while no radiative contribution to the energy loss is taken into account. This model provides two different predictions, corresponding to two evaluations of the b quark transport coefficients, one using Hard Thermal Loop perturbative calculations (HTL) and the other one computed from lattice QCD (LatQCD). The uncertainty associated with the LT1 model shown on Fig.1 (right) corresponds to variations of the nuclear shadowing taken into account in the initial heavy-quark production. The CUJET uncertainties correspond to variations of the running QCD coupling and of the magnetic screening mass.

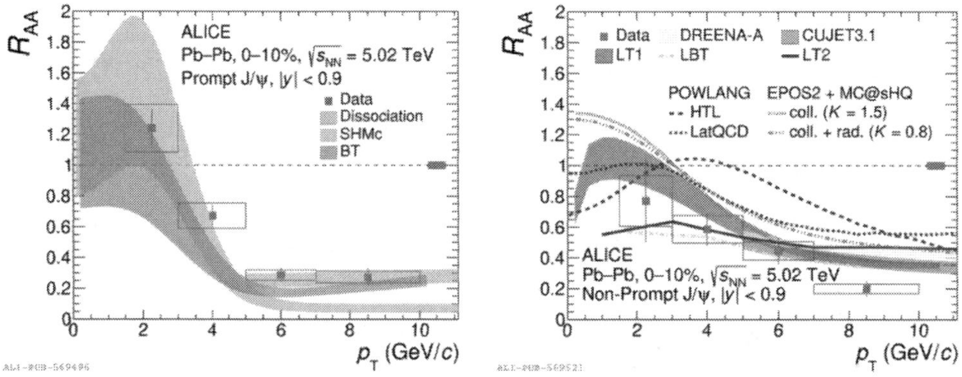

Figure 2. Prompt (left) and non-prompt J/ψ R_{AA} as a function of p_T in the $0-10\%$ centrality class, compared with models [7]. Error bars and boxes represent statistical and uncorrelated systematic uncertainties, respectively.

Using the measured cross sections in Pb–Pb and in pp, the nuclear modification factor R_{AA} can be computed. Figure 2 (left) shows comparisons of the prompt R_{AA} with models predictions. For $p_T > 5$ GeV/c, the BT model lies on the lower side of the data, while the SHMc model underestimates the data. Both models correctly describe the measured R_{AA} for $p_T < 5$ GeV/c. A dissociation model is also shown [11], based on a rate equation using a decay rate computed from momentum broadening of the charm quarks. It is in good agreement with the data at high p_T. The uncertainty on the predictions from the dissociation model shown on Fig.2 (left) corresponds to variations of the formation time of the charmonium state.

The non-prompt R_{AA} is shown of Fig.2 (right) and compared with models. The POWLANG model overpredicts the data, for both estimates of the transport coefficients. The EPOS2+MC@sHQ model tends to overpredict the data for $p_T < 3$ GeV/c, while it best encapsulates the high-p_T region by including only collisional energy loss. Additional mod-

© The Authors, published by EDP Sciences. This is an open access article distributed under the terms of the Creative Commons Attribution License 4.0 (https://creativecommons.org/licenses/by/4.0/).

els are also shown. The LT2 model [16] is another Langevin transport approach. The LBT model [17] is based on linear Boltzmann transport, which includes both short and long-range confining interactions in the heavy-quark potential. LT1, LT2 and LBT all include both radiative and collisional energy loss, and are in agreement with the data over the whole p_{T} range. Finally, DREENA [18] is based on a state-of-the-art energy-loss framework coupled with hydrodynamic description of the medium and provides predictions for $p_{\mathrm{T}} > 5$ GeV/c. DREENA and CUJET3.1 both overpredict the data for $p_{\mathrm{T}} > 7$ GeV/c.

3 Summary and outlook

New ALICE results were shown for prompt/non-prompt J/ψ separation in Pb–Pb collisions down to $p_{\mathrm{T}} = 1.5$ GeV/c. The prompt data are described by models including regeneration of uncorrelated $c\bar{c}$ pairs at low p_{T} and charmonium suppression through color screening and dynamical dissociation at high p_{T}. The non-prompt J/ψ data are described by several models. For $p_{\mathrm{T}} > 5$ GeV, all calculations are consistent with the data except for POWLANG, while at low p_{T}, EPOS2+MC@sHQ overpredits the data. The models which are in best agreement over the full p_{T} range are the ones which include both collisional and radiative energy loss mechanisms for the beauty quark.

For Run 3, ALICE underwent several important upgrades [19]. In particular, new detectors were installed, such as the Muon Forward Tracker (MFT). This new detector allows to perform precise vertexing for the muon spectrometer, and, as a consequence, to separate prompt and non-prompt J/ψ at forward rapidity in the dimuon decay channel. A first evaluation of the performance for prompt/non-prompt separation at forward rapidity was carried out in pp collisions. It showed that this separation is achievable down to $p_{\mathrm{T}} = 0$, thanks to the longitudinal boost of particles detected in this region, and opens the door to future prompt/non-prompt J/ψ analysis in Pb–Pb at forward rapidities, a phase-space region yet unexplored at LHC.

References

[1] A. Rothkopf, Phys. Rept. **858** (2020), 1-117
[2] P. Braun-Munzinger and J. Stachel, Phys. Lett. B **490** (2000), 196-202
[3] R. L. Thews, M. Schroedter and J. Rafelski, Phys. Rev. C **63** (2001), 054905
[4] F. Arleo, Phys. Rev. Lett. **119** (2017) no.6, 062302 [arXiv:1703.10852 [hep-ph]].
[5] L. Apolinário, Y. J. Lee and M. Winn, Prog. Part. Nucl. Phys. **127** (2022), 103990
[6] K. Aamodt *et al.* [ALICE], JINST **3** (2008), S08002
[7] S. Acharya *et al.* [ALICE], JHEP **02** (2024), 066
[8] S. Acharya *et al.* [ALICE], Phys. Lett. B **849** (2024), 138451
[9] A. Andronic *et al.*Phys. Lett. B **797** (2019), 134836
[10] B. Chen, Chin. Phys. C **43** (2019) no.12, 124101
[11] S. Aronson *et al.* Phys. Lett. B **778** (2018), 384-391
[12] M. Yang *et al.*Phys. Rev. C **107** (2023) no.5, 054917
[13] A. Beraudo, A. De Pace, M. Monteno, M. Nardi and F. Prino, JHEP **05** (2021), 279
[14] S. Shi, J. Liao and M. Gyulassy, Chin. Phys. C **43** (2019) no.4, 044101
[15] M. Nahrgang *et al.*Phys. Rev. C **93** (2016) no.4, 044909
[16] S. Q. Li *et al.*, Eur. Phys. J. C **81** (2021) no.11, 1035
[17] W. J. Xing, G. Y. Qin and S. Cao, Phys. Lett. B **838** (2023), 137733
[18] D. Zigic *et al.*, Front. in Phys. **10** (2022), 957019
[19] S. Acharya *et al.* [ALICE], JINST **19** (2024) no.05, P05062

© The Authors, published by EDP Sciences. This is an open access article distributed under the terms of the Creative Commons Attribution License 4.0 (https://creativecommons.org/licenses/by/4.0/).

Recent conventional and exotic charmonia results from LHCb

Youen Kang* on behalf of the LHCb Collaboration

[1]Department of Engineering Physics, Tsinghua University, Beijing 100084, China

Abstract. Quarkonia production in heavy-ion collisions is an important experimental probe that sheds light on the heavy quark interaction with the nuclear medium. The bound quarkonium states undergo dissociation and recombination in PbPb collisions, where they can also experience the initial and final state effects such as shadowing and co-mover breakup. With the large datasets of pp and pPb collisions, and excellent vertexing capabilities allowing separation of the prompt and b-decay components, LHCb performs precise measurements of J/ψ, $\psi(2S)$ and, for the first time at the LHC, χ_c production and modification in small collision systems. This contribution will discuss these results, along with the first measurement of the nuclear modification factor of the exotic $\chi_{c1}(3872)$.

1 Introduction

In ultra-relativistic heavy-ion collisions, quarkonium production is suppressed when the temperature exceeds its binding energy. Above the critical temperature of deconfinement, Debye color screening prevents quarks and anti-quarks from forming quarkonium in Quark-Gluon Plasma (QGP) [1]. The quarkonium can also go through a collisional dissociation and regeneration process inside QGP [2]. While in small collision systems like pp and pPb collisions, in the absence of Hot Nuclear Matter (HNM) effect, quarkonium can also dissociate by interacting with co-moving particles in the final state [3]. The excited quarkonium states are loosely bound compared to the ground state, leading to preferential suppression, especially in a high-multiplicity environment. The investigation of conventional and exotic charmonia production across different collision systems helps us understand the production mechanism of quarkonium, and study as well the evolution of medium properties with system size.

The LHCb detector [4] is a single-arm spectrometer with a forward pseudorapidity coverage ($2 < \eta < 4.8$). Its tracking system is comprised of the VErtex LOcator (VELO) and four planar tracking stations. The VELO provides precise measurements of track coordinates close to the interaction region. The number of tracks reconstructed by the VELO is often used as a proxy for charged particle multiplicity. The LHCb detector has recorded pp and pPb collisions data at $\sqrt{s} = 13$ TeV for pp collisions and $\sqrt{s_{NN}} = 5.02$ and 8.16 TeV in both the proton-going (forward) and lead-going (backward) pPb collisions, providing a transition environment from small collision systems to heavy-ion collisions.

*e-mail: youen.kang@cern.ch

© The Authors, published by EDP Sciences. This is an open access article distributed under the terms of the Creative Commons Attribution License 4.0 (https://creativecommons.org/licenses/by/4.0/).

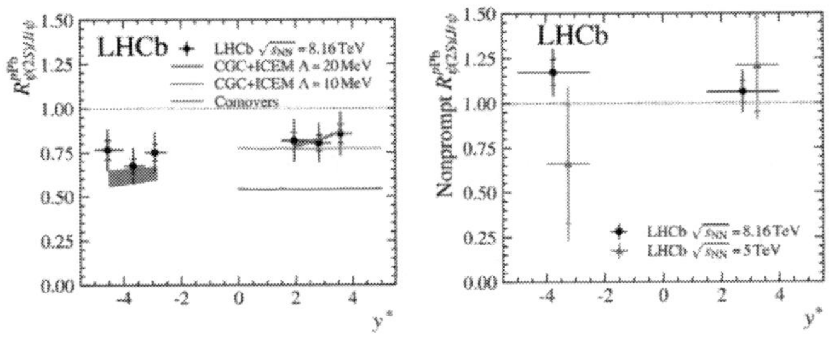

Figure 1. Cross section double ratios $R^{p\text{Pb}}_{\psi(2S)/J/\psi}$ as a function of rapidity in center-of-mass frame y^*, for (left) prompt and (right) nonprompt production. The horizontal bars are bin widths, vertical error bars represents statistical and total uncertainties.

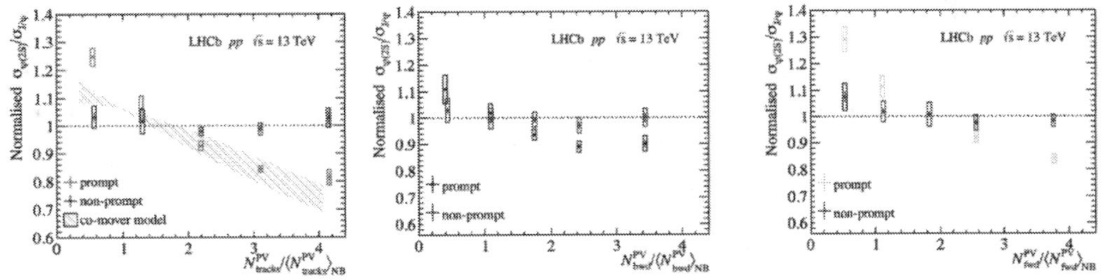

Figure 2. Self-normalized production ratio as a function of (left) charged-tracks density $N^{\text{PV}}_{\text{tracks}}/\langle N^{\text{PV}}_{\text{tracks}}\rangle_{NB}$, (middle) forward charged-tracks density $N^{\text{PV}}_{\text{fwd}}/\langle N^{\text{PV}}_{\text{fwd}}\rangle_{NB}$ and (right) backward charged-tracks density $N^{\text{PV}}_{\text{bwd}}/\langle N^{\text{PV}}_{\text{bwd}}\rangle_{NB}$. The error bars represent statistical uncertainties and heights of the boxes represent systematic uncertainties.

2 Conventional charmonium production

The LHCb collaboration measured the cross-section double ratio $R^{p\text{Pb}}_{\psi(2S)/J/\psi} = \frac{(\sigma_{\psi(2S)}/\sigma_{J/\psi})_{p\text{Pb}}}{(\sigma_{\psi(2S)}/\sigma_{J/\psi})_{pp}}$ in pPb collisions at $\sqrt{s_{NN}} = 8.16$ TeV [5], as shown in the Fig 1. The reference $\psi(2S)$ to J/ψ cross-section ratio is measured in 7 TeV pp collisions at LHCb [6]. The prompt and nonprompt production are separated using pseudo-proper time $t_z = \frac{z_X - z_{\text{PV}}}{p_z}$, where z_X and z_{PV} are the coordinates of charmonium and primary vertex (PV) along the beam axis. The ratio of nonprompt $\psi(2S)$ and J/ψ nuclear modification factors are compatible with one as expected since the nuclear matter effects affect $b-$hadron production but not their decays. The prompt double ratio is lower than one, especially in the backward rapidity where a higher multiplicity is achieved. The $\psi(2S)$ has a larger radius and lower binding energy, leading to a preferential dissociation when interacting with co-moving particles. The results agree with co-mover model calculations [3] and the predictions from Color-Glass Condensate (CGC) with Improved Color Evaporation Model (ICEM) models [7].

The LHCb Collaboration also measured the multiplicity-dependent self-normalized cross-section ratio of $\psi(2S)$ to J/ψ production $\frac{(\sigma_{\psi(2S)}/\sigma_{J/\psi})_n}{(\sum_n \sigma_{\psi(2S)}/\sum_n \sigma_{J/\psi})_n}$ at 13 TeV pp collisions [8], n stands for the multiplicity bin index. The prompt ratio decreases with increasing charged track multiplicity $N^{\text{PV}}_{\text{tracks}}$, as shown in the Fig 2, measured by the number of tracks used in PV reconstruction and normalized by that in No-Bias data, which is selected with a random trigger. The prompt ratio shows a decreasing trend as a function of forward charged track multiplicity while decreasing much slower as a function of backward-charged track multiplicity. This is consistent with expectation that charmonium can dissociate due to interactions with co-moving particles in the same direction.

The LHCb collaboration measured the fraction of χ_c (including χ_{c1} and χ_{c2}) decays in the prompt J/ψ yield $F_{\chi_c \to J/\psi} = \frac{\sigma_{\chi_c \to J/\psi}}{\sigma J/\psi}$ in pPb collisions at $\sqrt{s_{NN}} = 8.16$ TeV [9]. The $F_{\chi_c \to J/\psi}$ results in pPb collisions have an overall consistency with the results in 7 TeV pp

© The Authors, published by EDP Sciences. This is an open access article distributed under the terms of the Creative Commons Attribution License 4.0 (https://creativecommons.org/licenses/by/4.0/).

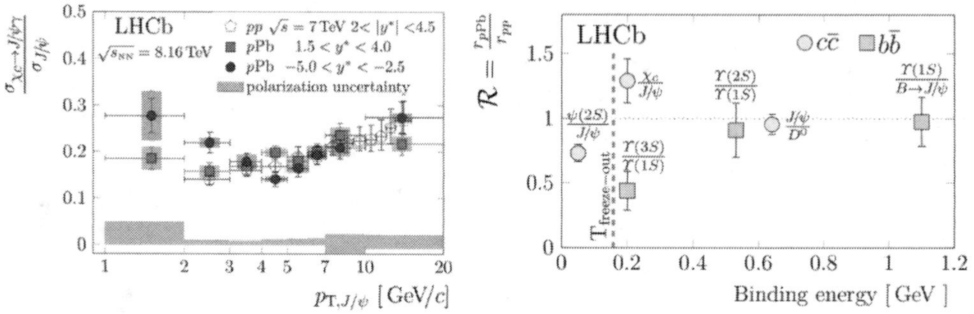

Figure 3. The (left) Fraction of χ_c decays to prompt J/ψ as a function of $p_{T,J/\psi}$, where the error bars (boxes) show the statistical (systematic) uncertainties and gray band is uncertainty from polarization effects, and (right) double ratio between quarkonium states vs. binding energy along with the estimated freeze-out temperature [11] in *p*Pb collisions.

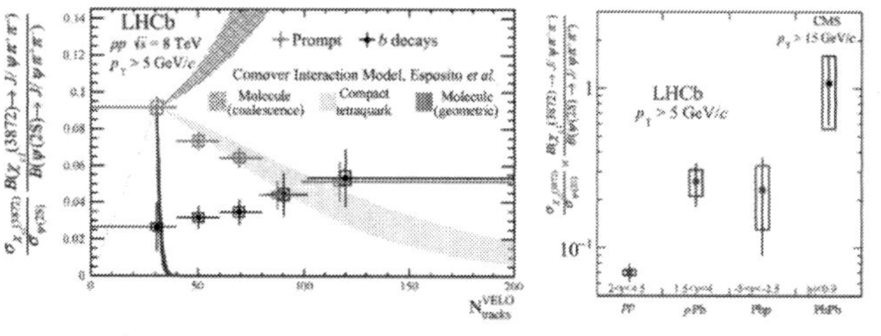

Figure 4. The $\chi_{c1}(3872)$ to $\psi(2S)$ cross-section ratio (left) as a function of number of VELO tracks in *pp* collisions, and (right) as a function of system sizes of collisions. The vertical error bars (boxes) are statistical (systematic) uncertainties. The horizontal error bars in the left plot are bin widths.

collisions [10], as shown in the Fig 3, suggesting that neither the prompt J/ψ mesons nor χ_c dissociate in the *p*Pb collisions. The non-dissociation of χ_c states suggests a constraint on the temperature of *p*Pb collision system of no more than 180 MeV, which is the smallest binding energy of χ_c states. This temperature is close to the estimated freeze-out temperature, 155-160 MeV, in *p*Pb collisions [11]. The double ratio for quarkonium states in the backward *p*Pb collisions is shown in Fig 3. The only quarkonium with binding energy higher than freeze-out temperature suppressed is $\Upsilon(3S)$. With similar binding energy, $\Upsilon(3S)$ is 2.9 times heavier than χ_c, which could favor its dissociation by interaction with co-moving particles.

3 Exotic charmonium

The LHCb collaboration measured the cross-section ratio of exotic state $\chi_{c1}(3872)$ to the conventional charmonium state $\psi(2S)$ with similar mass in 8.16 TeV *p*Pb collisions and 8 TeV *pp* collisions, as shown in the Fig. 4 [12, 13]. The result in *pp* collisions [13] shows a decreasing trend with increasing multiplicity for prompt $\chi_{c1}(3872)$ to $\psi(2S)$ cross-section ratio. Predictions from the co-mover interaction model suggest that, if $\chi_{c1}(3872)$ is a pure molecular with a corresponding higher radius, it will quickly dissociate with increasing multiplicity [3]. If coalescence provides an additional formation mechanism, the cross-section ratio will increase with increasing multiplicity. Neither of these calculations matches the data. If $\chi_{c1}(3872)$ is a compact tetraquark with a radius slightly larger than $\psi(2S)$, it will go through a slightly preferential suppression which matches the measured trend. The comparison between the *pp* and *p*Pb results from LHCb, together with the PbPb result from CMS [14], shows an increasing trend in the prompt $\chi_{c1}(3872)$ to $\psi(2S)$ cross-section ratio with increasing size of collisions, which can result from a combination of effects. Given that the $\chi_{c1}(3872)$ to

© The Authors, published by EDP Sciences. This is an open access article distributed under the terms of the Creative Commons Attribution License 4.0 (https://creativecommons.org/licenses/by/4.0/).

$\psi(2S)$ cross-section ratio decreases with increasing multiplicity in pp collisions, the increasing trend with collision size may indicate that the hadronic density in pPb collisions allows the quark coalescence to be a dominant mechanism for $\chi_{c1}(3872)$ production.

4 Conclusion

The LHCb collaboration measured the $\psi(2S)$ to J/ψ cross-section double ratio in pPb collisions at $\sqrt{s_{NN}} = 8.16$ TeV with a reference of cross-section ratio in 7 TeV pp collisions, and the self-normalized cross-section ratio in pp collisions. Results agree that $\psi(2S)$ is preferentially suppressed in high-multiplicity events. In pPb collisions, $\psi(2S)$ is more suppressed than in pp collisions. The relatively larger radius and lower binding energy for $\psi(2S)$ leads to an easier dissociation by interaction with final-state particles. The measurement of the fraction of χ_c decaying to prompt J/ψ in 8.16 TeV pPb collisions and 7 TeV pp collisions suggest that neither χ_c nor J/ψ are suppressed in pPb collisions, indicating that the temperature reached in 8.16 TeV pPb collisions is not able to inhibit the formation of quarkonium states with binding energy as small as 180 MeV.

The LHCb collaboration measured the relative production of exotic $\chi_{c1}(3872)$ to $\psi(2S)$ in 8 TeV pp collisions and 8.16 TeV pPb collisions. An increasing cross-section ratio with increasing size of collisions is observed, indicating the exotic $\chi_{c1}(3872)$ production modification in pPb and PbPb collisions is different from that of the $\psi(2S)$.

References

[1] Quigg. C, and R. Jonathan L. Phys. Rept. **56**, 167-235 (1979). https://doi.org/10.1016/0370-1573(79)90095-4

[2] J. Zhao, K. Zhou, S. Chen, and P. Zhuang, Prog. Part. Nucl. Phys. **114**, 103801 (2022). https://doi.org/10.1016/j.ppnp.2020.103801

[3] F. Elena. G. and Lansberg. J, JHEP. **10**, 094 (2018). https://doi.org/10.1007/JHEP10(2018)094

[4] R. Aaij *et al.* [LHCb], Int. J. Mod. Phys. A **30**, no.07, 1530022 (2015). https://doi.org/10.1142/S0217751X15300227

[5] R. Aaij *et al.* [LHCb], JHEP. **04**, 111 (2014), https://doi.org/10.1007/JHEP04(2024)111

[6] R. Aaij *et al.* [LHCb], Eur. Phys. J. C **80**, no.3, 185 (2020). https://doi.org/10.1140/epjc/s10052-020-7638-y

[7] Ma. Y, Venugopalan. R, Watanabe. K and Zhang. H, Phys. Rev. C **97**, no.1, 014909 (2018), https://doi.org/10.1103/PhysRevC.97.014909

[8] R. Aaij *et al.* [LHCb], JHEP. **05**, 243 (2024), https://doi.org/10.1007/JHEP05(2024)243

[9] R. Aaij *et al.* [LHCb], Phys. Rev. Lett. **132**, no.10, 102302 (2024), https://doi.org/10.1103/PhysRevLett.132.102302

[10] R. Aaij *et al.* [LHCb], Phys. Lett. B **718**, 431-440 (2012), https://doi.org/10.1016/j.physletb.2012.10.068

[11] Bazavov. A, and others, Phys. Lett. B **795**, 15-21 (2019), https://doi.org/10.1016/j.physletb.2019.05.013

[12] R. Aaij *et al.* [LHCb], Phys. Rev. Lett. **132**, no.24, 242301 (2024), https://doi.org/10.1103/PhysRevLett.132.242301

[13] R. Aaij *et al.* [LHCb], Phys. Rev. Lett. **126**, no.9, 092001 (2021), https://doi.org/10.1103/PhysRevLett.126.092001

[14] Sirunyan, Albert M. and others, Phys. Rev. Lett. **128**, no.3, 032001 (2022), https://doi.org/10.1103/PhysRevLett.128.032001

© The Authors, published by EDP Sciences. This is an open access article distributed under the terms of the Creative Commons Attribution License 4.0 (https://creativecommons.org/licenses/by/4.0/).

Probing a new regime of ultra-dense gluonic matter using high-energy photons with the CMS experiment

Pranjal Verma[1] for the CMS Collaboration

[1]Indian Institute of Technology Madras, Chennai, India

Abstract. In ultraperipheral collisions (UPCs) involving relativistic heavy ions, the production of heavy-flavor coherent vector mesons through photon-nuclear interactions is a key focus due to its direct sensitivity to the nuclear gluon density. Experimental measurements, however, face a two-way ambiguity as the collision involves two lead ions which can act as both a photon-emitter projectile and a target. This ambiguity hinders the separation of contributions from high and low-energy photon-nucleus interactions, restricting our ability to probe the extremely small-x regime where nonlinear QCD effects are anticipated. This writeup unveils the measurement of coherent heavy quarkonium photoproduction, addressing the two-way ambiguity by employing a forward neutron tagging technique in UPC PbPb collisions at 5.02 TeV. Overall these studies focus on the dominance of gluons in nuclear matter probed at higher energies.

1 Introduction

The electromagnetic fields surrounding relativistic heavy ions are highly Lorentz contracted and can be effectively treated as linearly polarized quasireal photons at a first approximation [2, 3]. These photons are powerful probes of the gluonic structure of nuclei [4]. Ultraperipheral collisions (UPCs) occur when two relativistic heavy ions pass each other at large impact parameters, reducing the probability of having the hadronic interactions. Cases like photon photon or photon nucleus interaction occurs in such collisions. Notably, vector meson photoproduction occurs when a photon from one nucleus fluctuates into a quark-antiquark pair, which then scatters off the other nucleus, emerging as a real meson. This phenomenon has been studied at high-energy facilities like the RHIC and LHC.

At energies accessible at the CERN LHC these studies can lead towards probing the poorly known nuclear gluonic structure and dynamics in the low Bjorken-x region. There is a possibility that at a point the increasing gluon density would reach saturation, along with the saturation of Vector meson cross section at the geometric limit of the nucleus. Another possibility can be nucleus target becoming totally absorptive to incoming photons leading to a "Black Disk Limit", where the internal structure of nucleus vanishes [1].

The photoproduction of heavy-flavor VMs is of particular interest since it can be studied in the perturbative quantum chromodynamic (QCD) framework with a large energy scale defined by the quark mass. In the leading-order QCD contribution, the coherent J/ψ photoproduction cross section is proportional to the square of the gluon density function. In photon-proton interactions, the J/ψ production cross section significantly increases with the photon-proton center-of-mass energy because of the rising gluon density at small x [1, 5].

© The Authors, published by EDP Sciences. This is an open access article distributed under the terms of the Creative Commons Attribution License 4.0 (https://creativecommons.org/licenses/by/4.0/).

For symmetric collision system, where either of the ions can be the photon emitter or target, the complications are even more increased. To solve this two-way ambiguity, we control the UPC impact parameter and thus the relative contributions of low and high energy photons to the measured J/ψ cross section can be separated [1].

2 Analysis Technique

The results shown in this writeup use Pb-Pb UPCs at a nucleon-nucleon center-of-mass energy $\sqrt{s_{NN}}$ = 5.02 TeV, which were collected by the CMS experiment at the LHC in 2018. The data correspond to an integrated luminosity of 1.52 nb^{-1}. The CMS detector has a highly sophisticated tracking system which captures particle trajectories with high precision. For muon detection, detectors containing 2 million cathode strip chamber wires are placed at the outer edges which detects the ionization trail left by muons. Additionally, the Zero Degree Calorimeters (ZDCs) made from tungusten embedded quartz fibers, detect neutrons produced during ultraperipheral collisions and helps us the identify the centrality in UPCs.

Events are selected using a hardware-based trigger system that requires at least one muon candidate with no explicit selection on its transverse momentum (p_T), coincident with a Pb-Pb bunch crossing [6]. At the trigger level, events with an energy deposit above the noise threshold in either of the forward calorimeters are vetoed. For the off-line analysis, events must have a primary interaction vertex, formed by using two or more tracks coming from the collision, that is within 20 cm along the beam axis and at a radius less than 2 cm in the transverse plane from the detector center. To suppress hadronic collisions, using empty bunch-crossing events only the forward calorimeters having their largest energy deposits below 7.3 and 7.6 GeV in the positive and negative rapidity sides, respectively are selected. [1]

Additionally, both the muon needs to be "soft muons" [7], lying in the range $|\eta| < 2.4$. The dimuon pair (ψ or J/ψ candidates) lies in mass region $2.6 < m_{\mu^+\mu^-} < 4.2$ GeV and rapidity region $1.6 < |y| < 2.4$. For each muon pair, at least one of the muon candidates must match the triggered muon [1].

3 Results

Based on the neutron peaks observed in the total ZDC energy distribution, events are classified with number of nuetrons fallen in ZDCs, 0n0n, 0nXn or XnXn. The relative fractions of J/ψ candidates in 0n0n, 0nXn, and XnXn classes are determined to be 73, 21, and 6% corrected from the corresponding raw fractions of 65, 28, and 7%, respectively [1].

Because of the symmetry of the Pb-Pb collision system,a J/ψ meson measured at rapidity y can result from two possible photon energies: $\omega_1 = (M_{J/\psi}/2$ exp(-y)) and $\omega_2 = (M_{J/\psi}/2$ exp(+y)). The measured differential cross section at a rapidity y for a neutron multiplicity class, injn (=0n0n, 0nXn, XnXn), of UPC events is the sum of both photon energy contributions [8]:

$$\frac{d\sigma^{in\,jn}_{J/\psi}(y)}{dy} = n^{in\,jn}_{\gamma A}(\omega_1)\,\sigma_{J/\psi}(\omega_1) + n^{in\,jn}_{\gamma A}(\omega_2)\,\sigma_{J/\psi}(\omega_2), \tag{1}$$

The measured coherent J/ψ photoproduction differential cross sections over the rapidity range 1.6 < |y| < 2.4 are reported in Fig. 1 (left) in three neutron multiplicity classes. The results with no neutron selection are given in the right plot of Fig. 1 [1].

The x of the associated parton and $W^{Pb}_{\gamma N}$ with a photon of energy ω_1 can be found using $x = (M_{J/\psi}/\sqrt{s_{NN}})\exp(+y)$ and $W^{Pb}_{\gamma N} = \sqrt{\sqrt{s_{NN}}M_{J/\psi}\exp(-y)}$. The corresponding values for a photon of energy ω_2 are obtained by changing the sign of y in each expression [1].

© The Authors, published by EDP Sciences. This is an open access article distributed under the terms of the Creative Commons Attribution License 4.0 (https://creativecommons.org/licenses/by/4.0/).

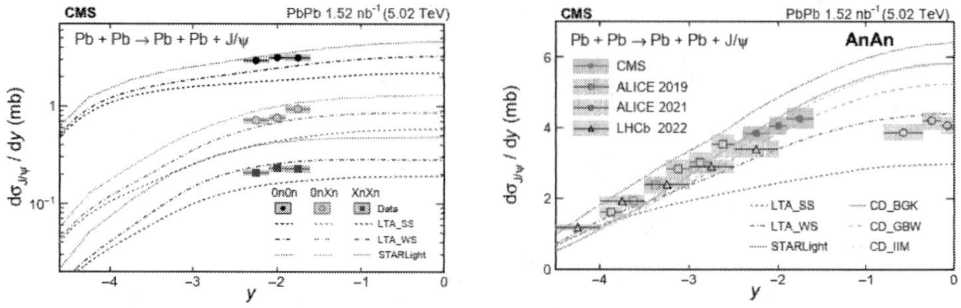

Figure 1. The differential coherent J/ψ photoproduction cross section as a function of rapidity, in different neutron multiplicity classes: 0n0n, 0nXn, and XnXn (left); AnAn (right) [1]

The measured total cross section has an unexpected energy dependence, approximately quadrupling as $W_{\gamma N}^{Pb}$ goes from 15 to 40 GeV, as seen in Fig. 2. This is consistent with the expectation of a fast growing gluon density at low x (e.g., from the IA model, where most of the nuclear effects are neglected, except coherence). However, this trend vanishes for $W_{\gamma N}^{Pb} > 40$ GeV, and instead the total cross section begins a slow linear rise with a slope of $(2.2 \pm 1.9) \times 10^{-5}$ mb/ GeV determined by a fit to CMS data with proper consideration of the covariance matrix of both statistical and systematic uncertainties. Considering the experimental uncertainties across the measured $W_{\gamma N}^{Pb}$ range, none of the theoretical models are consistent with the measurements, although data is close to Impulse Approximation (without nuclear effects) at low W and with the CD-BGK model having the best p value of 1.6×10^{-8}. Here, the photon-nucleus cross section approaches the unitarity limit allowed by the geometric size of the nucleus but the periphery of the nucleus won't become fully black [1].

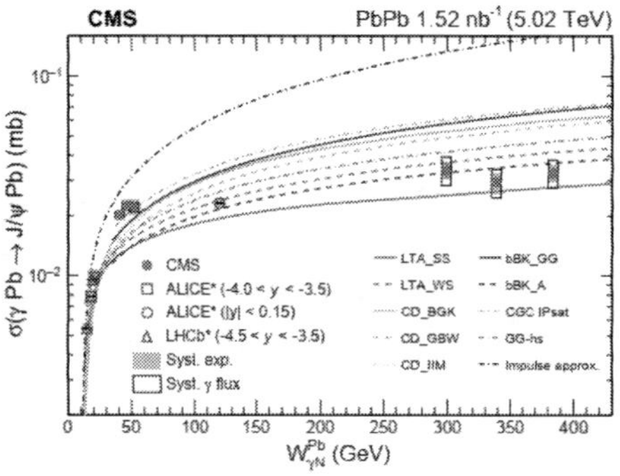

Figure 2. The total coherent J/ψ photoproduction cross section as a function of $W_{\gamma N}^{Pb}$ from the CMS measurement in Pb-Pb UPCs at = 5.02. Approximated results (implied by the asterisk) from the ALICE [9] and LHCb [10] experiments are displayed for specific rapidity regions, where the two-way ambiguity effect is expected to be negligible [1]

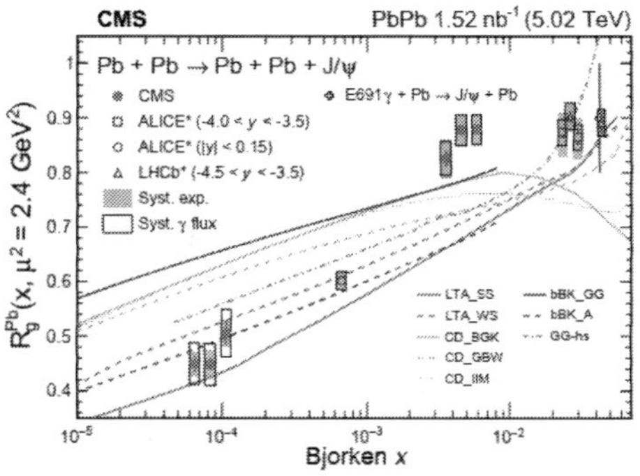

Figure 3. The nuclear gluon suppression factor R_g^{Pb} as a function of Bjorken x extracted from the coherent J/ψ photoproduction in Pb-Pb UPCs. Approximated results (implied by the asterisk) from the ALICE [9] and LHCb [10] experiments are displayed for specific rapidity regions, with negligible two-way ambiguity effect [1]

To quantify the nuclear effects on the observed gluon density function in a Pb nucleus, a nuclear gluon suppression factor, $R_g^{Pb}(x, \mu^2 = 2.4GeV^2)$ as a function of x and with $\mu =$

© The Authors, published by EDP Sciences. This is an open access article distributed under the terms of the Creative Commons Attribution License 4.0 (https://creativecommons.org/licenses/by/4.0/).

$M_{J/\psi}/2$, where μ is the energy scale, is defined as $R_g^{Pb} = \sqrt{\sigma^{\text{Meas}}/\sigma^{\text{IA}}}$, where σ^{Meas} is the measured cross section. The extracted R_g^{Pb} values are shown in Fig.3 as a function of x. The suppression in the relatively high-x (low $W_{\gamma N}^{Pb}$) region of $x > 5 \times 10^{-3}$ is approximately 0.8–0.9. At smaller x values, R_g^{Pb} starts dropping rapidly to 0.4–0.5 for $x \approx 6 \times 10^{-5}$. Similar to the cross section measurement, this is also beyond model expectation in full phase space. At high-x region,a flat trend is observed which quickly decrease towards lower x region [1].

4 Summary

A comprehensive analysis of the coherent J/ψ photoproduction cross section off lead nuclei has been presented, with a focus on its behavior as a function of the photon-nuclear center-of-mass energy per nucleon ($W_{\gamma N}^{Pb}$). Utilizing ultraperipheral Pb-Pb collision data and employing the forward neutron tagging method, the results span a wide range of $W_{\gamma N}^{Pb}$. At lower energies, the cross section shows a steep increase, but it levels off around $W_{\gamma N}^{Pb} \approx 40$ GeV and remains constant up to 400 GeV. This energy range explores a new regime of gluon momentum fraction, with Bjorken $x \approx 6 \times 10^{-5}$, within a heavy nucleus. Interestingly, this behavior is not aligned with predictions from existing theoretical models. It may either indicate the first direct observation of nuclear gluon saturation or suggest that the cross section is approaching the black-disk limit.

References

[1] CMS, Probing small Bjorken-x nuclear gluonic structure via coherent J/ψ photoproduction in ultraperipheral Pb-Pb collisions at $\sqrt{S_{NN}} = 5.02$ TeV. Phys. Rev. Lett. **131**, 262321 (2023). https://doi.org/10.1103/PhysRevLett.131.262301

[2] E. J. Williams, "Nature of the high energy particles of penetrating radiation and status of ionization and radiation formulae". Phys. Rev. **45**, 729 (1934) https://doi.org/10.1103/PhysRev.45.729

[3] Weizsäcker, C.F.v. Ausstrahlung bei Stößen sehr schneller Elektronen. Z. Physik 88, 612–625 (1934). https://arxiv.org/pdf/1806.07238

[4] S. R. Klein and P. Steinberg, "Photonuclear and two-photon interactions at high-energy nuclear colliders", Annu. Rev. Nucl. Part. Sci. **70**, 323 (2020). https://doi.org/10.1146/annurev-nucl-030320-033923

[5] ALICE Collaboration, "Exclusive J/ photoproduction off protons in ultra-peripheral p-Pb collisions at sNN = 5.02 TeV", Phys. Rev. Lett. **113**, 232504 (2014). http://10.1103/PhysRevLett.113.232504.

[6] CMS Collaboration, "The CMS trigger system" JINST **12** P01020 (2017), https://doi.org/10.1088/1748-0221/12/01/P01020

[7] CMS Collaboration, "Performance of CMS muon reconstruction in pp collision events at $\sqrt{s} = 7$ TeV" JINST **7** (2012) P10002, http://dx.doi.org/10.1088/1748-0221/7/10/P10002

[8] V. Guzey, M. Strikman, and M. Zhalov, "Disentangling coherent and incoherent quasielastic J/ photoproduction on nuclei by neutron tagging in ultraperipheral ion collisions at the LHC", Eur. Phys. J. C **74** (2014) 2942, https://doi.org/https://doi.org/10.1140/epjc/s10052-014-2942-z.

[9] Alice, Signals for black body limit in coherent ultraperipheral heavy ion collisions Phys. Lett. B **537**, 51 (2022). https://doi.org/10.48550/arXiv.hep-ph/0204175

[10] LHCb, Study of exclusive photoproduction of charmonium in ultra-peripheral lead-lead collisions. JHEP **06**, 146 (2023). https://doi.org/10.1007/JHEP06%282023%29146

© The Authors, published by EDP Sciences. This is an open access article distributed under the terms of the Creative Commons Attribution License 4.0 (https://creativecommons.org/licenses/by/4.0/).

First measurement of heavy flavour femtoscopy in Au+Au collisions at $\sqrt{s_{\mathrm{NN}}}$ = 200 GeV by STAR

Priyanka Roy Chowdhury[1,*] (for the STAR Collaboration)

[1]Warsaw University of Technology, Faculty of Physics, Koszykowa 75, 00-662 Warsaw, Poland

Abstract. In the very beginning stages of heavy-ion collisions, hard partonic scatterings yield heavy quarks, which go through the entire Quark-Gluon Plasma medium evolution. Femtoscopic correlation is characterized as a two particle correlation at low relative momentum that depends on the size of the region from which the correlated particles are emitted and on the final-state interaction. Such correlations between identifiable charged hadrons and charmed mesons could provide insights into their interactions with the medium and charm quarks in the hadronic phase. We describe the first femtoscopic correlation measurements made by the STAR experiment between $D^0/\overline{D^0} - \pi^{\pm}$, $D^0/\overline{D^0} - K^{\pm}$ and $D^0/\overline{D^0} - p^{\pm}$ pairs at mid-rapidity in Au+Au collisions at $\sqrt{s_{\mathrm{NN}}}$ = 200 GeV. The discussion of the physics consequences involves a comparison to theory calculations.

1 Introduction

Heavy quarks such as charm (c) and its charge conjugate ($\overline{c}$) are created in the early stages of ultra-relativistic collisions of heavy ions. Therefore, they probe all stages of the evolution of heavy-ion collisions - Quark-Gluon Plasma (QGP), hadronization, chemical and kinetic freeze-out. [1] One heavy ($c/\overline{c}$) and one light (($\overline{u}/u$) quark make up $D^0/\overline{D}^0$ mesons (lifetime, $c\tau \approx 123\,\mu$m [2]). These charmed mesons are useful to probe how heavy quarks interact with QGP medium. The STAR experiment at RHIC (Relativistic Heavy Ion Collider) observed significant D^0 elliptic flow [3] and D^0 suppression at high p_{T} [4] in heavy-ion collisions. Numerous theoretical models with varying assumptions were able to quantitatively describe these findings. In order to improve our comprehension of heavy quarks (c, b) interactions with the medium, new observables like the two-particle momentum correlation functions are of particular interest. Phase-space evolution of emission source and final-state interactions can be both studied with such measurements in heavy-ion collisions [5]. According to theory, phase-space cloud of outgoing correlated pairs or so-called area of homogeneity is responsive to QGP dynamics, such as collective flow [5, 6]. Size of area of homogeneity is substantially lower than the size of fireball in cases of strong correlation [5].

According to Koonin-Pratt equation [5], femtoscopic correlation function, $C(k^*) = \int S(r)|\psi(k^*, r)|^2 d^3r$, is a combination of emission source function, $S(r)$ and pair-wave function, $\psi(k^*, r)$. Here k^* represents the reduced momentum difference between correlated particle pairs emitting from a source of size r. We computed $C(k^*)$ by taking the k^* ratio of

*e-mail: priyanka.roy_chowdhury.dokt@pw.edu.pl

© The Authors, published by EDP Sciences. This is an open access article distributed under the terms of the Creative Commons Attribution License 4.0 (https://creativecommons.org/licenses/by/4.0/).

correlated [$A(k^*)$] and uncorrelated [$B(k^*)$] D^0-hadron pairs in the rest frame of their center of mass given in Eq. 1 [5].

$$C(k^*) = N\frac{A(k^*)}{B(k^*)} \text{ and } k^* = \frac{1}{2}(p_1 - p_2), \tag{1}$$

where N is normalization factor and p_1, p_2 are momenta of $D^0/\overline{D}^0$ and $K/\pi/p$ tracks in the pair-rest frame. Two tracks were chosen from the same event in order to compute $A(k^*)$. $B(k^*)$ was calculated using event mixing technique to select a pair of tracks from different events chosen within a similar primary z-vertex position (V_z) and centrality range as the same-event pairs.

2 $D^0/\overline{D}^0$ reconstruction and signal purity estimation

STAR is made up of multiple detectors with distinct functions [7]. The two primary detectors for tracking and identifying charge particles are the Time Projection Chamber (TPC) and Time of Flight (TOF). The HFT detector with outstanding track pointing resolution was used for the reconstruction of $D^0/\overline{D}^0$ mesons via the $K^{\mp}\pi^{\pm}$ decay channel using the set of topological criteria [4]. Fig. 1 shows the reconstructed D^0 and $\overline{D^0}$ invariant mass distributions using STAR data of Au+Au collisions at $\sqrt{s_{NN}}$ = 200 GeV [4]. We selected D^0 candidates (and their charge conjugates) based on two criteria: $p_T > 1$ GeV/c and the ratio of signal-to-combinatorial background under D^0 peak (S/B) had to exceed 30% in the lowest p_T bin. D^0 ($\overline{D^0}$) signal peak lies between 1.82 to 1.91 GeV/c^2 mass range. With rising transverse momentum, p_T, the S/B ratio increases. The D^0 signal purity ($\frac{S}{S+B}$) was calculated for each p_T bin (1-2, 2-3, 3-5, and 5-10 GeV/c). In the lowest p_T bin, the purity is 25% , gradually increasing to 80% in the highest p_T bin. D^0 purity is used to calculate the pair purity as discussed in detail in Sec. 3.

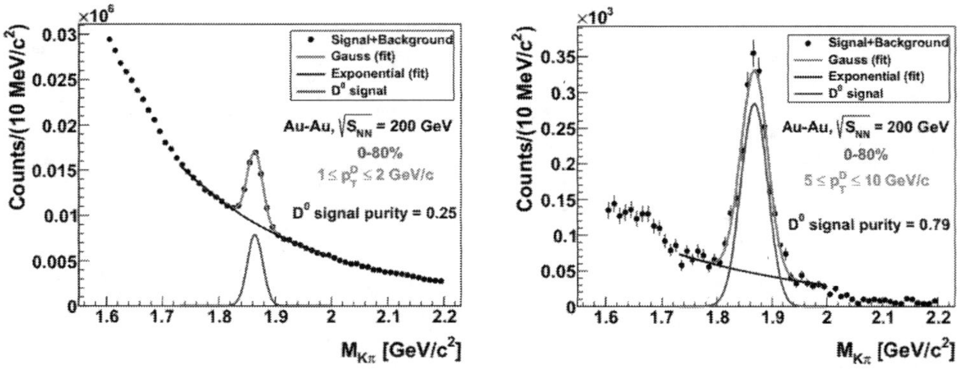

Figure 1. Invariant mass ($M_{K\pi}$) of D^0 and $\overline{D}^0$ candidates using STAR Run 2014 data. The black solid circles depict the invariant mass of unlike-sign (US) $K\pi$ pairs. The red curve corresponds to a Gaussian fit, while the black line shows an exponential fit to the US pairs. The blue curve illustrates the fit to the D^0 ($\bar{D}^0$) signal within the mass range of 1.73 to 2.0 GeV/c^2.

3 Correlation function calculation

The computed value of the correlation function can be affected by incorrectly identified correlated pairs. We used both TPC and TOF to select the primary tracks of considered charged hadrons (π, K and p). We eliminated splitted hadron tracks and excluded self-correlation

© The Authors, published by EDP Sciences. This is an open access article distributed under the terms of the Creative Commons Attribution License 4.0 (https://creativecommons.org/licenses/by/4.0/).

between D^0 daughters in order to reduce the potential effects of the TPC detector. Combining two different tracks into one is another possible detector impact, however our investigation showed that the contribution from merged tracks was negligible. To account for the contribution from the background under D^0 signal peak and impurity of the selected π, K and p samples with other hadrons and electrons present in the system, we implemented a purity correction for D^0-hadron pairs using a specific formula [8]:

$$C(k^*) = \frac{C_{\text{measured}}(k^*) - 1}{\text{Pair Purity}} + 1, \tag{2}$$

where $C(k^*)$ is the final correlation function after purity correction, $C_{\text{measured}}(k^*)$ represents the correlation function after correcting possible detector effects, while Pair Purity is determined by multiplying the D^0 signal purity with the average purity of the hadron sample. The purity of the hadron sample was determined using standard approach in STAR [9]. Hadrons were selected with momentum criteria, $p_\pi < 1$ GeV/c, $p_p < 1.2$ GeV/c and $p_K < 1$ GeV/c as they become difficult to distinguish from electrons and other hadrons beyond the mentioned thresholds respectively. Within selected momentum range, average purity of π and p are $(99.5 \pm 0.5)\%$ while for K sample, purity is $(97 \pm 3)\%$. Systematic uncertainty studies were made by variations in the topological cuts used for D^0 reconstruction [4], and by including the uncertainties in the purity estimation of D^0-hadron pairs.

4 Results and discussions

Figure 2 presents the final correlation function of $D^0/\overline{D^0} - \pi^\pm$, $D^0/\overline{D^0} - p^\pm$ and $D^0/\overline{D^0} - K^\pm$ pairs using STAR data from minimum bias Au+Au collision at $\sqrt{s_{\text{NN}}} = 200$ GeV within $|\eta| < 1$. The $C(k^*)$ distribution is consistent with unity, indicating an absence of significant correlation, with large statistical fluctuations. The strength of the correlation is linked to the size of the source [5, 6]. A weak or absent correlation suggests that the emission source

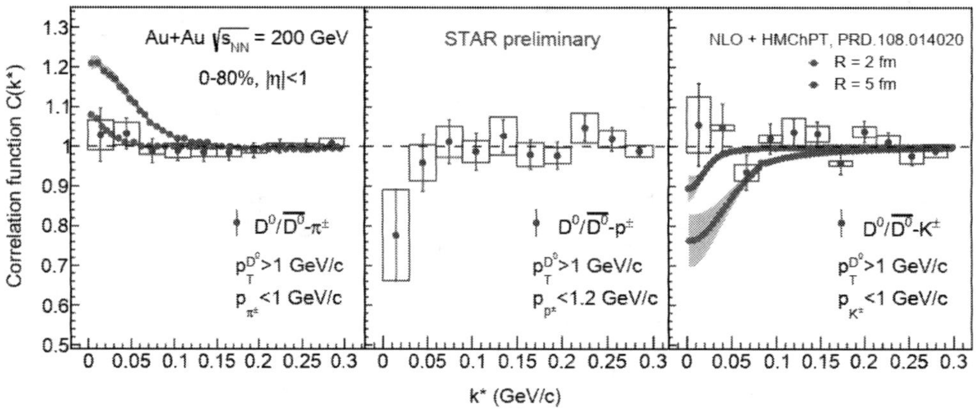

Figure 2. $C(k^*)$ for (left) $D^0/\overline{D^0} - \pi^\pm$, (mid) $D^0/\overline{D^0} - p^\pm$ and (right) $D^0/\overline{D^0} - K^\pm$ pairs. The blue solid circles, along with the boxes, indicate the STAR data and the associated systematic uncertainties. The green and pink bands illustrate the $C(k^*)$ values predicted by NLO + HMChPT model for source sizes of 2 fm and 5 fm, respectively.

from which the D-hadron pairs are produced is relatively large in size. In left and right panel, STAR data are compared with $C(k^*)$ calculated using NLO (next-to-leading order)-HMChPT (Heavy Meson Chiral Perturbation Theory) scheme [10–12]. In the left panel, the

© The Authors, published by EDP Sciences. This is an open access article distributed under the terms of the Creative Commons Attribution License 4.0 (https://creativecommons.org/licenses/by/4.0/).

theory predictions are for $D^0 - \pi^+$ and $D^+ - \pi^0$ pairs while in the right panel, for $D^0 - K^+$ pairs [10]. Neither of these channels includes Coulomb interaction. The STAR data and theoretical predictions align with an emission source size of 5 fm or more. The D^0K^+ channel exhibits a threshold near 0.083 GeV where a cusp effect is observed. The prediction shows a marked depletion around k^* between 0 to 0.05 GeV/c, which is attributed to the $D_S^*(2317)^{\pm}$ bound state and becomes more pronounced as the source radius decreases [10]. However, the resonance effect of this state is not evident in the STAR results due to either the large source size or significant experimental uncertainties. As a next step, we plan to update these results by combining data from Runs 2014 and 2016. This approach is expected to improve the precision of the correlation function measurements and offer more definitive conclusions about the source size. Additionally, theoretical inputs are necessary for a better interpretation of these data.

5 Acknowledgement

Priyanka Roy Chowdhury expresses gratitude for the financial support provided by the National Science Centre, Poland (NCN) under grant no. 2018/30/E/ST2/0008, as well as partial support from the U.S. Department of Energy (DOE).

References

[1] X. Dong *et al.*, Open Heavy-Flavor Production in Heavy-Ion Collisions, Ann. Rev. Nucl. Part. Sci. **69**, 417-445 (2019).

[2] R. L. Workman *et al.* [Particle Data Group], Review of Particle Physics, PTEP **2022**, 083C01 (2022).

[3] L. Adamczyk *et al.* [STAR], Measurement of D^0 Azimuthal Anisotropy at Midrapidity in Au+Au Collisions at $\sqrt{s_{NN}}$=200 GeV, Phys. Rev. Lett. **118**, no.21, 212301 (2017).

[4] J. Adam *et al.* [STAR], Centrality and transverse momentum dependence of D^0-meson production at mid-rapidity in Au+Au collisions at $\sqrt{s_{\rm NN}} = 200\,$GeV, Phys. Rev. C **99**, no.3, 034908 (2019).

[5] M. A. Lisa *et al.*, Femtoscopy in relativistic heavy ion collisions, Ann. Rev. Nucl. Part. Sci. **55**, 357-402 (2005).

[6] S. V. Akkelin and Y. M. Sinyukov, The HBT interferometry of expanding sources, Phys. Lett. B **356**, 525-530 (1995).

[7] Special Issue on RHIC and Its Detectors, edited by M. Harrison, T. Ludlam and S. Ozaki, Nucl. Instrum. Methods Phys. Res., Sect. A **499**, no.2–3 (2003).

[8] J. Adams *et al.* [STAR], Proton - lambda correlations in central Au+Au collisions at $\sqrt{S_{NN}}$ = 200-GeV, Phys. Rev. C **74**, 064906 (2006).

[9] L. Adamczyk *et al.* [STAR], Phys. Rev. C **95**, no. 3, 034907 (2017).

[10] M. Albaladejo, J. Nieves and E. Ruiz-Arriola, Femtoscopic signatures of the lightest S-wave scalar open-charm mesons, Phys. Rev. D **108**, no.1, 014020 (2023).

[11] F. K. Guo, C. Hanhart and U. G. Meissner, Interactions between heavy mesons and Goldstone bosons from chiral dynamics, Eur. Phys. J. A **40**, 171-179 (2009).

[12] L. S. Geng, N. Kaiser, J. Martin-Camalich and W. Weise, Low-energy interactions of Nambu-Goldstone bosons with D mesons in covariant chiral perturbation theory, Phys. Rev. D **82**, 054022 (2010).

© The Authors, published by EDP Sciences. This is an open access article distributed under the terms of the Creative Commons Attribution License 4.0 (https://creativecommons.org/licenses/by/4.0/).

EPJ Web of Conferences **316**, 04013 (2025) https://doi.org/10.1051/epjconf/202531604013
SQM 2024

Measurement of D^0 meson tagged jets in Au+Au collisions at $\sqrt{s_{NN}}$ = 200 GeV

Ondrej Lomicky[1] on behalf of the STAR Collaboration

[1]Czech Technical University in Prague, FNSPE, Prague 115 19, Czech Republic

Abstract. In this contribution, we report the measurement of D^0 meson tagged jets in Au+Au collisions at $\sqrt{s_{NN}} = 200$ GeV by the STAR experiment at RHIC. We present the nuclear modification factor R_{CP} as a function of the jet transverse momentum and transverse momentum fraction of the jet, carried by D^0 meson along the jet axis z_{Jet}, and the radial profile of the D^0 meson. These results are compared to theoretical predictions provided by a hybrid transport model. Additionally, we show the raw measurement of several generalized angularities λ_α^κ, which describe the jet substructure. These results may help distinguish between different models describing jet quenching and heavy flavor quark in-medium energy loss.

1 Introduction

The Quark-Gluon Plasma (QGP) produced in heavy-ion collisions can be studied using hard probes, such as c- and b-quarks created at the initial collision stage before the QGP is formed. The heavy quarks lose energy via the radiation of gluons and scattering with other partons present in the QGP. The radiation creates a cone-shaped spray of particles generally called a jet. The presence of the medium broadens the parton shower compared to p+p collisions or peripheral heavy-ion collisions where the creation of the QGP is not expected.

The interaction of quarks with the medium is mass-dependent. This is due to a phenomenon known as the dead-cone effect, predicted by the theory of quantum chromodynamics (QCD) and later observed by ALICE collaboration [1]. The dead-cone effect suppresses the emission of gluons within a cone of angular size m_q/E_q where m_q and E_q are mass and energy of a given quark, respectively.

The broadening of the jets in the medium in Pb+Pb collision to p+p collisions was observed at the LHC. The centrality-dependent modification of the jet shape nuclear modification factor as a function of radial distance from the jet axis reveals the different shower evolution in the presence of the QGP [2]. The other observable characterizing the jet substructure are the so-called generalized jet angularities λ_α^κ where the different choices of κ and α parameters tune the sensitivity of the observable to various jet aspects. These modifications of the jet shape contain information on the mechanism of the energy loss in the medium.

The study of D^0 and $\overline{D}^0$ meson tagged jets, consisting of the related c- and u-quarks in heavy-ion collisions opens a pathway to investigate the aforementioned modifications for the charm quark in the presence of a hot and dense medium.

© The Authors, published by EDP Sciences. This is an open access article distributed under the terms of the Creative Commons Attribution License 4.0 (https://creativecommons.org/licenses/by/4.0/).

2 Analysis details

The D^0 and $\overline{D}^0$ meson candidates were reconstructed from Au+Au collisions at $\sqrt{s_{NN}} = 200$ GeV in central (0-10%), midcentral (10-40%), and peripheral (40-80%) collisions, collected by the STAR experiment at Relativistic Heavy Ion Collider (RHIC) in 2014. The D^0 meson candidates with transverse momentum $1 < p_T < 10$ GeV/c are reconstructed via decay channel $D^0 \rightarrow K^- + \pi^+$ with branching ratio (3.947 ± 0.030) % [3]. The daughter particles were tracked and identified using Time Projection Chamber (TPC), Heavy-Flavor Tracker (HFT), and Time-of-Flight (TOF) detectors. The combinatorial D^0 background contribution is subtracted using a technique called $_s\mathcal{P}lot$ [4], which is part of the native class in RooStats. The inclusive D^0 meson tagged jets are reconstructed with C++ software package FastJet [5] employing anti-k_T algorithm with radius $R = 0.4$ within pseudorapidity range $|\eta| < 1 - R$. The neutral particle contribution was measured by the Barrel Electro-Magnetic Calorimeter (BEMC). The relatively soft background contribution and detector resolution effects are treated using the event-by-event background subtraction and Bayesian unfolding.

3 Nuclear modification factor

The nuclear modification factor for central (C) and peripheral (P) collisions as a function of jet transverse momentum $p_{T,\,Jet}$ is expressed as

$$R_{CP}(p_{T,\,Jet}) = \frac{\langle N_{coll}\rangle_P}{\langle N_{coll}\rangle_C}\frac{N_P^{events}}{N_C^{events}}\frac{d^2N_C/(dp_{T,\,Jet}dy)}{d^2N_P/(dp_{T,\,Jet}dy)}, \tag{1}$$

where $\langle N_{coll}\rangle$ and N^{events} are the corresponding number of nucleon-nucleon binary collisions and number of events, respectively.

The z_{Jet} observable is related to the fragmentation function in the DGLAP equation [6] and is defined as

$$z_{Jet} = \frac{\vec{p}_{T,\,Jet}\cdot\vec{p}_{T,\,D^0}}{|\vec{p}_{T,\,Jet}|^2}, \tag{2}$$

where $p_{T,\,jet}$ and $p_{T,\,D^0}$ are transverse momenta of the jet and D^0 meson, respectively. The nuclear modification factors R_{CP} as a function of $p_{T,\,Jet}$ and z_{Jet} for central and midcentral collision to the peripheral collision are presented in Fig. 1. The results of $R_{CP}(p_{T,\,Jet})$ show a hint of suppression of D^0-jets transverse momenta in central events. For midcentral collisions, the ratio is consistent with unity, which represents no significant impact of the QGP relatively to peripheral collisions. The second nuclear modification factor $R_{CP}(z_{Jet})$ shows a hint of D^0-jet suppression in central collisions, primarily from hard fragmented jets (high z_{Jet}). The ratios for the soft fragmented jets show no significant modification across both centrality ranges.

The charm quark diffusion can be studied by the radial distribution measurement of the D^0 meson inside the jet defined as

$$\Delta r = \sqrt{(\eta_{Jet} - \eta_{D^0})^2 + (\phi_{Jet} - \phi_{D^0})^2}, \tag{3}$$

where η and ϕ are the pseudorapidity and azimuthal angle, respectively. The measured ratios of per-jet normalized distributions $1/N_{Jet}\cdot\Delta r(p_{T,\,Jet})$ of different centralities are shown in Fig. 2 left. The measured ratios do not reveal any significant modification of them at the top RHIC energy of 200 GeV.

The model predictions in Fig. 1 and 2 left represented by magenta areas are provided by the LIDO model which is a hybrid transport model for the evolution of heavy quarks in

© The Authors, published by EDP Sciences. This is an open access article distributed under the terms of the Creative Commons Attribution License 4.0 (https://creativecommons.org/licenses/by/4.0/).

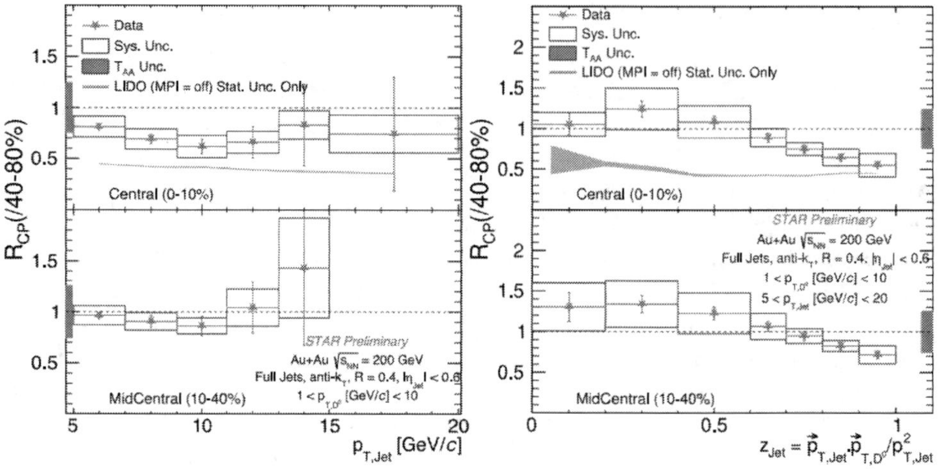

Figure 1. The nuclear modification factor R_{CP} for central (0-10 %) and midcentral (10-40%) collisions relative to peripheral (40-80%) collisions as a function of jet transverse momentum $p_{T,\,Jet}$ (left) and fragmentation function z_{Jet} (right) for D^0 meson tagged jets in Au+Au collisions at $\sqrt{s_{NN}} = 200$ GeV. The magenta areas represent the LIDO model prediction [7].

a QGP medium [7]. The model slightly underpredicts the yield in central collisions but it is consistent with the yield in peripheral events. This results in the underprediction of R_{CP} as a function of $p_{T,\,Jet}$ and z_{Jet}. Regarding the Δr distributions, LIDO qualitatively well explains the radial profile trends in both central and peripheral collisions.

4 Generalized angularities

One of the jet substructure observables that can discriminate between quark- and gluon-initiated jets or mass of the initiated parton are generalized angularities that depend on an angular exponent α and an energy weighting factor κ, defined as

$$\lambda_\alpha^\kappa = \sum_{i\in \mathrm{Jet}} \left(\frac{p_{T,i}}{p_{T,\,jet}}\right)^\kappa \left(\frac{\Delta R_i}{R_{Jet}}\right)^\alpha, \tag{4}$$

where $p_{T,\,Jet}$ and R_{Jet} are transverse momentum and jet radius, and $p_{T,i}$ and ΔR_i are transverse momentum and radial distance from the jet axis of the i-th track-based jet constituent (charged particles + D^0 meson), respectively. These observables are infrared and collinear safe only for sets where $\kappa = 1$ and $\alpha > 0$ [8]. In such cases, their distributions for p+p collisions can be calculated using perturbative QCD. For $\kappa = 1$, lower values of α parameter ($0 < \alpha < 1$) increase sensitivity to mass effects by emphasizing the influence of the dead-cone effect, i.e. contribution of jet constituent particles close to the jet axis. Conversely, higher values of α ($\alpha > 1$) suppresses the mass effects and increases the sensitivity to Casimir color effects.

In general, several generalized angularities have specific names due to their significance, such as $\lambda_{0.5}^1$ (Les Houches Angularity), which is sensitive to flavor discrimination, λ_1^1 (girth or width), which describes jet constituent broadening, and λ_2^1 (mass or thrust), which relates to the mass-squared at fixed jet energy [8]. The raw data results after background subtraction for the λ_1^1 and λ_3^1 in central collisions are shown in Fig. 2 right. The unphysical results, i.e., $\lambda_{\alpha\geq 1}^1 \notin \langle 0;1\rangle$, are caused by the jet background fluctuations and detector resolution. To obtain the true distributions, 2D or higher-dimensional unfolding is required, and work on this is currently ongoing.

© The Authors, published by EDP Sciences. This is an open access article distributed under the terms of the Creative Commons Attribution License 4.0 (https://creativecommons.org/licenses/by/4.0/).

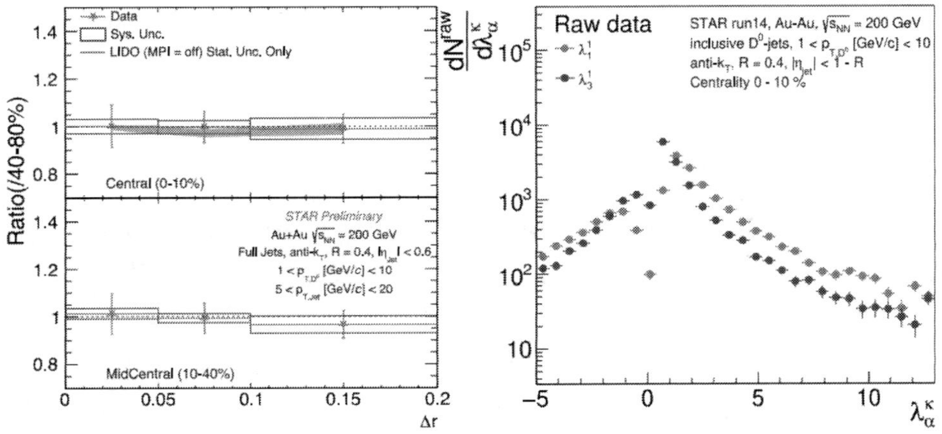

Figure 2. The ratios of the per-jet normalized distributions of radial profiles of D^0 mesons Δr for central (0-10%) and midcentral (10-40%) collisions relative to peripheral (40-80%) collisions (left), and the raw distributions of the generalized angularities λ^1_1 and λ^1_3 for central collisions (right) for D^0 meson tagged jets in Au+Au collisions at $\sqrt{s_{NN}} = 200$ GeV. The magenta area represents the LIDO model prediction [7].

5 Summary

We presented results of nuclear modification factors $R_{CP}(p_{T,\,Jet})$ and $R_{CP}(z_{Jet})$ of D^0 meson tagged jets in Au+Au collisions at $\sqrt{s_{NN}} = 200$ GeV. We observe suppression of D^0-jet transverse momenta in central collisions and hard fragmented jets in central and midcentral collisions. Next, we showed the ratios of the per-jet normalized distributions of radial profiles Δr of D^0 mesons, which, within uncertainties, are consistent with unity at $\sqrt{s_{NN}} = 200$ GeV. All three results were compared with the LIDO model. Overall, this model accurately predicts the trends of p_T, z_{Jet}, and Δr distributions, although it underpredicts the yield in the central collisions.

The future work will focus on calculation of D^0-jet angularity modification in central heavy-ion collisions in the form of $R_{CP}(\lambda^{\kappa}_{\alpha})$, to reveal the jet substructure modification in the presence of the QGP medium. In order to draw physical conclusions, unfolding, D^0 meson reconstruction efficiency, and other factors need to be included. This measurement may help to constrain several model predictions focused on the energy loss of partons in the QGP.

Acknowledgement: This work was supported by a grant from The Czech Science Foundation, grant number: 23-07499S.

References

[1] ALICE Collaboration, Nature **605**, 440–446 (2022).
[2] CMS Collaboration, Phys. Lett. B **730**, 243-263, (2014).
[3] R. L. Workman et al. (PDG), Prog. Theor. Exp. Phys. **2022**, 083C01 (2022).
[4] M. Pivk, F. R. Le Diberder, Nucl. Instrum. Methods Phys. Res. A **555**, 356-369 (2005).
[5] M. Cacciari, G. P. Salam, G. Soyez, Eur. Phys. J. C **72**, 1896 (2012).
[6] G. Altarelli, G. Parisi, Nucl. Phys. B **126**, 298–318 (1977).
[7] W. Ke, Y. Xu, S. A. Bass, Phys. Rev. C **98**, 064901 (2018).
[8] A. J. Larkoski, J. Thaler, W. J. Waalewijn, High Energ. Phys. **2014**, 129 (2014).

© The Authors, published by EDP Sciences. This is an open access article distributed under the terms of the Creative Commons Attribution License 4.0 (https://creativecommons.org/licenses/by/4.0/).

R_{AA} and v_n: relativistic transport approach for charm and bottom toward a more solid phenomenological determination of $D_s(T)$

Maria Lucia Sambataro[1,2*], *Salvatore* Plumari[1,2], *Vincenzo* Minissale[1,3], *Gabriele* Parisi[1,2], and *Vincenzo* Greco[1,2]

[1]Department of Physics and Astronomy, University of Catania, Via S. Sofia 64, 1-95123 Catania, Italy
[2]Laboratori Nazionali del Sud, INFN-LNS, Via S. Sofia 62, I-95123 Catania, Italy
[3]INFN Sezione di Catania, Via Santa Sofia, 64 - 95123 Catania, Italy

Abstract. Quasi-Particle Model (QPM) allows for a good description of the main features of thermodynamics computed via lattice QCD (lQCD) and for charm quark supplies a satisfying description of R_{AA} and $v_{2,3}$ of D mesons. Within an event-by-event full Boltzmann transport approach followed by a hybrid hadronization via coalescence plus fragmentation, we investigate the extension to bottom quark dynamics discussing predictions for R_{AA} and $v_{2,3}$ of B mesons comparing to the available experimental data by ALICE collaboration on electrons from B meson decay. Furthermore, we extend our QPM approach to a more realistic model in which partonic propagators explicitly depend on quark momentum (QPMp) following Dyson-Schwinger studies. The QPM_p improves the description of quark susceptibilities computed via lQCD, underestimated in the standard QPM approach and entails a D_s with a stronger non-perturbative behaviour near to T_c which leads to a better agreement with the recent lQCD data.

1 Introduction

In the investigation of Quark-Gluon Plasma (QGP), charm and bottom quarks are regarded as highly effective probes for observing the entire evolution of the system formed in ultra-relativistic heavy ion collisions (uRHIC) [1, 2]. These particles are generated out of equilibrium during the initial stages of the collisions via pQCD processes and due to their large mass, their thermalization time, in particular for charm quarks, is comparable to the lifetime of the QGP itself. Within this framework, two observables have been widely utilized as probes of QGP: the nuclear suppression factor R_{AA} and v_2 [3, 4]. Specifically, the R_{AA} provides a quantitative estimation of the interaction between heavy quarks and the bulk medium, defined as the ratio of the spectra of heavy flavor hadrons observed in nucleus-nucleus collisions with respect to proton-proton collisions. On the other hand, the v_2 is related to the anisotropy in the particle angular distribution due to the HQ partecipation in the collective motion of the medium. We have developed an event-by-event transport model that combines hybrid coalescence and fragmentation hadronization, also incorporating initial state fluctuations to investigate the HQs main observables in Pb+Pb collisions obtaining a good description of of

*e-mail: sambataro@lns.infn.it

© The Authors, published by EDP Sciences. This is an open access article distributed under the terms of the Creative Commons Attribution License 4.0 (https://creativecommons.org/licenses/by/4.0/).

R_{AA} and $v_{2,3}$ of D mesons [5–12]. The non perturbative interaction has been taken into account within a Quasi-particle model (QPM) with thermal masses which gives the possibility to quantitatively describe the Equation of State computed via lattice QCD (lQCD) even in the vicinity of phase transition [13, 14]. Recently, our approach has been extended to investigate the key observables in the bottom quark sector, offering predictions for centralities where experimental data are not yet available [15]. Furthermore, in order to simultaneously describe the Equation of State (EoS) from lQCD and the quark number susceptibilities, which are understimated in the standard QPM, we explore the possibility of using massive quasi particles that explicitly depend on the three-momentum of the medium approaching to the current quark masses predicted by perturbative QCD (pQCD) at high momenta [16]. This model extension, which in the following is named QPM_p, also incorporate the latest lattice results for the QCD Equation of State with 2 + 1 + 1 dynamical flavors (including the charm quark in the bulk). In this context, we estimate the heavy quark (HQs) spatial diffusion coefficient $D_s(T)$ especially in the infinite mass limit in order to ensure the same condition of the lQCD data.

2 Transport equation and extended Quasi-Particle model

In order to investigate the evolution of both the bulk and heavy quarks in the QGP, we solve the relativistic Boltzmann equations:

$$\{p_k^\mu \partial_\mu + m^*(x)\partial_\mu m^*(x)\partial_p^\mu\} f_k(x, p_k) = C[f_q, f_g](x, p_k) \tag{1}$$

$$p^\mu \partial_\mu f_Q(x, p) = C[f_q, f_g, f_Q](x, p) \tag{2}$$

Here, $C[f_q, f_g, f_Q](x, p)$ represents the relativistic Boltzmann-like collision integral, with $f_k(x, p)$ the on-shell phase space one-body distribution function for the k-th parton. We write the Boltzmann equation for bulk evolution in the *Mean Field Approximation* viewing the QPM masses as arising from a scalar field which has the effect to generate effective space-time dependent masses m*(x). The collision integral is also gauged to fixed $\eta/s \approx 0.1$, matching viscous hydrodynamics. In our calculations, we distribute the plasma particles in the coordinate space according to a Monte-Carlo Glauber model. The momentum distribution of light partons is a Boltzmann-Juttner distribution for momenta up to $p_T = 2\ GeV$, above this momentum we also include mini-jet production distributed according to pQCD calculation at NLO. Furthermore, we use FONLL distributions in momentum space for heavy quark. For further details about the initial conditions, see Refs. [12, 15]. To account for the non-perturbative effects in bulk-HQs scattering, we evaluate the collision integral using the QPM approach in which the coupling g is a temperature-dependent function, with its parameterized form derived by fitting to the energy density obtained in lQCD. For additional details, refer to Ref. [13]. Recently, new lattice results for the QCD equation of state with 2+1+1 dynamical flavors have been released [17]. Therefore, we extend our QPM from 2 + 1 to 2 + 1 + 1 flavors, incorporating the charm quark and, additionally, we investigate an upgraded version of the QPM, referred to as QPM_p, where partonic propagators explicitly depend on three-momentum to provide a more realistic model. Building on Ref. [18, 19], the key innovations in this extended approach compared to the standard QPM include the incorporation of momentum-dependent factors in the masses, inspired by Dyson-Schwinger studies. This adjustment ensures that the model approaches the pQCD limit as $p \to \infty$. The functional forms for the parton masses as functions of finite temperature T, quark chemical potential μ_q and the squared three-momentum p^2 can be found in Ref. [16, 18, 19].

© The Authors, published by EDP Sciences. This is an open access article distributed under the terms of the Creative Commons Attribution License 4.0 (https://creativecommons.org/licenses/by/4.0/).

3 Model results

Within the standard QPM, we firstly present our predictons for B meson R_{AA} at 0-10 % centrality class (left plot of Fig. 1) and $v_{2,3}$ at 30-50 % centrality class (right plot of Fig. 1). In particular, we can compare result for electrons from semi-leptonic B meson decay obtaining a good agreement with the available experimental data by ALICE collaboration. Further details are provided in Ref. [15], where we also present our prevision for the v_3 at top LHC energies whose corresponding experimental data are not yet available.

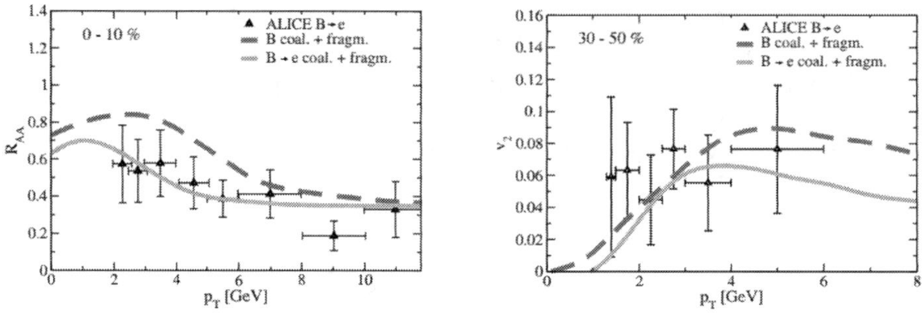

Figure 1. Nuclear modification factor $R_{AA}(p_T)$ (left panel) and elliptic flow $v_2(p_T)$(right panel) for B mesons and electrons from B mesons decay in PbPb collisions at $\sqrt{s} = 5.02\ ATeV$ for different centrality class. The data for electrons by ALICE collaboration are taken from Ref. [20, 21]

Moving on the QPM extension, in the left panel of Fig. 2 we present our result on light quark susceptibility $\chi_u = \frac{T}{V}\frac{\partial^2 lnZ}{\partial\mu^2}$ making a comparison between the standard QPM with 2+1 flavours which clearly undestimate the lQCD data and the QPM_p with 2+1+1 flavors which instead gives a good description of the data. In the calculation, we use a constant mass for charm quark $m_c = 1.5\,GeV$. For more details about quark suscetibilities in the QPM_p see Ref. [16] where also charm susceptibility is discussed for different temperature parametrization of charm mass. The most significant result of this work, is presented in the right plot of Fig. 2 where we compare the temperature dependence of $D_s(T)$ in both QPM and QPM_p to lQCD data. Our results indicate that the QPM_p, which correctly describe both EoS and susceptibilities, shows considerably better agreement with the recent lQCD data from Ref. [22] which include dynamical fermions, compared to the standard QPM.

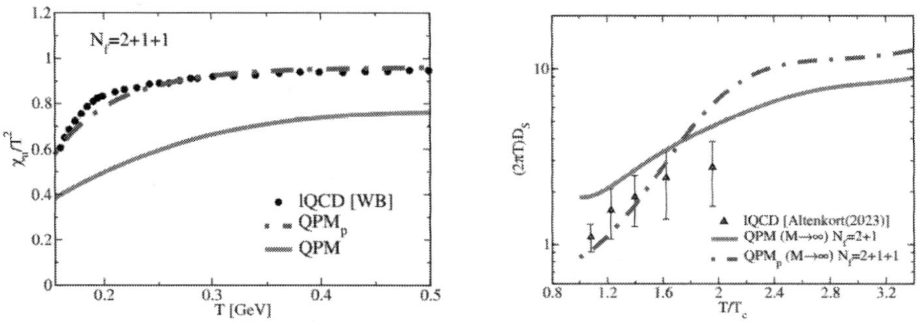

Figure 2. (left plot) Light quark susceptibility $\chi_u = \frac{T}{V}\frac{\partial^2 lnZ}{\partial\mu^2}$ in both QPM and QPM_p with $N_f = 2+1+1$ flavors compared to lQCD data from Ref. [17]. (right plot) The spatial diffusion coefficient $D_s(T)$ plotted as a function of T/T_c in both QPM and QPM_p in the infinite mass limit for HQs. The data points represent lQCD results taken from ref. [22].

© The Authors, published by EDP Sciences. This is an open access article distributed under the terms of the Creative Commons Attribution License 4.0 (https://creativecommons.org/licenses/by/4.0/).

4 Conclusion

We have examined the propagation of bottom quark in the QGP at top LHC energies using a relativistic, event-by-event Boltzmann transport approach which incorporates non-perturbative interaction effects through the standard QPM. We find a good agreement between our result on $R_{AA}(p_T)$ and $v_2(p_T)$ for electrons from B meson decay with the available experimental data by ALICE collaboration. Furthermore, we have explored a possible extension of the QPM approach to a more realistic model, where partonic masses explicitly depend on the three-momentum of plasma particles, allowing the model to reach the pQCD limit in the high-momentum regime. Given that the QPM_p more accurately describes the latest lattice QCD data for EoS, susceptibilities and the $D_s(T)$ in this model approaches to the new lQCD data in a better agreement with respect to the standard QPM, it will be important to investigate whether it offers an improved description of the main heavy flavour observables within a realistic simulation.

References

[1] X. Dong and V. Greco, Prog. Part. Nucl. Phys. **104** (2019), 97-141.
[2] M. He, H. van Hees and R. Rapp, Prog. Part. Nucl. Phys. **130** (2023), 104020.
[3] B. I. Abelev *et al.* [STAR], Phys. Rev. Lett. **98**, 192301 (2007) [erratum: Phys. Rev. Lett. **106**, 159902 (2011)].
[4] A. Adare *et al.* [PHENIX], Phys. Rev. Lett. **98**, 172301 (2007).
[5] S. Plumari, G. L. Guardo, F. Scardina and V. Greco, Phys. Rev. C **92**, no.5, 054902 (2015).
[6] F. Scardina, S. K. Das, V. Minissale, S. Plumari and V. Greco, Phys. Rev. C **96**, no.4, 044905 (2017).
[7] S. Plumari, G. Coci, V. Minissale, S. K. Das, Y. Sun and V. Greco, Phys. Lett. B **805**, 135460 (2020)
[8] L. Oliva, S. Plumari and V. Greco, JHEP **05**, 034 (2021)
[9] S. Plumari, A. Puglisi, F. Scardina and V. Greco, Phys. Rev. C **86**, 054902 (2012)
[10] Y. Sun, S. Plumari and V. Greco, Eur. Phys. J. C **80**, no.1, 16 (2020).
[11] S. Plumari, Eur. Phys. J. C **79**, no.1, 2 (2019).
[12] M. L. Sambataro, Y. Sun, V. Minissale, S. Plumari and V. Greco, Eur. Phys. J. C **82**, no.9, 833 (2022).
[13] S. Plumari, W. M. Alberico, V. Greco and C. Ratti, Phys. Rev. D **84** (2011), 094004.
[14] M. L. Sambataro, S. Plumari and V. Greco, Eur. Phys. J. C **80**, no.12, 1140 (2020).
[15] M. L. Sambataro, V. Minissale, S. Plumari and V. Greco, Phys. Lett. B **849**, 138480 (2024).
[16] M. L. Sambataro, V. Greco, G. Parisi and S. Plumari, Eur. Phys. J. C **84**, no.9, 881 (2024).
[17] S. Borsanyi *et al.* Nature **539**, no.7627, 69-71 (2016).
[18] H. Berrehrah, W. Cassing, E. Bratkovskaya and T. Steinert, Phys. Rev. C **93**, no.4, 044914 (2016).
[19] H. Berrehrah, E. Bratkovskaya, T. Steinert and W. Cassing, Int. J. Mod. Phys. E **25**, no.07, 1642003 (2016).
[20] S. Acharya *et al.* [ALICE], Phys. Rev. C **108**, no.3, 034906 (2023).
[21] S. Acharya *et al.* [ALICE], Phys. Rev. Lett. **126**, no.16, 162001 (2021).
[22] L. Altenkort *et al.* [HotQCD], Phys. Rev. Lett. **130**, no.23, 231902 (2023).

© The Authors, published by EDP Sciences. This is an open access article distributed under the terms of the Creative Commons Attribution License 4.0 (https://creativecommons.org/licenses/by/4.0/).

J/ψ photoproduction and polarization in peripheral Pb–Pb collisions with ALICE

Dukhishyam Mallick for the ALICE Collaboration[1,*]

[1]IJCLab Orsay, CNRS/IN2P3, Université Paris-Saclay, France

Abstract. Ultrarelativistic heavy-ion collisions generate a powerful electromagnetic field that produces photonuclear reactions. These processes have been extensively studied in ultraperipheral collisions, in which the impact parameter is larger than twice the nuclear radius. Recently, coherent photoproduction of J/ψ has been observed in nucleus–nucleus (A–A) collisions with nuclear overlap, based on the measurement of an excess of production with respect to hadron-production expectations in the very low transverse momentum (p_T). Such quarkonium measurements provide valuable information about the nuclear gluon distribution at low Bjorken x and high energy. In addition, they can constrain the theory describing photon-induced reactions in A–A collisions with nuclear overlap, including possible interactions of the measured probes with the formed and fast expanding Quark-Gluon Plasma. In order to confirm the photoproduction origin of the very low p_T J/ψ yield excess, polarization measurements represent a suitable observable. It is expected that the produced quarkonium inherits the polarization of the incoming photon because of the s-channel helicity conservation. ALICE can measure inclusive and exclusive quarkonium production down to $p_T = 0$, at forward rapidity ($2.5 < y < 4$) and midrapidity ($|y| < 0.9$). In this contribution, new preliminary measurement of the y-differential cross section and the first new polarization analysis at the LHC of coherently photoproduced J/ψ in peripheral Pb–Pb collisions will be presented. These measurements are conducted at forward rapidity in the dimuon decay channel. These results will be discussed together with the recent results on coherent photoproduction as a function of centrality at both mid and forward rapidities. Comparison with models will be shown whenever available.

1 Introduction

Photon-induced processes play an important role in ultraperipheral collisions (UPCs), i.e., collisions in which nuclei are separated by an impact parameter (b) larger than the sum of their radii. In such collisions, hadronic interactions are strongly suppressed, and the cross section of produced vector mesons (VM, e.g., J/ψ) is expected to be significant due to the strong electromagnetic field of the nuclei [1]. In this process, a photon emitted by one of the colliding nuclei fluctuates into a $q\bar{q}$ pair (acts as a color dipole), which can interact with the second nucleus via two gluon exchange at Leading Order (LO). The color dipole finally recombines into a VM. J/ψ photoproduction is one of the cleanest experimental tools for probing the gluon distributions inside the nucleus [2], which are poorly known at low

*e-mail: dukhishyam.mallick@cern.ch

© The Authors, published by EDP Sciences. This is an open access article distributed under the terms of the Creative Commons Attribution License 4.0 (https://creativecommons.org/licenses/by/4.0/).

Bjorken-x [3]. Photonuclear reactions in UPCs have been extensively studied in ALICE (see for instance a recent measurement in Ref. [4]).

The first experimental signature of photonuclear reactions in peripheral Pb–Pb collisions (PCs) at $\sqrt{s_{\rm NN}}$ = 2.76 TeV was reported by the ALICE Collaboration via the measurement of an excess in the J/ψ yield at very low $p_{\rm T}$ with respect to expectations from hadroproduction [5]. Similar observations were confirmed in Ref. [6, 7], interpreting the excess as the evidence of coherent J/ψ photoproduction in Pb–Pb collisions with nuclear overlap. Recently, the ALICE Collaboration measured Pb–Pb collisions at $\sqrt{s_{\rm NN}}$ = 5.02 TeV and observed a significant J/ψ yield excess at low $p_{\rm T}$ in collisions with nuclear overlap, even in the 30–50% centrality range [8]. Theoretical models describing coherent photoproduction, considering effective photon flux and photonuclear cross sections, were compared with these measurements [9–12]. Polarization measurements of coherently photoproduced VM are important, according to the s-channel helicity conservation (SCHC) hypothesis, the VM should retain the photon's transverse polarization [13]. Recent J/ψ polarization measurements in UPCs [14] confirmed consistency with SCHC. In these proceedings, we present the first rapidity (y) differential cross section measurement of coherent J/ψ photoproduction, and the first polarization measurement of inclusive J/ψ for $p_{\rm T}$ < 0.3 GeV/c, at forward rapidity (2.5 < y < 4.0) in the 70–90% centrality class in Pb–Pb collisions at $\sqrt{s_{\rm NN}}$ = 5.02 TeV.

2 Analysis details

A detailed description of the ALICE detector is provided in Ref. [15]. This analysis measures J/ψ production at forward rapidity using opposite-sign dimuon pairs in the 70–90% centrality range for Pb–Pb collisions at $\sqrt{s_{\rm NN}}$ = 5.02 TeV. The data, collected in 2015 and 2018, corresponds to an integrated luminosity of 756 ± 19 μb^{-1}. The J/ψ polarization is studied with the angular distribution of opposite sign (OS) dimuon pairs. The polarization is measured in the helicity frame [16], and dimuon angular distribution are expressed as :

$$W(\cos\theta,\varphi) \propto \frac{1}{3+\lambda_\theta}[1+\lambda_\theta\cos^2\theta+\lambda_\phi\sin^2\theta\cos(2\phi)+\lambda_{\theta\phi}\sin(2\theta)\cos(\phi)]. \quad (1)$$

If (λ_θ, λ_ϕ, $\lambda_{\theta\phi}$) = (0, 0, 0), the particle is unpolarized, if (λ_θ, λ_ϕ, $\lambda_{\theta\phi}$) = (-1, 0, 0), the particle has a longitudinal polarization, and if (λ_θ, λ_ϕ, $\lambda_{\theta\phi}$) = (1, 0, 0), the particle is transversely polarized. In both y-differential cross section and polarization analyses, J/ψ is reconstructed from the invariant mass of opposite-sign dimuon pairs. The signal extraction is performed by fitting the invariant mass distribution with a combined fit function that describes the signal and background components. A likelihood approach is adopted during the signal extraction due to the low statistics.

3 Results and discussion

The J/ψ $p_{\rm T}$ spectra are evaluated for six rapidity intervals in the centrality range 70–90% in Pb–Pb collisions at $\sqrt{s_{\rm NN}}$ = 5.02 TeV. An excess in the J/ψ raw yield is observed for $p_{\rm T}$ < 0.3 GeV/c in all y intervals with respect to hadroproduction yield. The contribution from hadronically produced J/ψ is subtracted using a data-driven model, as discussed in Ref. [8]. Figure 1 shows the coherent photoproduction cross section as a function of y for $p_{\rm T}$ < 0.3 GeV/c in the 70–90% centrality range for Pb–Pb collisions at $\sqrt{s_{\rm NN}}$ = 5.02 TeV. The data are compared with model predictions: GG-hs (hot-spot model using the Gribov–Glauber approach) [10], Zha (photon–pomeron coupling scenario) [11], and GBW/IIM (color dipole model) [12]. In the right panel of Fig. 1, three GBW/IIM scenarios are compared to data. The

© The Authors, published by EDP Sciences. This is an open access article distributed under the terms of the Creative Commons Attribution License 4.0 (https://creativecommons.org/licenses/by/4.0/).

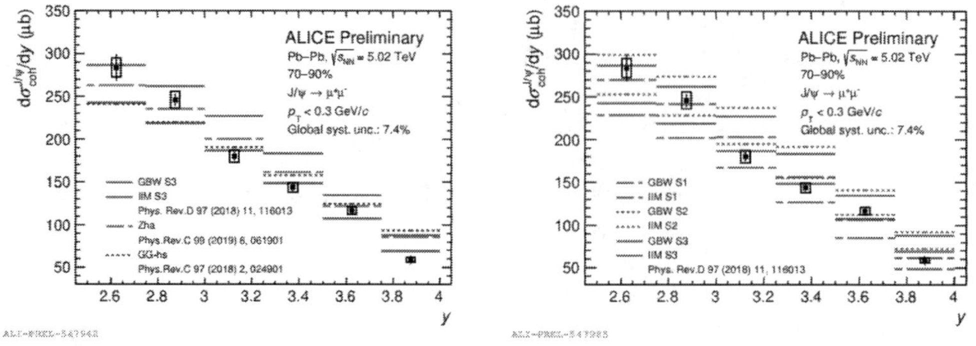

Figure 1. Cross section of coherently photoproduced J/ψ with $p_T < 0.3$ GeV/c as a function of y in the centrality range 70–90% in Pb–Pb collisions at $\sqrt{s_{NN}} = 5.02$ TeV. Results are compared with model predictions from Refs. [10–12].

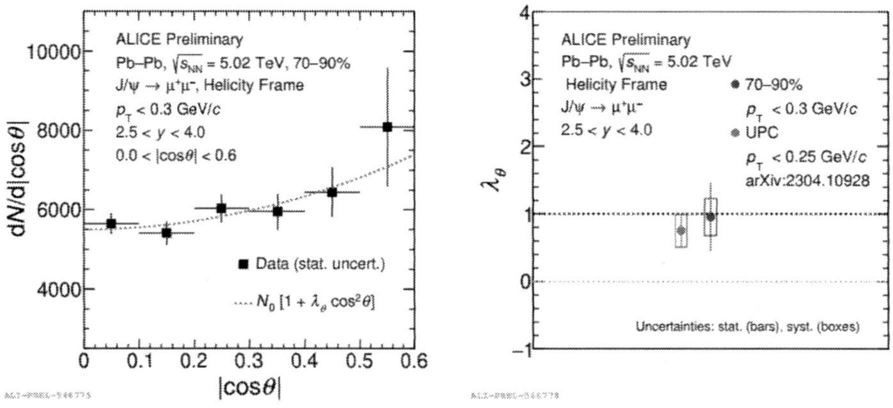

Figure 2. Left: Distribution of J/ψ cosθ in the helicity frame measured in 70–90% Pb–Pb collisions at $\sqrt{s_{NN}} = 5.02$ TeV. Reconstructed J/ψ were having $p_T < 0.3$ GeV/c and $2.5 < y < 4.0$. The distribution is fitted with Eq. 1 to extract λ_θ. Right: Comparison of the λ_θ from UPCs (red marker) and PCs (blue marker) in Pb–Pb collisions at $\sqrt{s_{NN}} = 5.02$ TeV.

S1 scenario does not differ from UPCs, S2 uses an effective photon flux outside the overlap region, and S3 is the same photon flux as S2 and a change in the effective photonuclear cross section to exclude the nuclear overlap. While all models reproduce the overall magnitude, they fail to capture the y-dependence of the photoproduced J/ψ. Similar results were seen in UPC measurements [17]. Figure 2 shows the cosθ angular distribution of J/ψ for $p_T <$ 0.3 GeV/c in Pb–Pb collisions with 70–90% centrality. A dependence on cosθ is observed and the distribution is fitted with Eq. 1 to extract the λ_θ parameter. The λ_θ value in PCs is compared with the value obtained in UPCs, as shown in the right panel of Fig. 2. The results suggest that inclusive J/ψ with $p_T < 0.3$ GeV/c have a transverse polarization and therefore obey the SCHC hypothesis as observed in UPC (see Ref. [14]). Even without subtracting the hadronic J/ψ contribution, this polarization measurement can be regarded as a

© The Authors, published by EDP Sciences. This is an open access article distributed under the terms of the Creative Commons Attribution License 4.0 (https://creativecommons.org/licenses/by/4.0/).

proxy for the photoproduced J/ψ polarization as it is established from Ref. [17] that coherent J/ψ photoproduction constitutes 78% of the inclusive J/ψ yield in this kinematic region. This polarization measurement serves as an additional method to test the J/ψ production mechanisms at very low p_T, and it is consistent with the picture of a coherent photoproduction origin for the measured J/ψ yield excess.

4 Summary

We report the first measurement of the y-differential coherent J/ψ photoproduction cross section and polarization for $p_T < 0.3$ GeV/c at forward rapidity in the 70–90% centrality range for Pb–Pb collisions at $\sqrt{s_{NN}} = 5.02$ TeV. A significant J/ψ yield excess is observed at low p_T in peripheral collisions, interpreted as coherent photoproduction. The y-dependence of the cross section is measured and compared with models, which qualitatively match the data but fail to describe the data across the entire y range. An angular $\cos\theta$ dependence is observed for inclusive J/ψ polarization measurement using the helicity frame for peripheral collisions, and the λ_θ value is compatible with unity within uncertainties. This result is consistent with a transverse polarization scenario and with the UPC results.

These measurements remain a challenge for theoretical studies since the implementation of coherent photoproduction in nuclear collisions is complex and all the aspects should be considered. The J/ψ cross section measurements as a function of y in UPCs and PCs, will be used for the understanding of the photon energy ambiguity as discussed in Ref. [18]. In Run 3 more differential and precise measurements of the coherent J/ψ photoproduction cross section (as well as other VMs) will be achieved in more central collisions, together with polarization measurements.

References

[1] C. A. Bertulani et al., *Ann. Rev. Nucl. Part. Sci. 55, 271–310 (2005).*
[2] V. Rebyakova, M. Strikman, and M. Zhalov, *Phys. Lett. B 710, 647 (2012).*
[3] K. J. Eskola, H. Paukkunen, and C. A. Salgado, *JHEP 04, 065 (2009).*
[4] S. Acharya et al., ALICE Collaboration, *Phys. Lett. B 817, 136280, (2021).*
[5] J. Adam et al., ALICE Collaboration, *Phys. Rev. Lett. 116, 222301 (2016).*
[6] J. Adam et al., STAR Collaboration, *Phys. Rev. Lett. 123, 132302 (2019).*
[7] R. Aaij et al., LHCb Collaboration, *Phys. Rev. C 105, L032201 (2022).*
[8] S. Acharya et al., ALICE Collaboration, *Phys. Lett. B 846, 137467 (2023).*
[9] M. Kłusek-Gawenda et al., *Phys. Rev. C 93, 044912 (2016).*
[10] J. Cepila et al., *Phys. Rev. C 97, 024901 (2018).*
[11] W. Zha et al., *Phys. Rev. C 97, 044910 (2018).*
[12] M. Gay Ducati et al., *Phys. Rev. D 97, 116013 (2018).*
[13] F. J .Gilman et al., *Phys. Lett. B 31, 387–390 (1970).*
[14] S. Acharya et al., ALICE Collaboration, *arXiv:2304.10928, (2023).*
[15] K. Aamodt et al., ALICE Collaboration, *JINST 3, S08002 (2008).*
[16] P. Faccioli et al., *Eur. Phys. J. C 69, 657–673 (2010).*
[17] S. Acharya et al., ALICE Collaboration, *Phys. Lett. B 798, 134926 (2019).*
[18] J. G. Contreras, *Phys. Rev. C 96, 015203 (2017).*

© The Authors, published by EDP Sciences. This is an open access article distributed under the terms of the Creative Commons Attribution License 4.0 (https://creativecommons.org/licenses/by/4.0/).

EPJ Web of Conferences **316**, 04016 (2025) https://doi.org/10.1051/epjconf/202531604016
SQM 2024

Charm and beauty production in hadronic collisions via muon measurements at forward rapidity with ALICE

Maolin Zhang[1,2] on behalf of the ALICE Collaboration,*

[1]Institute Of Particle Physics, CCNU-Wuhan, China
[2]Université Clermont-Auvergne, CNRS, LPCA, Clermont-Ferrand, France

Abstract. The latest results from the ALICE Collaboration on the measurements of open heavy flavours in the (di)muon channel at forward rapidity are presented. In the LHC Run 3, the new Muon Forward Tracker (MFT) adds vertex capabilities to the muon spectrometer, providing a unique way to discriminate muons from charm- and beauty-hadron decays. The analysis strategy is discussed and the status of the measurements in pp collisions at $\sqrt{s}$ = 13.6 TeV collected with the upgraded ALICE detector is reported.

1 Introduction

Heavy quarks (charm and beauty) are excellent probes to study the properties of the Quark-Gluon Plasma (QGP), a state of strongly-interacting matter at high temperature and energy density, formed in high-energy heavy-ion collisions. Indeed, due to their large masses, heavy quarks are produced in hard partonic scattering processes before the formation of the QGP, and they experience the full evolution of the hot and dense nuclear medium. Charm and beauty measurements in small collision systems, pp and p–Pb, serve as an important test of perturbative quantum chromodynamics (pQCD). They also represent the baseline to study the QGP and quantify hot-medium effects in heavy-ion collisions and provide the possibility to investigate cold nuclear matter effects. Measurements in these small collision systems gained additional interest with the observation, in high-multiplicity collisions, of final-state effects typically attributed to the QGP in Pb–Pb collisions. The origin of these effects in small collision systems is still under investigation.

Open heavy-flavour hadrons are measured with ALICE through the hadronic decay channels and with electrons from semielectronic decays of charm and/or beauty at midrapidity ($|y| < 0.8$), and in the semimuonic decay of charm and beauty hadrons at forward rapidity ($2.5 < y < 4.0$). A detailed description of the ALICE apparatus and its performance can be found in Refs. [1, 2]. In the following, a particular emphasis is placed on the recent measurements at forward rapidity using the muon spectrometer. These include the measurements of the muon azimuthal anisotropies in p–Pb collisions at $\sqrt{s_{\rm NN}}$ = 8.16 TeV, and the charm and beauty production cross sections in pp collisions at $\sqrt{s}$ = 13 TeV via dimuons.

A major upgrade of the ALICE apparatus has been successfully carried out for the LHC Run3 [3]. ALICE is equipped with several new subdetectors and the readout of all existing subdetectors is upgraded due to the continuous readout. In particular, the new Muon Forward

*e-mail: Maolin.Zhang@cern.ch

© The Authors, published by EDP Sciences. This is an open access article distributed under the terms of the Creative Commons Attribution License 4.0 (https://creativecommons.org/licenses/by/4.0/).

Tracker (MFT) provides vertexing to the muon spectrometer and allows us to distinguish muons from charm- and beauty-hadron decays in the single muon channel. The MFT analysis procedure is described and the status of the analysis of pp collisions at $\sqrt{s}$ = 13.6 TeV is presented.

2 Azimuthal anisotropies with muons in high-multiplicity p–Pb collisions at $\sqrt{s_{\rm NN}}$ = 8.16 TeV

The second-order Fourier coefficient of the inclusive muon azimuthal distribution (v_2) which quantifies the azimuthal anisotropy of produced muons, is studied as a function of transverse momentum $p_{\rm T}$ in high-multiplicity (0–20%) p–Pb collisions at both forward and backward rapidities using two-particle cumulants and forward-central two-particle correlations [4]. The results are obtained in a wide $p_{\rm T}$ interval, $0.5 < p_{\rm T} < 10$ GeV/c, where for $p_{\rm T} > 2$ GeV/c the contribution of muons from heavy-flavour hadron decays is dominant. Figure 1 (left) shows the $p_{\rm T}$-differential muon v_2 at forward rapidity after the subtraction of nonflow effects. The results from two-particle correlations and two-particle cumulants are consistent within uncertainties. A positive v_2 is measured with a significance reaching values of 4.6–12σ at intermediate $p_{\rm T}$ ($2 < p_{\rm T} < 6$ GeV/c). A comparison with the string-melting version of A MultiPhase Transport (AMPT) model [5, 6] and the colour glass condensate (CGC) predictions [7] is also displayed in Fig. 1 (middle and right). The anisotropies are generated from parton interactions in the early stage of the collisions in the CGC framework. The AMPT model addresses non-equilibrium dynamics and provides a microscopic evolution of parton interactions. The AMPT generates a positive v_2 for all particles and is in fair agreement with the measured inclusive muon v_2, suggesting that the v_2 is driven by the anisotropic parton escape mechanism. In the high-$p_{\rm T}$ region where the contribution of muons from heavy-flavour hadron decays dominates, both AMPT and CGC-based calculations describe the measured inclusive muon v_2. That indicates that possible contributions from initial-state effects are not excluded.

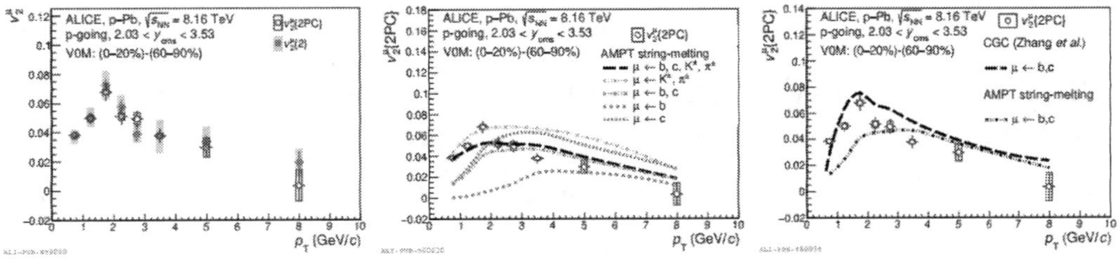

Figure 1. Left: inclusive muon v_2 measured at forward rapidity (p-going) in high-multiplicity p–Pb collisions at $\sqrt{s_{\rm NN}}$ = 8.16 TeV with two-particle cumulants ($v_2^{\mu}\{2{\rm PC}\}$) and two-particle correlations ($v_2^{\mu}\{2\}$) . Vertical bars and boxes represent statistical and systematic uncertainties. Middle: comparison with AMPT calculations. Right: comparison with AMPT and CGC-based calculations.

3 Charm and beauty production cross section at forward rapidity in the dimuon channel

During the LHC Run 2, differential measurements of the heavy-flavour production cross sections have been performed at forward rapidity in the single muon channel without separating the charm and beauty contributions (see Ref. [8] and references therein). Recently, ALICE also measured, for the first time, the $p_{\rm T}$-integrated charm and beauty production cross sections

© The Authors, published by EDP Sciences. This is an open access article distributed under the terms of the Creative Commons Attribution License 4.0 (https://creativecommons.org/licenses/by/4.0/).

at forward rapidity via the unlike-sign muon pair production in pp collisions at $\sqrt{s}$ = 13 TeV. The procedure consists in exploiting the high-mass region of the invariant mass distribution of $\mu^+\mu^-$ pairs ($m_{\mu^+\mu^-}$), namely the continuum region between the charmonium and bottomonium resonances ($4 < m_{\mu^+\mu^-} < 9$ GeV/c^2) which is dominated by the semimuonic decays of hadron pairs containing charm or beauty quarks. In this region, the combinatorial background (light-hadron decays) is found to be negligible. A simultaneous unbinned fit of the measured invariant mass and transverse momentum of $\mu^+\mu^-$ pairs ($p_{\mathrm{T},\,\mu^+\mu^-}$) up to $p_{\mathrm{T},\,\mu^+\mu^-}$ = 10 GeV/c with a cocktail of the different heavy-flavour components is performed to estimate the charm and beauty production cross sections. The templates are obtained from PYTHIA 8 Monte Carlo simulations. Figure 2 presents the measured charm and beauty production cross sections at forward rapidity in pp collisions at $\sqrt{s}$ = 13 TeV. The results at forward rapidity complement previous measurements by ALICE in the dielectron channel. The FONLL predictions [9] with CTEQ6 parton distribution functions are compatible with data within the experimental and theoretical uncertainties, although the measured charm (beauty) production cross section lies at the upper (lower) limit of the model calculations.

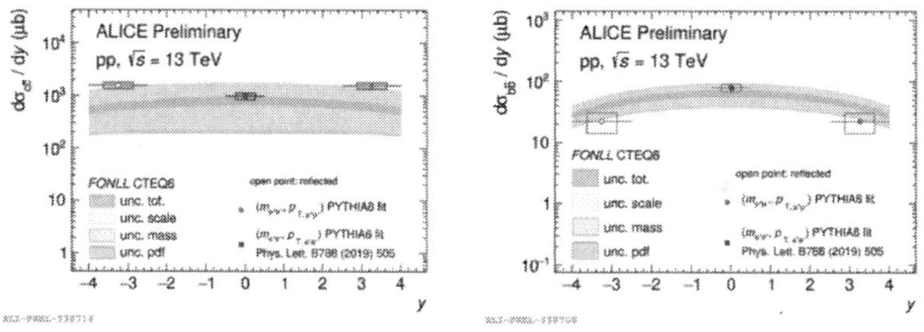

Figure 2. Charm (left) and beauty (right) production cross sections measured at forward rapidity via unlike-sign dimuons in pp collisions at $\sqrt{s}$ = 13 TeV. The vertical bars represent the statistical uncertainties and are smaller than the symbols, while the empty boxes correspond to the systematic uncertainties. The published results obtained at midrapidity in the dielectron channel are also reported. The production cross sections are compared with FONLL calculations.

4 Charm and beauty production at forward rapidity via the semimuonic decay channel in pp collisions at $\sqrt{s}$ = 13.6 TeV

In the Run 3, the presence of the MFT, designed to add vertexing to the muon spectrometer enhances the capabilities of ALICE for measuring muons in the forward rapidity. This new subdetector improves the tracking performance of the muon spectrometer and adds the possibility to separate muons originating from charm- and beauty-hadron decays in the single muon channel. The inclusive muon p_{T} distribution obtained at forward rapidity with a fraction of the pp sample collected in 2022, presented in Fig. 3 (left), indicates that muons can be measured in a wide p_{T} range and study the region where muons from W-boson decays are clearly observed. This is very promising to study the charm/beauty production with high statistical precision in a broad p_{T} interval.

The analysis strategy to distinguish the different muon sources is based on the distance of closest approach to the primary vertex in the transverse plane (DCA_{xy}) of global muon tracks (MFT tracks matched with tracks reconstructed in the muon tracking chambers and further matched with the muon identifier). Such study is performed exploiting the different decay lengths of beauty hadrons with respect to charm hadrons and other sources. Figure 3 (right) presents the typical DCA_{xy} distribution of the main muon sources estimated by means of simulations using PYTHIA 8 and the GEANT4 transport code. One notices that the signals and

© The Authors, published by EDP Sciences. This is an open access article distributed under the terms of the Creative Commons Attribution License 4.0 (https://creativecommons.org/licenses/by/4.0/).

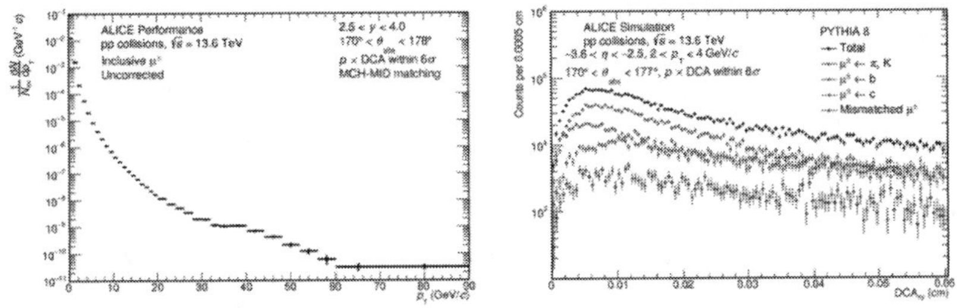

Figure 3. Left: inclusive muon p_T distribution measured in minimum bias collisions with a fraction of the sample of pp collisions at $\sqrt{s}$ = 13.6 TeV. Right: DCA distribution of the main muon sources obtained from a PYTHIA 8 simulation of pp collisions at $\sqrt{s}$ = 13.6 TeV.

the background exhibit different shapes, indicating that the separation of charm and beauty contributions can be achieved with the MFT. The performance for charm/beauty separation in pp collisions is first evaluated by means of dedicated Monte Carlo simulations. The DCA_{xy} distributions of the signals and background are parametrised with a variable-width Gaussian function. They further define the Monte Carlo templates which are combined after proper normalisation and employed to fit the total DCA_{xy}. The results indicate that the charm and beauty components can be measured separately in the semimuonic channel at forward rapidity in pp collisions at $\sqrt{s}$ = 13.6 TeV with the MFT coupled with the muon spectrometer. Such measurements can be achieved down to $p_T \simeq 0.5$ GeV/c and $p_T \simeq 1 - 2$ GeV/c for charm and beauty, respectively.

5 Conclusion

Measurements of azimuthal anistropies with muons in p–Pb collisions introduce new constraints in the interpretation of the collective behaviour in small systems and to model calculations. The p_T-integrated $c\bar{c}$ and $b\bar{b}$ production cross sections measured for the first time at forward rapidity via dimuons are in fair agreement with FONLL predictions. The MFT opens the path to new measurements such as charm and beauty production in the single muon channel. The performance study shows that the measurements can be performed differentially down to low p_T in pp collisions. High expectations are placed in the future charm/beauty analysis in Pb–Pb collisions at large rapidities, which will be unique at the LHC.

References

[1] K. Aamodt *et al.* [ALICE], JINST **3** (2008), S08002
[2] B. B. Abelev *et al.* [ALICE], Int. J. Mod. Phys. A **29** (2014), 1430044.
[3] S. Acharya *et al.* [ALICE], JINST **19** (2024) no.05, P05062.
[4] S. Acharya *et al.* [ALICE], Phys. Lett. B **846** (2023), 137782.
[5] Z. W. Lin, C. M. Ko, B. A. Li, B. Zhang and S. Pal, Phys. Rev. C **72** (2005), 064901.
[6] Z. W. Lin and L. Zheng, Nucl. Sci. Tech. **32** (2021) no.10, 113.
[7] C. Zhang, C. Marquet, G. Y. Qin, Y. Shi, L. Wang, S. Y. Wei and B. W. Xiao, Phys. Rev. D **102** (2020) no.3, 034010.
[8] S. Acharya *et al.* [ALICE], Eur. Phys. J. C **84** (2024) no.8, 813.
[9] M. Cacciari, S. Frixione, N. Houdeau, M. L. Mangano, P. Nason and G. Ridolfi, JHEP **10** (2012), 137.

© The Authors, published by EDP Sciences. This is an open access article distributed under the terms of the Creative Commons Attribution License 4.0 (https://creativecommons.org/licenses/by/4.0/).

Gluon decay into heavy quark pair under a strong magnetic field

Shile Chen[1,*], *Jiaxing* Zhao[2,3], and *Pengfei* Zhuang[1]

[1]Physics Department, Tsinghua University, Beijing 100084, China
[2]Helmholtz Research Academy Hesse for FAIR (HFHF), GSI Helmholtz Center for Heavy Ion Physics, Campus Frankfurt, 60438 Frankfurt, Germany
[3]Institut für Theoretische Physik, Johann Wolfgang Goethe-Universität,Max-von-Laue-Straße 1, D-60438 Frankfurt am Main, Germany

Abstract. Due to the extreme large magnetic field produced in the initial stage of non-central heavy-ion collision, the dynamical process of gluon decay into heavy quark pair will take place under an external field rather than in vacuum. Unlike in the vacuum case, where the process is forbidden by energy momentum conservation, under the external field, a process emerges considering the background energy which recovers the conservation. We calculate the gluon decay rate at leading order under a uniform magnetic field.

It is widely accepted that the strongest electromagnetic field in nature is generated in high energy nuclear collisions [1–4]. The maximum of the magnetic field can reach $5m_\pi^2 \sim 0.1$ GeV^2 in Au+Au collisions at top RHIC energy and almost $70m_\pi^2 \sim 1$ GeV^2 in Pb+Pb collisions at LHC energy [3, 4], where m_π is the pion mass. The strong magnetic field will cause many novel phenomena like chiral magnetic effect(CME) [1], photoproduction of dileptons and quarkonia in peripheral and ultra-peripheral collisions [5], splitting of D^0 and $\bar{D}^0$ directed flows [6], and spin-polarized difference between Λ and $\bar{\Lambda}$ [7].

In our previous study [8], we calculate the heavy quark production with the elementary process $gg \to Q\bar{Q}$ under the extreme strong magnetic field and show the analytical results with the Lowest Landau Level approximation (LLL). Because of the dimension reduction in phase space, the cross section of this process has a divergence around the incoming energy threshold. Then the heavy quark pair production in nucleus-nucleus collisions will be extremely enhanced at low transverse momentum and suppressed at high transverse momentum. The existence of external magnetic field breaks the rotational invariance and only the momentum along the magnetic field is conserved [8]. The production process becomes anisotropic and strongly dependent on the direction of motion.

To verify the validity of this LLL approximation, we put our calculation a step forward, extending to the next Landau level. From the result, we could conclude that if the partons mainly distributed in the small x region and the heavy quark produced at low p_T region with external magnetic field $\sim 10GeV^2$ then the process could be well described at LLL. In the calculation, we found that the cross section of such an elementary process will encounter a divergence for $u-$ and $t-$ channel if we set the final quarks up to $n = 1$ state. The inducement of this divergence is not from the threshold of the initial energy but the *on−shell* of the internal quark, which means the Feynman diagram could be cut, and the sub-process is just the decay

*e-mail: csl18@tsinghua.org.cn

© The Authors, published by EDP Sciences. This is an open access article distributed under the terms of the Creative Commons Attribution License 4.0 (https://creativecommons.org/licenses/by/4.0/).

of the on shell gluon to quark anti-quark pair. In vacuum, the elementary process like Fig. 1 is forbidden due to the energy-momentum conservation because the massless gluon has no rest mass. However, with the background field, this process will no longer be forbidden any more because the strong magnetic field could compensate the energy. This process has been widely studied in the astro physics when they consider the QED processes on a magnetars [12–14]. In heavy-ion collision, when considering the pre equilibrium stage with magnetic field, the gluon decay has barely contributed to the quark statistics. In this proceeding paper, we wi,ll calculate and discuss the process $g \to Q\bar{Q}$ under LLL.

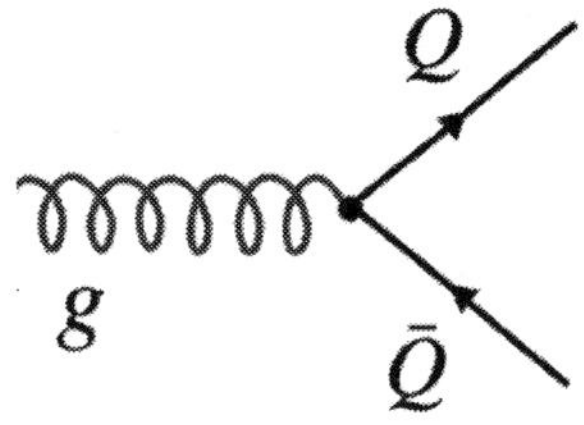

Figure 1. Elementary process of the gluon decay into quark anti-quark pair.

The Feynman rules associated with quarks are controlled by Dirac equation

$$\left[i\gamma^{\mu}\left(\partial_{\mu} + iqA_{\mu}\right) - m\right]\psi = 0, \tag{1}$$

where m is the quark mass, q the quark electric charge, and A_{μ} the electromagnetic potential. We consider an external magnetic field B in the direction of z-axis and choose the Landau gauge with $A_0 = 0$ and $\boldsymbol{A} = Bx\boldsymbol{e}_y$. In this case, the momentum along the x-axis is not conserved. Taking into account the Landau energy levels for a fermion moving in an external magnetic field, the stationary solution of the Dirac spinor can be written as

$$\psi^{-}_{n,\sigma}(x,p) = e^{-ip\cdot x}u_{n,\sigma}(\boldsymbol{x},p), \quad \psi^{+}_{n,\sigma}(x,p) = e^{ip\cdot x}v_{n,\sigma}(\boldsymbol{x},p) \tag{2}$$

where the quantum number n is Landau level, and the four-momentum is defined as $p_{\mu} = (\epsilon, 0, p_y, p_z)$ with $p_y = aqB$ controlled by the center of gyration a.

With the stationary solution of fermion field above, we could obtain the S-matrix of the decay process $g \to q\bar{q}$ as

$$S_{fi} = -ig\int d^4x\bar{\psi}^{(\sigma_-)}_{-}(x)_{p_{z-},0,a_-}\gamma_{\mu}t^{c}A^{\mu}_{c}(x)\psi^{(\sigma_+)}_{+}(x)_{p_{z+},0,a_+} \tag{3}$$

when we proceed our calculation under LLL, we could obtain

$$\begin{aligned} S_{fi} &= t^{c}\frac{g}{2\sqrt{2\omega}L^{7/2}}e^{-\frac{i}{4}\omega\sin\theta[4a_-\cos\phi+\lambda_B^2\omega\sin\theta(-i+\sin(2\phi))]} \\ &\times\ \epsilon_z\frac{p_{z-}(-p_{z-}+\omega\cos\theta)+(m+E_-)(m+E_+)}{\sqrt{m(m+E_-)}\sqrt{m(m+E_+)}}(2\pi)^3\delta^3_{fi}(E,p_y,p_z) \end{aligned} \tag{4}$$

where $\omega\hat{k} = \omega(1, \sin\theta\cos\phi, \sin\theta\sin\phi, \cos\theta)$ is the momentum of the initial gluon and ϵ_z is the z-component of its polarization vector.

The two independent polarization modes of gluon is O mode and $\mathcal{E}$ mode, with $\epsilon_O = (-\cos\theta\cos\phi, -\cos\theta\sin\phi, \sin\theta)$ and $\epsilon_{\mathcal{E}} = (\sin\phi, -\cos\phi, 0)$. Since the S-matrix is proportional to ϵ_z, only O-mode remains. If here we set $\theta = \pi/2$, then $\epsilon_z = 1$ and $p_{z+} = -p_{z-}$, $E_+ =$

© The Authors, published by EDP Sciences. This is an open access article distributed under the terms of the Creative Commons Attribution License 4.0 (https://creativecommons.org/licenses/by/4.0/).

$E_- = E$, we could obtain a simplified formulation

$$S_{fi} = t^c \frac{gm}{E\sqrt{2\omega}L^{7/2}} e^{-\frac{1}{4}\lambda_B^2\omega^2 + i\Phi}(2\pi)^3 \delta^3_{fi}(E, p_y, p_z) \tag{5}$$

where Φ is a phase term which will not contribute to the decay rate. And this S-matrix has the exact the same formula with [13, 14] except the color factor. The decay rate of the process $g \to Q\bar{Q}$ with $\theta = \pi/2$

$$\begin{aligned} \Gamma &= g^2 \sum_c \frac{1}{T} \int L\frac{dp_{z+}}{2\pi} \int L\frac{dp_{z-}}{2\pi} \int L\frac{da_+}{2\pi\lambda_B^2} \int L\frac{da_-}{2\pi\lambda_B^2} |S_{fi}|^2 \\ &= g^2 \frac{4q^2m^2}{12\pi E^2\omega\lambda_B^2} e^{-\lambda_B^2\omega^2/2} \int \frac{da_-}{L} \int dp_{z-}\delta(\omega - E_+ - E_-) \\ &= g^2 \frac{4qBm^2}{3\pi\omega^2\sqrt{\omega^2 - 4m^2}} e^{-\lambda_B^2\omega^2/2} \end{aligned} \tag{6}$$

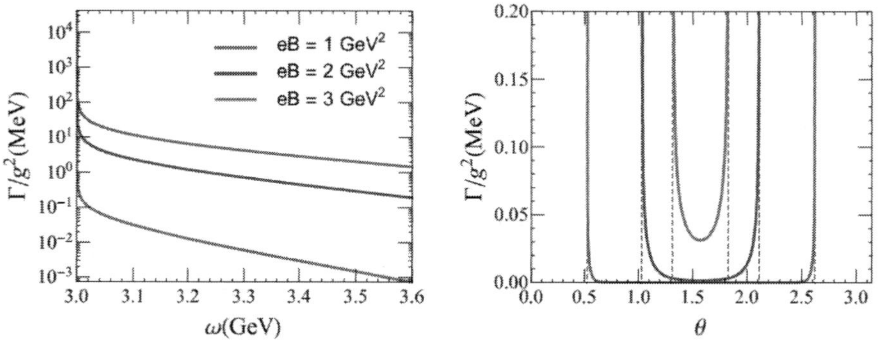

Figure 2. (Left)The decay rate of $g \to Q\bar{Q}$ with final quark mass $m = 1.5$GeV and $\theta = \pi/2$ changes with initial gluon energy. (Right)–Starting form Eq.(4). The decay rate changes with polarization angle at $eB = 1$ GeV^2 and different lines represent $\omega = 3.1$ GeV(Red), 3.5 GeV(Blue) and 6.0 GeV(Green) respectively. Dashed lines are the thresholds of angle window determined by the energy conservation condition.

From Fig.2, we could see that the decay rate decreasing with incoming energy has the same property with the scattering cross section $gg \to Q\bar{Q}$, which has a threshold depending on the rest mass of final fermions and increasing with magnetic field. The difference appears when considering the polarization angle dependence. The energy conservation condition $\sqrt{p_{z+}^2 + m^2} + \sqrt{p_{z-}^2 + m^2} = \omega$ is satisfied only with polarization angle is around $\pi/2$ which means the gluon is moving perpendicular to the magnetic field especially when the gluon energy approaches threshold. And when the energy grows, the angle window becomes larger and along with the exponential decrease of the magnitude of decay width. On the contrary, when the incoming gluon is moving along the magnetic field, it is obvious to see that the energy conservation is violated and this process is forbidden.

In summary, we derived the gluon decay rate under the external magnetic field in the strong field limit. This process is barely considered in the initial stage quark production since the magnetic field is believed to decay at a very fast speed when the system enters the QGP phase. From the result, we could conclude that the decay rate of gluon increases with the magnitude of magnetic field which means a stronger magnetic field can induce a faster decay

© The Authors, published by EDP Sciences. This is an open access article distributed under the terms of the Creative Commons Attribution License 4.0 (https://creativecommons.org/licenses/by/4.0/).

of the gluon even if the gluon has no interaction with magnetic field. As our previous work mentioned, the LLL is more suitable for a lighter quark, and we believed that the light quark production will also be affected by this process.

Acknowledgement: The work is supported by the NSFC grants No. 12075129, the Guangdong Major Project of Basic and Applied Basic Research No.2020B0301030008, the funding from the European Union's Horizon 2020 research, and the innovation program under grant agreement No. 824093 (STRONG-2020).

References

[1] D. E. Kharzeev, L. D. McLerran, and H. J. Warringa, *The Effects of topological charge change in heavy ion collisions: 'Event by event P and CP violation'*, *Nucl. Phys. A* **803** (2008) 227–253, [arXiv:0711.0950].

[2] V. Voronyuk, V. D. Toneev, W. Cassing, E. L. Bratkovskaya, V. P. Konchakovski, and S. A. Voloshin, *(Electro-)Magnetic field evolution in relativistic heavy-ion collisions*, *Phys. Rev. C* **83** (2011) 054911, [arXiv:1103.4239].

[3] W.-T. Deng and X.-G. Huang, *Event-by-event generation of electromagnetic fields in heavy-ion collisions*, *Phys. Rev. C* **85** (2012) 044907, [arXiv:1201.5108].

[4] K. Tuchin, *Particle production in strong electromagnetic fields in relativistic heavy-ion collisions*, *Adv. High Energy Phys.* **2013** (2013) 490495, [arXiv:1301.0099].

[5] **ALICE** Collaboration, J. Adam et al., *Measurement of an excess in the yield of J/ψ at very low p_T in Pb-Pb collisions at $\sqrt{s_{NN}}$ = 2.76 TeV*, *Phys. Rev. Lett.* **116** (2016) 222301, [arXiv:1509.08802].

[6] S. K. Das, S. Plumari, S. Chatterjee, J. Alam, F. Scardina, and V. Greco, *Directed Flow of Charm Quarks as a Witness of the Initial Strong Magnetic Field in Ultra-Relativistic Heavy Ion Collisions*, *Phys. Lett. B* **768** (2017) 260–264, [arXiv:1608.02231].

[7] Y. Guo, S. Shi, S. Feng, and J. Liao, *Magnetic Field Induced Polarization Difference between Hyperons and Anti-hyperons*, *Phys. Lett. B* **798** (2019) 134929, [arXiv:1905.12613].

[8] S. Chen, J. Zhao and P. Zhuang, "Heavy flavor production under a strong magnetic field," [arXiv:2401.17559 [hep-ph]].

[9] Z. Wang, J. Zhao, C. Greiner, Z. Xu, and P. Zhuang, *Incomplete electromagnetic response of hot QCD matter*, *Phys. Rev. C* **105** (2022) L041901, [arXiv:2110.14302].

[10] L. Yan and X.-G. Huang, *Dynamical evolution of magnetic field in the pre-equilibrium quark-gluon plasma*, arXiv:2104.00831.

[11] Y. Chen, X.-L. Sheng, and G.-L. Ma, *Electromagnetic fields from the extended Kharzeev-McLerran-Warringa model in relativistic heavy-ion collisions*, *Nucl. Phys. A* **1011** (2021) 122199, [arXiv:2101.09845].

[12] I. Caiazzo, J. Heyl and R. Turolla, Astrophys. Space Sci. Libr. **460**, 301-336 (2019) doi:10.1007/978-3-030-19715-5_12

[13] A. Kostenko and C. Thompson, *QED Phenomena in an Ultrastrong Magnetic Field. I. Electron–Photon Scattering, Pair Creation, and Annihilation*, *Astrophys. J.* **869** (2018) 44, [arXiv:1904.03324].

[14] A. Kostenko and C. Thompson, *QED Phenomena in an Ultrastrong Magnetic Field. II. Electron-Positron Scattering, $e^{\pm}$-Ion Scattering, and Relativistic Bremsstrahlung*, *Astrophys. J.* **875** (2019) 23, [arXiv:1904.03325].

© The Authors, published by EDP Sciences. This is an open access article distributed under the terms of the Creative Commons Attribution License 4.0 (https://creativecommons.org/licenses/by/4.0/).

Simulating Charm Quarks in IP-Glasma Initial Stage and Quark-Gluon Plasma: A Hybrid Approach for charm quark phenomenology

Charles Gale[1], *Sangyong* Jeon[1], *Manu* Kurian[2,*], *Björn* Schenke[3], and *Mayank* Singh[4]

[1]Department of Physics, McGill University, 3600 University Street, Montreal, QC, H3A 2T8, Canada
[2]RIKEN BNL Research Center, Brookhaven National Laboratory, Upton, New York 11973, USA
[3]Physics Department, Brookhaven National Laboratory, Upton, New York 11973, USA
[4]Department of Physics and Astronomy, Vanderbilt University, Nashville, TN 37240, USA

Abstract. We present phenomenological findings on charm quark transport while including its energy loss in both pre-equilibrium and hydrodynamic stages of the evolution. We employed the MARTINI event generator for the production and evolution of heavy quarks in the relativistic heavy-ion collisions. The sensitivity of the heavy meson nuclear modification factor and flow coefficient to the early stage of heavy-ion collisions and bulk medium evolution is analyzed for Pb+Pb collisions at 5.02 TeV. Our study provides insights into the interaction strength of charm quarks during the early phase and within the quark-gluon plasma.

1 Introduction

The accurate modeling of quark-gluon plasma evolution generated in relativistic nucleus-nucleus collisions relies on the understanding of the dynamics in the collision's early phase. Being generated during the early phase of the collisions, heavy quarks serve as effective probes of the early stages of the collision event. Heavy quark transport in the medium can therefore provide information about temperature, viscosity, and other transport properties. Efforts have been made to understand the Brownian transport of heavy-flavor particles in the quark-gluon plasma (QGP) and to estimate key experimental observables such as the nuclear suppression factor R_{AA} and elliptic flow v_2 at RHIC and LHC energies. Despite the existence of many studies on heavy-flavor dynamics in an expanding viscous QGP medium and the estimation of non-equilibrium effects on the heavy quark transport coefficients, major challenges persist in the phenomenological study of heavy quarks. The effects of the initial stage and the pre-equilibrium evolution of heavy quarks in the initial background are usually ignored.

We model relativistic heavy-ion collisions at LHC energies with a hybrid dynamical approach consisting of a fluctuating IP-Glasma initial state [1], followed by the viscous hydrodynamic approach MUSIC [2]. We evolve charm quarks in both the IP-Glasma initial stage and the hydrodynamically expanding QGP phase using the MARTINI event generator [3]. By integrating charm quark energy loss in the fluctuating initial state, in addition to that in

*e-mail: manukurian10406@gmail.com

© The Authors, published by EDP Sciences. This is an open access article distributed under the terms of the Creative Commons Attribution License 4.0 (https://creativecommons.org/licenses/by/4.0/).

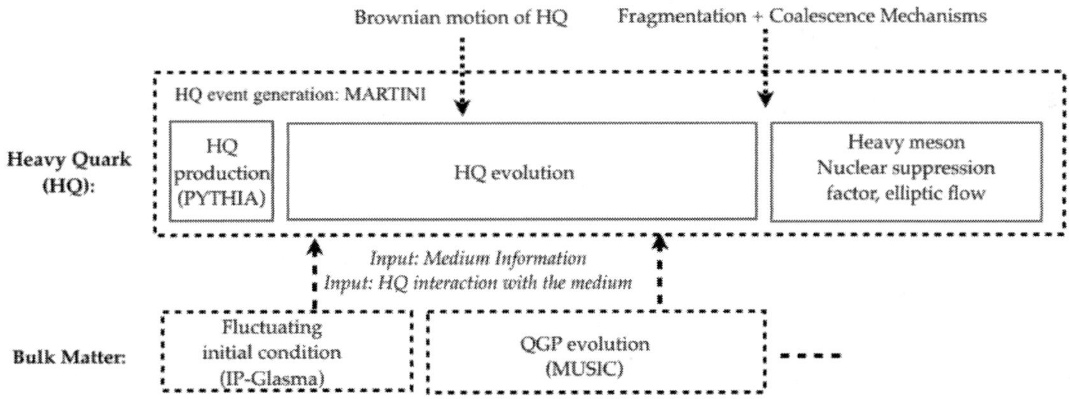

Figure 1. Multi-stage model for heavy quark evolution in heavy-ion collisions

the QGP phase, we present the phenomenological implications of charm quark evolution in the early stages of heavy-ion collisions.

2 Charm quark multi-stage evolution

We describe heavy-quark evolution with a multi-stage model as described in figure 1. We initially generate charm quarks using PYTHIA8.2, which involves sampling the 6-dimensional momentum distributions of $Q\bar{Q}$ systems. The initial spatial distribution of charm quarks is obtained from the IP-Glasma initial state, with charm quarks randomly sampled on the background IP-Glasma. The precise determination of charm quark interactions in the expanding space-time geometry, which captures the gluon fields in the fluctuating initial state, is nontrivial and not yet fully developed. The phenomenological implication of the impact of charm quark evolution in the pre-equilibrium stage is an interesting aspect to explore. In the present analysis, as a first step toward capturing the impact of the fluctuating initial state on charm quark evolution, we approximate the initial state as an evolving QCD medium dominated by gluons. We assume that the gluons form a thermalized medium with a conformal equation of state where a temperature can be defined. The transport of heavy quarks is studied as Brownian motion in such a medium, where the interaction strength of the quarks in the medium can be measured in terms of the appropriate diffusion coefficient.

The next phase is the evolution of the charm quark in the hydrodynamically expanding QGP. Relativistic second-order viscous hydrodynamics is employed to describe the evolution of the QGP medium. The change in charm quark momentum, dp_i, over a time interval dt, can be quantified numerically using the Langevin equations in the local rest frame of the medium as follows [4],

$$\begin{aligned} dp_i &= -\eta(p_i)p_i\,dt + \xi_i(\mathbf{p})\,dt, \\ \langle \xi_i(t)\xi_j(0)\rangle &= \left(\delta_{ij} - \frac{p_i p_j}{|\mathbf{p}|^2}\right)\kappa_T(|\mathbf{p}|) + \frac{p_i p_j}{|\mathbf{p}|^2}\,\kappa_L(|\mathbf{p}|), \end{aligned} \tag{1}$$

where $\eta(p_i)$ is the drag coefficient, $\kappa_{L/T}$ is the longitudinal/transverse momentum diffusion coefficient, and ξ_i denotes the stochastic force acting on the charm quark. We utilized recently

© The Authors, published by EDP Sciences. This is an open access article distributed under the terms of the Creative Commons Attribution License 4.0 (https://creativecommons.org/licenses/by/4.0/).

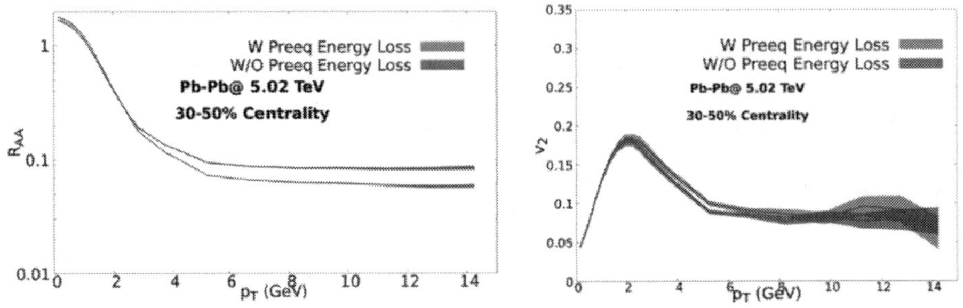

Figure 2. The impact of charm quark energy loss in the initial stage of the collision on the D-meson nuclear suppression factor R_{AA} and elliptic flow v_2. The temperature-dependent lattice estimation of spatial diffusion coefficient is used to quantify the interaction strength of charm quark in the medium

estimated $N_f = 2+1$ lattice spatial diffusion coefficient $2\pi DT$ [6] to describe the charm quark interaction strength in the medium. The spatial diffusion coefficient can be related to $\eta(p_i)$ and $\kappa_{L/T}$ [4, 5].

The current analysis specifically focuses on open charm mesons. We use a combination of fragmentation and heavy-light coalescence mechanisms for the hadronization process. In the fragmentation approach, a charm quark with momentum p fragments into a D-meson with momentum zp. We also consider the coalescence model, where the recombination probability is described by the momentum-space Wigner function [7].

To quantify the impact of charm quark evolution in the early stages of the collision, we analyze charm quark observables with and without accounting for its pre-equilibrium energy loss, as shown in figure 2. The switching time from the Glasma to plasma phase is fixed at $\tau_w = 0.4$ fm. Here, we considered only the fragmentation mechanism for the hadronization process. It is observed that the R_{AA} is modified by the pre-equilibrium energy loss of the charm quark. Elliptic flow is another key observable that measures the azimuthal anisotropy of charm mesons. The elliptic flow, v_2, is estimated using the scalar product method. It is shown that the fluctuating initial state modifies the D-meson v_2 in two ways. In comparison with a smooth background, fluctuations increase local pressure gradients, which in turn increase v_2. Additionally, a fluctuating background causes a decorrelation between the event planes of light and heavy-flavor mesons, which suppresses v_2. Refer [8] for details. The evolution of charm quarks in the fluctuating initial state further modifies D-meson v_2. Despite the relatively short lifetime of the initial phase, we observe that the energy loss of charm quarks in the IP-Glasma phase has a non-negligible effect in the phenomenological study.

In figure 3, we consider momentum-dependent heavy quark transport coefficients and the heavy-light coalescence mechanism in addition to the fragmentation process for hadronization. This allows for a comparison of our results with experimental data. A parameterized momentum dependence, motivated by perturbative QCD estimates of heavy quark coefficients, is introduced into the temperature-dependent 2 + 1 flavor lattice rates. It is observed that both the momentum dependence and the coalescence hadronization mechanism play significant roles in the phenomenological study of heavy quarks.

© The Authors, published by EDP Sciences. This is an open access article distributed under the terms of the Creative Commons Attribution License 4.0 (https://creativecommons.org/licenses/by/4.0/).

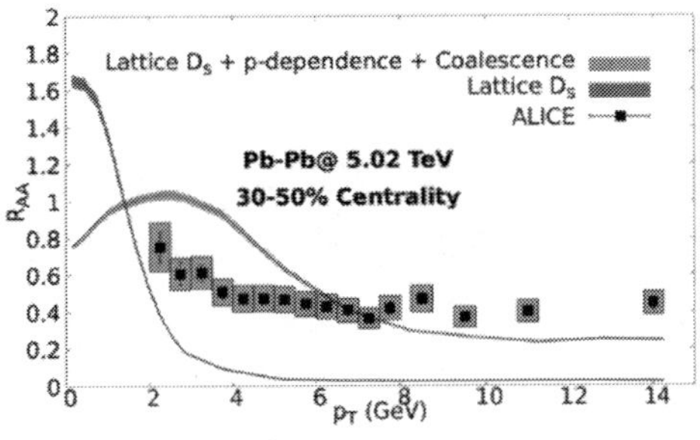

Figure 3. Comparison of D meson R_{AA} with the experimental data [9] for Pb+Pb at 5.02 TeV.

3 Summary

One of the interesting questions we aim to address is whether charm quarks can act as efficient probes of the initial phase despite their shorter lifetime and fluctuating energy density profile. An exact solution to this aspect requires a thorough understanding of charm quark interactions with the non-Abelian fields, which is highly non-trivial. In this study, we provide a qualitative estimation of the effect of pre-equilibrium charm quark evolution on charm meson observables by approximating the pre-equilibrium phase as a gluonic system with a temperature. Our analysis shows that charm quarks demonstrate their potential to provide penetrating tomographic information about the initial stage of heavy-ion collisions.

Acknowledgments: CG and SJ is supported by the Natural Sciences and Engineering Research Council of Canada (Grant numbers SAPIN-2018-00024 and SAPIN-2020-00048). MK acknowledges the RIKEN SPDR program. BS is supported by the U.S. Department of Energy, Office of Science, Office of Nuclear Physics, under DOE Contract No. DE-SC0012704 and within the framework of the Saturated Glue (SURGE) Topical Theory Collaboration. MS is supported by U.S. DOE Grants No. DE-SC0024347.

References

[1] B. Schenke, P. Tribedy and R. Venugopalan, Phys. Rev. Lett. **108**, 252301 (2012).
[2] B. Schenke, S. Jeon and C. Gale, Phys. Rev. C **82**, 014903 (2010).
[3] C. Young, B. Schenke, S. Jeon and C. Gale, Phys. Rev. C **86**, 034905 (2012).
[4] G. D. Moore and D. Teaney, Phys. Rev. C **71**, 064904 (2005).
[5] M. Kurian, M. Singh, V. Chandra, S. Jeon and C. Gale, Phys. Rev. C **102**, 044907 (2020).
[6] L. Altenkort *et al.* [HotQCD], Phys. Rev. Lett. **130**, no.23, 231902 (2023).
[7] J. Zhao, J. Aichelin, P. B. Gossiaux, *et al.* Phys. Rev. C **109**, no.5, 054912 (2024).
[8] M. Singh, M. Kurian, S. Jeon and C. Gale, Phys. Rev. C **108**, no.5, 054901 (2023).
[9] S. Acharya *et al.* [ALICE], JHEP **01**, 174 (2022).

© The Authors, published by EDP Sciences. This is an open access article distributed under the terms of the Creative Commons Attribution License 4.0 (https://creativecommons.org/licenses/by/4.0/).

EPJ Web of Conferences **316**, 04019 (2025) https://doi.org/10.1051/epjconf/202531604019
SQM 2024

Investigation of early magnetic field and angular momentum in ultrarelativistic heavy-ion collisions via D^{*+} spin alignment with ALICE

Himanshu Sharma[1] on belalf of the ALICE collaboration,*

[1]INFN Padova

Abstract. Heavy quarks, i.e. charm and beauty, are produced at the initial stages of heavy-ion collisions, on a time scale shorter than the medium formation time, and are sensitive to the initial angular momentum of the system and the magnetic field produced perpendicular to the reaction plane (defined by the impact parameter direction and beam direction) in non-central heavy-ion collisions. In the presence of a large angular momentum and initial magnetic field, the charm quark can be polarized. Experimentally, the heavy-flavor polarization can be probed by measuring the spin density matrix element (ρ_{00}) of spin-1 mesons such as D^{*+} mesons. Any deviation of the parameter from 1/3 can be attributed to the spin alignment of the D^{*+} meson.
In these proceedings, we present the first measurement of the ρ_{00} parameter of prompt D^{*+} mesons in Pb–Pb collisions at $\sqrt{s_{\rm NN}}$= 5.02 TeV, exploiting the data sample collected by the ALICE Collaboration during LHC Run 2. A comparison with the J/ψ polarization measurement is also reported to investigate the effect of the magnetic field. Moreover, the ρ_{00} parameter of D^{*+} mesons measured in pp collisions is also presented.

1 Introduction

In ultrarelativistic nuclear collisions, the produced system evolves through various phases. In non-central collisions, charged spectator motion produces a magnetic field ($B \approx 10^{15}$ T) in very early stages of the collisions, which decreases with time [1]. Additionally, a highly vortical system with orbital angular velocity ($\omega \approx 10^{22}$ s^{-1}) is produced due to the spectator motion [2] which results in a large orbital angular momentum (L). The two effects can preferentially align particle's spin projection along the spin quantization axis through spin-orbit coupling. Charm quarks and antiquarks are produced in the early stages (~ 0.1 fm/c) of the collision. Therefore, they are more sensitive to the high intensity of the electromagnetic fields compared to light quarks [3]. In presence of the large L and strong B fields, the charm quarks and antiquarks become polarized. Eventually, this polarization is reflected in the spin alignment of vector mesons (J/ψ, D^{*+}) formed by these quarks [3].

Vector meson spin alignment is quantified by measuring the spin density matrix element (ρ_{00}) which is extracted by parameterizing the angular distribution of D^{*+} decay products using the following expression:

*e-mail: himanshu.sharma@cern.ch

© The Authors, published by EDP Sciences. This is an open access article distributed under the terms of the Creative Commons Attribution License 4.0 (https://creativecommons.org/licenses/by/4.0/).

$$\frac{dN}{d\cos\theta^*} = N_0\left[(1-\rho_{00}) + (3\rho_{00}-1)\cos^2\theta^*\right], \quad (1)$$

Here, θ^* is the angle between the momentum of the daughter pion and the polarization axis in the D^{*+} rest frame, and N_0 is the normalization constant. In Pb–Pb collisions, ρ_{00} is measured in the direction of L and B fields, i.e. the direction orthogonal to the reaction plane. In pp collisions, it is measured in the helicity frame, which is defined as the direction of a vector meson in the laboratory frame. The spin alignment of vector meson is influenced by two hadronization mechanisms, which dominate charm-meson formation at different p_T. At low p_T, spin alignment depends on the magnitude of polarization of the both recombining quark and antiquark. At high p_T, it depends on the magnitude of polarization of the fragmenting quark (or antiquark) and its correlation with the constituent antiquark (or quark) [4, 5].

The ALICE apparatus [7] allows for the reconstruction of D^{*+} mesons at midrapidity ($|y| < 0.8$) through the decay channel $D^{*+} \to D^0\,(K^-\pi^+)\,\pi^+$ and respective charge conjugates. The D^{*+} mesons formed in charm-quark hadronization, or in decays of excited charm states, are defined as prompt, while those produced in beauty hadron decays are called non-prompt D^{*+}. The inclusive charmonium states are reconstructed down to $p_T = 0$ at forward rapidity ($2.5 < y < 4$) through the dimuon decay channel. The Time Projection Chamber (TPC) and Inner Tracking System (ITS) detectors are used for precise tracking of the pions and kaons from the D^{*+} decay at midrapidity. The TPC offers outstanding particle identification capabilities down to very low p_T (around 200 MeV/c) via the measurement of specific energy loss in a gaseous medium.

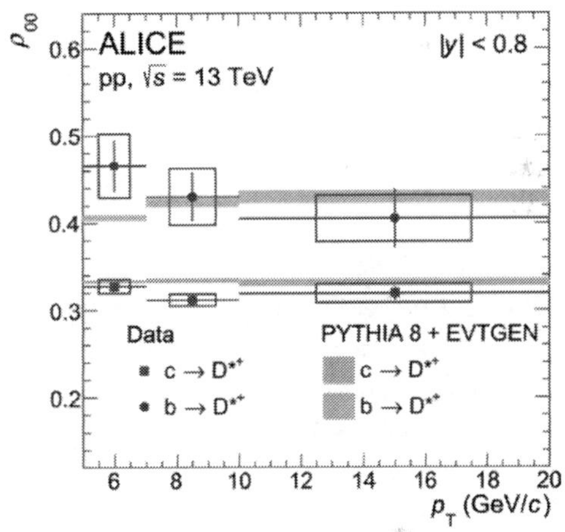

Figure 1. ρ_{00} of prompt and non-prompt D^{*+} as a function of p_T measured in the helicity frame in pp collisions at $\sqrt{s} = 13$ TeV [6].

The ITS consists of six layers of silicon detectors. The two innermost layers, equipped with pixel detectors, provide the spatial resolution needed to distinguish between prompt and non-prompt D^{*+} mesons at midrapidity in Pb–Pb collisions. Boosted Decision Trees are used to reduce the combinatorial background and separate the prompt and non-prompt components. The muon spectrometer is employed to trigger on and reconstruct muon tracks at forward pseudorapidity.

2 Results

Figure 1 shows the first measurement of the prompt and non-prompt D^{*+} spin alignment at the LHC. The prompt and non-prompt ρ_{00} are shown as a function of p_T in the helicity frame at midrapidity in pp collisions at $\sqrt{s} = 13$ TeV. The measured ρ_{00} for prompt D^{*+} is consistent with 1/3. Thus, there is no evidence of spin alignment of prompt D^{*+} with respect to the helicity axis. In contrast, a significant spin alignment is observed for non-prompt D^{*+}, as expected from the helicity conservation in weak decays of beauty hadrons. The model predictions, obtained from PYTHIA 8 + EvtGen [8, 9], are consistent with these measurements. These results serve as a benchmark for D^{*+} spin alignment measurements in Pb–Pb collisions.

© The Authors, published by EDP Sciences. This is an open access article distributed under the terms of the Creative Commons Attribution License 4.0 (https://creativecommons.org/licenses/by/4.0/).

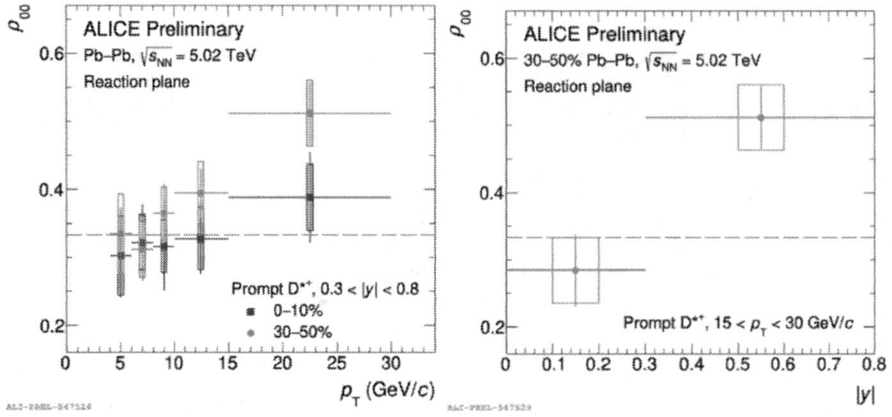

Figure 2. (Left) ρ_{00} of prompt D^{*+} as a function of p_T measured with respect to reaction plane in the 0–10% and 30–50% centrality ranges, and (Right) as a function of rapidity in $15 < p_T < 30$ GeV/c in the 30–50% centrality range in Pb–Pb collisions.

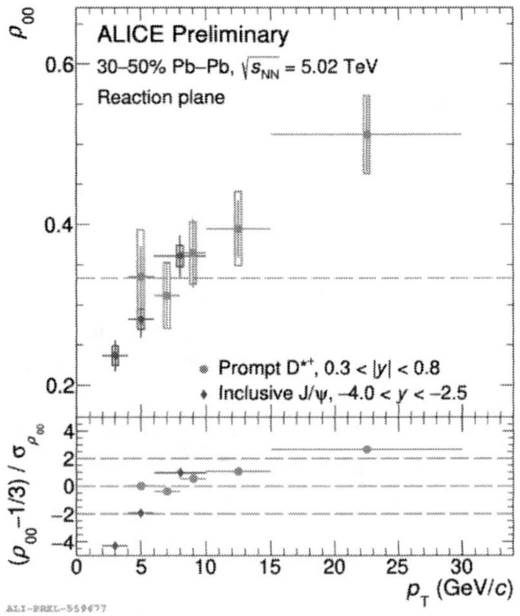

Figure 3. ρ_{00} of prompt D^{*+} and inclusive J/ψ as a function of p_T measured with respect to the reaction plane in 30–50% centrality range at forward rapidity in Pb–Pb collisions.

Figure 2 demonstrates the first measurement of D^{*+} spin alignment with respect to the reaction plane in the 0–10% and 30–50% centrality intervals in Pb–Pb collisions. The left panel of Fig. 2 shows prompt D^{*+} ρ_{00} as a function of p_T in the $0.3 < |y| < 0.8$ rapidity region. The data show a hint of rising trend with p_T in 30–50% centrality interval. In the 0–10% centrality interval, ρ_{00} is consistent with 1/3. In the 30–50% centrality interval, it is larger than 1/3 at high p_T, where hadronization is expected to occur mainly through quark fragmentation. This case is further investigated as a function of rapidity. In the right panel of Figure 2, ρ_{00} of prompt D^{*+} is shown in two different rapidity ranges: at midrapidity ($|y| < 0.3$) and at slightly more forward rapidity ($0.3 < |y| < 0.8$), measured at $15 < p_T < 30$ GeV/c in 30–50% centrality. At midrapidity, ρ_{00} does not show any significant deviation from 1/3, while at forward rapidity it is significantly larger than this value. At large rapidity, the electromagnetic fields decrease slowly in time and charm quarks with large momentum might be more influenced by those fields [3]. Theoretical predictions are necessary to make conclusive remarks on these findings.

Depending on the hadronization mechanism, quark polarization is reflected in the hadrons. Both J/ψ and D^{*+} possess a charm quark (or an antiquark), therefore, polarization of these hadrons might be governed by similar processes. Figure 3 shows a comparison

© The Authors, published by EDP Sciences. This is an open access article distributed under the terms of the Creative Commons Attribution License 4.0 (https://creativecommons.org/licenses/by/4.0/).

of the prompt D^{*+} ρ_{00} with that of inclusive J/ψ as a function of p_T measured with respect to reaction plane at forward rapidity in 30–50% centrality interval in Pb–Pb collisions. Similar to prompt D^{*+}, the ρ_{00} of inclusive J/ψ rises with p_T. The ρ_{00} for prompt D^{*+} and inclusive J/ψ are compatible within uncertainties in the overlapping p_T range. The inclusive J/ψ ρ_{00} is significantly smaller (by 4σ) than 1/3 at p_T smaller than 5 GeV/c, where J/ψ is dominantly produced by charm and anti-charm recombination.

3 Summary

The first results of prompt D^{*+} spin alignment with respect to the reaction plane in Pb–Pb collisions were presented. A significant spin alignment is observed for prompt D^{*+} at high p_T in 30–50% centrality interval at rapidity $0.3 < |y| < 0.8$ in Pb–Pb collisions. This observation supports the hypothesis that electromagnetic fields can polarize the charm quarks at the large momentum produced during the early stages of collisions. The prompt D^{*+} results are consistent with inclusive J/ψ polarization in the overlapping p_T region measured with respect to the reaction plane. A comparison with theoretical predictions is necessary to obtain a robust interpretation of the data.

Since 2022, ALICE has been collecting data in a continuous readout mode in pp and Pb–Pb collisions. Eventually, the total size of this data sample from minimum-bias pp and Pb–Pb collisions will be significantly larger than that collected in LHC Run 2 and will allow more differential spin alignment measurements at high p_T, up to 100 GeV/c.

References

[1] P. Christakoglou et al. *Eur. Phys. J. C* 81.8 (2021), 717.
[2] STAR Collaboration. *Nature* 548 (2017), 62–65.
[3] S. K. Das et al. *Phys. Lett. B* 768 (2017), 260–264.
[4] Y. Yang et al. *Phys. Rev. C* 97.3 (2018), 034917.
[5] Z. Liang et al. *Phys. Lett. B* 629 (2005), 20–26.
[6] ALICE Collaboration. *Phys. Lett. B* 846 (2023), 137920.
[7] ALICE Collaboration. *JINST* 3 (2008), S08002.
[8] P. Skands et al. *Eur. Phys. J. C* 74.8 (2014), 3024.
[9] D. J. Lange. *Nucl. Instrum. Meth. A* 462 (2001), 152–155.

© The Authors, published by EDP Sciences. This is an open access article distributed under the terms of the Creative Commons Attribution License 4.0 (https://creativecommons.org/licenses/by/4.0/).

EPJ Web of Conferences **316**, 05001 (2025) https://doi.org/10.1051/epjconf/202531605001
SQM 2024

Investigating the system size dependence of hypernuclei production with A < 5 using the ALICE detector

Yuanzhe Wang[1,*] for the ALICE Collaboration

[1]Key Laboratory of Nuclear Physics and Ion-beam Application (MOE), Institute of Modern Physics, Fudan University, 200433, Shanghai, China

Abstract. The production of (anti)hypernuclei is among the most promising probes for studying the production mechanism of light nuclei in high-energy hadronic collisions. According to coalescence predictions, the production of ${}^{3}_{\Lambda}$H, ${}^{4}_{\Lambda}$H, and ${}^{4}_{\Lambda}$He is sensitive to their internal wave function. In contrast, the yields predicted with the Statistical Hadronitazion Models (SHM) are based on the mass of the (hyper)nuclei and the freeze-out temperature, with no explicit dependence on the nuclear structure. In these proceedings, the measurements of ${}^{3}_{\Lambda}$H, ${}^{4}_{\Lambda}$H, and ${}^{4}_{\Lambda}$He from pp to central Pb–Pb collisions are presented. The results are based on the data samples collected by ALICE during the LHC Run 2 and Run 3. For the ${}^{3}_{\Lambda}$H, in addition, an innovative method to extract its properties based on the system size dependency of its production yield is also presented.

1 Introduction

Hypernuclei are nuclei that contain at least one hyperon. Since this unique system was first discovered in the 1950s, hypernuclei have gathered substantial interests as natural probes to study the hyperon-nucleon (Y-N) interaction, which is expected to play an important role in the innermost core of neutron stars [1]. Relativistic hadronic collisions provide a favourable environment for the formation of hypernuclei and antihypernuclei. A basic understanding of the physical properties of hypernuclei has already been established by experimental measurements. The ${}^{3}_{\Lambda}$H is found to be a loosely bound structure with a size around 10 fm due to the small Λ separation energy (B_{Λ}) about one hundred keV [2]. Moreover, the B_{Λ} of both ${}^{4}_{\Lambda}$H and ${}^{4}_{\Lambda}$He are of the order of several MeV [3], which is similar to the binding energy per nucleon of ordinary nucleus.

Recently, the production of hypernuclei is proposed as an effective method to study the nucleosynthesis mechanism in hadronic collisions. The coalescence model [4] suggests that the probability to produce a nucleus relies on the overlap of the nucleus Wigner function and the nucleon distribution function in the emission source. On the other hand, the Statistical hadronization Model (SHM) [5] assumes that the yield of particles is mainly dominated by their mass and the chemical freeze-out temperature but do not depend on their internal structure. Furthermore, it is also possible that the elliptic flow of hypernuclei could distinguish between SHM and the coalescence models, because the emission source of fireball differs in different directions in non-central collisions [6].

*e-mail: yuanzhe.wang@cern.ch

© The Authors, published by EDP Sciences. This is an open access article distributed under the terms of the Creative Commons Attribution License 4.0 (https://creativecommons.org/licenses/by/4.0/).

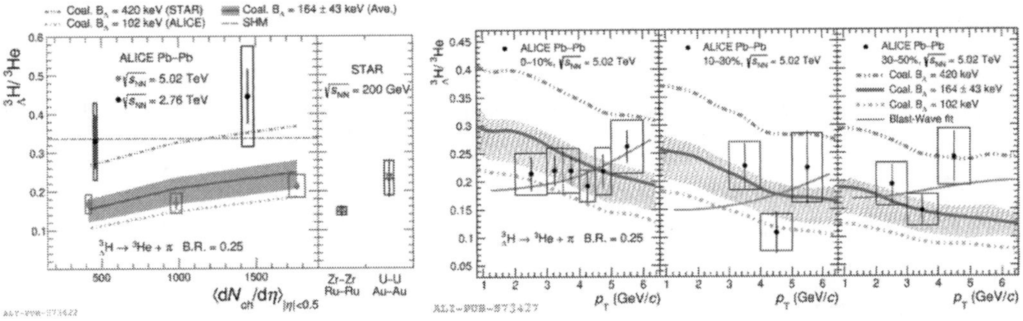

Figure 1. Yield ratio of $^3_\Lambda$H to ^{3}He in Pb–Pb collisions together with theoretical predictions. Left panel: $^3_\Lambda$H/ ^{3}He as a function of charged-particle multiplicity density $\langle dN_{ch}/d\eta \rangle$. The recent experimental results of STAR are shown alongside for comparison. Right panel: $^3_\Lambda$H/ ^{3}He versus p_T in different centrality intervals.

In these proceedings, we present the measurements of hypernuclei production ($^3_\Lambda$H, $^4_\Lambda$H, $^4_\Lambda$He and their charge conjugates) obtained by analysing the data sample collected by ALICE during the LHC Run 2 and Run 3. In the following, we use the notation $^3_\Lambda$H, $^4_\Lambda$H, and $^4_\Lambda$He for both particles and antiparticles.

2 Results

2.1 Production of $^3_\Lambda$H in Pb–Pb collisions at $\sqrt{s_{NN}}$ = 5.02 TeV

In order to test the two nuclear production models, namely the SHM and the coalescence models, the yield ratio of $^3_\Lambda$H to ^{3}He was calculated based on the p_T differential measurements of $^3_\Lambda$H in different centrality intervals [7]. The $^3_\Lambda$H candidates are reconstructed in Pb–Pb collisions via its two-body decay channel $^3_\Lambda$H → ^{3}He + π^- using 210 million collision events collected by ALICE in 2018. Fig. 1 shows measured $^3_\Lambda$H/^{3}He yield ratios with theoretical predictions [8, 9] and experimental results from STAR Collaboration [10] for comparison. The multiplicity dependence of $^3_\Lambda$H/^{3}He, shown in left the panel reflects the dependence on the size of the emission source. The expectation of SHM stays constant at large multiplicities, while all the three coalescence predictions with different B_Λ hypotheses are more sensitive to the system size. Compared with the prior results which were measured at $\sqrt{s_{NN}}$ = 2.76 TeV, the uncertainties of this work have been improved due to larger statistics, further understanding of properties of $^3_\Lambda$H, and the new Machine Learning analysis method [11] with a gradient-boosted decision tree classifier (BDT). These two results are consistent within a 2σ confidence interval. T or the one measured by ALICE [2]. In addition, the suppression of $^3_\Lambda$H yield in smaller collision systems has also been measured by STAR [10]. This approach offers a novel method to probe properties of $^3_\Lambda$H through its production dependence on system size. The right panel of Fig. 1 presents the transverse momentum (p_T) dependence of the $^3_\Lambda$H/^{3}He yield ratio, together with the coalescence predictions and a hydrodynamic description of the observable based on the Blast-Wave model [12]. The coalescence predicts larger suppression of $^3_\Lambda$H production at higher p_T, while the Blast-Wave picture provides an increasing trend of $^3_\Lambda$H/^{3}He. However, the current experimental uncertainties do not support drawing any conclusion.

© The Authors, published by EDP Sciences. This is an open access article distributed under the terms of the Creative Commons Attribution License 4.0 (https://creativecommons.org/licenses/by/4.0/).

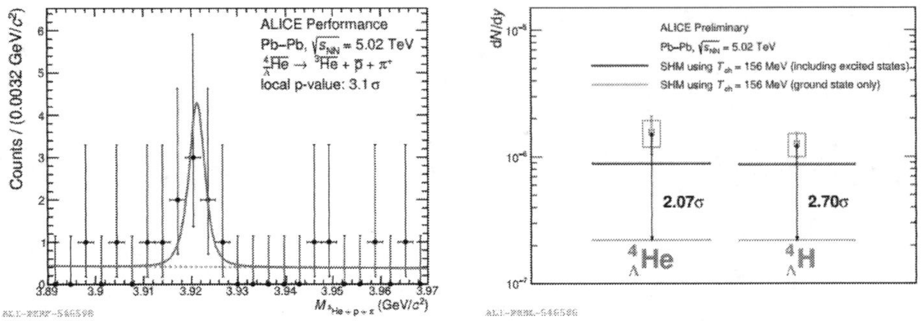

Figure 2. Left panel: Invariant mass distribution of the $^{4}_{\bar{\Lambda}}\overline{\mathrm{He}}$ in 0–10% most central Pb–Pb collisions at $\sqrt{s_{\mathrm{NN}}}$ = 5.02 TeV. The data points are fitted with a function which is a combination of KDE signal and an exponential background. Right panel: Integrated yields of $^{4}_{\Lambda}$He and $^{4}_{\Lambda}$H in 0–10% most central Pb–Pb collisions at $\sqrt{s_{\mathrm{NN}}}$ = 5.02 TeV with SHM predictions.

2.2 Production of A = 4 hypernuclei in Pb–Pb collisions at $\sqrt{s_{\mathrm{NN}}}$ = 5.02 TeV

The search of A = 4 hypernuclei ($^{4}_{\Lambda}$H and $^{4}_{\Lambda}$He) was also performed in Pb–Pb collisions at $\sqrt{s_{\mathrm{NN}}}$ = 5.02 TeV. The $^{4}_{\Lambda}$H and $^{4}_{\Lambda}$He are reconstructed via the decay channels $^{4}_{\Lambda}\mathrm{H} \to {}^{4}\mathrm{He} + \pi^{-}$ and $^{4}_{\Lambda}\mathrm{He} \to {}^{3}\mathrm{He} + \mathrm{p} + \pi^{-}$ respectively. The first evidence of $^{4}_{\bar{\Lambda}}\overline{\mathrm{He}}$ is shown in the left panel of Fig. 2. The right panel presents the integrated yield of A = 4 hypernuclei, where both production yields deviate by at least 2σ from the SHM prediction which only includes ground states. However, the effect of the excited states is observed as an enhancement in the yields of $^{4}_{\Lambda}$H and $^{4}_{\Lambda}$He. Additionally, taking into account the spin degeneracy, the feed-down effects originating from these excited states can increase the expected yields by a factor up to 4 in total. With the contribution from the excited states, the yields of both $^{4}_{\Lambda}$H and $^{4}_{\Lambda}$He are consistent with the SHM predictions. It appears that the SHM works well for compact states such as $^{4}_{\Lambda}$H and $^{4}_{\Lambda}$He, but it becomes less effective when the structure is larger, as in the case of $^{3}_{\Lambda}$H.

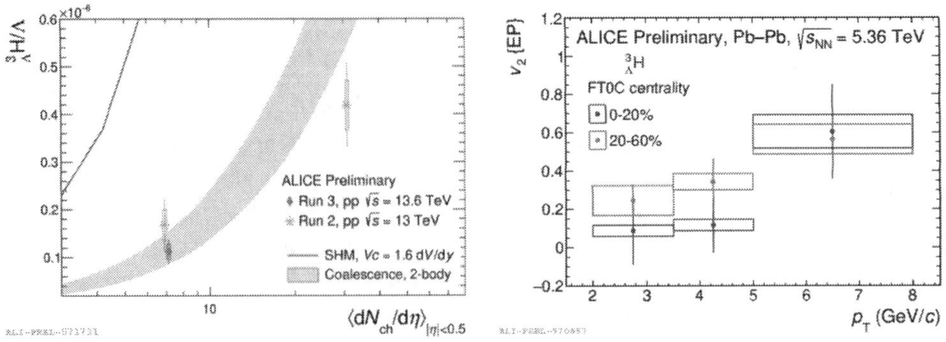

Figure 3. Left panel: Yield ratio of $^{3}_{\Lambda}$H to Λ in pp collisions with theoretical predictions. The results of $^{3}_{\Lambda}$H are obtained from antimatter only. Right panel: v_2 of $^{3}_{\Lambda}$H in Pb–Pb collision at $\sqrt{s_{\mathrm{NN}}}$ = 5.36 TeV in different centrality classes.

© The Authors, published by EDP Sciences. This is an open access article distributed under the terms of the Creative Commons Attribution License 4.0 (https://creativecommons.org/licenses/by/4.0/).

2.3 Measurement of $^{3}_{\Lambda}$H in Run3

As previously discussed, there is a notable discrepancy between the SHM and the coalescence model in their predictions regarding the yield of $^{3}_{\Lambda}$H. In the small collision systems where the emission size is comparable to that of $^{3}_{\Lambda}$H, the $^{3}_{\Lambda}$H yield is expected to be more significantly suppressed in the coalescence picture. Therefore, it would be more decisive to test these two models at low multiplicities. Such studies are also performed during the LHC Run3 period. The $^{3}_{\Lambda}$H p_{T} differential production in pp collision at $\sqrt{s}$ = 13.6 TeV was measured using the data collected in 2022. The left panel of Fig. 3 shows the p_{T} integrated yield ratio of $^{3}_{\Lambda}$H to Λ. The new results obtained in Run3 is compatible with the ones from Run2. From a comparison with the model predictions, the SHM [5] is found to be inconsistent with the results, while the Run3 $^{3}_{\Lambda}$H/ Λ ratio aligns well with the 2-body coalescence model [4] which assumes that $^{3}_{\Lambda}$H is formed from the coalescence of d and Λ.

The elliptic flow of $^{3}_{\Lambda}$H was measured for the first time in Pb–Pb collisions at $\sqrt{s_{\mathrm{NN}}}$ = 5.36 TeV, using the dataset collected by ALICE in 2023. The results are shown in the right panel of Fig. 3. In general, the v_2 of $^{3}_{\Lambda}$H increases with both centrality and p_{T}. In future, uncertainties on the measurement will be improved from the full event statistics of LHC Run 3.

3 Summary

The production of A < 5 hypernuclei has been investigated by ALICE. The $^{3}_{\Lambda}$H yield is measured among different collision systems and appears to be in favor of the coalescence picture. Notably, based on the predictions of the coalescence model, the system size dependence of $^{3}_{\Lambda}$H production is sensitive to its B_{Λ}. However, in the case of $^{4}_{\Lambda}$H and $^{4}_{\Lambda}$He where the Λ separation energy is larger, SHM is found to be effective while taking into account the feed-down effects from the excited states, while no coalescence prediction is currently available for them. It indicates that there may still be unknown factors in nuclear production. The extensive data sample collected during the ongoing LHC Run3 period will enable more precise measurements and differential analyses of hypernuclei.

The author is supported by the National Natural Science Foundation of China (Nos. 123B2083, 12322508, 12061141008) and the Science and Technology Commission of Shanghai Municipality (No. 23590780100).

References

[1] L. Tolos, L. Fabbietti, Prog. Part. Nucl. Phys. **112**, 103770 (2020).
[2] S. Acharya et al. (ALICE), Phys. Rev. Lett. **131**, 102302 (2023).
[3] M. Abdallah et al. (STAR), Phys. Lett. B **834**, 137449 (2022).
[4] K.J. Sun, C.M. Ko, B. Dönigus, Phys. Lett. B **792**, 132 (2019).
[5] V. Vovchenko, B. Dönigus, H. Stoecker, Phys. Lett. B **785**, 171 (2018).
[6] R. Vozabova, B. Tomasik, Phys. Rev. C **109**, 064908 (2024).
[7] S. Acharya et al. (ALICE), arXiv:2405.19839 (2024).
[8] V. Vovchenko, H. Stoecker, Comput. Phys. Commun. **244**, 295 (2019).
[9] D.N. Liu et al., Phys. Lett. B **855**, 138855 (2024).
[10] M. Abdulhamid et al. (STAR), Nature **632**, 1026 (2024).
[11] L. Barioglio, F. Catalano, M. Concas, P. Fecchio, F. Grosa, F. Mazzaschi, M. Puccio, hipe4ml/hipe4ml (2021)
[12] E. Schnedermann, J. Sollfrank, U.W. Heinz, Phys. Rev. C **48**, 2462 (1993).

© The Authors, published by EDP Sciences. This is an open access article distributed under the terms of the Creative Commons Attribution License 4.0 (https://creativecommons.org/licenses/by/4.0/).

Measurements of ${}^{4}_{\Lambda}$H and ${}^{4}_{\Lambda}$He Production in $\sqrt{s_{NN}}$ = 3.0 - 3.5 GeV Au+Au Collisions at RHIC

Chenlu Hu[1,*] for STAR collaboration

[1]University of Chinese Academy of Science

Abstract. Hypernuclei, which are bound states of nuclei with at least one hyperon, serve as excellent experimental probes for studying the hyperon-nucleon (Y-N) interaction. In these proceedings, the measurements of A=4 hypernuclei (${}^{4}_{\Lambda}$H and ${}^{4}_{\Lambda}$He) production from the RHIC-STAR experiment utilizing the fixed target datasets are presented. The measured yields dN/dy of ${}^{4}_{\Lambda}$H and ${}^{4}_{\Lambda}$He as a function of rapidity are shown from $\sqrt{s_{NN}}$ = 3.0, 3.2 and 3.5 GeV Au+Au collisions. Additionally, the energy dependencies of the ratio of ${}^{4}_{\Lambda}$H/Λ and ${}^{4}_{\Lambda}$He/Λ are examined to explore isospin effects. The mass dependence of the mean transverse momentum $<p_T>$ are also discussed. Furthermore, calculations from PHQMD, thermal model and transport model JAM plus coalescence afterburner are compared to these results and the relevant physics implications are discussed.

1 Introduction

Relativistic heavy ion collisions are an abundant source of strangeness. As strange quarks have to be newly produced during the hot and dense stage of the collision, they are thought of carrying information on the properties of the matter that was created [1, 2]. Hypernuclei, which consist of at least one hyperon, serve as an excellent experimental tools for studying the hyperon-nucleon (Y-N) interaction. Y-N interactions, especially at high baryon density, are not only essential for understanding the inner structure of compact stars [3, 4], but also for describing the hadronic phase of heavy-ion collisions. Heavy-ion collisions provide an environment where it is possible to study the Y-N interaction under finite temperature and density conditions through measurements of hypernuclei properties, such as their collective flow and production yields.

A=4 mirror hypernuclei (${}^{4}_{\Lambda}$H and ${}^{4}_{\Lambda}$He) are substantially tighter bound states compared to the hypertriton (${}^{3}_{\Lambda}$H). The existence of the spin-1 excited states (${}^{4}_{\Lambda}$H$^{*}(1^{+})$ and ${}^{4}_{\Lambda}$He$^{*}(1^{+})$) [5] may also enhance the measured yields through feed-down. As such, their yields allow us to gain insight on the effects of hypernuclear binding, spin and isospin content on their production in heavy-ion collisions. In these proceedings, the yields dN/dy and mean transverse momentum p_T spectra of ${}^{4}_{\Lambda}$H and ${}^{4}_{\Lambda}$He in Au+Au collisions at $\sqrt{s_{NN}}$ = 3.0 - 3.5 GeV will be discussed.

*e-mail: huchenlu@ucas.ac.cn

© The Authors, published by EDP Sciences. This is an open access article distributed under the terms of the Creative Commons Attribution License 4.0 (https://creativecommons.org/licenses/by/4.0/).

2 Experimental and Data Analysis

The STAR experiment carried out the Beam Energy Scan (BES) program in order to study the properties of quark-gluon plasma (QGP) and search for quantum chromodynamics (QCD) critical point. In BES-II program, in fixed-target (FXT) mode, the center of mass collision energy extends from 7.7 GeV down to 3.0 GeV.

In this analysis, we used the dataset of Au + Au collisions at $\sqrt{s_{\mathrm{NN}}}$ = 3.0 - 3.5 GeV collected using the FXT setup at RHIC of the STAR experiment. We mainly used Time Projection Chamber (TPC) detector for particle identification. The hypernuclei ${}^{4}_{\Lambda}$H and ${}^{4}_{\Lambda}$He are reconstructed with the following decay channels: ${}^{4}_{\Lambda}\mathrm{H} \rightarrow {}^{4}\mathrm{He} + \pi^{-}$, ${}^{4}_{\Lambda}\mathrm{He} \rightarrow {}^{3}\mathrm{He} + \mathrm{p} + \pi^{-}$. The secondary decay topology is reconstructed by the KFParticle program which is based on a Kalman filter method [6]. In the program, error-matrices are used to enhance the reconstruction significance. A set of cuts on topological variables are applied to the hypernuclei candidates to optimize the signal significance.

3 Results and Discussions

3.1 Particle Yields

The p_{T}-integrated yields dN/dy for ${}^{4}_{\Lambda}$H and ${}^{4}_{\Lambda}$He are calculated from the p_{T} spectra by combining data in the measured p_{T} range and function-fitting extrapolation in the unmeasured p_{T} range. Figure 1 presents the rapidity dependence of the dN/dy in 0-10% and 10-40% central Au + Au collisions across a range of $\sqrt{s_{\mathrm{NN}}}$ from 3.0 to 3.5 GeV. The rapidity distributions show slight variations, with a downward trend in central collisions and an increase towards backward rapidity in mid-central collisions. In 0-40% centrality, the ${}^{4}_{\Lambda}$He yield at mid-rapidity is comparable to that of ${}^{4}_{\Lambda}$H. The prediction from Jet AA Microscopic (JAM) plus Coalescence [7, 8] is plotted for comparison. In the JAM+Coalescence model, the JAM transport model generates hadron phase space distributions at freezeout, followed by a coalescence procedure that forms (hyper)nuclei when the relative momentum and spatial distance of their constituents fall within defined limits. JAM+Coalescence calculations could qualitatively describe the rapidity dependence of dN/dy for ${}^{4}_{\Lambda}$H in 0-40% centrality.

Figure 2 shows dN/dy for various particles, scaled by their spin degeneracy factor (2J+1). The measured dN/dy exhibits an approximate exponential dependence on mass, but the yields of A=4 hypernuclei exceed this trend, as shown by grey lines. Here, the yields of A=4 particles contains feed down from excited state (J=1), which enhance the yields by a factor of 3. Even if we remove the excited state contribution, the A=4 hypernuclei yields remain above the exponential trend. Dashed lines are calculations from JAM+Coalescence afterburner, which include contributions from the excited states and show a similar exponential dependence of dN/dy/(2J+1) vs mass. Here, Λ is weighted to the data, and different coalescence parameters for ${}^{3}_{\Lambda}$H and ${}^{4}_{\Lambda}$H (${}^{4}_{\Lambda}$He) are needed to describe the data. The ΔR is 4.8 fm for both ${}^{3}_{\Lambda}$H and ${}^{4}_{\Lambda}$H (${}^{4}_{\Lambda}$He), and ΔP is 0.24 GeV/c for ${}^{3}_{\Lambda}$H and 0.38 GeV/c for ${}^{4}_{\Lambda}$H (${}^{4}_{\Lambda}$He). The larger ΔP may reflect the tighter binding of A=4 hypernuclei. The Parton Hadron Quantum Molecular Dynamics (PHQMD) [9] approach could describe the yields of Λ, ${}^{4}_{\Lambda}$H and ${}^{4}_{\Lambda}$He, but overestimates that of ${}^{3}_{\Lambda}$H, and does not include contributions from excited states. The measured yields have been corrected by their corresponding branching ratios, which are listed in the figure.

3.2 Particle Yield Ratios

Figure 3 presents the particle ratios of hypernuclei to hyperon (${}^{4}_{\Lambda}$H/Λ and ${}^{4}_{\Lambda}$He/Λ) as a function of collision energy. It shows the similar decreasing trend of ${}^{4}_{\Lambda}$H/Λ and ${}^{4}_{\Lambda}$He/Λ with

© The Authors, published by EDP Sciences. This is an open access article distributed under the terms of the Creative Commons Attribution License 4.0 (https://creativecommons.org/licenses/by/4.0/).

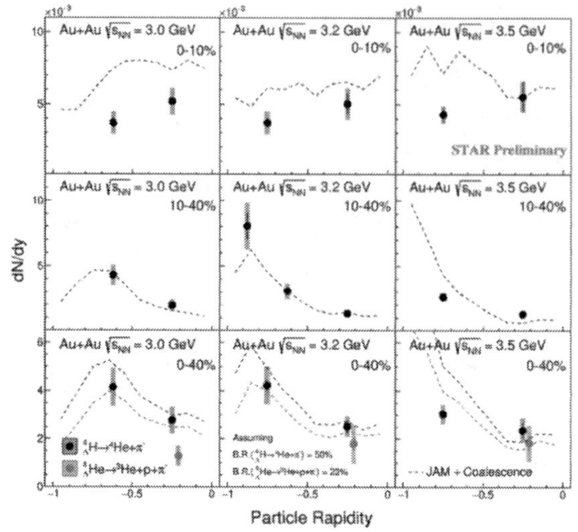

Figure 1. Rapidity distribution of $^4_\Lambda$H and $^4_\Lambda$He in 0-10% and 10-40% Au + Au collision at $\sqrt{s_{NN}}$ = 3.0 - 3.5 GeV. The symbols represent measurements while the lines represent JAM+Coalescence calculations.

the increasing energy. $^4_\Lambda$H/Λ is systematically larger than $^4_\Lambda$He/Λ probably because there are more neutrons than protons in the colliding system. The measured data are well described with JAM+Coalescence calculations, while overestimated by the Thermal-Fist [10].

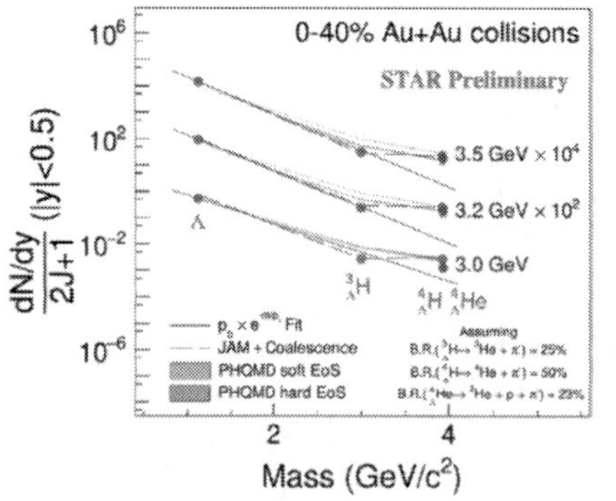

Figure 2. Mass dependence of measured dN/dy scaled by the spin degeneracy factor (2J+1). The symbols represent measurements while the lines represent different model calculations.

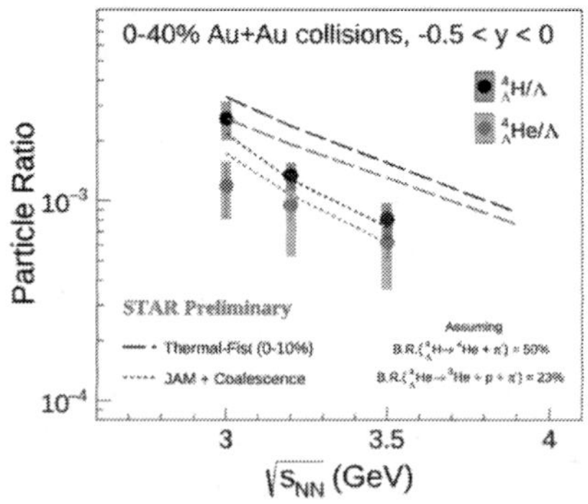

Figure 3. $^4_\Lambda$H/Λ and $^4_\Lambda$He/Λ at mid-rapdity in 0-40% Au+Au collisions as function of the center of mass collision energy. The symbols represent measurements while the lines represent different model calculations.

3.3 Mean Transverse Momentum

Figure 4 presents the mass dependence of mid-rapidity $<p_T>$ for Λ, $^3_\Lambda$H, $^4_\Lambda$H and $^4_\Lambda$He, from the $\sqrt{s_{NN}}$ = 3.0 - 3.5 GeV in 0-10% and 0-40% Au+Au collisions. The measured $<p_T>$ follow the linear mass scaling up to 3.5 GeV, which is consistent with coalescence as the dominant process for hypernuclei production at mid-rapidity. Both JAM+Coalescence and PHQMD model could reproduce the mass dependence of $<p_T>$ qualitatively.

© The Authors, published by EDP Sciences. This is an open access article distributed under the terms of the Creative Commons Attribution License 4.0 (https://creativecommons.org/licenses/by/4.0/).

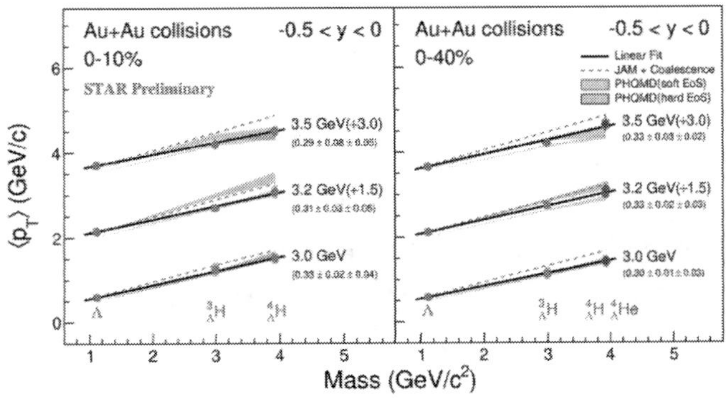

Figure 4. Mass dependence of the mid-rapidity $<p_T>$ for Λ, ${}^{3}_{\Lambda}$H, ${}^{4}_{\Lambda}$H and ${}^{4}_{\Lambda}$He, from the $\sqrt{s_{NN}}$ = 3.0 - 3.5 GeV in 0-10% and 0-40% Au+Au collisions. The symbols represent measurements while the lines represent JAM+Coalescence and PHQMD model calculations. The numbers in parentheses next to the energy represent the vertical shifts, while the numbers in the lower parentheses indicate the slopes of the linear fits.

4 Summary and Outlook

In summary, we carry out the rapidity and centrality dependence measurement of ${}^{4}_{\Lambda}$H and ${}^{4}_{\Lambda}$He yields in Au+Au collisions from $\sqrt{s_{NN}}$ = 3.0 to 3.5 GeV in the high-baryon-density region. JAM+Coalescence model could qualitatively reproduce the rapidity and centrality dependence of ${}^{4}_{\Lambda}$H production. The yields of Λ, ${}^{3}_{\Lambda}$H, ${}^{4}_{\Lambda}$H and ${}^{4}_{\Lambda}$He do not strictly follow an exponential scaling with mass when divided by spin degeneracy, suggesting significant contributions from feed-down of excited A=4 hypernuclei. The ratio of ${}^{4}_{\Lambda}$H/Λ and ${}^{4}_{\Lambda}$He/Λ are well described with JAM+Coalescence calculations, while overestimated by Thermal-Fist. The linear mass scaling is observed in the mass dependence of mid-rapidity $<p_T>$ up to 3.5 GeV, which is well described by JAM+Coalescence afterburner and PHQMD calculations qualitatively. This is consistent with coalescence as the dominant process for hypernuclei production at mid-rapidity.

The results presented in these proceedings are based on a subset of the BES-II datasets. In Run 21, STAR collected 2 billion events at $\sqrt{s_{NN}}$ = 3 GeV. This larger dataset will enable measurements of heavier hypernuclei (A>4) and may help us gain valuable insights into the mass dependence of hypernuclei production.

References

[1] P. Koch, B. Muller and J. Rafelski, Phys. Rept. **142**, 167 (1986)
[2] J. Chen et al., (2024), **arXiv:2407.02935** [nucl-ex]
[3] D. Gerstung, N. Kaiser, and W. Weise, Eur. Phys. J. **A56**, 175 (2020)
[4] D. Lonardoni et al., Phys. Rew. Lett. **114**, 092301 (2015)
[5] F. Schulz et al. (A1 Collaboration), Nucl. Phys. A **954** (2016), 149-160
[6] X. Y. Ju, Y. H. Leung, S. Radhakrishnann et al., Nucl. Sci. Tech. **34** (2023) 15
[7] Y. Nara, EPJ Web Conf. **208** (2019), 11004
[8] J. Steinheimer et al., Phys. Lett. B **714** (2012), 85-91
[9] S. Gläßel et al., Phys. Rev. C **105** (2022) no.1, 014908
[10] T. Reichert et al, Phys. Rev. C **107** (2023) no.1, 014912

© The Authors, published by EDP Sciences. This is an open access article distributed under the terms of the Creative Commons Attribution License 4.0 (https://creativecommons.org/licenses/by/4.0/).

Energy dependence of ϕ(1020) meson production in nucleus-nucleus collisions at the CERN SPS

Łukasz Rozpłochowski[1,*] for NA61/SHINE Collaboration

[1]H. Niewodniczański Institute of Nuclear Physics, PAS, Radzikowskiego 152, 31-342 Kraków, Poland

Abstract. This paper presents preliminary results from the NA61/SHINE experiment on ϕ meson production in central Ar+Sc collisions at beam momenta of 150*A*, 75*A*, and 40*A* GeV/*c* (corresponding to $\sqrt{s_{NN}}$ = 16.8, 11.9, and 8.8 GeV, respectively). These results include double differential distributions in rapidity (y) and transverse momentum (p_T), as well as p_T-integrated rapidity distributions and total ϕ yields. Additionally, the widths of the rapidity distributions of ϕ and ϕ/π yield ratios are compared to those from Pb+Pb and p+p reactions. Notably, this work represents the first-ever measurements of ϕ production in a mid-sized system at SPS energies.

1 Introduction

The ϕ meson, a resonance particle with a rest mass m_0 = 1019.461 ± 0.016 MeV and a mass width Γ = 4.249 ± 0.013 MeV, consists almost entirely of strange and anti-strange quarks. As the lightest hidden-strangeness particle, it is valuable for studying the Quark-Gluon Plasma. In a hot and dense QGP environment, ϕ production may be particularly sensitive to strangeness enhancement effects, potentially behaving like a "doubly strange" particle.

The methods applied to investigate ϕ production in Ar+Sc reactions are largely consistent with those used in the analysis of ϕ production in p+p collisions [1, 2]. Direct observation of the ϕ meson is not feasible experimentally, so the K^+K^- decay channel is used as an indirect method to study ϕ production. The entire available phase space was divided into bins, and the tag-and-probe method [3] was used to extract the number of ϕ mesons in each bin.

2 Results

Fig. 1 shows double differential (y, p_T) distributions of ϕ mesons produced in central Ar+Sc collisions at 150*A* GeV/*c* beam momentum. Integrating these distributions over p_T, yields the rapidity distribution (Fig. 2, left). A correction (up to 5%) for the tail of the double differential distribution is calculated from the fit of the function $f(p_T) \propto p_T \cdot \exp(-m_T/T)$ visible as blue curves in Fig. 1.

Rapidity distributions for all three energies considered here, are presented in Fig. 2. These distributions are fitted with the sum of two Gaussian functions:

$$g(\mathrm{y}) \propto \exp\left(\frac{-(x-\mu)^2}{2\sigma^2}\right) + \exp\left(\frac{-(x+\mu)^2}{2\sigma^2}\right). \quad (1)$$

*e-mail: lukasz.rozplochowski@ifj.edu.pl

© The Authors, published by EDP Sciences. This is an open access article distributed under the terms of the Creative Commons Attribution License 4.0 (https://creativecommons.org/licenses/by/4.0/).

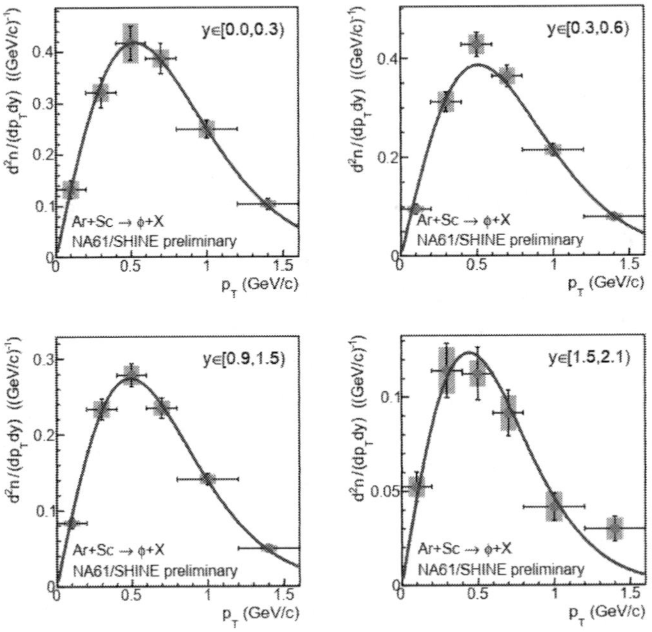

Figure 1. Double differential spectra (y, p_T) of ϕ produced in central Ar+Sc collisions at p_{beam} = 150A GeV/c. Statistical and systematic uncertainties are represented by vertical bars and red rectangles, respectively. The blue curve is a fit used to obtain the integral of the unmeasured tail of the distribution.

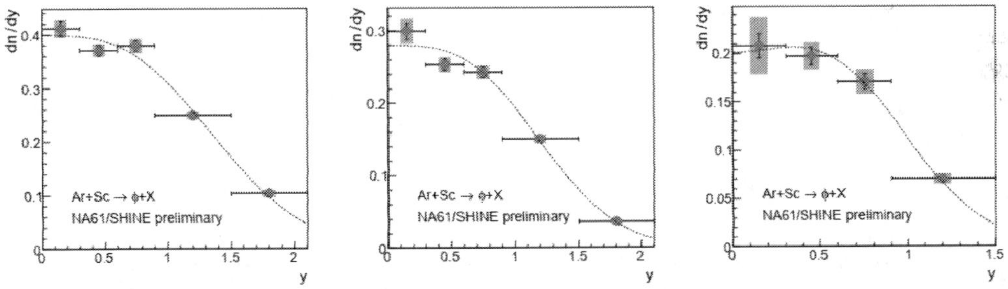

Figure 2. Rapidity distributions of ϕ mesons produced in central Ar+Sc collision at beam momenta equal to 150A, 75A and 40A GeV/c (from left to right). The dotted lines are the double Gaussian functions (see text) fitted to the experimental points. Statistical uncertainties are represented by vertical bars and systematic uncertainties by red rectangles.

From these fits, the widths RMS = $\sqrt{\sigma^2 + \mu^2}$ of the distributions are obtained, and the corrections (up to 2.5%) for the tails of the spectra. By integrating the rapidity distributions, one determines the total yields $\langle\phi\rangle$.

Fig. 3 shows the widths of the rapidity distributions as a function of the beam rapidity for various particles from three different systems. Pale background consist of π^-, K^+, K^- and $\overline{\Lambda}$ produced in Pb+Pb [4–6] and p+p [7, 8] collisions, which seem to follow a similar trend with increasing beam rapidity. Overlaid on this background are the widths of the ϕ rapidity distributions for Pb+Pb [9], p+p [1, 10], and Ar+Sc (this analysis). The points representing

© The Authors, published by EDP Sciences. This is an open access article distributed under the terms of the Creative Commons Attribution License 4.0 (https://creativecommons.org/licenses/by/4.0/).

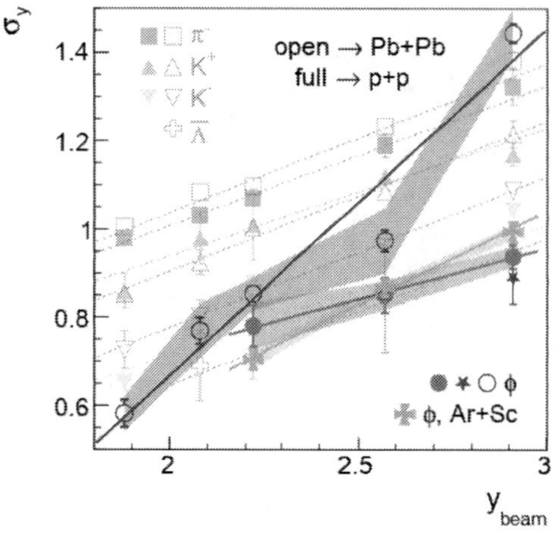

Figure 3. Widths of rapidity distributions for various particles as a function of beam rapidity. Data come from three different systems: central Pb+Pb [4–6, 9], minimum bias $p+p$ [1, 7, 8, 10] and central Ar+Sc. Vertical bars represent the statistical uncertainties, and the bands systematic ones.

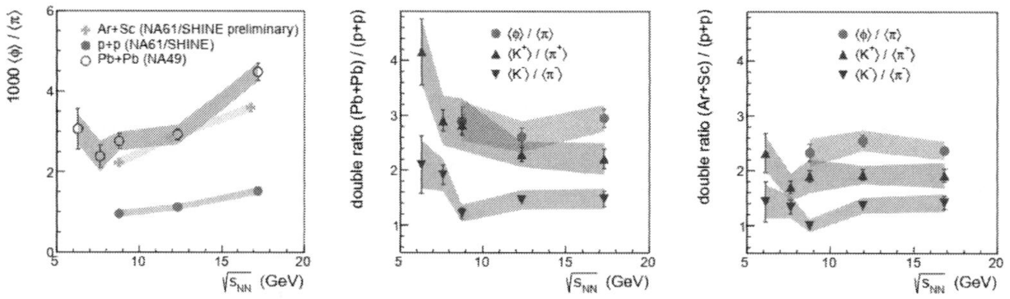

Figure 4. Left: energy dependence of the ϕ/π ratio for the minimum bias $p+p$, central Ar+Sc, and central Pb+Pb collisions. Middle and right: enhancement of ϕ and charged kaon production relative to pion production in Pb+Pb [4, 9] and Ar+Sc [12] collisions over the $p+p$ [1, 11] reactions, as a function of the collision energy. Vertical bars denote statistical uncertainties, while shaded bands represent systematic uncertainties.

ϕ from Pb+Pb deviate from the overall trend, whereas the ϕ from $p+p$ does not. Therefore, examining ϕ production in intermediate systems, such as Ar+Sc, may help to clarify this puzzling behaviour. As visible in Fig. 3, the widths for ϕ mesons from Ar+Sc collisions are very similar to those from $p+p$ reactions.

The left panel of Fig. 4 presents the energy dependence of the ϕ/π ratio for three systems: $p+p$, Ar+Sc, and Pb+Pb. The ratio for Ar+Sc is slightly lower than for Pb+Pb but much higher than for $p+p$ collisions. The middle panel shows the double ratios for ϕ/π, defined as:

$$\text{double ratio (Pb+Pb)}/(p+p) = \frac{(\langle\phi\rangle/\langle\pi\rangle)_{\text{Pb+Pb}}}{(\langle\phi\rangle/\langle\pi\rangle)_{p+p}}\,. \tag{2}$$

This quantity is defined correspondingly for other particles. The right panel of Fig. 4 also displays the double ratio, but instead of comparing Pb+Pb, it compares Ar+Sc to $p+p$. The enhancement in both Ar+Sc and Pb+Pb reactions over the $p+p$ collisions, of particle production relative to pion production, is slightly higher for ϕ mesons than that for kaons, and is independent of the collision energy within the considered range. The data for $p+p$ collisions is taken from Refs. [1, 11], for Pb+Pb from Refs. [4, 9] and for Ar+Sc from Refs. [12].

© The Authors, published by EDP Sciences. This is an open access article distributed under the terms of the Creative Commons Attribution License 4.0 (https://creativecommons.org/licenses/by/4.0/).

3 Summary

Preliminary results on the $\phi(1020)$ meson production in central Ar+Sc reactions at beam momenta of $150A$, $75A$ and $40A$ GeV/c were shown. They include double differential (y, p_T) spectra, p_T-integrated rapidity distributions, and total yields. The rapidity widths of ϕ mesons produced in Ar+Sc collisions are similar to those from p+p collisions and different than in Pb+Pb. On the other hand, the ratio of ϕ to π mesons production in Ar+Sc collisions is slightly lower than for Pb+Pb reactions and significantly higher than for p+p. Additionally, the enhancement of ϕ/π mesons production in both Ar+Sc and Pb+Pb collisions over the p+p, is slightly higher than that for charged kaons relative to pions and independent of the collision energy in the considered range.

This work was supported by the Polish National Agency for Academic Exchange NAWA under the Programme STER - Internationalisation of doctoral schools, Project no. PPI/STE/2020/1/00020

References

[1] A. Aduszkiewicz et al. (NA61/SHINE), Eur. Phys. J. C **80**, 199 (2020). 10.1140/epjc/s10052-020-7675-6
[2] A.J. Marcinek, ϕ meson production in proton-proton collisions in the NA61/SHINE experiment at CERN SPS (2016), presented 18 Nov 2016, `https://cds.cern.ch/record/2234262`
[3] R. Aaij et al., Physics Letters. Section B: Nuclear, Elementary Particle and High-Energy Physics **703**, 267 (2011). 10.1016/j.physletb.2011.08.017
[4] S.V. Afanasiev et al. (The NA49 Collaboration), Phys. Rev. C **66**, 054902 (2002). 10.1103/PhysRevC.66.054902
[5] C. Alt et al. (NA49 Collaboration), Phys. Rev. C **77**, 024903 (2008). 10.1103/PhysRevC.77.024903
[6] T. Anticic et al. (The NA49 Collaboration), Phys. Rev. Lett. **93**, 022302 (2004). 10.1103/PhysRevLett.93.022302
[7] N. Abgrall et al., The European Physical Journal C **74** (2014). 10.1140/epjc/s10052-014-2794-6
[8] S. Pulawski, Identification and analysis of charged hadrons in p+p interactions from NA61/SHINE experiment at CERN SPS energies (2015), `https://cds.cern.ch/record/2240156`
[9] C. Alt et al. (NA49 Collaboration), Phys. Rev. C **78**, 044907 (2008). 10.1103/PhysRevC.78.044907
[10] S. Afanasiev et al., Physics Letters B **491**, 59 (2000). https://doi.org/10.1016/S0370-2693(00)01023-6
[11] A. Aduszkiewicz et al. (NA61/SHINE), Eur. Phys. J. C **77**, 671 (2017). 10.1140/epjc/s10052-017-5260-4
[12] H. Adhikary et al. (NA61/SHINE), Eur. Phys. J. C **84**, 416 (2024). 10.1140/epjc/s10052-024-12602-2

© The Authors, published by EDP Sciences. This is an open access article distributed under the terms of the Creative Commons Attribution License 4.0 (https://creativecommons.org/licenses/by/4.0/).

Clusters as a probe of the equation-of-state of strongly interacting matter

Susanne Gläßel[1], *Vadim* Voronyuk[2], *Michael* Winn[3], *Viktar* Kireyeu[2], *Gabriele* Coci[4], *Jörg* Aichelin[3,5], *Christoph* Blume[1,6,8], and *Elena* Bratkovskaya[6,7,8*],

[1]Institut für Kernphysik, Max-von-Laue-Str. 1, 60438 Frankfurt, Germany
[2]Joint Institute for Nuclear Research, Joliot-Curie 6, 141980 Dubna, Moscow region, Russia
[3]SUBATECH, Université de Nantes, IMT Atlantique, IN2P3/CNRS 4 rue Alfred Kastler, 44307 Nantes cedex 3, France
[4]Dipartimento di Fisica e Astronomia "E. Majorana", Università degli Studi di Catania, Via S. Sofia, 64, I-95125 Catania, Italy
[5]Frankfurt Institute for Advanced Studies, Ruth Moufang Str. 1, 60438 Frankfurt, Germany
[6]GSI Helmholtzzentrum für Schwerionenforschung GmbH, Planckstr. 1, 64291 Darmstadt, Germany
[7]Institut für Theoretische Physik, Johann Wolfgang Goethe University, Max-von-Laue-Str. 1, 60438 Frankfurt, Germany
[8]Helmholtz Research Academy Hessen for FAIR (HFHF), GSI Helmholtz Center for Heavy Ion Physics. Campus Frankfurt, 60438 Frankfurt, Germany

Abstract. We investigate the influence of the equation-of-state (EoS) of strongly interacting matter created in heavy-ion collisions on the light cluster and hypernuclei production within the Parton-Hadron-Quantum-Molecular Dynamics (PHQMD) microscopic transport approach. In earlier PHQMD calculations, nucleon interactions were modeled using a static, density-dependent potential corresponding to the soft and hard equation-of-state. In this study, we incorporate a momentum-dependent potential for the baryon-baryon interaction, derived from the soft EoS. We study the influence of momentum dependent potential on light cluster production.

1 Introduction

The study of light baryonic clusters in the central rapidity region of ultra-relativistic heavy-ion collisions has garnered significant theoretical and experimental interest. A key focus is understanding the mechanisms behind the production and survival of these loosely bound objects in the extreme heat and density of the collision zone. Additionally, a major challenge lies in accurately identifying and calculating these clusters in dynamical simulations of heavy-ion reactions. At low beam energies, the cluster production is directly linked to the nucleon dynamics, which is highly sensitive to baryon-baryon potential interactions related to nuclear matter equation-of-state (cf. a review and refs. therein [1]). One of the key question is related to study of the sensitivity of different observables to the equation-of-state of nuclear matter, which is a subject of this study.

*speaker at the 'Strange Quark Matter 2024'

© The Authors, published by EDP Sciences. This is an open access article distributed under the terms of the Creative Commons Attribution License 4.0 (https://creativecommons.org/licenses/by/4.0/).

2 Cluster production within the PHQMD

The Parton-Hadron-Quantum-Molecular Dynamics is a microscopic n-body transport model based on the QMD propagation of the baryonic degrees of freedom, where the clusters are formed via 'potential' and "kinetic" mechanisms [2–7].

The interaction between nucleons in the hadronic phase leads to the formation of bound clusters of varying sizes, with their multiplicity depending on the expansion dynamics and composition of the hot interaction zone. Modeling this cluster formation through nucleon interactions involves propagating the n-body phase space density, a process used in methods like Quantum Molecular Dynamics (QMD). Cluster identification throughout the system's evolution is achieved using the advanced Minimum Spanning Tree (aMST) method, which combines MST with a stabilization procedure, as outlined in Ref. [3]. It is important to note that MST is a recognition method for clusters, not a mechanism for their formation, as QMD propagates baryons rather than pre-formed clusters.

The identification of clusters at different stages of the system's dynamical evolution is carried out using the advanced Minimum Spanning Tree procedure, i.e. MST followed by the stabilization procedure, detailed in Ref. [6]. It is crucial to emphasize that MST serves as a cluster recognition method rather than a 'cluster-building' mechanism, given that the QMD transport approach propagates baryons rather than pre-formed clusters.

Additionally, 'kinetic' mechanisms for deuteron production are incorporated by catalytic hadronic reactions accounting for all isospin channels of the various $\pi NN \leftrightarrow \pi d$, $NNN \leftrightarrow Nd$ reactions which enhances deuteron production. It also considers the quantum nature of the deuteron by means of its finite size modelled by the finite-size excluded volume effect in coordinate space and a projection of the relative momentum of the interacting pair of nucleons on the deuteron wave-function in momentum space, leading to a strong reduction of deuteron production, especially at target/projectile rapidities.

In Refs. [8–11], we conducted a comparative analysis of the coalescence, potential and kinetic mechanisms for cluster formation within the PHQMD model and compared this also to the coalescence of UrQMD models. We have found that there are observables, which are sensitive to the deuteron production mechanism: the rapidity distribution has a different form and the transverse momentum distribution has a different slope at low p_T. These differences are large enough to be measurable and will allow, therefore, for discriminating between the different mechanisms for deuteron production when confronting these results with data [11].

In the previous PHQMD calculations we employed a static interaction between nucleons in terms of density dependent potential, which corresponds to the soft and hard EoS. Here have incorporated the momentum dependent potential for the baryon-baryon interaction which is evaluated from the soft EoS. The parameters of momentum dependent potential are fitted to the "optical" potential (i.e. Schrödinger equivalent potential U_{SEP}), extracted from elastic scattering data in pA reactions [12, 13]. The potential grows up to the total momentum of proton $p \sim 1.5$ GeV$/c$ and then decreases. The extended publication of the influence of momentum dependent potential is in preparation now. Here we report our preliminary results from this study.

3 Results

We have studied the influence of equation-of-state realized via different baryon-baryon potential: static soft (S), static hard (H) and soft momentum dependent (SM) potentials, on different observables such as rapidity distributions, transverse momentum spectra of protons and light clusters as well as collective observables - the flow coefficients v_1 and v_2 as a func-

© The Authors, published by EDP Sciences. This is an open access article distributed under the terms of the Creative Commons Attribution License 4.0 (https://creativecommons.org/licenses/by/4.0/).

tion of rapidity and p_T transverse momentum for different energies - from SIS to low BES RHIC.

In Figure 1 we show the PHQMD results for the transverse momentum p_T-spectra of protons, deuterons, tritons, ^{3}He, ^{4}He for Au+Au 0-10% central collisions at $\sqrt{s_{NN}} = 3$ GeV for different rapidity bins calculated with 3 different EoS: soft EoS -indicated as dashed lines, hard EoS - presented by the dotted lines and soft with momentum dependent potential - shown by solid lines. While at the low SIS energies the p_T-spectra show a pronounced splitting depending on the particular EoS chosen, at low BES-RHIC energies of $\sqrt{s_{NN}} = 3$ GeV the difference between static hard potential and momentum dependent soft potential are relatively small due to the fact that the potential decreases with increasing proton energy. A comparison of the PHQMD results to the STAR data [14] shows that the soft momentum dependent EoS provides a better description of STAR data. However, the hard static potential spectra are only slightly softer then those calculated with the momentum dependent soft EoS.

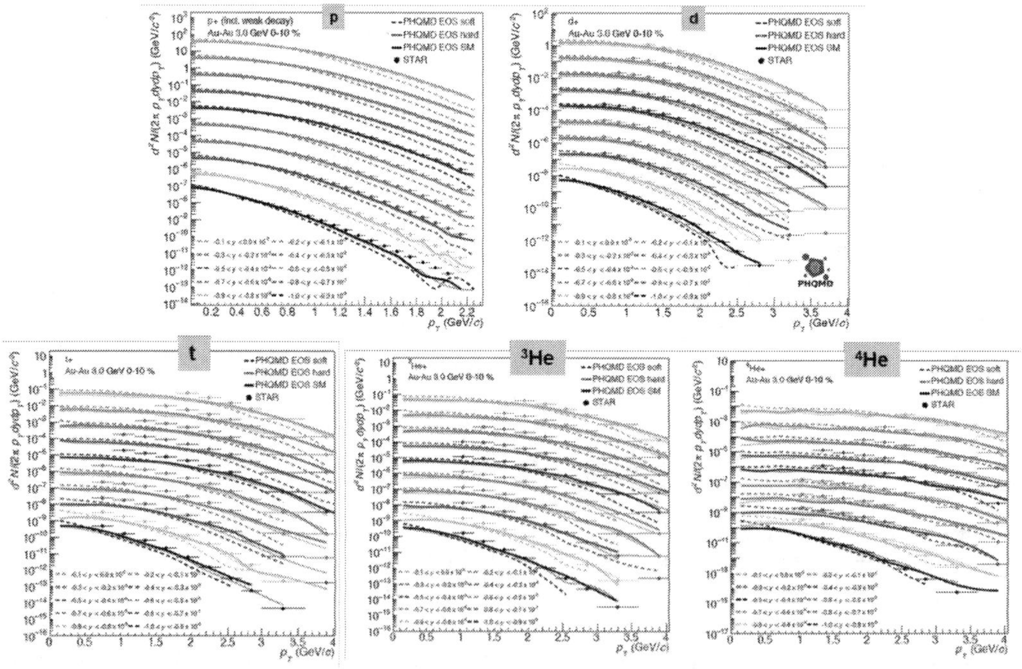

Figure 1. The PHQMD results for the transverse momentum p_T-spectra of protons, deuterons, tritons, ^{3}He, ^{4}He for Au+Au 0-10% central collisions at $\sqrt{s_{NN}} = 3$ GeV for different rapidity bins (scaled with 10^n factors for distinguishing lines) calculated with 3 different EoS: soft EoS - indicated as dashed lines, hard EoS - presented by the dotted lines and soft with momentum dependent potential - shown by solid lines, in comparison to the STAR data [14].

4 Conclusions

In this contribution we have discussed the influence of EoS on the cluster observables. The comparison of the PHQMD results to the HADES and STAR data show a strong sensitivity

© The Authors, published by EDP Sciences. This is an open access article distributed under the terms of the Creative Commons Attribution License 4.0 (https://creativecommons.org/licenses/by/4.0/).

of the transverse momentum spectra and the flow coefficient v_1 and especially elliptic flow coefficient v_2 on momentum dependence of the potential at SIS energies while this sensitivity decreases with increasing energies due to the decrease of potential as demonstrated in the example figure presented in this contribution. The extended study within the PHQMD on this issue is following up.

Acknowledgements: The authors acknowledge the partial support by the Deutsche Forschungsgemeinschaft (DFG, German Research Foundation), by the GSI-IN2P3 agreement under contract number 13-70 as well as by the European Union's Horizon 2020 research and innovation program under grant agreement STRONG–2020 – No 824093.

References

[1] A. Sorensen et al., Prog. Part. Nucl. Phys. **134**, 104080 (2024), 2301.13253

[2] J. Aichelin, Phys. Rept. **202**, 233 (1991)

[3] J. Aichelin, E. Bratkovskaya, A. Le Fèvre, V. Kireyeu, V. Kolesnikov, Y. Leifels, V. Voronyuk, G. Coci, Phys. Rev. C **101**, 044905 (2020), 1907.03860

[4] S. Gläßel, V. Kireyeu, V. Voronyuk, J. Aichelin, C. Blume, E. Bratkovskaya, G. Coci, V. Kolesnikov, M. Winn, Phys. Rev. C **105**, 014908 (2022), 2106.14839

[5] V. Kireyeu, Phys. Rev. C **103**, 054905 (2021), 2103.10542

[6] G. Coci, S. Gläßel, V. Kireyeu, J. Aichelin, C. Blume, E. Bratkovskaya, V. Kolesnikov, V. Voronyuk, Phys. Rev. C **108**, 014902 (2023), 2303.02279

[7] V. Kireyeu, G. Coci, S. Glaessel, J. Aichelin, C. Blume, E. Bratkovskaya (2023), 2304.12019

[8] V. Kireyeu, J. Steinheimer, J. Aichelin, M. Bleicher, E. Bratkovskaya, Phys. Rev. C **105**, 044909 (2022), 2201.13374

[9] E. Bratkovskaya, S. Glässel, V. Kireyeu, J. Aichelin, M. Bleicher, C. Blume, G. Coci, V. Kolesnikov, J. Steinheimer, V. Voronyuk, EPJ Web Conf. **276**, 03005 (2023), 2208.11802

[10] V. Kireyeu, G. Coci, S. Gläßel, J. Aichelin, C. Blume, V. Voronyuk, E. Bratkovskaya, EPJ Web Conf. **296**, 12003 (2024)

[11] V. Kireyeu, G. Coci, S. Gläßel, J. Aichelin, C. Blume, E. Bratkovskaya, Phys. Rev. C **109**, 044906 (2024)

[12] B.C. Clark, E.D. Cooper, S. Hama, Phys. Rev. C **73**, 024608 (2006)

[13] E.D. Cooper, S. Hama, B.C. Clark, R.L. Mercer, Phys. Rev. C **47**, 297 (1993)

[14] H. Liu (STAR), Acta Phys. Polon. Supp. **16**, 1 (2023), 2208.04650

© The Authors, published by EDP Sciences. This is an open access article distributed under the terms of the Creative Commons Attribution License 4.0 (https://creativecommons.org/licenses/by/4.0/).

Multiplicity dependence of hyperon and hypertriton production in Zr+Zr and Ru+Ru collisions at $\sqrt{s_{NN}}$ = 200 GeV

Dongsheng Li[1,*] (for the STAR collaboration)

[1]University of Science and Technology of China

Abstract. We present yield measurements on hyperons ($\Lambda, \overline{\Lambda}$ and $\Xi^-, \overline{\Xi}^+$) and hypertriton (${}^3_\Lambda$H, ${}^3_{\overline{\Lambda}}\overline{\text{H}}$) in four different centrality classes of Zr+Zr and Ru+Ru collisions at $\sqrt{s_{\rm NN}}$ = 200 GeV. The yield ratios of Λ/π^-, Ξ^-/π^-, ${}^3_\Lambda$H$/\Lambda$ and S_3 = (${}^3_\Lambda$H$/\Lambda$)/(^{3}He$/$p) are reported as a function of multiplicity, while the ratio of ${}^3_\Lambda$H$/{}^3$He is reported as a function of $p_{\rm T}$ in each centrality. The comparisons between data and models are discussed. These results provide insights on particle production mechanisms in heavy-ion collisions.

1 Introduction

In analogy to the Big Bang theory, which describes the origin and evolution of the universe, high energy heavy-ion collisions (HIC) can be dubbed the Little Bang. Although they are different systems, measurements on particle production in HIC would possibly help us understand the first few minutes of the evolution of our universe.

Recently, the dependence of particle production in HIC on system size has drawn a lot of attention [1–10, 15]. For hadron production, such dependence mainly reflects the properties of the hot medium. While for large nuclear clusters like hypertriton (${}^3_\Lambda$H), their internal structures might also leave some non-negligible fingerprints on the yields, because their nuclear size are of the same magnitude as the size of the fireball created in the collisions and may be even larger than the fireball size for some small collision systems. These studies aim to verify our understanding of particle production mechanisms in HIC.

2 Results and discussions

In system size studies, the charged-particle multiplicity within unit of pseudo-rapidity $\langle {\rm dN_{ch}}/{\rm d}\eta \rangle$ is usually suggested as a measure of system size. In this work, we report the multiplicity dependence of hyperon and ${}^3_\Lambda$H production in Zr+Zr and Ru+Ru collisions at $\sqrt{s_{\rm NN}}$ = 200 GeV, by analysing about 3.7 billion minimum bias events. These short-lived particles are reconstructed in STAR with the KFParticle Package [16], using the 2-body decay channels including $\Lambda \to {\rm p} + \pi^-$, $\Xi^- \to \Lambda + \pi^-$, ${}^3_\Lambda{\rm H} \to {}^3{\rm He} + \pi^-$ and their charge conjugates within the mid-rapidity ($|y| < 0.5$ for (anti)hyperon, $|y| < 0.8$ for (anti)hypertriton). The daughter particle tracks including $\pi^\pm$, p($\overline{\rm p}$), ^{3}He(${}^3\overline{\rm He}$) are identified with energy loss $\langle {\rm dE/dx} \rangle$ and momentum information measured by the Time Projection Chamber (TPC). The production yields are measured and the systematic uncertainties include components from tracking efficiency, topological cut variation, MC weighting and extrapolation.

*e-mail: erl@mail.ustc.edu.cn

© The Authors, published by EDP Sciences. This is an open access article distributed under the terms of the Creative Commons Attribution License 4.0 (https://creativecommons.org/licenses/by/4.0/).

2.1 Hyperon-to-pion ratio

In Fig. 1, hyperon-to-pion ratios are shown as a function of multiplicity, where the solid markers show the results from this study, while open markers show results from other collision systems [1–5, 11, 12]. In this analysis, the feed-down contributions from weak decay have been subtracted for Λ and $\overline{\Lambda}$, while for Ξ^- and $\overline{\Xi}^+$ such contributions are negligible. The yield ratios have similar values for the systems with similar multiplicity, which means that the strangeness production mechanisms are similar despite differences in the collision energies and the beam particle species. These yield ratios show a slightly increasing trend from small to large systems and such enhancement is usually considered as a signature of QGP formation in large systems.

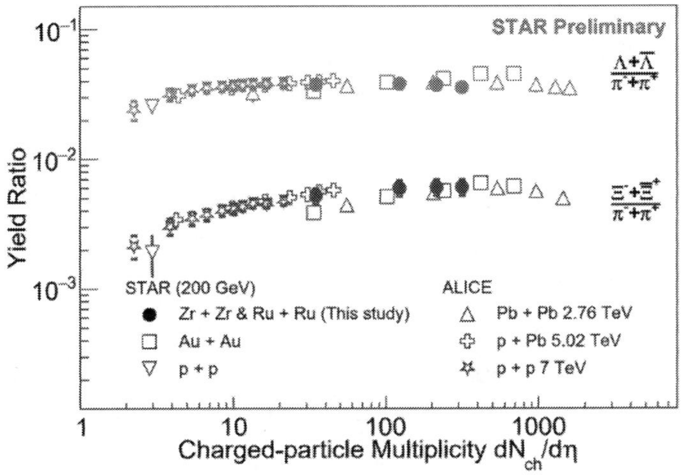

Figure 1. The hyperon-to-pion ratios as a function of charged-particle multiplicity. Yields of particles are combined with those of anti-particles. Red markers show the $(\Lambda + \overline{\Lambda})/(\pi^- + \pi^+)$ ratio, while blue markers show the $(\Xi^- + \overline{\Xi}^+)/(\pi^- + \pi^+)$ ratio, respectively [1–5, 11, 12].

2.2 $^3_\Lambda$H/Λ and S_3 ratios

The light hypernuclei production mechanisms in HIC are still not fully understood. The yield ratios of $^3_\Lambda\mathrm{H}/\Lambda$ and $S_3 = (^3_\Lambda\mathrm{H}/\Lambda)/(^3\mathrm{He}/p)$ are suggested as probes to distinguish different production mechanisms. Measurements on these hypernuclei yield ratios are shown in Fig. 2. In this analysis, the yields of Λ, proton and ^{3}He are corrected for feed-down from weak decay channels. Predictions from several popular theoretical models are also shown for comparison. The thermal model calculations are generated with canonical ensemble using the Thermal-Fist package [17]. The thermal model parameters for Zr+Zr and Ru+Ru collisions are obtained by fitting the yields of light hadrons, including $\pi^\pm$, $K^\pm$, p, $\Lambda(\bar{\Lambda})$ and $\Xi^-(\bar{\Xi}^+)$, in each centrality interval. The correlation volume (V_c) is varied from $V_c = dV/dy$ to $3dV/dy$. Other thermal model parameters are varied by 1-σ to generate the uncertainty bands. The thermal model predictions for LHC energies are directly taken from [15]. The analytical coalescence model calculations are generated with several assumptions of a thermalized hadron emission source, using both 2-body and 3-body Wigner functions to treat the nucleon coalescence process [18]. We also compared it with another coalescence model that applies the MUSIC and UrQMD models to simulate the QGP evolution and hadronic rescattering before the coalescence. Subsequently, it uses the 3-body Wigner function with different inputs of the Λ binding energy (B_Λ) of $^3_\Lambda$H for the coalescence afterburner [19].

The S_3 ratio in this study is roughly consistent with that in Au+Au and U+U collisions [13]. The $^3_\Lambda\mathrm{H}/\Lambda$ and S_3 ratios from this study and from the ALICE collaboration

© The Authors, published by EDP Sciences. This is an open access article distributed under the terms of the Creative Commons Attribution License 4.0 (https://creativecommons.org/licenses/by/4.0/).

measurements [14, 15] show similar trends. These results strongly deviate from the thermal model predictions, while agreeing with calculations by coalescence model of certain configurations. In particular, we find that the MUSIC + UrQMD + Coal. model with B_Λ = 0.42 MeV [20] can simultaneously describe both yield ratios well. However, we note that the B_Λ averaged over all current measurements is 0.164 ± 0.043 MeV, which is significantly smaller than 0.42 MeV. An increasing trend from small to large collision systems is observed in the ${}^{3}_{\Lambda}\mathrm{H}/\Lambda$ ratio, which is understood as a result of canonical suppression and possibly also the large nuclear size of ${}^{3}_{\Lambda}\mathrm{H}$. While for the S_3 ratio, a much weaker multiplicity dependence is observed compared to the ${}^{3}_{\Lambda}\mathrm{H}/\Lambda$ ratio, which is possibly due to the nuclear size effect since the conserved charges like baryon number and strangeness number all cancel out.

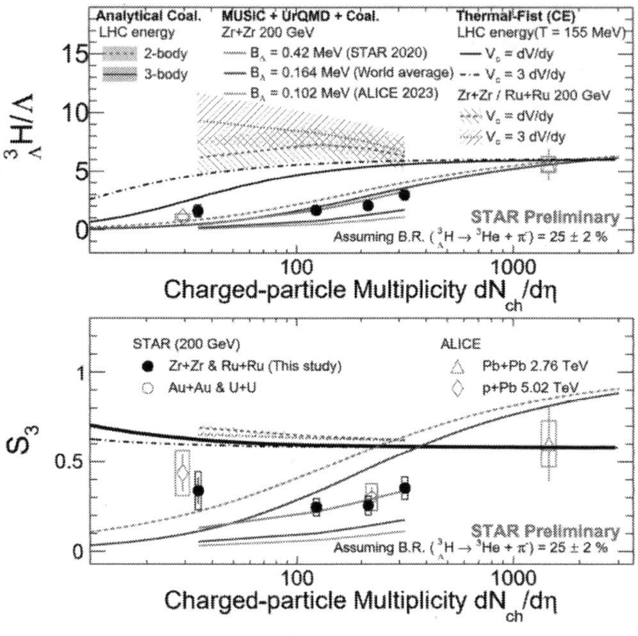

Figure 2. The multiplicity dependence of ${}^{3}_{\Lambda}\mathrm{H}/\Lambda$ and S_3. Solid circles show results from 200 GeV Zr+Zr and Ru+Ru collisions by this study, while open markers show results from other collision systems. Different model calculations are shown for comparison, including the analytical coalescence model (red and blue shaded bands), MUSIC + UrQMD + Coal. model of different Λ binding energy inputs (magenta, violet and grey lines) and Thermal-Fist model with the canonical ensemble assumption (black lines, green and brown shaded bands).

2.3 ${}^{3}_{\Lambda}\mathrm{H}/{}^{3}\mathrm{He}$ ratio

The constituents of ${}^{3}_{\Lambda}\mathrm{H}$ would be the same as those of ${}^{3}\mathrm{He}$ if one substitutes the Λ with a proton. Although their masses are very similar, they have very different sizes. The nuclear radius of ${}^{3}_{\Lambda}\mathrm{H}$ is ~ 5 fm, while that of ${}^{3}\mathrm{He}$ is ~ 2 fm, thus the coalescence model expects a strong multiplicity dependence and a p_T softening of the ${}^{3}_{\Lambda}\mathrm{H}/{}^{3}\mathrm{He}$ ratio [19]. While in the thermal model, all particles are generally treated point-like with no internal structure.

In Fig. 3, the ${}^{3}_{\Lambda}\mathrm{H}/{}^{3}\mathrm{He}$ ratio is measured in different centrality classes of Zr+Zr and Ru+Ru collisions, showing a weak p_T and multiplicity dependence. By comparing with the model predictions, it seems our data points are roughly described by the coalescence model with B_Λ= 0.42 MeV, while overestimated by the thermal model. Again, we note that this value seems to be large compared with the world average value.

3 Summary and outlook

The yield measurements on hyperons ($\Lambda, \overline{\Lambda}$ and $\Xi^-, \overline{\Xi}^+$) and hypertriton (${}^{3}_{\Lambda}\mathrm{H}, {}^{3}_{\overline{\Lambda}}\overline{\mathrm{H}}$) in Zr+Zr and Ru+Ru collisions at $\sqrt{s_{NN}}$ = 200 GeV are reported. The multiplicity dependence of

© The Authors, published by EDP Sciences. This is an open access article distributed under the terms of the Creative Commons Attribution License 4.0 (https://creativecommons.org/licenses/by/4.0/).

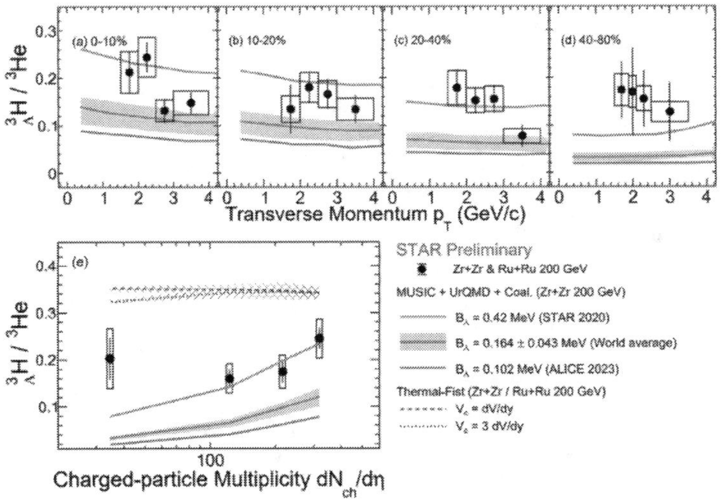

Figure 3. Fig.(a) - Fig.(d) show the p_T dependence of ${}^{3}_{\Lambda}H/{}^{3}He$, while Fig(e) shows the multiplicity dependence of the p_T integrated ratio. Black points show the measurement in this study. Model predictions from MUSIC + UrQMD + Coal. model of different Λ binding energy inputs are shown as orange, magenta and green lines, while Thermal-Fist are shown as the green and brown shaded bands.

hyperon and light hypernuclei production are investigated. It is found that the hyperon production mechanisms are similar in systems with the same multiplicity, despite differences of the collision energies and the beam particle species. For light hypernuclei production, the measurements roughly agree with certain coalescence model predictions and strongly deviate from the thermal model calculations. However, the coalescence model with the world average measured B_Λ does not describe the data well. Thus, more efforts from the theoretical side, as well as higher precision measurements in small systems will be necessary to elucidate the role of the nuclear size on the formation of light hypernuclei in HIC.

References

[1] J. Adams et al. (STAR collaboration), Phys. Rev. Lett. **98** (2007), 062301. https://doi.org/10.1103/PhysRevLett.98.062301

[2] B. B. Abelev et al. (ALICE collaboration), Phys. Lett. B **728** (2014), 216-227. https://doi.org/10.1016/j.physletb.2014.05.052 [erratum: Phys. Lett. B **734** (2014), 409-410. https://doi.org/10.1016/j.physletb.2014.05.052]

[3] B. B. Abelev et al. (ALICE collaboration), Phys. Lett. B **728** (2014), 25-38. https://doi.org/10.1016/j.physletb.2013.11.020]

[4] J. Adam et al. (ALICE collaboration), Phys. Lett. B **758** (2016), 389-401. https://doi.org/10.1016/j.physletb.2016.05.027

[5] J. Adam et al. (ALICE collaboration), Nature Phys. **13** (2017), 535-539. https://doi.org/10.1038/nphys4111

[6] S. Acharya et al. (ALICE collaboration), Eur. Phys. J. C **80** (2020) no.9, 889. https://doi.org/10.1140/epjc/s10052-020-8256-4

[7] S. Acharya et al. (ALICE collaboration), Phys. Rev. C **101** (2020) no.4, 044906. https://doi.org/10.1103/PhysRevC.101.044906

[8] S. Acharya et al. (ALICE collaboration), Phys. Lett. B **800** (2020), 135043. https://doi.org/10.1016/j.physletb.2019.135043

[9] S. Acharya et al. (ALICE collaboration), Phys. Lett. B **794** (2019), 50-63. https://doi.org/10.1016/j.physletb.2019.05.028

[10] J. H. Chen, X. Dong, X. H. He, H. Z. Huang, F. Liu et al. arXiv:2407.02935. https://doi.org/10.48550/arXiv.2407.02935

© The Authors, published by EDP Sciences. This is an open access article distributed under the terms of the Creative Commons Attribution License 4.0 (https://creativecommons.org/licenses/by/4.0/).

[11] B. I. Abelev et al. (STAR collaboration), Phys. Rev. C **75** (2007), 064901. https://doi.org/10.1103/PhysRevC.75.064901

[12] B. I. Abelev et al. (STAR collaboration), Phys. Rev. C **79** (2009), 034909. https://doi.org/10.1103/PhysRevC.79.034909

[13] M. I. Abdulhamid, et al. (STAR collaboration), Nature **632** (2024), 1026–1031. https://doi.org/10.1038/s41586-024-07823-0

[14] J. Adam et al. (ALICE collaboration), Phys. Lett. B **754** (2016), 360-372. https://doi.org/10.1016/j.physletb.2016.01.040

[15] S. Acharya et al. (ALICE collaboration), Phys. Rev. Lett. **128** (2022) no.25, 252003. https://doi.org/10.1103/PhysRevLett.128.252003

[16] X. Y. Ju, Y. H. Leung, S. Radhakrishnann et al., Nucl. Sci. Tech. **34** (2023) 158. https://doi.org/10.1007/s41365-023-01320-1

[17] V. Vovchenko and H. Stoecker, Comput. Phys. Commun. **244** (2019), 295-310. https://doi.org/10.1016/j.cpc.2019.06.024

[18] K. J. Sun, C. M. Ko and B. Dönigus, Phys. Lett. B **792** (2019), 132-137. https://doi.org/10.1016/j.physletb.2019.03.033

[19] D. N. Liu, C. M. Ko, Y. G. Ma, F. Mazzaschi, M. Puccio, Q. Y. Shou, K. J. Sun and Y. Z. Wang, Phys. Lett. B **855** (2024), 138855. https://doi.org/10.1016/j.physletb.2024.138855

[20] J. Adam et al. (STAR Collaboration) Nature Phys. **16** (2020) 4, 409-412. https://doi.org/10.1038/s41567-020-0799-7

© The Authors, published by EDP Sciences. This is an open access article distributed under the terms of the Creative Commons Attribution License 4.0 (https://creativecommons.org/licenses/by/4.0/).

Production of Light Nuclei in Au+Au Collisions with the STAR BES-II Program

Yixuan Jin[1,*] (For the STAR Collaboration)

[1]Institute of Particle Physics and Key Laboratory of Quark & Lepton Physics (MOE), Central China Normal University, Wuhan, 430079, China.

Abstract. The yields and ratios of light nuclei in heavy-ion collisions offer a method to distinguish between the thermal and coalescence models. Ratios such as $N_t \times N_p/N_d^2$ and $N_{^3He} \times N_p/N_d^2$ are suggested as potential probes to investigate critical phenomena within the QCD phase diagram. The significantly larger datasets from STAR BES-II compared to BES-I, combined with enhanced detector capabilities, allow for more precise measurements. In this proceeding, we present the centrality and energy dependence of transverse momentum spectra and particle yields of (anti-)proton, (anti-)deuteron, and ^{3}He at BES-II energies ($\sqrt{s_{NN}}$ = 7.7 – 27 GeV), as well as the light nuclei to proton yield ratios and coalescence parameters (B_2(d) and $B_3(^3$He)).

1 Introduction

In Quantum Chromodynamics (QCD) framework, quarks and gluons are confined within hadrons at low temperatures and baryon densities. Conversely, at higher temperatures and/or high baryon densities, hadronic matter transitions into quark-gluon plasma (QGP), where the dominant degrees of freedom are quarks and gluons. Experimentally, as the initial temperature (T) and the baryon chemical potential (μ_B) vary with the centre-of-mass collision energy, the phase diagram can be explored by adjusting the beam energy [1–3].

The Beam Energy Scan (BES) program at the Relativistic Heavy-Ion Collider (RHIC) aims to investigate QGP signatures, understand the nature of the phase transition, and identify the conjectured critical point. The second phase of the Beam Energy Scan (BES-II), with about ten times more data than the BES-I phase and enhanced detector capabilities, is driven by the need for more precise experimental results, allowing more firm physics conclusions.

Based on the coalescence model, the light nuclei compound ratios such as $N_t \times N_p/N_d^2$ and $N_{^3He} \times N_p/N_d^2$ are sensitive to the neutron density fluctuations. Therefore, the ratios have been suggested as a potential probe to search for the critical phenomena in the QCD phase diagram [4]. The STAR BES-I and fixed-target program have measured the light nuclei production in Au+Au collisions [5–7].

This proceeding presents the transverse momentum spectra, particle yields, and yield ratios for light nuclei in Au+Au collisions at $\sqrt{s_{NN}}$ = 7.7 – 27 GeV from the STAR BES-II program. Light nuclei to proton yield ratios decrease with increasing collision energy. Furthermore, the coalescence parameters (B_2(d) and $B_3(^3$He)) demonstrate centrality and p_T dependence, indicating the influence of collective expansion.

*e-mail: jyx@mails.ccnu.edu.cn

© The Authors, published by EDP Sciences. This is an open access article distributed under the terms of the Creative Commons Attribution License 4.0 (https://creativecommons.org/licenses/by/4.0/).

2 Analysis details

Particles are identified using the Time Projection Chamber (TPC) at low p_T, with Time of Flight (TOF) added at higher p_T. After signal extraction, the spectra are corrected for energy loss, TPC tracking efficiency, TOF matching efficiency, and knock-out effects specifically for protons and deuterons. The feed-down contributions of proton and anti-proton from the weak decay of strange baryons are also applied.

Point-to-point systematic uncertainties on the spectra are estimated by varying track selection and analysis cuts. The systematic uncertainties of dN/dy are assessed by considering the point-to-point uncertainties in spectra, and the extrapolated range from the spectra fitting by examining the deviation between the Blast-Wave model fit and the double-p_T ($p_0 e^{-p_T^2/p_1^2} + p_2 e^{-p_T^2/p_3^2}$) function.

3 Results

3.1 Corrected transverse momentum spectra

The transverse momentum (p_T) spectra of deuterons and ^{3}He in Au+Au collisions from the STAR BES-II program are shown in Fig. 1. Due to higher statistics, the BES-II spectra cover more centrality bins, and remain consistent with those from BES-I within the uncertainties. Additionally, thanks to the inner TPC (iTPC), the p_T acceptance is significantly enhanced, and the extended p_T ranges in BES-II contribute to a reduction in the systematic uncertainties of the particle yields.

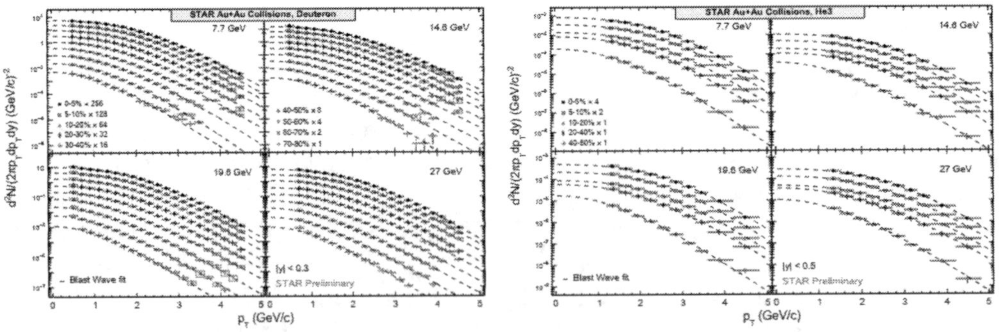

Figure 1. The transverse momentum spectra of deuteron (left) and ^{3}He (right) at mid-rapidity (with $|y| < 0.3$ for deuteron and $|y| < 0.5$ for ^{3}He) for different centrality bins in Au+Au collisions from the STAR BES-II program. The dashed lines represent the corresponding Blast-Wave fits. Statistical and systematic uncertainties are indicated by vertical lines and boxes, respectively.

3.2 Particle Yields

The particle yields (dN/dy) are extracted from p_T spectra, with measurement ranges integrated over p_T and extrapolations carried out using Blast-Wave model fits. Figure 2 shows dN/dy for different particles with the scaling of their corresponding spin degeneracy factor $2J + 1$. These scaled yields of light nuclei follow an exponential distribution of the form p_0/P^{A-1}, where P is the penalty factor determined by the Boltzmann factor $e^{(m_N - \mu_B)/T}$ in the thermal model. The penalty factor increases with higher beam energy, indicating that it becomes more challenging to form high-mass objects.

© The Authors, published by EDP Sciences. This is an open access article distributed under the terms of the Creative Commons Attribution License 4.0 (https://creativecommons.org/licenses/by/4.0/).

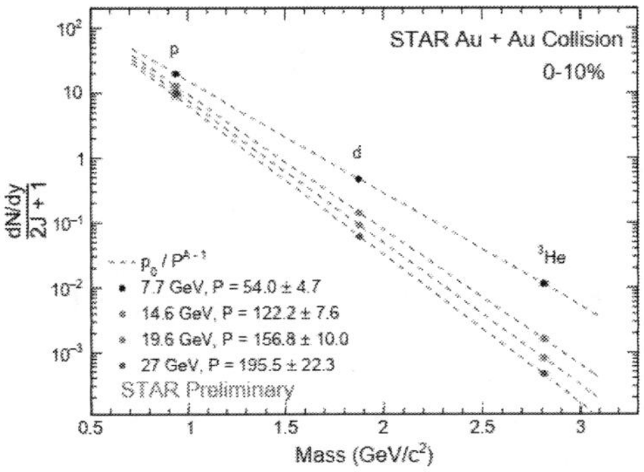

Figure 2. The mass m_A (where A is the mass number) dependence of $\mathrm{d}N/\mathrm{d}y$ for protons, deuterons, and ^{3}He, scaled by the spin degeneracy factor $2J+1$, is shown for Au+Au collisions at BES-II energies of $\sqrt{s_{\mathrm{NN}}} = 7.7 - 27$ GeV. Statistical uncertainties are indicated by vertical lines, while systematic uncertainties are shown by boxes. The yields are fitted using dashed lines representing exponential functions.

3.3 Particle Yield Ratios

Figure 3 presents the light nuclei to proton yield ratios consistent with BES-I results within uncertainties. The $\mathrm{N_d/N_p}$ and $\mathrm{N_{^3He}/N_p}$ ratios decrease monotonically with increasing collision energy, and the difference between these ratios becomes smaller at lower energies. The thermal model, shown by dashed lines, describes the $\mathrm{N_d/N_p}$ ratios well. However, it overestimates the $\mathrm{N_t/N_p}$ and $\mathrm{N_{^3He}/N_p}$ ratios by about factor 2, likely due to the effects of hadronic re-scattering during the hadronic expansion stage.

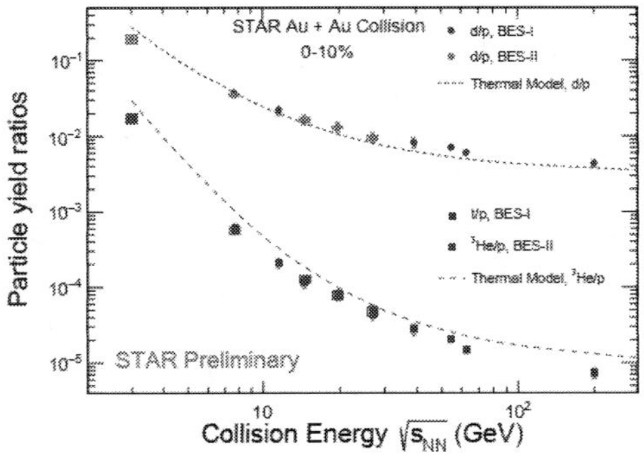

Figure 3. The energy dependence of $\mathrm{N_d/N_p}$ (the red circles) and $\mathrm{N_{^3He}/N_p}$ (the blue squares) in the most central 0-10% collisions from BES-II. The gray points represent $\mathrm{N_d/N_p}$ and $\mathrm{N_t/N_p}$ from BES-I. Vertical lines and boxes represent statistical and systematic uncertainties, respectively. The lines show the light nuclei to proton yield ratios from the thermal model.

3.4 Coalescence Parameters

In the coalescence picture:

$$E_A \frac{\mathrm{d}^3 N_A}{\mathrm{d}p_A^3} = B_A \left(E_p \frac{\mathrm{d}^3 N_p}{\mathrm{d}p_p^3} \right)^Z \left(E_n \frac{\mathrm{d}^3 N_n}{\mathrm{d}p_n^3} \right)^{A-Z} \approx B_A \left(E_p \frac{\mathrm{d}^3 N_p}{\mathrm{d}p_p^3} \right)^A \tag{1}$$

© The Authors, published by EDP Sciences. This is an open access article distributed under the terms of the Creative Commons Attribution License 4.0 (https://creativecommons.org/licenses/by/4.0/).

where A is the mass number and Z is the charge number of the light nucleus. p_n, p_p, p_A are the momenta of the neutron, proton, and nucleus.

The invariant yield of light nuclei is proportional to the invariant yield of nucleons, with the coalescence parameter B_A reflecting the probability of nucleon coalescence and relating to the local nucleon density. Figure 4 shows coalescence parameters as a function of p_T/A across different centralities. B_A increases with transverse momentum, suggesting an expanding collision system, and also increases from central to peripheral collisions, likely due to a decreasing source volume.

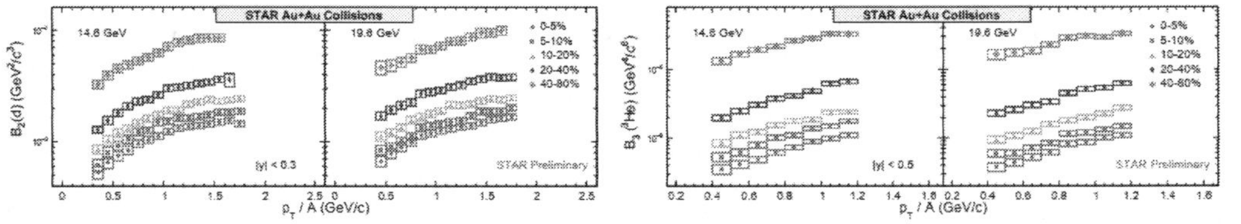

Figure 4. Number of constituent nucleon (NCN) scaling for p_T dependence of B_2 for deuteron (left) and B_3 for ^{3}He (right) in centrality 0-5%, 5-10%, 10-20%, 20-40% and 40-80% in Au+Au collisions at $\sqrt{s_{NN}}$ = 14.6 and 19.6 GeV. The boxes represent the systematic uncertainties.

4 Conclusions

The production of primordial proton, deuteron, and ^{3}He in Au+Au collisions at $\sqrt{s_{NN}} = 7.7 - 27$ GeV from RHIC STAR BES-II is reported in these proceedings. The particle ratios N_d/N_p and $N_{^3He}/N_p$ show a monotonic decrease with collision energy. The thermal model over-predicts N_t/N_p and $N_{^3He}/N_p$ by a factor of about 2. The coalescence parameter B_A increase from low to high p_T due to collective expansion, and decreasing source volume results in a rise of B_A from central to peripheral collisions. The analysis of BES-II data at 9.2, 11.5, and 17.3 GeV is currently underway, promising further insights into the energy-dependent behavior of the system.

5 Acknowledgement

This work was supported by National Key Research and Development Program of China (No.2022YFA1604900, No.2020YFE0202002), National Natural Science Foundation of China (No.12122505) and the Fundamental Research Funds for the Central Universities (CCNU220N003).

References

[1] A. Bzdak, S. Esumi, V. Koch, J. F. Liao, M. Stephanov, N. Xu. Phys. Rep. **853**, 1-87 (2020).
[2] X. F. Luo, Q. Wang, N. Xu, P. F. Zhuang, *Properties of QCD matter at high baryon density* (Springer, Singapore, 2022) 1-75
[3] J. H. Chen, X. Dong, X. H. HE, H. Z. Huang, F. Liu *et al.* arXiv:2407.02935[nucl-ex].
[4] K. J. Sun, L. W. Chen, C. M. Ko, Z. B. Xu. Phys. Lett. B. **774**, 103-107 (2017).
[5] J. Adam *et al.* [STAR Collaboration]. Phys. Rev. C 99 (2019) 6, 064905.
[6] M. I. Abdulhamid *et al.* [STAR Collaboration]. Phys. Rev. Lett. 130 (2023) 202301.
[7] STAR Collaboration. arXiv:2311.11020[nucl-ex].

© The Authors, published by EDP Sciences. This is an open access article distributed under the terms of the Creative Commons Attribution License 4.0 (https://creativecommons.org/licenses/by/4.0/).

EPJ Web of Conferences **316**, 06001 (2025) https://doi.org/10.1051/epjconf/202531606001
SQM 2024

Bayesian constraints on the high density QCD EoS for nuclear matter from Heavy-ion collision data

Manjunath Omana Kuttan[1,*], *Jan* Steinheimer[1,**], *Kai* Zhou[1,2], and *Horst* Stoecker[1,3,4]

[1]Frankfurt Institute for Advanced Studies, Ruth-Moufang-Str. 1, D-60438 Frankfurt am Main, Germany
[2]School of Science and Engineering, The Chinese University of Hong Kong, Shenzhen (CUHK-Shenzhen), Guangdong, 518172, China.
[3]Institut für Theoretische Physik, Goethe Universität Frankfurt, Max-von-Laue-Str. 1, D-60438 Frankfurt am Main, Germany
[4]GSI Helmholtzzentrum für Schwerionenforschung GmbH, Planckstr. 1, D-64291 Darmstadt, Germany

Abstract. A Bayesian framework is introduced to constrain the density dependence of the Equation of State (EoS) for nuclear matter. The EoS is inferred using existing measurements of elliptic flow and the mean transverse kinetic energy of protons in the beam energy range of $\sqrt{s_{NN}}$ = 2–10 GeV. Tight constraints are obtained for densities up to 4 times the nuclear saturation density. However, the results are highly sensitive to the choice of observables, highlighting the need for consistent, high-precision measurements in this energy range.

1 Introduction

Relativistic nuclei-nuclei collisions generate extremely hot and/or dense nuclear matter, which cools, expands and eventually freezes out into stable hadrons, traversing the phase diagram of quantum chromodynamics (QCD), providing a unique opportunity for laboratory exploration of the phase structure of strongly interacting nuclear matter. In theoretical calculations, the dynamics and evolution of such a hot and/or dense system are governed by the Equation of State (EoS), which relates the system's pressure to its energy density and net-baryon density. The EoS decides the location of phase boundaries, the nature of transitions between the phases and the critical regions in the QCD phase diagram, making its extraction one of the primary objectives of heavy-ion collision (HIC) experiments worldwide.

At vanishing chemical potential, first-principles lattice QCD calculations predict a smooth crossover for both chiral and deconfinement transitions [1–3]. However, at finite baryon densities, the fermionic sign problem prevents direct lattice QCD calculations, necessitating a more comprehensive approach which would involve comparing experimental data with QCD inspired, effective model calculations to infer the EoS [4–8]. Due to the lack of a definitive observable for the unambiguous extraction of the EoS, a global fit to multiple EoS-sensitive observables is necessary to systematically constrain or infer the EoS. Several machine learning methods have also been proposed to study the EoS [9–12].

For the beam energies used to study high baryon density matter ($\sqrt{s_{NN}}$ < 15 GeV), another challenge has been the absence of a consistent theoretical framework that describes

*e-mail: manjunath@fias.uni-frankfurt.de
**e-mail: steinheimer@fias.uni-frankfurt.de

© The Authors, published by EDP Sciences. This is an open access article distributed under the terms of the Creative Commons Attribution License 4.0 (https://creativecommons.org/licenses/by/4.0/).

both the early compression and subsequent expansion phases of the collision using a consistent EoS. This issue was recently addressed in [13, 14], where a realistic chiral mean-field EoS (CMF) was incorporated into both the non-equilibrium and equilibrium phases of the collision within the microscopic UrQMD [15, 16] model. By extending this method, arbitrary density-dependent EoSs can be implemented in UrQMD. Leveraging this approach, the present work [17] constrains the high-density EoS through a Bayesian inference procedure, utilizing existing experimental data. This not only provides the first quantitative analysis of the constraints that current data impose on the density dependence of the EoS but also establishes a robust analysis framework that can be employed for future experimental data.

2 Methodology

To consistently describe the entire evolution using a consistent EoS, we employ the microscopic UrQMD model, which has been augmented to incorporate a density-dependent potential. In the model, the density-dependent potential energy V enters the QMD equations of motion, and can be directly related to the pressure of the system [18], and thus to the EoS.

For our problem, we parameterise the potential energy V as a 7th degree polynomial for densities above 2 times saturation density (n_0), below which we fix it to be the CMF EoS,

$$i.e., V(n_B) = \sum_{i=1}^{7} \theta_i \left(\frac{n_B}{n_0} - 2\right)^i + h \quad (1)$$

where h=-22.07 MeV is set to ensure that the potential energy is a continuous function at $2n_0$.

Given model parameters $\boldsymbol{\theta}$ and experimental data $\mathbf{D}$, Bayes' theorem estimate the posterior $P(\boldsymbol{\theta}|\mathbf{D})$ by scaling our prior knowledge $P(\boldsymbol{\theta})$ based on the likelihood $P(\mathbf{D}|\boldsymbol{\theta})$ which quantifies how effectively the parameters describe the data. The parameters of the model to be constrained are the polynomial coefficients $\boldsymbol{\theta} = \{\theta_1, \theta_2, ..., \theta_7\}$. In order to constrain them, we will be using the experimental data $\mathbf{D}$ comprising elliptic flow (v_2) of mid rapidity protons from mid-central Au-Au collisions at $\sqrt{s_{\mathrm{NN}}}$ = 2.24, 2.32, 2.4, 2.42, 2.51, 3.0, 3.32, 3.84, 4.23 and 4.72 GeV [19–25] and the mean transverse kinetic energy ($\langle m_T \rangle - m_0$) in central Au-Au collisions at $\sqrt{s_{\mathrm{NN}}}$ = 3.83, 4.29, 6.27, 7.7 and 8.86 GeV [26–28].

To calculate the likelihood, numerous UrQMD simulations must be performed with a wide range of EoSs to generate predictions for comparison with the data. However, this approach can be computationally slow and infeasible. To address this, we train Gaussian Process (GP) models as fast emulators for the UrQMD simulations. Once trained, the GP emulator accurately predicts the observables, v_2 and $\langle m_T \rangle - m_0$.

3 Results and Conclusions

The experimental observations and the trained GP models are now employed to construct the posterior distributions for the EoS. The left panel of Figure 1 illustrates the posterior distribution obtained when all 15 data points were included in the inference. As shown, tight constraints can be established for baryon densities up to 4 n_0. The "MEAN" EoS constructed by averaging the sampled EoS at different densities, suggests a stiff EoS at these densities. To evaluate the sensitivity of the inference to the choice of observables, we performed an additional analysis excluding the $\langle m_T \rangle - m_0$ values for $\sqrt{s_{\mathrm{NN}}}$ = 3.83 and 4.29 GeV in the inference. The resulting posterior distribution, displayed in the right panel of Figure 1, shows significant differences for densities above 3 n_0. Notably, the exclusion of these two data points results in a softening of the EoS in the density range of 3 - 5 n_0, which is consistent with signals of a first-order phase transition.

© The Authors, published by EDP Sciences. This is an open access article distributed under the terms of the Creative Commons Attribution License 4.0 (https://creativecommons.org/licenses/by/4.0/).

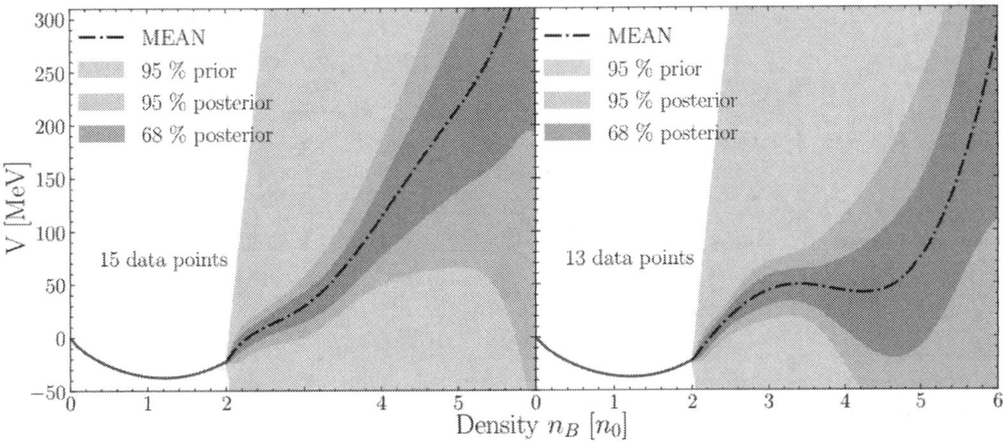

Figure 1. (Left) Posterior distribution for the EoS inferred using all 15 experimental observations. (Right) Posterior distribution for the EoS obtained excluding the $\langle m_T \rangle - m_0$ values for $\sqrt{s_{\mathrm{NN}}} = 3.83$ and 4.29 GeV (13 data points). The violet curve up to $2n_0$ represents the CMF EoS.

To investigate the significant deviation observed with the exclusion of just two data points, the MEAN EoSs extracted from both cases were implemented in UrQMD, and the v_2 and $\langle m_T \rangle - m_0$ values were calculated and compared with the experimental data. It was found that both EoSs reproduce the v_2 values reasonably well at low beam energies, with small deviations at higher energies ($\sqrt{s_{\mathrm{NN}}} > 3.5$ GeV). Interestingly, the $\langle m_T \rangle - m_0$ values for $\sqrt{s_{\mathrm{NN}}} > 6$ GeV showed better agreement with the experimental data when the two data points were excluded. This also resulted in an improved fit for v_2 for $\sqrt{s_{\mathrm{NN}}} > 3.5$ GeV. Conversely, when the two $\langle m_T \rangle - m_0$ data points were included in the inference, the fit to these specific points improved but led to a worse fit for v_2 at beam energies greater than 3.5 GeV. This indicates a potential tension between the v_2 and $\langle m_T \rangle - m_0$ data at around 4 GeV. Such a tension could either be a result of inconsistencies in the experimental measurements or is a limitation in the model's ability to accurately describe the physics at higher energies. For more details on the analysis and results, see [17].

In summary, using the bayesian framework, we find that tight constraints can be obtained for the density dependence of the EoS for densities up to 4 n_0 using the available heavy-ion collision data in in the beam energy range $\sqrt{s_{\mathrm{NN}}} = 2 - 10$ GeV. However, the results demonstrate significant sensitivity to the choice of observables, indicating that consistent data is critical for tightly constraining the EoS of nuclear matter at high densities. High-precision, consistent measurements are necessary for the unambiguous extraction of the EoS. Future data from the STAR-FXT program at RHIC, the CBM experiment at FAIR, and upcoming experiments at HIAF and NICA will be pivotal in achieving this goal. In future, the framework can be extended by incorporating additional observables (e.g., higher-order flow coefficients, differential flow observables) and introducing more parameters into the inference, such as a momentum dependence to the EoS, for a more comprehensive analysis.

Acknowledgment. The authors thank Volker Koch, Behruz Kardan and Shuzhe Shi for the insightful discussions. This work is supported by the Helmholtzzentrum für Schwerionenforschung GSI (M.O.K), the BMBF under the ErUM-Data and KISS project (M.O.K, K.Z), the SAMSON AG (M.O.K, J.S, K.Z) and the Walter Greiner Gesellschaft zur Förderung der physikalischen Grundlagenforschung e.V. through the Judah M. Eisenberg Laureatus Chair at the Goethe Universität Frankfurt am Main (H.S).

© The Authors, published by EDP Sciences. This is an open access article distributed under the terms of the Creative Commons Attribution License 4.0 (https://creativecommons.org/licenses/by/4.0/).

References

[1] Y. Aoki, G. Endrodi, Z. Fodor, S.D. Katz, K.K. Szabo, Nature **443**, 675 (2006). 10.1038/nature05120

[2] S. Borsanyi, Z. Fodor, C. Hoelbling, S.D. Katz, S. Krieg, K.K. Szabo, Phys. Lett. B **730**, 99 (2014). 10.1016/j.physletb.2014.01.007

[3] A. Bazavov et al., Phys. Rev. D **90**, 094503 (2014). 10.1103/PhysRevD.90.094503

[4] H. Stoecker, W. Greiner, Phys. Rept. **137**, 277 (1986). 10.1016/0370-1573(86)90131-6

[5] J. Hofmann, H. Stoecker, U.W. Heinz, W. Scheid, W. Greiner, Phys. Rev. Lett. **36**, 88 (1976). 10.1103/PhysRevLett.36.88

[6] H. Stoecker, Nucl. Phys. A **750**, 121 (2005). 10.1016/j.nuclphysa.2004.12.074

[7] M.A. Stephanov, K. Rajagopal, E.V. Shuryak, Phys. Rev. Lett. **81**, 4816 (1998). 10.1103/PhysRevLett.81.4816

[8] Y. Hatta, M.A. Stephanov, Phys. Rev. Lett. **91**, 102003 (2003), [Erratum: Phys.Rev.Lett. 91, 129901 (2003)]. 10.1103/PhysRevLett.91.102003

[9] L.G. Pang, K. Zhou, N. Su, H. Petersen, H. Stöcker, X.N. Wang, Nature Commun. **9**, 210 (2018). 10.1038/s41467-017-02726-3

[10] J. Steinheimer, L. Pang, K. Zhou, V. Koch, J. Randrup, H. Stoecker, JHEP **12**, 122 (2019). 10.1007/JHEP12(2019)122

[11] Y.L. Du, K. Zhou, J. Steinheimer, L.G. Pang, A. Motornenko, H.S. Zong, X.N. Wang, H. Stöcker, Eur. Phys. J. C **80**, 516 (2020). 10.1140/epjc/s10052-020-8030-7

[12] M. Omana Kuttan, K. Zhou, J. Steinheimer, A. Redelbach, H. Stoecker, JHEP **21**, 184 (2020). 10.1007/JHEP10(2021)184

[13] M. Omana Kuttan, A. Motornenko, J. Steinheimer, H. Stoecker, Y. Nara, M. Bleicher, Eur. Phys. J. C **82**, 427 (2022). 10.1140/epjc/s10052-022-10400-2

[14] J. Steinheimer, M. Omana Kuttan, A. Motornenko, A. Sorensen, Y. Nara, V. Koch, M. Bleicher, H. Stoecker, EPJ Web Conf. **276**, 01021 (2023). 10.1051/epjconf/202327601021

[15] S.A. Bass et al., Prog. Part. Nucl. Phys. **41**, 255 (1998). 10.1016/S0146-6410(98)00058-1

[16] M. Bleicher et al., J. Phys. G **25**, 1859 (1999). 10.1088/0954-3899/25/9/308

[17] M. Omana Kuttan, J. Steinheimer, K. Zhou, H. Stoecker, Phys. Rev. Lett. **131**, 202303 (2023). 10.1103/PhysRevLett.131.202303

[18] J. Steinheimer, A. Motornenko, A. Sorensen, Y. Nara, V. Koch, M. Bleicher, Eur. Phys. J. C **82**, 911 (2022). 10.1140/epjc/s10052-022-10894-w

[19] C. Pinkenburg et al., Phys. Rev. Lett. **83**, 1295 (1999). 10.1103/PhysRevLett.83.1295

[20] D. Adamova et al., Nucl. Phys. A **698**, 253 (2002). 10.1016/S0375-9474(01)01371-9

[21] A. Andronic et al., Phys. Lett. B **612**, 173 (2005). 10.1016/j.physletb.2005.02.060

[22] L. Adamczyk et al., Phys. Rev. C **86**, 054908 (2012). 10.1103/PhysRevC.86.054908

[23] J. Adam et al., Phys. Rev. C **103**, 034908 (2021). 10.1103/PhysRevC.103.034908

[24] J. Adamczewski-Musch et al., Phys. Rev. Lett. **125**, 262301 (2020). 10.1103/PhysRevLett.125.262301

[25] M.S. Abdallah et al., Phys. Lett. B **827**, 137003 (2022). 10.1016/j.physletb.2022.137003

[26] L. Ahle et al., Phys. Rev. C **60**, 064901 (1999). 10.1103/PhysRevC.60.064901

[27] C. Alt et al., Phys. Rev. C **73**, 044910 (2006). 10.1103/PhysRevC.73.044910

[28] L. Adamczyk et al., Phys. Rev. C **96**, 044904 (2017). 10.1103/PhysRevC.96.044904

© The Authors, published by EDP Sciences. This is an open access article distributed under the terms of the Creative Commons Attribution License 4.0 (https://creativecommons.org/licenses/by/4.0/).

4D-TExS: A new 4D lattice-QCD equation of state with extended density coverage

Johannes Jahan[1,*], *Ahmed* Abuali[1], *Szabolcs* Borsányi[2], *Micheal* Kahangirwe[1], *Paolo* Parotto[3], *Attila* Pásztor[4,5], *Claudia* Ratti[1], *Hitansh* Shah[1], and *Seth A.* Trabulsi[1]

[1]Department of Physics, University of Houston, Houston, TX 77204, USAl
[2]Department of Physics, Wuppertal University, Gaussstr. 20, D-42119, Wuppertal, Germany
[3]Dipartimento di Fisica, Università di Torino and INFN Torino, Via P. Giuria 1, I-10125 Torino, Italy
[4]Institute for Theoretical Physics, ELTE Eötvös Loránd University, Pázmány P. sétány 1/A, H-1117 Budapest, Hungary
[5]HUN-REN-ELTE Theoretical Physics Research Group, Pázmány Péter sétány 1/A,1117 Budapest, Hungary

Abstract. Although calculations of QCD thermodynamics from first-principle lattice simulations are limited to zero net-density due to the fermion sign problem, several methods have been developed to extend the equation of state (EoS) to finite values of the B, Q, S chemical potentials. Taylor expansion around $\mu_i = 0$ ($i = B, Q, S$) enables to cover with confidence the region up to $\mu_i/T < 2.5$. Recently, a new method has been developed to compute a 2D EoS in the (T, μ_B) plane. It was constructed through a T-expansion scheme (TExS), based on a resummation of the Taylor expansion, and is trusted up to densities around $\mu_B/T = 3.5$. We present here the new 4D-TExS EoS, a generalization of the TExS to all 3 chemical potentials, expected to offer a larger coverage than the 4D Taylor expansion EoS. After explaining the basics of the T-Expansion Scheme and how it is generalized to multiple dimensions, we will present results for thermodynamic observables as functions of temperature and both finite baryon and strangeness chemical potentials.

1 Introduction

The exploration of the phase diagram of nuclear matter at finite density, to understand how matter behaves in extreme conditions, is a very active field of research nowadays [1]. In order to interpret experimental results and bridge the gap with theory, heavy-ion collision and neutron star merger simulations require a nuclear EoS as input. With the recent development of new frameworks taking into account local charge conservation and diffusion, mainly to study the existence of a critical point in the nuclear phase diagram, or the baryon stopping for instance, the need for a four-dimensional EoS with independent chemical potentials is growing. Since lattice QCD simulations are hindered from going to finite density because of the sign problem [2], one has to use expansion methods to reach this purpose from first-principle QCD calculations. The most successful attempt so far was achieved through a Taylor expansion in several conserved-charge chemical potentials [3, 4], which is, however, limited to $\hat{\mu}_i$ ($\equiv \mu_i/T$) $\leqslant 2 - 2.5$, with $i = B, Q, S$ [5]. We present here the generalization to several dimensions of a new expansion method developed in Refs. [5, 6].

*e-mail: jjahan@central.uh.edu

© The Authors, published by EDP Sciences. This is an open access article distributed under the terms of the Creative Commons Attribution License 4.0 (https://creativecommons.org/licenses/by/4.0/).

2 T-Expansion Scheme

The EoS construction we present is based on the T-expansion scheme (TExS), initially proposed in [5] as a 2-dimensional construction, and extended in [6] through the addition of a correction to take into account the Stefan-Boltzmann limit at high temperature. Noticing that $\lim_{\hat{\mu}_X \to 0} \frac{\chi_1^X(T,\hat{\mu}_X)}{\hat{\mu}_X} = \chi_2^X(T,0)$, one can start from the ansatz [6]:

$$\frac{\chi_1^X(T,\hat{\mu}_X)}{\overline{\chi}_1^X(\hat{\mu}_X)} = \frac{\chi_2^X(T',0)}{\overline{\chi}_2^X(0)}\,, \tag{1}$$

with χ_j^X the susceptibilities of conserved charge X, computed on the lattice at $\hat{\mu} = 0$. The rescaled temperature T' is expanded as $T' = T\left(1 + \lambda_2^X(T)\hat{\mu}_X^2 + (\mathcal{O}^4)\right)$, thus giving to this expansion scheme a dependence on both T and $\hat{\mu}_X = \mu/T$, whereas the Taylor expansion provides only an expansion in $\hat{\mu}_X$ at constant T [4].

3 4D-TExS - Generalization to 4 dimensions

One can generalize the approach defined in Sec. 2 from a two-dimensional phase space $(T,\hat{\mu}_B)$, with one conserved charge X, to a four-dimensional phase space $(T,\hat{\mu}_B,\hat{\mu}_Q,\hat{\mu}_S)$, with B the baryon number, Q the electric charge and S the strangeness number. By a simple change of variables to spherical coordinates, with the following definitions:

$$\begin{aligned}
\hat{\mu}_B &= \hat{\mu}\,.\,\cos(\theta) & & \hat{\mu} = \sqrt{\hat{\mu}_B^2 + \hat{\mu}_Q^2 + \hat{\mu}_S^2}\,,\\
\hat{\mu}_Q &= \hat{\mu}\,.\,\sin(\theta)\cos(\varphi) & \Longleftrightarrow\quad & \varphi = \arccos\left(\hat{\mu}_Q \Big/ \sqrt{\hat{\mu}_Q^2 + \hat{\mu}_S^2}\right)\,,\\
\hat{\mu}_S &= \hat{\mu}\,.\,\sin(\theta)\sin(\varphi) & & \theta = \arccos\left(\hat{\mu}_B/\hat{\mu}\right)\,,
\end{aligned} \tag{2}$$

one maps the system to a $(T,\hat{\mu},\theta,\varphi)$ space, hence using the same two-dimensional TExS approach described in Sec. 2, but now allowing to explore three finite chemical potentials at the same time by varying the angles θ and φ. One needs thus to define the so-called generalized second-order susceptibility:

$$\begin{aligned}
X_2^{\theta,\varphi}(T) = \left.\frac{\partial^2 P/T^4}{\partial\hat{\mu}^2}\right|_{\hat{\mu}=0} &= c_\theta^2 \cdot \chi_2^B(T) + s_\theta^2 c_\varphi^2 \cdot \chi_2^Q(T) + s_\theta^2 s_\varphi^2 \cdot \chi_2^S(T)\\
&+ 2c_\theta s_\theta c_\varphi \cdot \chi_{11}^{BQ}(T) + 2c_\theta s_\theta s_\varphi \cdot \chi_{11}^{BS}(T) + 2s_\theta^2 c_\varphi s_\varphi \cdot \chi_{11}^{QS}(T)\,,
\end{aligned} \tag{3}$$

as a linear combination of the second-order charge susceptibilities, where $c_{\theta/\varphi} \equiv \cos(\theta/\varphi)$ and $s_{\theta/\varphi} \equiv \sin(\theta/\varphi)$. In the same way, one obtains by further derivation the generalized fourth-order cumulant $X_4^{\theta,\varphi}(T)$.

This leads to rewrite the main identity from Eq. (1) using $X_1^{\theta,\varphi}$ and $X_2^{\theta,\varphi}$, where the expanded temperature is written $T'^{\theta,\varphi}(T,\hat{\mu}) = T\left(1 + \lambda_2^{\theta,\varphi}(T) \times \hat{\mu}_B^2 + ...\right)$. The second-order expansion coefficient is expressed as:

$$\lambda_2^{\theta,\varphi}(T) = \frac{1}{6T}\frac{1}{X_2'^{\theta,\varphi}(T)} \times \left(X_4^{\theta,\varphi}(T) - \frac{\overline{X}_4^{\theta,\varphi}}{\overline{X}_2^{\theta,\varphi}} X_2^{\theta,\varphi}(T)\right)\,. \tag{4}$$

© The Authors, published by EDP Sciences. This is an open access article distributed under the terms of the Creative Commons Attribution License 4.0 (https://creativecommons.org/licenses/by/4.0/).

4 Results and thermodynamics

In this section, we show as a proof of concept some preliminary results obtained for an expansion along a trajectory with $(\theta, \varphi) = (90^\circ, 135^\circ)$, where $\mu_B = -\mu_S$. All the lattice QCD data used in our construction have been computed by the Wuppertal-Budapest collaboration, and have been matched with hadron resonance gas (HRG) calculation for $T < 130$ MeV [7].

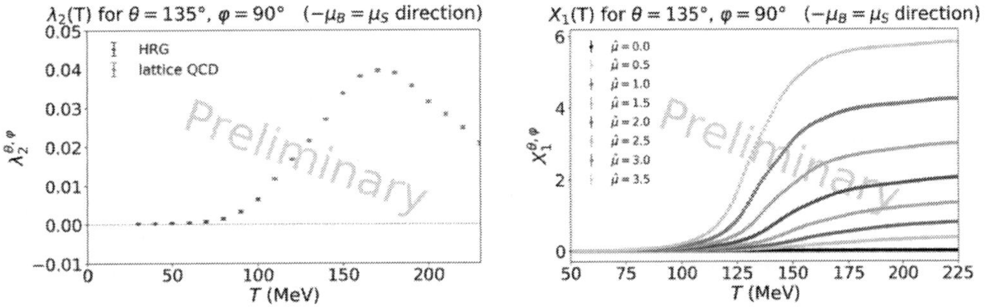

Figure 1. *Left panel:* T' expansion coefficient $\lambda_2^{\theta,\varphi}$ as a function of T, for $\theta = 135^\circ$ and $\varphi = 90^\circ$. We display in blue the range calculated with HRG data, and in red the range from lattice QCD calculations. *Right panel:* generalized charge density X_1 as a function of T, for different values of $\hat{\mu}$ from 0 to 3.5.

We start the construction of the EoS by calculating the generalized second-order and fourth-order susceptibilities for a given direction (θ, φ) in the phase diagram. From there, one obtains the expansion parameter $\lambda_2^{\theta,\varphi}$ as a function of T, shown on the left panel of Fig.1 for the chosen direction. Note that, for consistency, we stop the expansion of T' at second order, since the $\lambda_4^{\theta,\varphi}$ coefficient relies on sixth-order susceptibilities which have not all been computed yet for mixed-charges B, Q, S. The generalized charged density $X_1^{\theta,\varphi}$, thus obtained using Eq. (1), is shown on the right panel of Fig.1 for different values of $\hat{\mu}$. From there, one can compute pressure by integrating $X_1^{\theta,\varphi}$ using $P^{\theta,\varphi}(T, \hat{\mu}) = P(T, 0) + \int_0^{\hat{\mu}} X_1^{\theta,\varphi}(T, \hat{\mu}')d\hat{\mu}'$, with the zero-density pressure $P(T, 0)$ obtained from lattice simulations. One can obtain entropy density directly from pressure as well, using $s^{\theta,\varphi}(T, \hat{\mu}) = s(T, 0) + \left[\partial P^{\theta,\varphi}(T, \hat{\mu})/\partial T\right]_{\hat{\mu}}$. Both quantities are displayed in Fig.2 as functions of T, for different values of $\hat{\mu}$. Although pressure looks smooth up to $\hat{\mu} = 3.5$, one can observe that the entropy density starts to display oscillations as a function of T from $\hat{\mu} = 2.5$. Even though it remains an increasing function of temperature, as physically expected, these oscillations are most likely due to a problem in the way lattice susceptibilities used to construct the EoS are fitted.

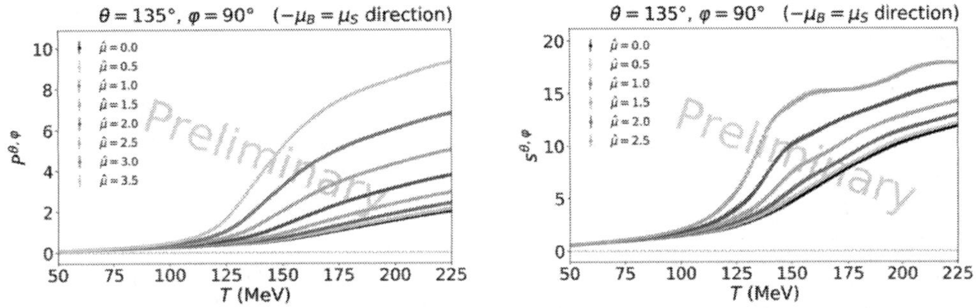

Figure 2. Pressure (*left panel*) and entropy density (*right panel*) as a function of T, for $\theta = 135^\circ$ and $\varphi = 90^\circ$. Different colors correspond to different values of $\hat{\mu}$ used, ranging from 0 to 3.5.

© The Authors, published by EDP Sciences. This is an open access article distributed under the terms of the Creative Commons Attribution License 4.0 (https://creativecommons.org/licenses/by/4.0/).

5 Summary & Outlook

We present a formalism to construct a four-dimensional EoS based on a generalization of the T-expansion scheme introduced in Refs. [5, 6], by means of a change of coordinates. This new expansion method, which consist of a resummation of the Taylor expansion, has been used to compute an equation of state at finite (T, μ_B, μ_S), for which thermodynamics quantities has been shown. Although this expansion scheme has been shown to be trusted up to $\hat{\mu}_B = 3.5$ when other chemical potentials are 0 [5], the values it can reach in other direction is still left to be determined. In the future, we will make calculation to finite chemical potentials for all three conserved charges B, Q and S. We will work on the modeling of the susceptibilities, as we need to ensure a smooth merging between HRG and lattice QCD regimes. At the same time, we will work as well on extending the range of the susceptiblities by extrapolating them to higher temperature, using lattice QCD data as a guidance and by ensuring a convergence to their respective Stefan-Boltzmann limits at $T \to \infty$. Finally, it is worth mentioning that all the work presented here being preliminary, no proper error estimate has been performed yet.

Acknowledgements

This material is based upon work supported in part by the National Science Foundation within the framework of the MUSES collaboration, under grant number No. OAC-2103680, as well as under grants No. PHY-2208724, PHY-1748958 and PHY-2116686, in part by the U.S. Department of Energy, Office of Science, Office of Nuclear Physics, under Award Number DE-SC0022023, as well as by the National Aeronautics and Space Agency (NASA) under Award Number 80NSSC24K0767.

References

[1] R. Kumar et al. (MUSES), Theoretical and experimental constraints for the equation of state of dense and hot matter, Living Rev. Rel. **27**, 3 (2024), 2303.17021. 10.1007/s41114-024-00049-6

[2] K. Splittorff, J.J.M. Verbaarschot, The QCD Sign Problem for Small Chemical Potential, Phys. Rev. D **75**, 116003 (2007), hep-lat/0702011. 10.1103/PhysRevD.75.116003

[3] A. Monnai, B. Schenke, C. Shen, Equation of state at finite densities for QCD matter in nuclear collisions, Phys. Rev. C **100**, 024907 (2019), 1902.05095. 10.1103/PhysRevC.100.024907

[4] J. Noronha-Hostler, P. Parotto, C. Ratti, J.M. Stafford, Lattice-based equation of state at finite baryon number, electric charge and strangeness chemical potentials, Phys. Rev. C **100**, 064910 (2019), 1902.06723. 10.1103/PhysRevC.100.064910

[5] S. Borsányi, Z. Fodor, J.N. Guenther, R. Kara, S.D. Katz, P. Parotto, A. Pásztor, C. Ratti, K.K. Szabó, Lattice QCD equation of state at finite chemical potential from an alternative expansion scheme, Phys. Rev. Lett. **126**, 232001 (2021), 2102.06660. 10.1103/PhysRevLett.126.232001

[6] S. Borsanyi, J.N. Guenther, R. Kara, Z. Fodor, P. Parotto, A. Pasztor, C. Ratti, K.K. Szabo, Resummed lattice QCD equation of state at finite baryon density: Strangeness neutrality and beyond, Phys. Rev. D **105**, 114504 (2022), 2202.05574. 10.1103/PhysRevD.105.114504

[7] V. Vovchenko, H. Stoecker, Thermal-FIST: A package for heavy-ion collisions and hadronic equation of state, Comput. Phys. Commun. **244**, 295 (2019), 1901.05249. 10.1016/j.cpc.2019.06.024

© The Authors, published by EDP Sciences. This is an open access article distributed under the terms of the Creative Commons Attribution License 4.0 (https://creativecommons.org/licenses/by/4.0/).

Critical Point Fluctuations in Heavy-Ion Collisions within Molecular Dynamics with Expansion

Volodymyr A. Kuznietsov[1,3,*], *Volodymyr* Vovchenko[1], *Volker* Koch[2], and *Mark I.* Gorenstein[3]

[1]Physics Department, University of Houston, 3507 Cullen Blvd, Houston, TX 77204, USA
[2]Nuclear Science Division, Lawrence Berkeley National Laboratory, 1 Cyclotron Road, Berkeley, CA 94720, USA
[3]Department of the high-density Energy Physics, Bogolyubov Institute for Theoretical Physics, 03680 Kyiv, Ukraine

Abstract. We analyze particle number fluctuations near the critical endpoint of a first-order phase transition using molecular dynamics simulations of the Lennard-Jones fluid. Building on our previous study [V.A. Kuznietsov et al., Phys. Rev. C **105**, 044903 (2022)], we incorporate longitudinal collective flow. The scaled variance of particle number distributions in various coordinate and momentum space acceptances is computed through ensemble averaging, confirming previous time-averaging results. We find that significant collective flow is crucial for observing large fluctuations from the critical point in momentum space. These results are discussed in the context of heavy-ion collisions.

1 Introduction

Identifying the QCD critical point (CP) at finite baryon density remains a primary objective in beam energy scans of heavy-ion collisions [1]. Proton number fluctuations and their cumulants are critical for CP detection [2, 3], as non-monotonic behavior in cumulants is expected near the CP [4]. Recent measurements [5] observed excess variance below 20 GeV, exceeding predictions from baryon conservation and repulsive interactions [6]. Prior molecular dynamics (MD) simulations of the Lennard-Jones (LJ) fluid demonstrated strong CP signals in coordinate space, but they disappeared in momentum space due to a lack of particle correlations in the uniform LJ system [7, 8]. In this work, we extend our analysis to heavy-ion conditions. Using ensemble averaging over multiple events, in line with the ergodic hypothesis [9], and incorporating longitudinal flow (correlating momenta with coordinates), we assess whether large fluctuations remain detectable within experimentally relevant acceptances.

2 Simulation setup

2.1 Lennard-Jones Fluid

The LJ fluid consists of classical non-relativistic particles interacting via the potential:

$$V_{\rm LJ}(r) = 4\varepsilon\left[\left(\frac{\sigma}{r}\right)^{12} - \left(\frac{\sigma}{r}\right)^{6}\right]. \quad (1)$$

*e-mail: vkuzniet@cougarnet.uh.edu

© The Authors, published by EDP Sciences. This is an open access article distributed under the terms of the Creative Commons Attribution License 4.0 (https://creativecommons.org/licenses/by/4.0/).

The first term represents short-range repulsion, while the second models intermediate-range attraction. Parameters σ and ε define the repulsive core size and attractive well depth, respectively, setting the system's length and energy scales.

Using dimensionless variables, we define the reduced temperature $\tilde{T} = T/(k_B\varepsilon)$, density $\tilde{n} = n\sigma^3$, and time $\tilde{t} = t\sqrt{\varepsilon/(m\sigma^2)}$. In these terms, most LJ fluid properties, including the phase diagram, become independent of σ and ε.

While the LJ equation of state is not exact, it has been extensively studied via molecular dynamics (MD). Its phase diagram includes a first-order liquid-gas transition with a critical point (CP) in the 3D-Ising universality class [10], located at $\tilde{T}_c = 1.321 \pm 0.007$ and $\tilde{n}_c = 0.316 \pm 0.005$ [11].

2.2 Molecular Dynamics

MD simulations integrate Newton's equations of motion using the Velocity-Verlet method for N particles with periodic boundary conditions [12]. In our previous work, we studied particle number fluctuations along the supercritical isotherm $\tilde{T} = 1.06\tilde{T}_c$ by simulating for extended periods and computing the moments of particle number distributions as time averages.

Here, we replicate the conditions of Ref. [8] using the same GPU-accelerated MD code [13]. The key difference is that we now calculate observables as ensemble averages by performing multiple simulations with random initial conditions at each density. This allows us to compare ensemble and time averaging, and to study equilibration dynamics. Our simulations use $N = 400$ particles, corresponding approximately to the number of baryons in central heavy-ion collisions when baryon-antibaryon pair production is negligible.

3 Incorporating longitudinal flow

In heavy-ion collisions, collective flow causes correlations between particle coordinates and momenta during the final stage of hydrodynamic evolution. To explore this, we incorporate longitudinal flow into our simulations to study fluctuations in rapidity acceptance, relevant to heavy-ion collisions. We focus on the longitudinal direction, integrating out transverse momenta. The longitudinal rapidity y for each particle is defined as the sum of collective and thermal components:

$$y = y^{\text{coll}} + y^{\text{th}}, \quad \text{where} \quad y^{\text{th}} = \sqrt{\frac{T_{\text{frz}}}{m_N \tilde{T}}}\,\tilde{v}_z^{\text{LJ}}. \tag{2}$$

In LJ fluid simulations, where there is no collective motion, the thermal rapidity component is $y^{\text{th}} \propto \tilde{v}_z^{\text{LJ}}$, with $\tilde{v}_z^{\text{LJ}}$ following a Maxwell-Boltzmann distribution with width $\sigma_{\tilde{v}_z^{\text{LJ}}} = \sqrt{\tilde{T}}$. In heavy-ion collisions, the thermal rapidity distribution width is $\sigma_y = \sqrt{T_{\text{frz}}/m_N}$, leading to the expression for y^{th}.

The collective rapidity component y^{coll} corresponds to space-time rapidity η_s in the Bjorken picture, where $y_{\text{coll}} = \eta_s$. The LJ longitudinal coordinate $\tilde{z}^{\text{LJ}}$ maps to η_s using:

$$y_{\text{cm}}^{\text{beam}}(\sqrt{s_{\text{NN}}}) = \ln\left[\frac{\sqrt{s_{\text{NN}}} + \sqrt{s_{\text{NN}} - 4m_N^2}}{2m_N}\right], \quad \text{thus} \quad y^{\text{coll}} = \frac{2y_{\text{cm}}^{\text{beam}}}{\tilde{L}}\,\tilde{z}^{\text{LJ}}. \tag{3}$$

© The Authors, published by EDP Sciences. This is an open access article distributed under the terms of the Creative Commons Attribution License 4.0 (https://creativecommons.org/licenses/by/4.0/).

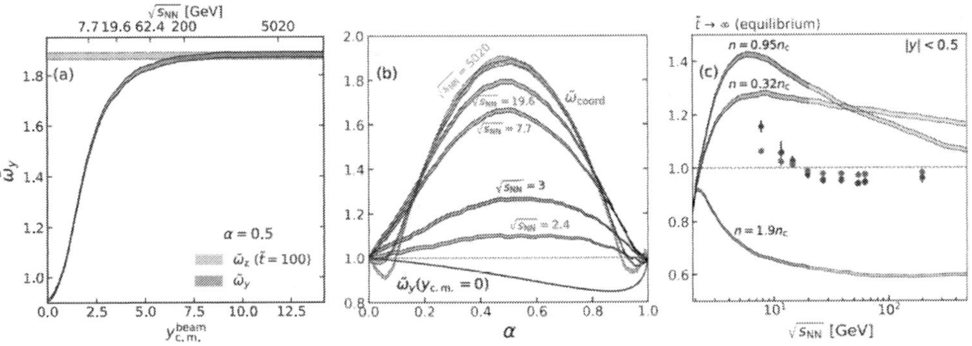

Figure 1. The corrected scaled variance with the effect of collective flow. Panels (a) and (b) stands for the density $n = 0.95n_c$ and temperature $T = 1.06T_c$ and panel (c) depicts $n = 0.32n_c, 0.95n_c$ and $1.9n_c$ for the same temperature. Panel (a) shows the flow at constant $\alpha = 0.5$ as a function of collision energy $\sqrt{s_{NN}}$. Panel (b) shows the same quantity for different α. Panel (c) depicts $\tilde{\omega}_y$ as a function of $\sqrt{s_{NN}}$ for fixed y_{cut}. The points correspond to the experimental data of the STAR Collaboration [5] for protons (green) and reconstructed baryons (blue).

We assume a flat particle density in space-time rapidity η_s, implying boost invariance up to the beam rapidity. Future work may explore non-uniform η_s distributions.

The described procedure is the simplest way to incorporate longitudinal flow, transforming a single fireball in a box into an expanding one. It follows the Bjorken picture and assumes no event-by-event fluctuations in longitudinal flow. For more precise applications, especially at lower RHIC beam energies, this model requires refinement. However, we use this simplified model to estimate the longitudinal flow effect under the most favorable conditions.

4 Results

4.1 Fluctuations at fixed α

We examine the fluctuations by the scaled variance defined as

$$\tilde{\omega}_y = \frac{1}{1-\alpha}\left[\frac{\langle N^2\rangle - \langle N\rangle^2}{\langle N\rangle}\right], \tag{4}$$

with baryon conservation correction [14] at different energies for a fixed α, adjusting y_{cut} at each energy to satisfy $\alpha = \langle N_{acc}\rangle/N$ as depicted on Fig. 1 in panels (a) and (b). As y_{beam} increases, $\tilde{\omega}_y$ rises and eventually saturates at a value consistent with coordinate space fluctuations. This shows that strong collective flow maps coordinate space fluctuations onto momentum space. For $y_{beam} \to 0$, the results match box simulations where momentum space fluctuations are absent.

4.2 Fluctuations at fixed y_{cut}

In heavy-ion collisions, measurements are usually performed within a fixed rapidity interval near midrapidity, such as in STAR measurements of proton number cumulants. Fixing y_{cut} leads to a smaller α at higher energies due to the increased total rapidity coverage.

In Fig. 1(c) we observe that baryon number fluctuations in a fixed rapidity interval behave as expected for systems with a critical point. In the case of $n = 0.95n_c$, the fluctuations reflect the assumption that freeze-out near the CP occurs at a specific collision energy. However, the interplay between the finite lifetime and system size can suppress fluctuations in real systems.

© The Authors, published by EDP Sciences. This is an open access article distributed under the terms of the Creative Commons Attribution License 4.0 (https://creativecommons.org/licenses/by/4.0/).

5 Discussion and summary

In this article, we explore particle number fluctuations near the critical endpoint of a first-order phase transition using molecular dynamics simulations of the Lennard-Jones fluid. We found that collective flow is crucial for detecting large fluctuations in momentum space, similar to how those are observed in heavy-ion collisions. This study shows that the best signals for the CP can be observed in the 5–7 GeV collision energy range.

Future studies in the framework of molecular dynamics on higher statistics could include obtaining of the next two cumulants with and without the collective flow, testing different types of cumulants as a CP indicator, mixed phase studies, cross-check the results with Yukawa potential and utilizing of the AI for the phase diagram mapping.

6 Acknowledgments

We thank Carsten Greiner, Roman Poberezhnyuk, Claudia Ratti and Yuri Sinyukov for discussions. The authors acknowledge the use of the PhysGPU Cluster and the support from the Research Computing Data Core at the University of Houston to carry out the research presented here. V.K. has been supported by the U.S. Department of Energy, Office of Science, Office of Nuclear Physics, under contract number DE-AC02-05CH11231.

References

[1] A. Bzdak, S. Esumi, V. Koch, J. Liao, M. Stephanov and N. Xu, Mapping the Phases of Quantum Chromodynamics with Beam Energy Scan, Physics Reports, **853** (2020)

[2] M. A. Stephanov, K. Rajagopal and E. Shuryak, Event-by-event fluctuations in heavy ion collisions and the QCD critical point, Phys. Rev. D **60** 114028 (1999)

[3] Y. Hatta and M. A. Stephanov, Proton number fluctuation as a signal of the QCD critical endpoint, Phys. Rev. Lett. **91** 102003 (2003)

[4] M. A. Stephanov, Non-Gaussian fluctuations near the QCD critical point, Phys. Rev. Lett. **102** 032301 (2009)

[5] M. Abdallah and al., Cumulants and correlation functions of net-proton, proton, and antiproton multiplicity distributions in Au+Au collisions at energies available at the BNL Relativistic Heavy Ion Collider, Phys. Rev. C **104** 024902 (2021)

[6] V. Vovchenko, QCD at finite temperature and density: Criticality, EPJ Web Conf. **296** 01017 (2024)

[7] V. A. Kuznietsov, O. Savchuk, R. V. Poberezhnyuk, V. Vovchenko, M. I. Gorenstein and H. Stöker, Molecular dynamics analysis of particle number fluctuations in the mixed phase of a first-order phase transition, Phys. Rev. C **107** 055206 (2023)

[8] V. A. Kuznietsov. O. Savchuk, M. I. Gorenstein, V. Koch and V. Vovchenko, Critical point particle number fluctuations from molecular dynamics, Phys. Rev. C **105** 044903 (2022)

[9] V. A. Kuznietsov, M. I. Gorenstein, V. Koch and V. Vovchenko, Coordinate versus momentum cuts and effects of collective flow on critical fluctuations, Phys. Rev. C **110** 015206 (2024)

[10] S. Stephan, J. Staubach and H. Hasse, Review and comparison of equations of state for the Lennard-Jones fluid, **523** 112772 (2020)

[11] S. Stephan, M. Thol, J. Vrabec and H. Hasse, Thermophysical Properties of the Lennard-Jones Fluid: Database and Data Assessment, Journal of Chemical Information and Modeling **59** 4248-4265 (2019)

[12] M. P. Allen and D. J. Tildesley, Computer Simulation of Liquids, Oxford University Press (2017)

[13] V. Vovchenko, Molecular dynamics simulation and visualization of the Lennard-Jones system utilizing CUDA-enabled GPUs (GitHub repository), https://github.com/vlvovch/lennard-jones-cuda

[14] V. Vovchenko, O. Savchuk, R. V. Poberezhnyuk, M. I. Gorenstein and V. Koch, Connecting fluctuation measurements in heavy-ion collisions with the grand-canonical susceptibilities, Phys. Lett. B **811** 135868 (2020)

© The Authors, published by EDP Sciences. This is an open access article distributed under the terms of the Creative Commons Attribution License 4.0 (https://creativecommons.org/licenses/by/4.0/).

Bayesian location of the QCD critical point: a holographic perspective

Claudia Ratti[1,*], *Joaquin* Grefa[1,2], *Mauricio* Hippert[3], *T. Andrew* Manning[4], *Jorge* Noronha[5], *Jacquelyn* Noronha-Hostler[5], *Israel* Portillo Vazquez[1], *Romulo* Rougemont[6], and *Michael* Trujillo[1]

[1]Physics Department, University of Houston, Houston TX 77204
[2]Department of Physics, Kent State University, Kent, Ohio 44243, USA
[3]Department of Theoretical Physics, Rio de Janeiro State University, 20550-013 Rio de Janeiro, Brazil
[4]National Center for Supercomputing Applications, University of Illinois Urbana-Champaign, Urbana, IL 61801, USA
[5]Illinois Center for Advanced Studies of the Universe and Department of Physics, University of Illinois Urbana-Champaign, Urbana, IL 61801-3003, USA
[6]Instituto de Física, Universidade Federal de Goiás, Av. Esperança - Campus Samambaia, CEP 74690-900, Goiânia, Goiás, Brazil

Abstract. We present a Bayesian analysis, based on holography and constrained by lattice QCD simulations, which leads to a prediction for the existence and location of the QCD critical point. We employ two different parametrizations of the functions that characterize the breaking of conformal invariance and the baryonic charge in the Einstein-Maxwell-dilaton holographic model. They lead to predictions for the critical point that overlap at one sigma. While some samples of the prior distribution do not predict a critical point, or produce critical points that cover large regions of the phase diagram, all posterior samples present a critical point at chemical potentials μ_{Bc} ~550-630 MeV.

1 Introduction

Mapping out the QCD phase diagram is an open challenge in nuclear physics, which is the main program component of experiments such as STAR at the Relativistic Heavy-Ion Collider (RHIC) and CBM at FAIR [1]. In particular, while the phase transition between hadrons and quarks is a smooth crossover at low densities [2], several models predict the existence of a second-order critical point (CP), beyond which the transition becomes first order (for recent reviews see e.g.[3, 4]). In our analysis, we use a holographic approach [5] to extrapolate lattice QCD information from zero to large baryon chemical potential μ_B. This model is a natural good candidate, since it can reproduce lattice QCD results at low densities, where they are available [6, 7], and naturally leads to the almost ideal-fluid behavior of the Quark-Gluon Plasma [8], as observed in experiments. We use a Bayesian analysis, and two functional forms for the potential and coupling functions that are the main unknown in our model, to perform a systematic scan of the parameters, constrained by lattice QCD results at zero density [9, 10], with a probability given by the respective error bars. Our predictions for the QCD critical point, based on the posterior parameter distributions, agree within one sigma for the two model realizations, and collapse to a very small region of the phase diagram, which is within the phase transition band from lattice QCD simulations [11].

*e-mail: cratti@uh.edu

© The Authors, published by EDP Sciences. This is an open access article distributed under the terms of the Creative Commons Attribution License 4.0 (https://creativecommons.org/licenses/by/4.0/).

2 Einstein-Maxwell-Dilaton holographic model

The gauge/gravity duality provides a way to describe the physics of a thermal, strongly coupled field theory in 3+1 Minkowski space time in terms of dual black holes in a 4+1 dimensional asymptotically anti-de Sitter bulk spacetime.

We employ the following action, in which a Maxwell field A_μ is used to provide a description of a conserved charge in QCD, in our case baryon number, while a dilaton scalar field ϕ breaks conformal invariance. The action of our theory is given by

$$S = \frac{1}{2\kappa_5^2}\int_{M_5} d^5\sqrt{-g}\left[R - \frac{(\partial_\mu\phi)^2}{2} - V(\phi) - \frac{f(\phi)F_{\mu\nu}^2}{4}\right], \tag{1}$$

where g is the determinant of the metric, R is the Ricci scalar, $F_{\mu\nu} = \partial_\mu A_\nu - \partial_\nu A_\mu$ is the Maxwell field strength, while $V(\phi)$ and $f(\phi)$ are functions tweaked to reproduce QCD physics at $\mu_B = 0$. The numerical solutions for the EMD fields in thermal equilibrium, generated by solving the bulk equations of motion for different pairs of initial conditions, are associated through the holographic dictionary with definite thermal states at the boundary QFT, corresponding to given values of temperature T and chemical potential μ_B. These functions are the central unknowns in our analysis, and we choose for them two ansätze as follows:

1. Polynomial-Hyperbolic Ansatz (PHA), a more common parametrization, similar to the one chosen in Ref. [6]

$$\begin{aligned} V(\phi) &= -12\cosh(\gamma\,\phi) + b_2\,\phi^2 + b_4\,\phi^4 + b_6\,\phi^6, \\ f(\phi) &= \frac{\mathrm{sech}(c_1\phi + c_2\phi^2 + c_3\phi^3)}{1+d_1} + \frac{d_1}{1+d_1}\mathrm{sech}(d_2\phi). \end{aligned} \tag{2}$$

2. Parametric Ansatz (PA), a parametrization where it is more straightforward to relate the parameters directly to plateaus and exponential slopes in the potential [11]

$$\begin{aligned} V(\phi) &= -12\cosh\left[\left(\frac{\gamma_1\,\Delta\phi_V^2 + \gamma_2\,\phi^2}{\Delta\phi_V^2 + \phi^2}\right)\phi\right], \\ f(\phi) &= 1-(1-A_1)\left[\frac{1}{2}+\frac{1}{2}\tanh\left(\frac{\phi-\phi_1}{\delta\phi_1}\right)\right] + -A_1\left[\frac{1}{2}+\frac{1}{2}\tanh\left(\frac{\phi-\phi_2}{\delta\phi_2}\right)\right]. \end{aligned} \tag{3}$$

3 Bayesian analysis and results

We will scan the parameter space for the two ansätze described above, in order to generate an ensemble of models distributed according to the error bars on lattice QCD results. We use Bayes theorem to find the posterior distributions for the model parameters, $\vec{\theta}$, given the lattice QCD constraints $\vec{d}$ as follows

$$P(\vec{\theta}|\vec{d}) = \frac{P(\vec{d}|\vec{\theta})P(\vec{\theta})}{P(\vec{d})} \tag{4}$$

where $P(\vec{\theta})$ is the prior probability distribution for the model parameters, and we treat $P(\vec{d})$, known as the evidence, as a normalization factor for the posterior.

To locate the critical point, we look at lines of constant ϕ_0 while increasing Φ_1, where ϕ_0 and Φ_1 are the values of the dilaton and $U(1)$ electric fields at the event horizon of a black hole solution, which correspond to the two initial conditions needed to numerically solve the

© The Authors, published by EDP Sciences. This is an open access article distributed under the terms of the Creative Commons Attribution License 4.0 (https://creativecommons.org/licenses/by/4.0/).

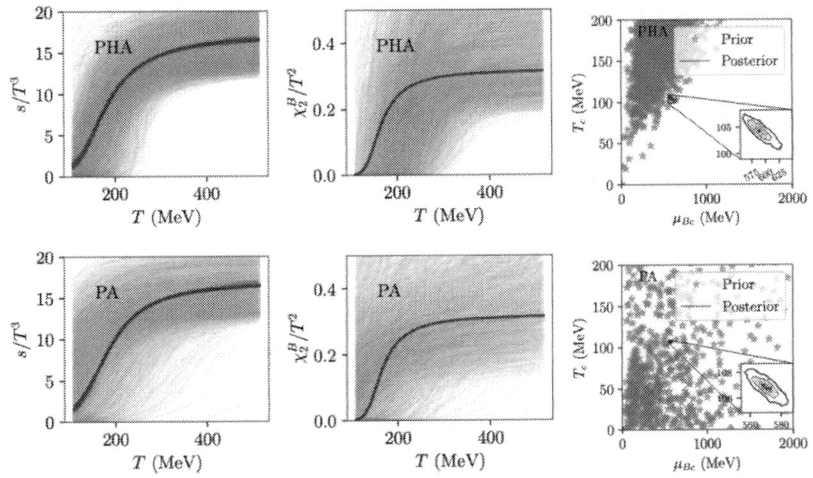

Figure 1. Results for sample equations of state from the prior (gray) and posterior (blue) distributions, for the PHA (top) and PA (bottom) Ansätze. Left: normalized entropy density vs temperature. Center: normalized baryon susceptibility vs temperature. Right: Predictions for the QCD critical point for the prior (grey stars) and posterior samples (blue histogram and contours). In the left and center panels, the red points with error bars are the lattice results from Refs. [9, 10]. In the rightmost panels, the blue lines visible in the inset represent 68% and 95% confidence levels for the posterior distribution.

bulk field equations. These lines are parallel at $\mu_B = 0$, but they cross at the CP and lead to a three-solution region beyond it.

The gray lines in the leftmost and middle panels of Fig. 1 display the prior equations of state, resulting from the functions $f(\phi)$ and $V(\phi)$ utilized in our analysis. The rightmost panels show the spatial distribution of critical points in the $(T,\ \mu_B)$ plane corresponding to these samples of the priors. The top and bottom panels correspond to the PHA and PA models, respectively. While $\sim 20\%$ of the prior sample does not produce a critical point at all for the PA model, critical points found in this sample are scattered over a very wide region in the phase diagram. On the other hand, the prior for the PHA version of the model comparatively produces critical points that are concentrated mainly in one region of the phase diagram.

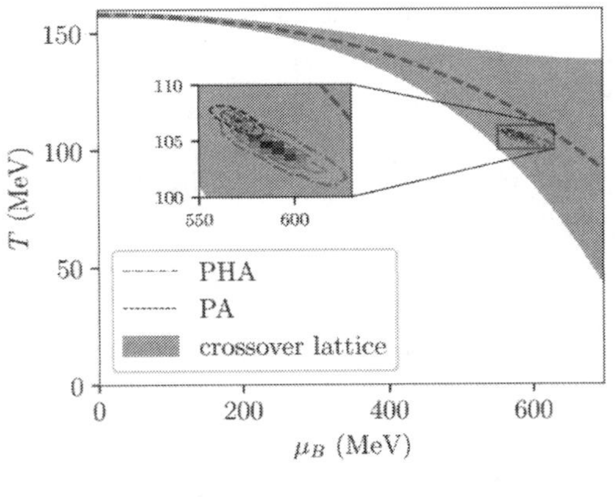

Figure 2. Predictions for the critical point location on the $(T,\ \mu_B)$-plane, based on the posterior distributions for the PHA model (red) and the PA model (blue). Also shown is the extrapolation of the lattice QCD transition line from Ref. [12] (green band), based on the peak of the chiral susceptibility. Lines around confidence regions for the critical point location represent 68% and 95% confidence levels.

© The Authors, published by EDP Sciences. This is an open access article distributed under the terms of the Creative Commons Attribution License 4.0 (https://creativecommons.org/licenses/by/4.0/).

Figure 2 shows that the predicted distributions for the critical point location from our Bayesian analysis for the PHA (red) and PA (blue) Ansätze, are located within a narrow region in T and μ_B. Moreover, the regions for the PA and PHA Ansätze agree, with overlapping 68% confidence regions. For a posterior distribution of the parameter values see Ref. [11].

4 Conclusions

We have presented results for the first Bayesian analysis of the phase diagram of QCD constrained by first-principles lattice QCD results at zero baryon density. The posterior distribution of CP locations was computed for two different parametrizations of a holographic EMD model. We find that imposing agreement with lattice QCD tightly constrains predictions for the QCD CP location, which were spread all around the phase diagram in the unconstrained prior. Moreover, bands for the CP location within each model overlap within one sigma, indicating the robustness of our results against parametrization choices.

5 Acknowledgements

This material is based upon work supported in part by the National Science Foundation within the framework of the MUSES collaboration, under grant number No. OAC-2103680, as well as under grants No. PHY-2208724, PHY-1748958 and PHY-2116686, in part by the U.S. Department of Energy, Office of Science, Office of Nuclear Physics, under Award Number DE-SC0022023, DE-SC0023861, as well as by the National Aeronautics and Space Agency (NASA) under Award Number 80NSSC24K0767. The authors also acknowledge support from the Illinois Campus Cluster, a computing resource that is operated by the Illinois Campus Cluster Program (ICCP) in conjunction with the National Center for Supercomputing Applications (NCSA), and which is supported by funds from the University of Illinois at Urbana-Champaign. R.R. acknowledges financial support by National Council for Scientific and Technological Development (CNPq) under grant number 407162/2023-2.

References

[1] A. Lovato, *et al.* [arXiv:2211.02224 [nucl-th]].

[2] Y. Aoki, G. Endrodi, Z. Fodor, S. D. Katz and K. K. Szabo, Nature **443** (2006), 675-678.

[3] A. Bzdak, S. Esumi, V. Koch, J. Liao, M. Stephanov N. Xu, Phys. Rept. **853** (2020), 1-87

[4] L. Du, A. Sorensen and M. Stephanov, [arXiv:2402.10183 [nucl-th]].

[5] R. Rougemont, J. Grefa, M. Hippert, J. Noronha, J. Noronha-Hostler, I. Portillo and C. Ratti, Prog. Part. Nucl. Phys. **135** (2024), 104093

[6] R. Critelli, J. Noronha, J. Noronha-Hostler, I. Portillo, C. Ratti and R. Rougemont, Phys. Rev. D **96** (2017) no.9, 096026

[7] J. Grefa, J. Noronha, J. Noronha-Hostler, I. Portillo, C. Ratti and R. Rougemont, Phys. Rev. D **104** (2021) no.3, 034002

[8] J. Grefa, M. Hippert, J. Noronha, J. Noronha-Hostler, I. Portillo, C. Ratti and R. Rougemont, Phys. Rev. D **106** (2022) no.3, 034024

[9] S. Borsanyi, Z. Fodor, C. Hoelbling, S. D. Katz, S. Krieg and K. K. Szabo, Phys. Lett. B **730** (2014), 99-104

[10] R. Bellwied, S. Borsanyi, Z. Fodor, S. D. Katz, A. Pasztor, C. Ratti and K. K. Szabo, Phys. Rev. D **92** (2015) no.11, 114505

[11] M. Hippert, J. Grefa, T. A. Manning, J. Noronha, J. Noronha-Hostler, I. Portillo Vazquez, C. Ratti, R. Rougemont and M. Trujillo, Phys. Rev. D **110**, no.9, 094006 (2024)

[12] S. Borsanyi, Z. Fodor, J. N. Guenther, R. Kara, S. D. Katz, P. Parotto, A. Pasztor, C. Ratti and K. K. Szabo, Phys. Rev. Lett. **125** (2020) no.5, 052001

© The Authors, published by EDP Sciences. This is an open access article distributed under the terms of the Creative Commons Attribution License 4.0 (https://creativecommons.org/licenses/by/4.0/).

Spinodal decomposition in Bjorken flow

Joseph Kapusta[1,*], *Mayank* Singh[1,2,**], and *Thomas* Welle[1,3,***]

[1]School of Physics and Astronomy, University of Minnesota, Minneapolis, MN 55455, USA
[2]Department of Physics and Astronomy, Vanderbilt University, Nashville, TN 37240, USA
[3]Applied Research Associates, 8537 Six Forks Road, Raleigh, NC 27615, USA

Abstract. The QCD first-order phase transition at large baryon densities is expected to proceed by spinodal decomposition. This spinodal phase is likely to leave its signatures on the experimental observables measured in heavy-ion collision experiments. Identifying these signatures requires phenomenological models integrating surface effects resulting from the phase transition into the hydrodynamical description of the expanding quark gluon plasma. We write the equations of relativistic hydrodynamics with spinodal decomposition and solve it in on a background of Bjorken flow relevant for heavy-ion collisions.

1 Introduction

Mapping the QCD phase diagram has been a priority for the heavy-ion collision community. Lattice calculations with non-zero quark masses have shown that the QCD matter changes phase from a hadron gas to the quark gluon plasma (QGP) by a smooth crossover when the baryon chemical potential is near zero [1]. At high baryon densities, the phase change is expected to proceed via a first-order phase transition [2]. The first-order phase transition curve ends in a critical point before reaching the zero chemical potential axis.

The Beam Energy Scan (BES) program at the Relativistic Heavy-Ion Collider (RHIC) and the upcoming experiments at the Facility for Antiproton and Ion Research (FAIR) aim to explore different regions of the QCD phase diagram with the goal of quantitatively mapping its features. Heavy nuclei are collided at a range of collision energies which cover a wide region of the phase diagram. This program relies on identifying the observable signatures of the critical point and the phase transition curve and doing extensive model-to-data comparisons [3]. We expect the phase transition curve to cover a bigger region of the phase diagram than the critical point and hence it is crucial that we understand its dynamics.

The phase transition in these collisions is expected to proceed via spinodal decomposition. Here we extend the equation of state to metastable and unstable regions and report on the equations describing spinodal decomposition in a relativistic QCD medium. We write and solve these equations in a fluid with Bjorken flow.

*e-mail: kapusta@umn.edu
**e-mail: mayank.singh@vanderbilt.edu
***e-mail: twelle2357@gmail.com

© The Authors, published by EDP Sciences. This is an open access article distributed under the terms of the Creative Commons Attribution License 4.0 (https://creativecommons.org/licenses/by/4.0/).

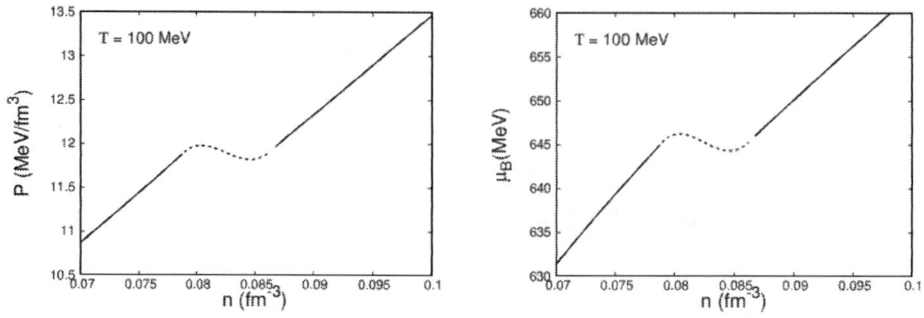

Figure 1. Pressure (left) and baryon chemical potential (right) are interpolated in metastable and unstable phases at fixed temperature.

2 Metastable and unstable regions

An important step towards describing the spinodal decomposition is extending the equation of state to the coexistence phase. This phase is often not described by the first principle approaches as it is composed of metastable and unstable regions. Given an equation of state for the stable region, we can interpolate the pressure in this phase as [4]

$$P_{\rm int}(n) = P_X(T) + \sum_{i=1}^{4} c_i(n - n_G)^i. \tag{1}$$

The subscript X denotes the pressure along the line where the phases coexist. This line is often defined by Maxwell construction. The baryon density is denoted by n while n_G denotes the baryon density in the hadronic gas phase along the coexistence line. Similarly, the chemical potential can be parameterized as

$$\mu_{\rm int}(n) = \mu_X(T) + d_0 \ln(n/n_G) + \sum_{i=1}^{3} d_i(n - n_G)^i, \tag{2}$$

The coefficients c_i and d_i can be obtained by applying the thermodynamic relations at $n = n_G$ and $n = n_L$ where the equation of state is stable. Further constraints are obtained by applying the continuity conditions on pressure and chemical potential and their derivatives on the border between the coexistence and the QGP and hadronic gas phases. Thermodynamic relations also hold at the boundaries and provide additional constraints. The interpolated pressure and baryon chemical potential for a temperature of 100 MeV are shown in figure 1. The stable QCD equation of state is taken from reference [5].

3 Hydrodynamics with phase separation

The Helmholtz free energy can be written as

$$F\{n(\mathbf{x}, t)\} = \int d^3x \left[\tfrac{1}{2}K(\boldsymbol{\nabla} n)^2 + f(T, n)\right], \tag{3}$$

where the first function on the right accounts for the free energy due to a gradient in baryon density across the phase boundary. The second term $f(T, n)$ is the bulk free energy. Cahn and Hilliard [6, 7] showed in their seminal papers that this leads to the stress tensor

$$T_{ij} = \tilde{P}\delta_{ij} + K(\partial_i n)(\partial_j n). \tag{4}$$

© The Authors, published by EDP Sciences. This is an open access article distributed under the terms of the Creative Commons Attribution License 4.0 (https://creativecommons.org/licenses/by/4.0/).

https://doi.org/10.1051/epjconf/202531606005

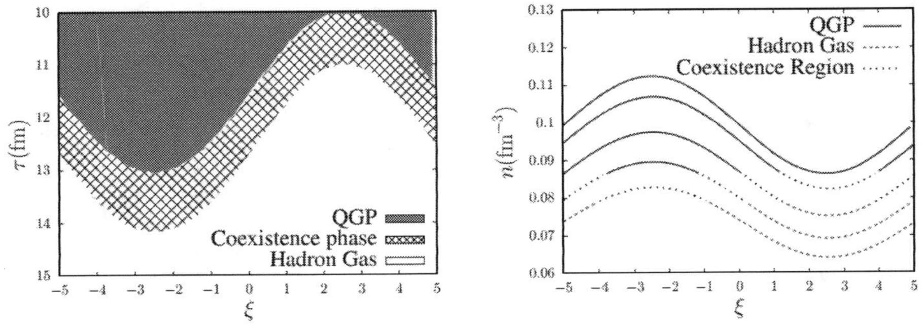

Figure 2. The phases of QCD matter as a function of proper time and rapidity (left). The spatial distribution of baryon density (right). The five curves from top to bottom correspond to τ = 10 fm, 10.5 fm, 11.5 fm, 12.5 and 13.5 fm respectively.

For a relativistic fluid, the covariant energy-momentum tensor can be given by [4]

$$T^{\mu\nu} = \tilde{P}(u^\mu u^\nu - g^{\mu\nu}) + \tilde{\epsilon} u^\mu u^\nu + K(D^\mu n)(D^\nu n), \tag{5}$$

where the local thermodynamic variables are pressure $\tilde{P} = P + KnD^2 n + \frac{1}{2}K(D^\mu n)(D_\mu n)$ and energy density $\tilde{\epsilon} = \epsilon - \frac{1}{2}K(D^\mu n)(D_\mu n)$. Here $D^\mu = \partial^\mu - u^\mu u^\alpha \partial_\alpha$ is the gradient orthogonal to the fluid velocity u^μ. We are using the mostly negative metric. In the Landau-Lifshitz frame baryon current has the form

$$J^\mu = nu^\mu + \sigma_B T D^\mu \left(\frac{\tilde{\mu}}{T}\right), \tag{6}$$

with baryon chemical potential $\tilde{\mu} = \mu + KD^2 n$ and baryon conductivity σ_B.

We solve equations for boost invariant hydrodynamics with the simplification that system is homogeneous in transverse direction. We start the evolution at proper time τ_x = 10 fm when energy density is kept boost invariant. Baryon density is initialized as $n(\tau_x) = n_0(\tau_x) + n_{fl}\sin(k\xi)$ where n_0 is the average value of baryon density. Here n_{fl} is the amplitude of sinusoidal fluctuations in baryon density in the rapidity direction ξ. The values of n_0 and n_{fl} are chosen such that the lowest point of baryon density distribution just touches the coexistence phase region. Energy density gradient terms are neglected for simplicity. We verified that the violation of boost invariance in energy density is less than 3%.

The evolution of different phases and that of baryon density can be seen in figure 2. The baryon density reduces as the system expands and cools with time. It is interesting to note that unlike the steady-state systems undergoing spinodal decomposition, which have been extensively studied, this is a rapidly expanding system. So the system always dilutes and ends up entirely in the hadronic gas phase. If the phase change is via a first-order transition, then it encounters the coexistence phase.

The baryon density chosen here is smoothly varying. This is a simplification and ensures that the fourth-order derivatives encountered in the Cahn-Hilliard model are simplified. We chose a relatively small value of $K = 5 \times 10^{-5}$ MeV^{-4} which ensured that this simplification holds for the duration of this evolution. In the future, this simplification can be relaxed for a more realistic baryon density distribution.

The energy density evolution is shown in figure 3. We show the results with $K = 5 \times 10^{-5}$ MeV^{-4} and $K = 0$. The later choice means that the phase surface has no energy contribution and is shown for reference. The energy increases immediately as the system enters the coexistence region. The energy ends up reducing in the negative rapidity when it goes in the

© The Authors, published by EDP Sciences. This is an open access article distributed under the terms of the Creative Commons Attribution License 4.0 (https://creativecommons.org/licenses/by/4.0/).

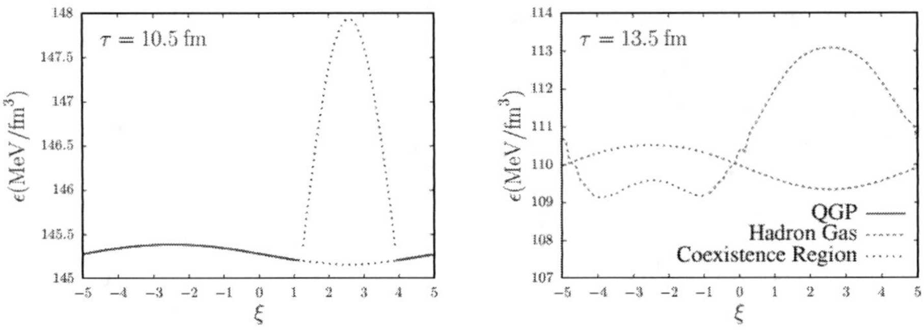

Figure 3. Energy density evolution for two different values of τ. The curves with higher maximum value of ϵ correspond to $K = 5 \times 10^{-5}$ MeV^{-4}. The other curves corresponds to $K = 0$.

coexistence phase. Whether the energy density increases or decreases depends on the sign of baryon density gradients. These surface energy contributions have the net effect of enhancing the fluctuations already present in the system.

4 Conclusions

The QCD phase diagram is postulated to have a first-order phase transition curve at large baryon densities. Experimental verification of such a phase transition relies on successful model-to-data comparisons with the models incorporating the phase transition dynamics. In heavy-ion collisions, such a phase transition is expected to proceed by spinodal decomposition. We wrote the equations of relativistic hydrodynamics with spinodal decomposition. The covariant energy-momentum tensor incorporating the surface effects is explicitly shown. We also provide an interpolation prescription to extend the QCD equation of state to metastable and unstable phases. We solved the equations for Bjorken flow.

The formalism can be included in the existing models of heavy-ion collisions to simulate the QCD phase transition. This will help us identify the observables containing the signatures of the phase transition and help map the QCD phase diagram.

Acknowledgements

This work was supported by the U.S. Department of Energy Grant Nos. DE-FG02-87ER40328 (JK, MS and TW) and DE-SC-0024347 (MS).

References

[1] Y. Aoki, G. Endrodi, Z. Fodor, S. D. Katz and K. K. Szabo, Nature **443**, 675-678 (2006)
[2] K. Fukushima and T. Hatsuda, Rept. Prog. Phys. **74**, 014001 (2011)
[3] X. An, *et al.* Nucl. Phys. A **1017**, 122343 (2022)
[4] J. I. Kapusta, M. Singh and T. Welle, [arXiv:2407.16963 [hep-ph]].
[5] J. Kapusta, T. Welle and C. Plumberg, Phys. Rev. C **106**, no.1, 014909 (2022)
[6] J. W. Cahn and J. E. Hilliard, J. Chem. Phys. **28**, 258 (1958)
[7] J. W. Cahn and J. E. Hilliard, J. Chem. Phys. **31**, 688 (1959)

© The Authors, published by EDP Sciences. This is an open access article distributed under the terms of the Creative Commons Attribution License 4.0 (https://creativecommons.org/licenses/by/4.0/).

Simulating collectivity in dense baryon matter with multiple fluids

Iurii Karpenko[1,*], *Jakub* Cimerman, *Pasi* Huovinen[2], and *Boris* Tomášik[1,3]

[1]Faculty of Nuclear Sciences and Physical Engineering, Czech Technical University in Prague, Břehová 7, Prague, Czech Republic
[2]Incubator of Scientific Excellence—Centre for Simulations of Superdense Fluids, University of Wrocław, Poland
[3]Univerzita Mateja Bela, Banská Bystrica, Slovakia

Abstract. We report on construction of a modern multi-fluid approach to heavy-ion collisions at FAIR/BES energies (MUFFIN) and show the reproduction of basic experimental observables in Au-Au collisions in the RHIC Beam Energy Scan program. We also show the p_T-differential and p_T-integrated polarization of (anti-)Λ hyperons. In MUFFIN simulations, we observe a strong splitting between polarizations of Λ and anti-Λ. The splitting is driven purely by a finite baryon chemical potential.

1 Introduction

One of the key objectives of heavy-ion collision experiments at center-of-mass energies ranging from a few to a hundred GeV is to explore the characteristics of the dense baryonic matter that is formed, particularly its equation of state (EoS) and transport properties. The fluid dynamic approach plays a crucial role in this endeavor, as it allows for the flexible incorporation of different equations of state.

This approach has proven highly effective in describing nucleus-nucleus collisions at high energies, specifically for $\sqrt{s_{\rm NN}} = 200$ GeV and above. At these energy scales, the evolution is typically divided into two stages: the initial state, where hard scatterings occur and lead to the isotropization or effective fluidization of the system, followed by the fluid phase, which is governed by fluid dynamical equations.

At lower energies, however, modeling heavy-ion collisions presents a distinct challenge. The Lorentz contraction of the colliding nuclei is weaker, requiring up to several fm/c for the nuclei to fully pass through one another and for all primary nucleon-nucleon (NN) scatterings to take place. In the meantime, a dense medium may already form in regions where the first NN scatterings have occurred, even while some nucleons are still approaching their first interactions.

Multi-fluid dynamics offers an elegant, albeit phenomenological, method to account for the complex space-time evolution of nucleus-nucleus collisions at intermediate energies. In this approach, the incoming nuclei are approximated as two blobs of cold, baryon-rich fluids. The collision is then modeled as an interpenetration of these fluids, which are decelerated by frictional forces. The energy and momentum lost due to friction is used to generate a third fluid, representing the particles produced during the reaction.

*e-mail: yu.karpenko@gmail.com

© The Authors, published by EDP Sciences. This is an open access article distributed under the terms of the Creative Commons Attribution License 4.0 (https://creativecommons.org/licenses/by/4.0/).

In the following, we present a small selection of the basic observables, as well as polarization observables, computed in the MUFFIN (MUlti Fluid simulation for Fast IoN collisions) model. For a detailed description of the model and a broader set of key observables — such as rapidity distributions and transverse momentum spectra of various hadron types — we direct the reader to [1].

2 Multi-fluid model

Dynamics of heavy-ion reactions in the multi-fluid model (3-fluid model specifically) is desrcibed by a set of coupled local energy-momentum and baryon charge (non-)conservation equations:

$$\begin{aligned} \partial_\mu T^{\mu\nu}_{\mathrm{p}}(x) &= -F^\nu_{\mathrm{p}}(x) + F^\nu_{\mathrm{fp}}(x), \\ \partial_\mu T^{\mu\nu}_{\mathrm{t}}(x) &= -F^\nu_{\mathrm{t}}(x) + F^\nu_{\mathrm{ft}}(x), \\ \partial_\mu T^{\mu\nu}_{\mathrm{f}}(x) &= F^\nu_{\mathrm{p}}(x) + F^\nu_{\mathrm{t}}(x) - F^\nu_{\mathrm{fp}}(x) - F^\nu_{\mathrm{ft}}(x), \\ \partial_\mu N^\mu_{\mathrm{p,t,f}}(x) &= 0, \end{aligned} \tag{1}$$

each equation represents evolution of a fluid with a source term describing local energy-momentum exchange with the two other fluids. The indices "p" and "t" stand for projectile and target fluids which represent the original nuclei, and "f" stands for the fireball fluid, which represents meson-dominated medium created in the collision. Projectile and target fluids are initialised as moving blobs with local rest frame energy ε_0 and baryon n_0 densities equal to those of ordinary nuclear matter, whereas the fireball fluid is empty at the beginning.

The first three equations (1) summed together correspond to exact energy-momentum conservation of the entire system:

$$\partial_\mu \left[T^{\mu\nu}_p(x) + T^{\mu\nu}_t(x) + T^{\mu\nu}_f(x)\right] = 0.$$

The functional form of the friction terms is derived from the energy and momentum transport in NN and Nπ collisions using kinetic theory. The friction terms are then scaled to take into account for shortcomings of the derivation e.g. emergence of the QGP phase. The scaling also allows for a better fit to the experimental data.

3 Results and discussion

Due to space limitations, only a selection of basic observables pertaining to collision energy range $\sqrt{s_{\mathrm{NN}}} = 7.7 \ldots 19.6$ GeV is presented here. For the basic observables in a broader collision energy range the Reader is again referred to [1]. Figure 1 shows pseudorapidity distributions of all charged hadrons, rapidity distributions of net protons and transverse momentum distributions of pions, kaons, protons and protons in Au-Au collisions at $\sqrt{s_{\mathrm{NN}}} = 19.6$ GeV. Due to the lack of experimental data from STAR, the rapidity distribution of net protons is compared to an older experimental data from NA49 experiment at SPS, with somewhat different collision system (Pb-Pb) at collision energy $\sqrt{s_{\mathrm{NN}}} = 17.3$ GeV per nucleon pair, which we consider being close enough to Au-Au collisions at $\sqrt{s_{\mathrm{NN}}} = 19.6$ GeV.

Overall, MUFFIN reproduces the shapes of both the $dN_{\mathrm{ch}}/d\eta$ and net proton distributions while slightly missing the magnitudes of both. The shape of net proton distribution from NA49 experiment is reproduced well, which indicates a correct amount of baryon stopping in the present version of the friction terms. The transverse momentum distributions are reproduced generally well, with a notable exception of anti-proton ones, where the spectra are flatter than the data.

© The Authors, published by EDP Sciences. This is an open access article distributed under the terms of the Creative Commons Attribution License 4.0 (https://creativecommons.org/licenses/by/4.0/).

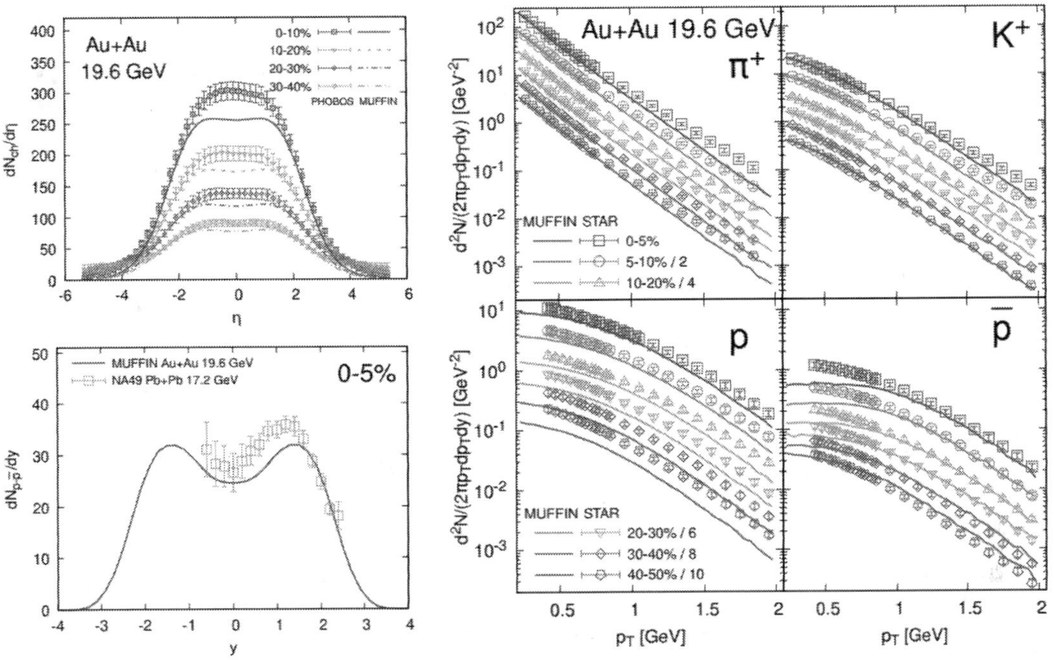

Figure 1: A selection of experimental observables from MUFFIN for Au-Au collisions at $\sqrt{s_{NN}}$ = 7.7 GeV: pseudorapidity distributions of all charged hadrons compared to PHOBOS data [2] (top left panel), rapidity distributions of net protons compared to NA49 data at 17.3 GeV [3] (bottom left panel) and transverse momentum spectra of positive pions, kaons, protons and anti-protons compared to STAR data [4] (right panels).

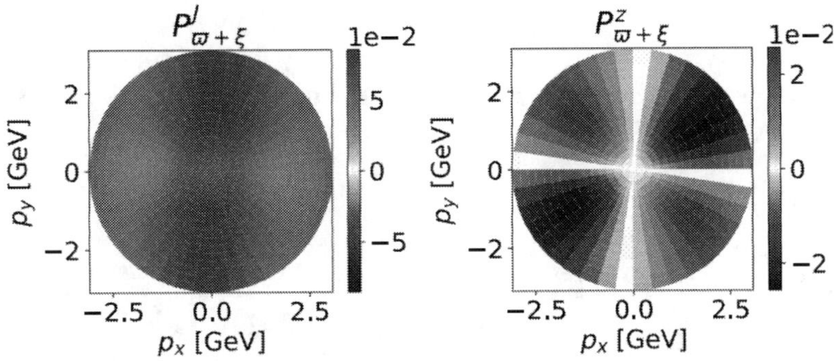

Figure 2: Transverse momentum dependent polarization of Λ hyperons from MUFFIN, computed at mid-rapidity for 20-50% central Au-Au collisions at $\sqrt{s_{NN}}$ = 7.7 GeV. The panels show the polarization component along the total angular momentum (left), and along the beam direction (right).

Next, we turn to hyperon polarization observables. Mean polarization of Λ hyperons at mid-rapidity is evaluated at the hypersurface of particlization according to the formula from [5], including shear-spin coupling introduced in [6]. No re-adjustment of the model parameters has been done for those observables, which can be therefore considered as a prediction from the model.

© The Authors, published by EDP Sciences. This is an open access article distributed under the terms of the Creative Commons Attribution License 4.0 (https://creativecommons.org/licenses/by/4.0/).

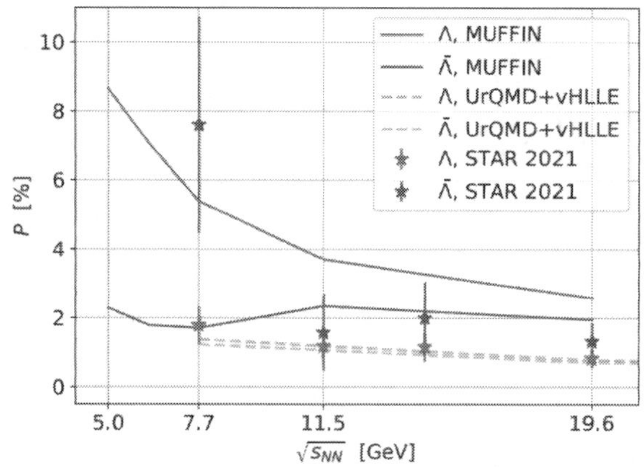

Figure 3. Momentum-averaged (global) polarization of Λ and anti-Λ hyperons at mid-rapidity in 20-50% central Au-Au collisions computed with MUFFIN (solid lines), compared to existing results from UrQMD+vHLLE (dashed lines) and experimental data from STAR [9] corrected for updated value of α_H decay parameter.

Figure 2 shows components of the polarization vector in the direction of the total angular momentum (on the left) and in the direction of the beam (on the right). MUFFIN simulation for this figure corresponds to 20-50% central Au-Au collisions at $\sqrt{s_{\rm NN}} = 7.7$ GeV.

Figure 3 shows momentum-averaged (sometimes referred to as "global") polarization of Λ and anti-Λ hyperons, computed for 20-50% central Au-Au collisions at different collision energies. The results from MUFFIN are shown with solid lines, whereas the results published in [7] using conventional "sandwich-type" hybrid model UrQMD+vHLLE [8]. MUFFIN simulations hit the ballpark of the experimental data points and result in a strong Λ - anti-Λ splitting, with stronger polarization of Λ hyperons. This seems to be in apparent contradiction to the original results from the STAR experiment [9]. The updated polarization measurements presented by STAR at the SQM2024 conference, however, show a splitting which is much more consistent with zero within the error bars.

Acknowledgements: IK and BT acknowledge support by the Czech Science Foundation under project No. 22-25026S. PH was supported by the program Excellence Initiative–Research University of the University of Wrocław of the Ministry of Education and Science. BT acknowledges support by VEGA 1/0521/22. Computational resources were provided by the e-INFRA CZ project (ID:90254), supported by the Ministry of Education, Youth and Sports of the Czech Republic.

References

[1] J. Cimerman, I. Karpenko, B. Tomasik and P. Huovinen, Phys. Rev. C **107**, no.4, 044902 (2023) [arXiv:2301.11894 [nucl-th]].

[2] B. Alver *et al.* [PHOBOS], Phys. Rev. C **83**, 024913 (2011) [arXiv:1011.1940].

[3] H. Appelshauser *et al.* [NA49], Phys. Rev. Lett. **82**, 2471-2475 (1999) doi:10.1103/PhysRevLett.82.2471 [arXiv:nucl-ex/9810014 [nucl-ex]].

[4] L. Adamczyk *et al.* [STAR], Phys. Rev. C **96**, no.4, 044904 (2017) [arXiv:1701.07065 [nucl-ex]].

[5] F. Becattini, V. Chandra, L. Del Zanna and E. Grossi, Annals Phys. **338**, 32-49 (2013) [arXiv:1303.3431 [nucl-th]].

[6] F. Becattini, M. Buzzegoli and A. Palermo, Phys. Lett. B **820**, 136519 (2021) [arXiv:2103.10917 [nucl-th]].

[7] I. Karpenko and F. Becattini, Eur. Phys. J. C **77**, no.4, 213 (2017)

[8] I. A. Karpenko, P. Huovinen, H. Petersen and M. Bleicher, Phys. Rev. C **91**, no.6, 064901 (2015) [arXiv:1502.01978 [nucl-th]].

[9] L. Adamczyk *et al.* [STAR], Nature **548**, 62-65 (2017)

© The Authors, published by EDP Sciences. This is an open access article distributed under the terms of the Creative Commons Attribution License 4.0 (https://creativecommons.org/licenses/by/4.0/).

Extracting the dependence of directed flow differences on conserved charges

Zi-Wei Lin[1,*], *Kishora* Nayak[2], and *Shusu* Shi[3]

[1]Department of Physics, East Carolina University, Greenville, NC, 27858, USA
[2]Department of Physics, Panchayat College, Sambalpur University, Bargarh 768028, Odisha, India
[3]Key Laboratory of Quark & Lepton Physics (MOE) and Institute of Particle Physics, Central China Normal University, Wuhan, 430079, China

Abstract. Recently, seven produced hadron species have been used to construct multiple hadron sets with given differences in the net electric charge (Δq) and strangeness (ΔS) between the two sides. A nonzero directed flow difference Δv_1 has been proposed as a consequence of the electromagnetic field produced in relativistic heavy ion collisions. Previously, we have shown with quark coalescence that Δv_1 and the slope difference $\Delta v_1'$ depend linearly on both Δq and ΔS with zero intercept. Here we emphasize that a two-dimensional function or fit is necessary for extracting the Δq- and ΔS-dependences of Δv_1. On the other hand, a one-dimensional fit gives a different value for the slope parameter of the Δq- or ΔS-dependence. Furthermore, a one-dimensional fit is incorrect because its slope parameter depends on the arbitrary scaling factor of a hadron set and is thus ill-defined. We use test data of Δv_1 to explicitly demonstrate these points.

1 Introduction

Direct flow analyses of produced hadrons that contain no u and d quarks were proposed [1, 2]. Seven hadron species, K^-, ϕ, $\bar{p}$, $\bar{\Lambda}$, $\bar{\Xi}^+$, Ω^- and $\bar{\Omega}^+$, were used [2] to construct multiple hadron sets with given differences in the net electric charge (Δq) and strangeness (ΔS) between the two sides. A nonzero directed flow difference Δv_1 (or the difference of the directed flow slopes at midrapidity $\Delta v_1'$) has been proposed as a consequence of the electromagnetic field produced in relativistic heavy ion collisions [2, 3], especially if Δv_1 or $\Delta v_1'$ increases with Δq.

In earlier studies [4, 5], we have examined the consequence of the coalescence sum rule (CSR) or quark coalescence on Δv_1 and $\Delta v_1'$ of the hadron sets. We found [4] that quark coalescence leads to $\Delta v_1' = c_q^* \Delta q + c_S^* \Delta S$, so in general $\Delta v_1' \neq 0$ for a hadron set with nonzero Δq and/or ΔS. Specifically, the CSR gives the following relation:

$$\Delta v_1' = \left(v_{1,\bar{d}}' - v_{1,\bar{u}}'\right) \Delta q + \left[\left(v_{1,\bar{s}}' - v_{1,s}'\right)/2 - \left(v_{1,\bar{d}}' - v_{1,\bar{u}}'\right)/3\right] \Delta S. \quad (1)$$

Therefore, the coefficients c_q^* and c_S^* reflect the v_1' difference of produced quarks with different electric charges. We then proposed two methods, the 5-set method and the 3-set method [4], to extract the coefficients for the Δq- and ΔS-dependences of Δv_1 or $\Delta v_1'$. Equivalently, we can write $\Delta v_1' = c_q^* \Delta q + c_B^* \Delta B$ that involves the difference in the net-baryon number (ΔB) between the two sides, where quark coalescence gives $c_B^* = -3 c_S^*$ [5].

*e-mail: linz@ecu.edu

© The Authors, published by EDP Sciences. This is an open access article distributed under the terms of the Creative Commons Attribution License 4.0 (https://creativecommons.org/licenses/by/4.0/).

Table 1 shows our choice of five hadron sets (Set 1-5) and the STAR Collaboration's choice (Index 1-5) [3]. For example, Δq_{ud} is the net electric charge in light quarks ($\bar{u}$ and $\bar{d}$ here) of the left side minus that of the right side including the weighting factor of each hadron (such as 1 or 1/2). For all the hadrons sets made of these seven produced hadron species, there are only five independent hadron sets [2, 4]. For example, each hadron index from the STAR Collaboration can be written as a linear combination of the hadron sets of our choice: Index1=-Set1, Index2=Set1+Set4+Set5, Index3=Set1+2*Set5, Index4=3*Set3+6*Set4, Index5=2*Set1-2*Set2+Set4+3*Set5.

Table 1. Set 1 to 5 are our choice of five independent hadron sets, while Index 1 to 5 are the STAR choice. Differences of electric charge, strangeness and baryon number between the two sides are given.

Set/Index #	Δq	Δq_{ud}	ΔS	ΔB	Left side	Right side
Set 1	0	0	0	0	$v_1[K^-(\bar{u}s)] + v_1[\overline{\Lambda}(\bar{u}\bar{d}\bar{s})]$	$v_1[\bar{p}(\bar{u}\bar{u}\bar{d})] + v_1[\phi(s\bar{s})]$
Set 2	0	0	0	0	$v_1[\overline{\Lambda}(\bar{u}\bar{d}\bar{s})]$	$\frac{1}{2}v_1[\overline{\Xi}^+(\bar{d}\bar{s}\bar{s})] + \frac{1}{2}v_1[\bar{p}(\bar{u}\bar{u}\bar{d})]$
Set 3	0	0	0	0	$\frac{1}{3}v_1[\Omega^-(sss)] + \frac{1}{3}v_1[\overline{\Omega}^+(\bar{s}\bar{s}\bar{s})]$	$v_1[\phi(s\bar{s})]$
Set 4	1/3	0	1	-1/3	$\frac{1}{2}v_1[\phi(s\bar{s})]$	$\frac{1}{3}v_1[\Omega^-(sss)]$
Set 5	2/3	1/3	1	-1/3	$\frac{1}{2}v_1[\phi(s\bar{s})] + \frac{1}{3}v_1[\bar{p}(\bar{u}\bar{u}\bar{d})]$	$v_1[K^-(\bar{u}s)]$
Index 1	0	0	0	0	$v_1[\bar{p}(\bar{u}\bar{u}\bar{d})] + v_1[\phi(s\bar{s})]$	$v_1[K^-(\bar{u}s)] + v_1[\overline{\Lambda}(\bar{u}\bar{d}\bar{s})]$
Index 2	1	1/3	2	-2/3	$v_1[\overline{\Lambda}(\bar{u}\bar{d}\bar{s})]$	$\frac{1}{3}v_1[\Omega^-(sss)] + \frac{2}{3}v_1[\bar{p}(\bar{u}\bar{u}\bar{d})]$
Index 3	4/3	2/3	2	-2/3	$v_1[\overline{\Lambda}(\bar{u}\bar{d}\bar{s})]$	$v_1[K^-(\bar{u}s)] + \frac{1}{3}v_1[\bar{p}(\bar{u}\bar{u}\bar{d})]$
Index 4	2	0	6	-2	$v_1[\overline{\Omega}^+(\bar{s}\bar{s}\bar{s})]$	$v_1[\Omega^-(sss)]$
Index 5	7/3	1	4	-4/3	$v_1[\overline{\Xi}^+(\bar{d}\bar{s}\bar{s})]$	$v_1[K^-(\bar{u}s)] + \frac{1}{3}v_1[\Omega^-(sss)]$

2 Extracting the Δq or ΔS dependences

We have shown earlier [4] that Δv_1 and $\Delta v_1'$ of the hadron sets from quark coalescence depend linearly on both Δq and ΔS, as shown in Eq.(1). Therefore, we suggest to analyze the data with a two-dimensional (2-D) linear function: $\Delta v_1' = c_0^* + c_q^* \Delta q + c_S^* \Delta S$. Since the quark coalescence predicts $c_0^* = 0$, a non-zero intercept parameter c_0^* would mean the breaking of the coalescence sum rule. To extract the coefficients for the dependences on Δq and ΔS, one can use the above function to fit the five independent data points; we call this the 5-set method [4]. In the following, we use this method to demonstrate the proper way of extracting the Δq and ΔS dependences. We use the STAR 10-40% Au+Au data at 27A GeV as an example, where we only take the central values for demonstration (i.e., without considering the experimental error bars). These data are given in the STAR27 row in Table 2, where column 2 to 6 give the $\Delta v_1'$ values (multiplied by 10^4) of the hadron index 1 to 5, respectively.

Table 2. Values of $\Delta v_1' \times 10^4$ for Set 1-5 or Index 1-5 of different test data. Values of the coefficients c_0^*, c_q^* and c_S^* from 2-D fits and the coefficients $K_{\Delta q}$ and $K_{\Delta S}$ from 1-D fits are also given.

Test data	#1	2	3	4	5	Fit	c_0^*	c_q^*	c_S^*	$K_{\Delta q}$	$K_{\Delta S}$
STAR27	3	41	39	83	64	w/o c_0^*	0	12.8	9.67	33.4	15.2
						with c_0^*	6.66	8.66	9.93	30.5	13.0
Our27	-3	8	-55/3	23	21	w/o c_0^*	0	-6.00	25.0	-	-
						with c_0^*	-4.44	-6.00	29.4	-	-
Our27ideal	0	0	0	23	21	w/o c_0^*	0	-6.00	25.0	-	-
						with c_0^*	0.00	-6.00	25.0	-	-
STAR27ideal	0	44	42	138	86	w/o c_0^*	0	-6.00	25.0	47.2	22.4
						with c_0^*	0.00	-6.00	25.0	49.0	22.9
STAR27ideal2	0	44	42	69	86	w/o c_0^*	0	-6.00	25.0	40.1	21.9
(Index4 scaled by 1/2)						with c_0^*	0.00	-6.00	25.0	34.5	21.9

© The Authors, published by EDP Sciences. This is an open access article distributed under the terms of the Creative Commons Attribution License 4.0 (https://creativecommons.org/licenses/by/4.0/).

We first use the 2-D function to fit the five STAR27 data points. We obtain the fit function as $6.66+8.66\Delta q+9.93\Delta S$ (or $12.8\Delta q+9.67\Delta S$ if we do not allow the intercept c_0^*). Figure 1 shows the 2-D fit (w/o the intercept) of the STAR27 test data over the $\Delta q-\Delta S$ plane. We also follow the method used by the STAR Collaboration [3] and perform a one-dimensional (1-D) fit of the STAR27 test data as shown in Fig. 1. When fit with the function $K_{\Delta q}\Delta q$, we obtain $33.4\Delta q$ (or $5.33 + 30.5\Delta q$ with intercept). When using the function $K_{\Delta S}\Delta S$, we obtain $15.2\Delta S$ (or $9.65 + 13.0\Delta S$ with intercept). Note that the $K_{\Delta q}$ and $K_{\Delta S}$ values without intercept are rather close to those extracted by the STAR Collaboration (central value of 29 and 19, respectively) [3]. However, we see in Table 2 that the Δq and ΔS slope coefficients from the 1-D fits can be quite different from those from the 2-D fits. For example, the $K_{\Delta q}$ values from the 1-D fit are much larger than the c_q^* values from the 2-D fit.

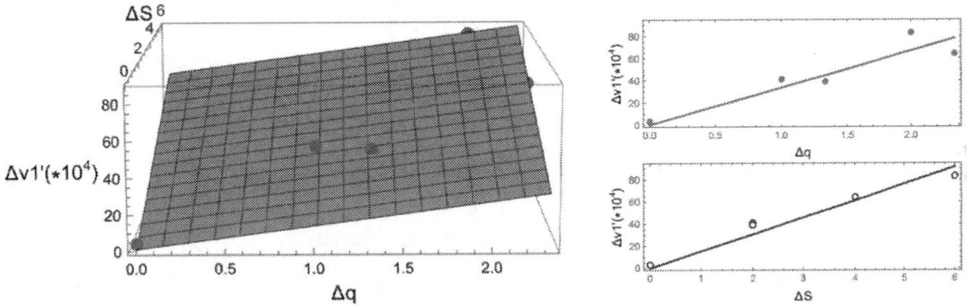

Figure 1. 2-D fit (left) and 1-D fits (right), all w/o intercept, of the STAR27 test data in Table 2.

We then convert the STAR27 test data into the five hadron sets of our choice: the Our27 row in Table 2. Fitting of Our27 test data with the 2-D function gives $-4.44-6.00\Delta q+29.4\Delta S$ (or $-6.00\Delta q+25.0\Delta S$ with no intercept). These c_q^* and c_S^* values are drastically different from those obtained from fitting the STAR27 test data. This difference may seem to be unexpected at first, since our hadron sets and the STAR hadron sets are related to each other linearly and thus equivalent. However, they are only equivalent when the Δv_1 data follow the coalescence sum rule. The left panel of Fig. 2 shows the 2-D fit (w/o intercept) of Our27 test data. We see that $\Delta v_1' \neq 0$ for Set 1, 2, 3 (all at $\Delta q = \Delta S = 0$); this clearly violates the coalescence sum rule.

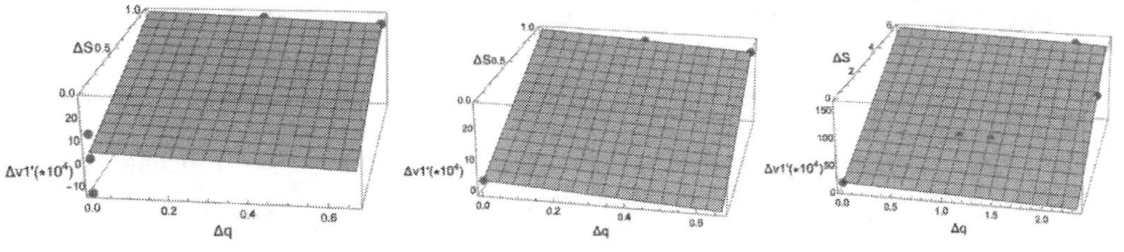

Figure 2. 2-D fit (w/o intercept) of Our27 test data (left). 2-D fits of Our27ideal test data (middle) and the STAR27ideal test data (right) give the same 2-D plane over the $\Delta q-\Delta S$ space.

To explicitly demonstrate this, we set the $\Delta v_1'$ values to zero for Set 1, 2, and 3 of Our27 test data to obtain the Our27ideal test data in Table 2. This test data thus satisfy the coalescence sum rule and serve as an ideal CSR case. A 2-D fit of Our27ideal test data gives $-6.00\Delta q+25.0\Delta S$; this result is obtained with or without an intercept, which is expected for an ideal CSR case. We then convert Our27ideal test data into the five hadron indices chosen

© The Authors, published by EDP Sciences. This is an open access article distributed under the terms of the Creative Commons Attribution License 4.0 (https://creativecommons.org/licenses/by/4.0/).

by STAR, labeled as STAR27ideal in Table 2. A 2-D fit of the STAR27ideal test data also gives $-6.00\Delta q + 25.0\Delta S$, as shown in Fig. 2. Therefore, the difference in the 2-D coefficients from fitting the STAR27 and Our27 test data is due to the breaking of CSR in the test data.

When we perform 1-D fits of the STAR27ideal test data, as shown in Fig. 3, we obtain $K_{\Delta q} = 47.2$ without intercept or 49.0 with intercept. They are drastically different from the Δq slope of -6.00 from the 2-D fits. Therefore, the coefficient for the Δq or ΔS dependence from 1-D fits can be quite different from that from 2-D fits even in the ideal CSR case and is thus incorrect. Note that the 2-D fit plane of the ideal CSR test data perfectly goes through each data point, as shown in Fig. 2. However, that is not the case for the 1-D fits (see Fig. 3).

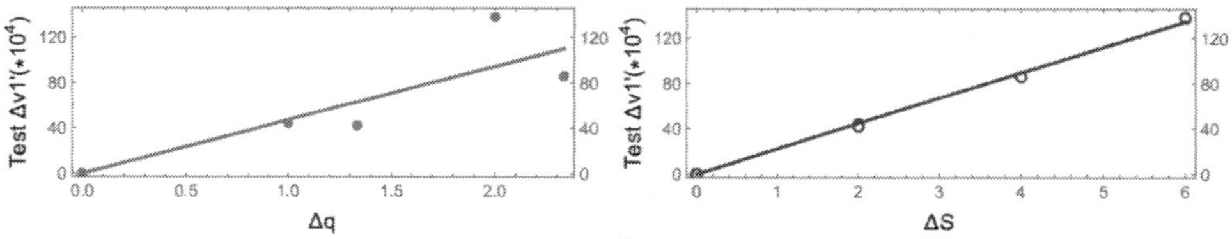

Figure 3. 1-D fits (w/o intercept) of the STAR27ideal test data that satisfy the coalescence sum rule.

In the ideal case where the Δv_1 data follow the CSR, the 2-D coefficients are unchanged when a given hadron set is scaled by a constant, e.g., changing the STAR Index 4 to $v_1[\overline{\Omega}^+(\bar{s}\bar{s}\bar{s})]/2$ (left side) and $v_1[\Omega^-(sss)]/2$ (right side). After all, the overall normalization of the hadron weighting factors for each hadron set is arbitrary. For demonstration, we scale Index4 of the STAR27ideal test data by 1/2 to get the STAR27ideal2 test data; the new Index4 now has $\Delta q = 1$ and $\Delta S = 3$ and STAR27ideal2 still represents the ideal CSR case. As expected, 2-D fits of the unscaled (STAR27ideal) and scaled (STAR27ideal2) test data give exactly the same coefficients. When we perform 1-D fits, however, the $K_{\Delta q}$ and $K_{\Delta S}$ values from fitting the unscaled and scaled test data are different, as shown in Table 2. This clearly shows that the coefficients from 1-D fits are ill-defined mathematically; therefore, 1-D fits should not be used for the analysis of the v_1 difference data.

3 Summary

On the direct flow difference of hadron sets consisting of produced hadrons, we show that 2-D fits give different Δq (or ΔS) slope values than 1-D fits, even for the ideal test data that follow the CSR relations perfectly. We also demonstrate that different choices of the five independent hadron sets are equivalent and give the same 2-D coefficient values when the v_1 difference data follow the CSR. On the other hand, if the extracted coefficients depend on the choice, it indicates the breaking of the CSR. Finally, we show that the coefficients from 1-D fits depend on the arbitrary scaling of a hadron set and are therefore ill-defined.

Acknowledgments. ZWL is supported by the U.S. NSF grants 2012947 and 2310021.

References

[1] L. Adamczyk *et al.* [STAR Collaboration], Phys. Rev. Lett. **120**, 062301 (2018).
[2] A.I. Sheikh, D. Keane, P. Tribedy, Phys. Rev. C **105**, 014912 (2022).
[3] STAR Collaboration, arXiv:2304.02831 (2023).
[4] K. Nayak, S. Shi, Z. W. Lin, Phys. Lett. B **849**, 138479 (2024).
[5] K. Nayak, S. Shi, Z. W. Lin, Universe **10**, 112 (2024).

© The Authors, published by EDP Sciences. This is an open access article distributed under the terms of the Creative Commons Attribution License 4.0 (https://creativecommons.org/licenses/by/4.0/).

Measurement of charge-dependent directed flow in STAR Beam Energy Scan (BES-II) Au+Au and U+U collisions

Muhammad Farhan Taseer[1,2,*] for the STAR Collaboration

[1]Institute of Modern Physics, Chinese Academy of Sciences, Lanzhou 730000, China
[2]University of Chinese Academy of Sciences, Beijing 100049, China

Abstract. We presented the rapidity dependence of directed flow (v_1) and its slope (dv_1/dy) for $\pi^{\pm}$, $K^{\pm}$ and $p(\bar{p})$ in Au+Au collisions at $\sqrt{s_{NN}} = 7.7 - 19.6$ GeV from the Beam Energy Scan Phase-II, as well as in U+U collisions at $\sqrt{s_{NN}} = 193$ GeV measured by the STAR experiment. The v_1 values are reported as a function of rapidity and centrality. Additionally, the dv_1/dy and the charge dependent difference, $\Delta(dv_1/dy)$, of the three identified particles in U+U collisions is compared to those in Au+Au and isobar (Ru+Ru and Zr+Zr) collisions. These findings offer insights into the initial electromagnetic field as well as baryon transport at various system sizes and beam energies.

1 Introduction

An ultra-strong magnetic field ($B \approx 10^{18}$ Gauss) is anticipated during the early stages of heavy-ion collisions. Such a strong magnetic field holds significant importance in exploring the properties of QCD matter under extreme conditions, including understanding topology of QCD vacuum and QCD phase transition. Indeed, magnetic fields are primarily generated by spectators and dissipate rapidly, with timescales similar to the duration of the colliding nuclei passage [1, 2]. The presence of such an extremely strong magnetic field significantly influences the properties of the particles produced in the initial stage.

The directed flow or the first harmonic flow coefficient (v_1) describes the collective sideward motion of produced particles and nuclear fragments, and carries information from the very early stages of collision [3]. Specifically, particles and antiparticles with opposite charges experience different contribution of the electromagnetic forces to their rapidity-odd directed flow, $v_1(y)$ [4]. In addition, v_1 can capture information from the initial geometry of the system and also offer means to understand baryon transport in the early phase of collision.

In the expanding quark-gluon plasma (QGP), quarks experience different electromagnetic effects. Specifically, the Lorentz force $\mathbf{F} = q\mathbf{v} \times \mathbf{B}$ (resulting from the Hall effect) acts in the opposite direction to the forces arising from Faraday induction and Coulomb effect [1, 2]. The $\Delta(dv_1/dy)$, difference in v_1 slope between positively and negatively charged particles, can be used to investigate the effects of electromagnetic fields in heavy-ion collisions. Moreover, model studies [5–7] suggest that transported quarks from the colliding nuclei to the mid-rapidity regions and the produced quarks in the final state may have different contributions due to differences in quark compositions, resulting a charge dependent v_1 and $\Delta(dv_1/dy)$ [4].

*e-mail: mfarhan_taseer@impcas.ac.cn

© The Authors, published by EDP Sciences. This is an open access article distributed under the terms of the Creative Commons Attribution License 4.0 (https://creativecommons.org/licenses/by/4.0/).

Previous studies by STAR Collaboration on $\Delta(dv_1/dy)$ in peripheral Au+Au collisions at $\sqrt{s_{NN}}$ = 27 GeV, as well as Au+Au and isobar collisions at top RHIC energy ($\sqrt{s_{NN}}$ = 200 GeV) are consistent with the expectation from the dominance of Faraday and Coulomb effects in peripheral collisions [4, 8]. Another STAR analysis [9] observed the positive $\Delta(dv_1/dy)$ splitting in the mid-central collisions using combinations of hadrons without transported quarks, which can be explained by the presence of an electromagnetic field predominantly influenced by the Hall effect. We extend these charge dependent v_1 measurements by different collision systems such as U+U, Au+Au and isobar (Ru+Ru and Zr+Zr).

2 Dataset and event selection

The data set analyzed in this work was recorded by the STAR experiment at the Relativistic Heavy Ion Collider (RHIC) in the year 2012 for U+U collisions at $\sqrt{s_{NN}}$ = 193 GeV. A total of 250 million good quality minimum-bias triggered events were analyzed, with each event required to have a primary vertex within $|V_z| \leq 50$ cm along the beam direction and $|V_r| \leq 2$ cm in the transverse direction. Time Projection Chamber (TPC) of STAR is used for charged particle tracking within pseudorapidity $|\eta| < 1$ and provides full 2π azimuthal coverage. Standard track selection and particle identification, utilizing TPC and Time-of-Flight (TOF) detectors, are performed as described in Ref. [4].

3 Results and discussion

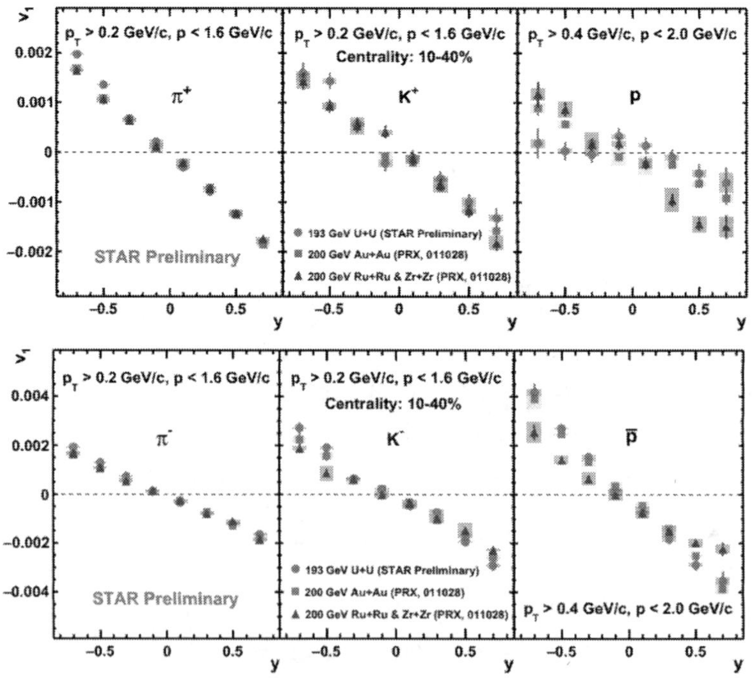

Figure 1. v_1 as a function of rapidity for $\pi^{\pm}$, $K^{\pm}$, protons and anti-protons in U+U, Au+Au and isobar (Ru+Ru and Zr+Zr) collisions at $\sqrt{s_{NN}}$ = 193 and 200 GeV [4]. Transverse momentum (p_T) and total momentum in collision center-of-mass frame (p) are specified in figure legends.

Figure 1 shows STAR measurement of directed flow as a function of rapidity in mid-central (10-40%) U+U collisions for $\pi^{\pm}$, $K^{\pm}$ and $p(\bar{p})$. The slope (dv_1/dy) is extracted using a linear fit within rapidity ($-0.8 < y < 0.8$). Moreover, $v_1(y)$ measured in this work for U+U is also compared with published results in Au+Au and isobar collisions at 200 GeV [4].

© The Authors, published by EDP Sciences. This is an open access article distributed under the terms of the Creative Commons Attribution License 4.0 (https://creativecommons.org/licenses/by/4.0/).

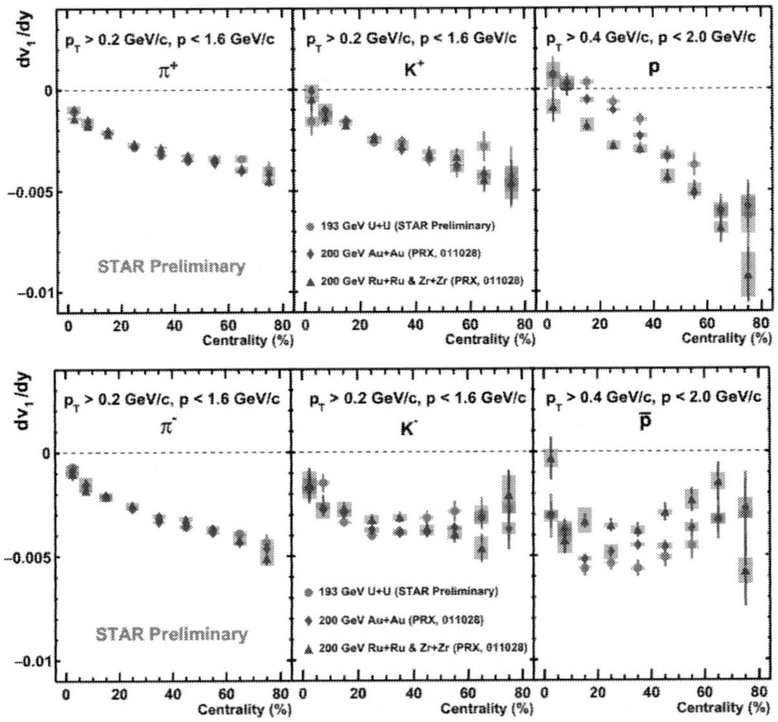

Figure 2. dv_1/dy as a function of centrality for $\pi^{\pm}$, $K^{\pm}$, protons and anti-protons in U+U, Au+Au and isobar (Ru+Ru and Zr+Zr) collisions at $\sqrt{s_{NN}}$ = 193 and 200 GeV [4].

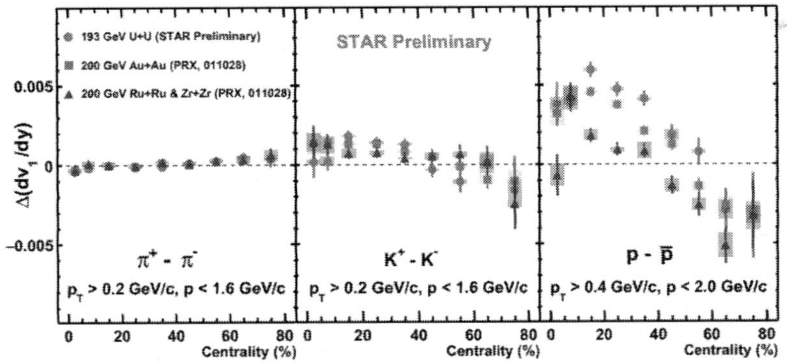

Figure 3. $\Delta(dv_1/dy)$ as a function of centrality for pions, kaons and protons in U+U, Au+Au and isobar (Ru+Ru and Zr+Zr) collisions at $\sqrt{s_{NN}}$ = 193 and 200 GeV [4].

Figure 2 presents (dv_1/dy) slope as a function of centrality for $\pi^{\pm}$, $K^{\pm}$ and $p(\bar{p})$. A significant splitting in (dv_1/dy) slope as well as dv_1/dy dependence on system size is observed for proton and antiproton in the mid-central (10-40%) collisions among the three different collision systems. In contrary, no significant dv_1/dy splitting is observed for pions and kaons and the data points are found consistent between different colliding species.

Figure 3 shows the difference of (dv_1/dy) slope between positively and negatively charged particles as a function of centrality. Proton $\Delta(dv_1/dy)$ shows a clear ordering in system size in the mid-central (10-40%) collisions as well as a sign change in peripheral (50-80%) collisions. This sign change for the proton $\Delta(dv_1/dy)$ is consistent with naive expectations of transported quarks + electromagnetic effects. No significant (dv_1/dy) splitting between π^+

© The Authors, published by EDP Sciences. This is an open access article distributed under the terms of the Creative Commons Attribution License 4.0 (https://creativecommons.org/licenses/by/4.0/).

and π^- is observed in U+U collisions at 193 GeV similar to Au+Au and isobar collisions at 200 GeV. A similar but less obvious trend, compared to pion $\Delta(dv_1/dy)$, is found for kaon $\Delta(dv_1/dy)$ and the data points are consistent within uncertainties.

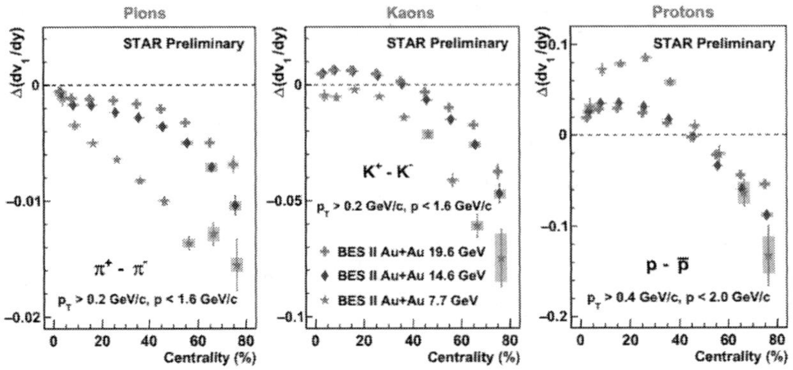

Figure 4. $\Delta(dv_1/dy)$ as a function of centrality for pions, kaons and protons in Au+Au collisions at $\sqrt{s_{NN}}$ = 19.6, 14.6 and 7.7 GeV.

Figure 4 presents the beam energy dependence of $\Delta(dv_1/dy)$ as a function of centrality for pions, kaons and protons in Au+Au collisions at $\sqrt{s_{NN}} = 7.7 - 19.6$ GeV. The $\Delta(dv_1/dy)$ for protons become more negative in the peripheral collisions as the beam energy is reduced, consistent with the expectation from the dominance of (Faraday + Coulomb) effect [8].

4 Summary

We have presented a measurement of v_1 and $\Delta(dv_1/dy)$ in U+U collisions at 193 GeV as well as in Au+Au collisions at 7.7 – 19.6 GeV using STAR experiment. The positive $\Delta(dv_1/dy)$ for kaons and protons in central collisions can be attributed to the contributions from transported quarks, while the notable negative values of $\Delta(dv_1/dy)$, observed in the peripheral collisions, are consistent with the effects of the electromagnetic field, primarily due to Faraday induction combined with the Coulomb effect [1, 2]. The v_1 dependence on system size can be due to the presence of extremely strong magnetic fields at very early stages of heavy-ion collisions, however it needs more exploration and theoretical input. Additionally, the stronger charge splitting observed in Au+Au collisions at lower BES-II energies [8] support the idea that the electromagnetic field decays more slowly at lower collision energies. Other mechanisms such as effect from baryon transport and its inhomogeneities are under investigation [10].

References

[1] U. Gürsoy et al. Phys. Rev. C 98,055201 (2018)
[2] U. Gürsoy et al. Phys. Rev. C 89 054905 (2014)
[3] H. Sorge. Phys. Rev. Lett. 78, 2309 (1997)
[4] M. I. Abdulhamid et al. (STAR Collaboration) Phys. Rev. X 011028 (2024)
[5] Y. Guo et al. Phys. Rev. C 86, 044901 (2012)
[6] K. Nayak et al. Phys. Rev. C 100, 054903 (2019)
[7] P. Bożek, Phys. Rev. C 106, L061901 (2022)
[8] A. P. Dash (for STAR Collaboration), EPJ Web of Conferences 296, 10007 (2024)
[9] STAR Collaboration, arXiv:2304.02831
[10] T. Parida et al. arXiv:2305.08806

© The Authors, published by EDP Sciences. This is an open access article distributed under the terms of the Creative Commons Attribution License 4.0 (https://creativecommons.org/licenses/by/4.0/).

EPJ Web of Conferences **316**, 06009 (2025) https://doi.org/10.1051/epjconf/202531606009
SQM 2024

Collision Energy Dependence of Hypertriton Production in Au+Au Collisions at RHIC

Xiujun Li[1,*] , for the STAR Collaboration

[1]University of Science and Technology of China

Abstract. Hypernuclei are bound states of nucleons and at least one hyperon. In particular, the hypertriton ${}^{3}_{\Lambda}$H, a bound state consisting of a proton, neutron and hyperon, is the lightest known hypernucleus. Precise measurements on the energy dependence of ${}^{3}_{\Lambda}$H production will give invaluable information on hypernuclei production mechanisms due to its unique intrinsic properties. The second phase of the Beam Energy Scan (BES-II) at RHIC provides a great opportunity to study hypernuclei production. In these proceedings, we present p_{T}−integrated yields (dN/dy) at mid-rapidity, yield ratios and average transverse momentum ($\langle p_T \rangle$) of ${}^{3}_{\Lambda}$H as a function of collision energy in Au+Au collisions from $\sqrt{s_{\mathrm{NN}}}$ = 3 to 27 GeV. These results are compared with phenomenological model calculations, and physics implications on production mechanism are discussed.

1 Introduction

Despite extensive measurements of light nuclei production in heavy-ion collisions, the precise mechanisms underlying their formation remain an open question. Unlike ordinary nuclei, hypernuclei, which contain strange quarks, offer an additional dimension to these studies, while their production mechanisms are not well understood. The equation-of-state (EoS) with the presence of strangeness is crucial for probing the dense neutron star interiors, but the hyperon-nucleon (Y−N) interaction, essential for constraining this EoS, remains poorly understood [1]. Hypernuclei serve as a natural laboratory to study the Y−N interaction. Precise measurements of hypernuclei properties and production yields can shed light on the production mechanisms and the role of Y−N interactions play at neutron stars densities.

The hypertriton ${}^{3}_{\Lambda}$H, a bound state of a proton, neutron, and Λ hyperon, is the lightest known hypernucleus [2]. A prominent enhancement of the strangeness population factor, $S_3 = {}^{3}_{\Lambda}\mathrm{H}/({}^{3}\mathrm{He} \times \frac{\Lambda}{\mathrm{p}})$, has been proposed as a probe for deconfinement [3]. Detailed studies of the energy dependence of ${}^{3}_{\Lambda}$H production could provide valuable insights into the mechanisms of hypernuclei formation.

Notably, light hypernuclei are predicted by thermal models to be abundantly produced at low collision energies due to the high baryon density [4]. A fixed target mode in the STAR (Solenoidal Tracker At RHIC) experiment was specifically implemented during the BES-II running period to reach lower center-of-mass energies, which corresponds to higher baryon density. The low $\sqrt{s_{\mathrm{NN}}}$, enhanced detector capabilities, and high statistics datasets collected during BES-II offer great opportunity for hypernuclei study.

*e-mail: lixiujun@mail.ustc.edu.cn

© The Authors, published by EDP Sciences. This is an open access article distributed under the terms of the Creative Commons Attribution License 4.0 (https://creativecommons.org/licenses/by/4.0/).

2 Analysis Details

The measurements of ${}^{3}_{\Lambda}$H production in this analysis are carried out by utilizing the Au+Au collision datasets in at $\sqrt{s_{\rm NN}}$ = 3.0, 3.2, 3.5, 3.9, 4.5 and 5.2 GeV collected with the FXT mode, and data at $\sqrt{s_{\rm NN}}$ = 7.7, 11.5, 14.6, 19.6 and 27 GeV collected in collider mode from STAR experiment during BES-II . The hypertriton ${}^{3}_{\Lambda}$H is reconstructed via its two-body decay channel (${}^{3}_{\Lambda}{\rm H} \rightarrow \pi^{-} + {}^{3}{\rm He}$) using the KFParticle package [5]. Particle identification of daughters (π^{-} and ^{3}He) is done by using energy loss information from the Time Projection Chamber (TPC). The decay topology is used for ${}^{3}_{\Lambda}$H reconstruction.

3 Results and discussion

The transverse momentum (p_T) spectra of ${}^{3}_{\Lambda}$H are measured in 0-10% and 10-40% central Au+Au collisions at $\sqrt{s_{\rm NN}}$ = 3 to 27 GeV. Figure 1 provides an example of the p_T spectra of ${}^{3}_{\Lambda}$H at $\sqrt{s_{\rm NN}}$ = 7.7 to 27 GeV, with the dotted lines indicating fits to the data.

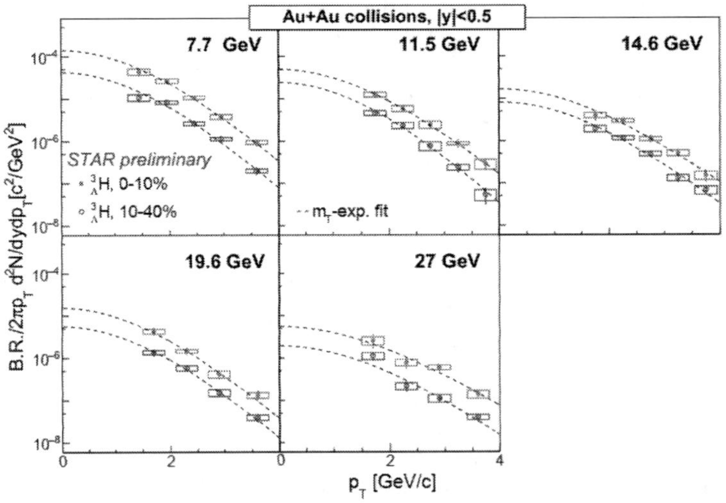

Figure 1. The ${}^{3}_{\Lambda}$H p_T spectra in 0-10% (red circles) and 10-40% (blue circles) central Au+Au collisions. Systematic uncertainties are represented by boxes. The dotted lines are m_T exponential fits to data.

The p_T-integrated yields dN/dy for hypertriton are calculated from the p_T spectra by combining data in the measured p_T range and function-fitting extrapolation in the unmeasured p_T range. Figure 2 presents the energy dependence of the dN/dy for ${}^{3}_{\Lambda}$H at mid-rapidity ($|y| < 0.5$) in 0-10% central Au+Au collisions across a range of $\sqrt{s_{\rm NN}}$ from 3 to 27 GeV. Two predictions, from Ultra-relativistic Quantum Molecular Dynamics (UrQMD)+coalescence [6] and the thermal model [6] are plotted for comparison. In the UrQMD+Coalescence model, the UrQMD transport model generates hadron phase space distributions at freeze-out. A coalescence procedure is then applied, where light nuclei and hypernuclei are formed if the relative momentum and spatial distance of their constituents are within specified thresholds [6]. The thermal model assumes that the chemical freeze-out of light nuclei and hypernuclei happens simultaneously with hadrons [6]. The dN/dy increases as the collision energy decreases, peaking around 3-4 GeV. Both the UrQMD+Coalescence and thermal model calculations capture the overall trend. While the thermal model significantly overestimates the experimental data at all energies, the UrQMD+Coalescence model matches the experimental data below 11.5 GeV.

© The Authors, published by EDP Sciences. This is an open access article distributed under the terms of the Creative Commons Attribution License 4.0 (https://creativecommons.org/licenses/by/4.0/).

Figure 3 presents the particle ratios of (hyper)nucleus to hadron (d/p, t/p, and ${}^3_\Lambda\mathrm{H}/\Lambda$) as a function of collision energy. The thermal model results, indicated by dashed/ dash-dotted lines, overestimates the ${}^3_\Lambda\mathrm{H}/\Lambda$ and t/p ratios by a factor of approximately 2, while it successfully describes the d/p ratio in the data. These indicate ${}^3_\Lambda\mathrm{H}$ and triton (t) yields are not in equilibrium and fixed at chemical freeze-out along with other light hadrons.

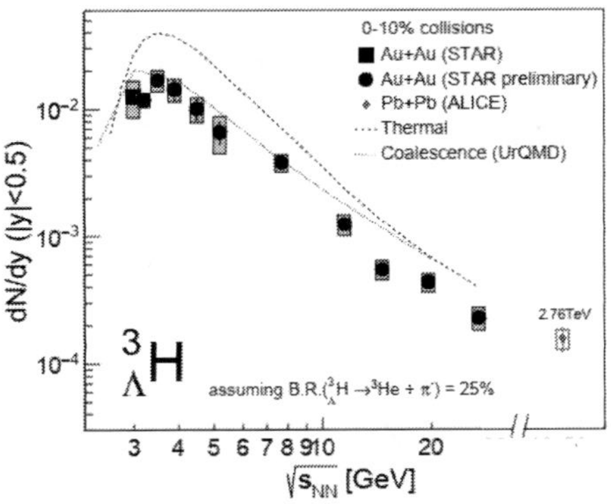

Figure 2. Measured mid-rapidity yields of ${}^3_\Lambda\mathrm{H}$ in 0-10% central Au+Au collisions as function of the center of mass collision energy [7, 8]. The dashed line is from thermal model calculations [6]. The solid line is from transport model calculations with coalescence as an afterburner [6].

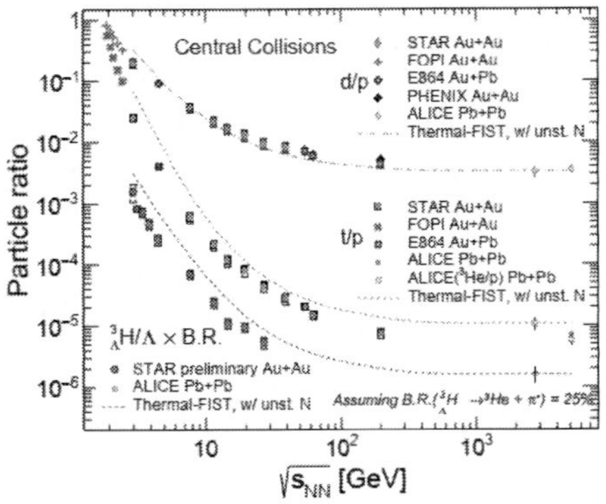

Figure 3. d/p, t/p, and ${}^3_\Lambda\mathrm{H}/\Lambda$ yield ratios in 0-10% central Au+Au collisions as function of the center of mass collision energy. The thermal model results are shown by dashed and dash-dotted lines [6]. Experimental measurements are shown as symbols [9, 10].

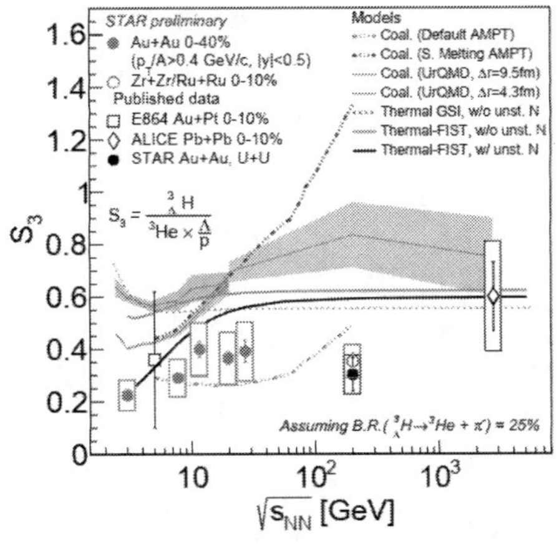

Figure 4. Energy dependence of S_3 from experiments [8, 11–13] and models [3, 4, 6]. The UrQMD+coalescence results with $\Delta r = 9.5$ fm are shown as orange line, while the results of $\Delta r=$ 4.3 fm are shown as magenta line [6]. The black and grey solid lines correspond to thermal model with/without feed down from unstable nuclei.

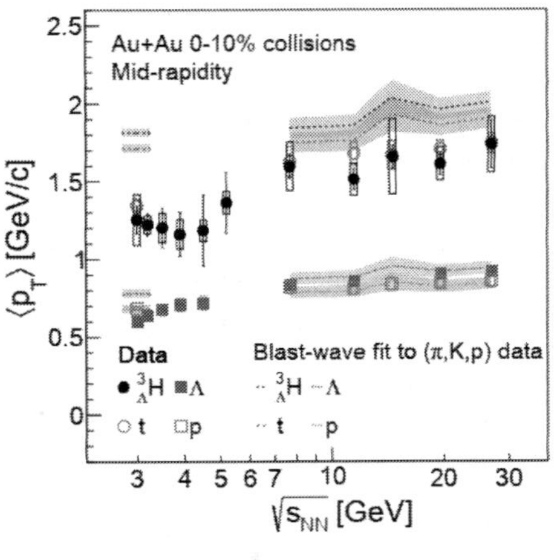

Figure 5. Energy dependence of $\langle p_T\rangle$ for ${}^3_\Lambda\mathrm{H}$, t, Λ, and p in 0-10% most central collisions. In addition, the calculations by fitting the experimental spectra of π, K, p with a blast-wave model are shown as the dashed/solid lines corresponding to the marker color.

© The Authors, published by EDP Sciences. This is an open access article distributed under the terms of the Creative Commons Attribution License 4.0 (https://creativecommons.org/licenses/by/4.0/).

Figure 4 presents the energy dependence of the strangeness population factor S_3 from experimental data and compares them with predictions from coalescence and thermal model approaches. The measured S_3 shows a mild increasing trend from $\sqrt{s_{\rm NN}}$ = 3 GeV to 2.76 TeV. If feed down from unstable nuclei to stable protons and ^{3}He is included in the thermal model, the S_3 is reduced and provides a better description of the experimental measurements. In the UrQMD+coalescence results, the energy dependence of S_3 exhibits sensitivity to the coalescence parameter source radius (Δr), possibly due to the difficulty in forming $^3_\Lambda$H with a large radius in smaller systems [6].

Figure 5 shows the energy dependence of $\langle p_T \rangle$ of hypertritons, tritons, Λ, and protons in 0-10% central Au+Au collisions from $\sqrt{s_{\rm NN}}$ = 3 to 27 GeV. The $\langle p_T \rangle$ of $^3_\Lambda$H and t shows a similar energy-dependence trend. The blast-wave fit using measured kinetic freeze-out parameters from light hadrons (π, K, p) slightly overestimates the $\langle p_T \rangle$ for both $^3_\Lambda$H and t. It suggests that $^3_\Lambda$H and t do not follow same collective expansion as light hadrons, indicating that $^3_\Lambda$H and t decouples at different times compared to light hadrons.

4 Summary and outlook

In summary, we carry out the energy dependence study of hypertriton yields in Au+Au collisions from $\sqrt{s_{\rm NN}}$ = 3 to 27 GeV in the high-baryon-density region. The UrQMD+Coalescence calculations are consistent with data, but the thermal model with the freeze-out parameters determined by the light hadron yields over-predicts hypertriton yields. The over-predictions of the $^3_\Lambda$H/Λ yield ratio by the thermal model, and the $\langle p_T \rangle$ of $^3_\Lambda$H by blast-wave model parametrized from light hadron data, suggest that the $^3_\Lambda$H is likely formed at or decouples from the system at a different temperature compared to the light hadrons.

The results in these proceedings utilize only a subset of the BES II datasets. STAR collected 2 billion Au+Au events at $\sqrt{s_{\rm NN}}$ = 3 GeV in Run 21, and is projected to collect 18 billion Au+Au events at $\sqrt{s_{\rm NN}}$ = 200 GeV in Run 23-25, much larger than the datasets presente in these proceedings. The huge datasets would enable precise light hypernuclei measurements and offer great opportunity to the measurements of heavier hypernuclei with A>3.

References

[1] D. Chatterjee, I. Vidana, Eur. Phys. J. A **52**, 1 (2016)
[2] J. Chen, X. Dong, Y.G. Ma, Z. Xu, Sci. Bull. **68**, 3252 (2023)
[3] S. Zhang et al., Phys. Lett. B **684**, 224 (2010)
[4] A. Andronic et al., Phys. Lett. B **697**, 203 (2011)
[5] X.Y. Ju et al., Nucl. Sci. Tech. **34**, 158 (2023)
[6] T. Reichert et al., Phys. Rev. C **107**, 014912 (2023)
[7] M.S. Abdallah et al. (STAR Collaboration), Phys. Rev. Lett. **128**, 202301 (2022)
[8] J. Adam et al. (ALICE Collaboration), Phys. Lett. B **754**, 360 (2016)
[9] M.I. Abdulhamid et al. (STAR Collaboration), Phys. Rev. Lett. **130**, 202301 (2023)
[10] M.I. Abdulhamid et al. (STAR Collaboration) (2023), `arXiv:2311.11020`
[11] T.A. Armstrong et al. (E864 Collaboration), Phys. Rev. C **70**, 024902 (2004)
[12] B. Abelev et al. (STAR Collaboration), Science **328**, 58 (2010)
[13] M.I. Abdulhamid et al. (STAR), Nature **632**, 1026 (2024)

© The Authors, published by EDP Sciences. This is an open access article distributed under the terms of the Creative Commons Attribution License 4.0 (https://creativecommons.org/licenses/by/4.0/).

Study of baryon-strangeness and charge-strangeness correlations in Pb–Pb collisions at $\sqrt{s_{\mathrm{NN}}}$ = 5.02 TeV with ALICE

Swati Saha[1,2,*] for the ALICE Collaboration

[1]School of Physical Sciences, National Institute of Science Education and Research, Jatni-752050, Odisha, India

[2]Homi Bhabha National Institute, Training School Complex, Anushaktinagar, Mumbai-400094, Maharastra, India

Abstract. In the quest to unravel the mysteries of the strong force and the underlying properties of the quark-gluon plasma, the ALICE collaboration at CERN has carried out a comprehensive study focusing on the correlations between net-conserved quantities such as net-baryon, net-charge and net-strangeness. These correlations play a crucial role in the study of QCD phase structure as they are closely related to the ratios of thermodynamic susceptibilities in lattice QCD calculations. This work mainly focuses on the correlations between net-kaon and net-proton, and net-kaon and net-charge in Pb–Pb collisions at $\sqrt{s_{\mathrm{NN}}}$ = 5.02 TeV using data recorded during LHC Run 2. The net-proton and net-kaon serve as proxies for net-baryon and net-strangeness, respectively, with measurements analyzed as a function of collision centrality. Theoretical predictions from the Thermal-FIST model are compared with experimental results, providing insights into the effects of resonance decays and charge conservation laws on the correlations.

1 Introduction

Quantum Chromodynamics (QCD) predicts that, at sufficiently high energy densities, nuclear matter transitions into a state called quark-gluon plasma (QGP), where quarks and gluons are no longer confined within hadrons [1]. Studying fluctuations and correlations of conserved charges in heavy-ion collisions provides a powerful tool for probing the QCD phase transition and the properties of the QGP medium [2, 3]. These observables are linked to thermodynamic susceptibilities in lattice QCD (LQCD) calculations and exhibit characteristic changes in the crossover region between low- and high-temperature phases of QCD [4, 5]. Comparing experimental measurements of these fluctuations and correlations with Hadron Resonance Gas (HRG) model predictions can enhance our understanding of freeze-out parameters and the thermodynamic properties of the medium formed in heavy-ion collisions [5]. However, it is important to recognize that, in addition to critical signals related to phase transition, other dynamical sources can also influence the measurements of conserved charge fluctuations and correlations, such as quantum number conservation [6], volume fluctuations [7, 8], resonance decays [9, 10], etc. Understanding the effects of these individual sources is crucial. This report will primarily focus on the impact of exact conservation laws–baryon number (B),

*e-mail: swati.saha@cern.ch

© The Authors, published by EDP Sciences. This is an open access article distributed under the terms of the Creative Commons Attribution License 4.0 (https://creativecommons.org/licenses/by/4.0/).

electric charge (Q), and strangeness (S)–on particle production, as well as on the role of resonance decays and final-state interactions in affecting the measurements.

For a system with pressure P and temperature T, the thermodynamic susceptibilities of conserved charges in the framework of LQCD are obtained from partial derivatives of P/T^4 with respect to the chemical potentials of the conserved charges (μ_x: x = Q, B, or S):

$$\chi^{l,m,n}_{\mathrm{B,Q,S}} = \frac{\partial^{(l+m+n)}(P/T^4)}{\partial^l(\mu_\mathrm{B}/T)\partial^m(\mu_\mathrm{Q}/T)\partial^n(\mu_\mathrm{S}/T)},$$

where $l, m, n = 1, 2, 3, ..., n$ are the order of derivative. These susceptibilities are studied experimentally in terms of the ratios of cumulants of net-conserved charge distributions [11]. Owing to the experimental challenges in detecting all baryons and strange hadrons, net-proton (p) and net-kaon (K) are employed as proxies for net-baryon and net-strangeness, respectively. The second-order ($l + m + n = 2$) diagonal and off-diagonal cumulants for net-charge, net-proton, and net-kaon multiplicity distributions can be expressed as: $\sigma^2_\alpha = \langle(\delta N_\alpha)^2\rangle$ and $\sigma^{11}_{\alpha,\beta} = \langle(\delta N_\alpha)(\delta N_\beta)\rangle$, where α, β can be Q, p, or K, $\delta N_\alpha = (N_{\alpha^+} - N_{\alpha^-}) - \langle(N_{\alpha^+} - N_{\alpha^-})\rangle$, and angular brackets $\langle ... \rangle$ denote average over all events.

In this report, the cumulant ratios $C_{\mathrm{p,K}} = \sigma^{11}_{\mathrm{p,K}}/\sigma^2_\mathrm{K}$ and $C_{\mathrm{Q,K}} = \sigma^{11}_{\mathrm{Q,K}}/\sigma^2_\mathrm{K}$ (proxies of baryon-strangeness correlation and charge-strangeness correlation respectively) [12, 13] in Pb–Pb collisions at $\sqrt{s_\mathrm{NN}}$ = 5.02 TeV measured using the data recorded by the ALICE [14, 15] detector are presented as a function of centrality.

2 Analysis details

The analysis is conducted using 80 million Pb–Pb collision events recorded during the 2015 LHC run, selected by a minimum-bias trigger that requires signals from both the V0A ($2.8 < \eta < 5.1$) and V0C ($-3.7 < \eta < -1.7$) detectors [14, 15]. The centrality of an event is determined from the signal amplitudes in these detectors. Charged particle tracks are reconstructed in the central barrel of ALICE using the Inner Tracking System (ITS) and Time Projection Chamber (TPC) [14, 15], which provide full azimuthal coverage within the pseudorapidity range $|\eta| < 0.8$. Particle identification for pions, kaons, and protons is achieved by analyzing specific energy loss ($\mathrm{d}E/\mathrm{d}x$) in the TPC and measuring flight time from the collision's primary vertex to the Time-of-Flight (TOF) [14, 15] detector. The p_T range for particle selection is 0.2 to 2.0 GeV/c for pions and kaons, and 0.4 to 2.0 GeV/c for protons. Measurements are corrected for detector efficiencies considering binomial response of detector as detailed in Ref. [16]. Track reconstruction efficiencies for protons, pions, and kaons (and their anti-particles) are determined via Monte Carlo simulations with HIJING [17] and GEANT3 [18], and these efficiencies are applied to correct the diagonal and off-diagonal cumulants on a track-by-track basis. Statistical uncertainties in the measurements are estimated using the bootstrap resampling method. Systematic uncertainties are evaluated by varying the criteria for event selection, track selection, and altering the conditions used for particle identification.

3 Results

The measured correlation between net-proton and net-kaon, denoted as $C_{\mathrm{p,K}}$, is shown in the left panel of Fig. 1 as a function of centrality, while the right panel displays the correlation between net-charge and net-kaon, $C_{\mathrm{Q,K}}$. If the event-by-event multiplicity distributions of pions, kaons, and protons follow a Poisson distribution, all the off-diagonal cumulants ($\sigma^{11}_{\mathrm{p,K}}$, $\sigma^{11}_{\pi,\mathrm{K}}$, and $\sigma^{11}_{\pi,\mathrm{p}}$) are zero. Consequently, the Poisson expectations for $C_{\mathrm{p,K}} = \sigma^{11}_{\mathrm{p,K}}/\sigma^2_\mathrm{K}$ and

© The Authors, published by EDP Sciences. This is an open access article distributed under the terms of the Creative Commons Attribution License 4.0 (https://creativecommons.org/licenses/by/4.0/).

$C_{\mathrm{Q,K}} = (\sigma^2_{\mathrm{K}} + \sigma^{11}_{\mathrm{p,K}} + \sigma^{11}_{\pi,\mathrm{K}})/\sigma^2_{\mathrm{K}} = 1 + \sigma^{11}_{\mathrm{p,K}}/\sigma^2_{\mathrm{K}} + \sigma^{11}_{\pi,\mathrm{K}}/\sigma^2_{\mathrm{K}}$ would be zero and one, respectively. Measurements of both the correlations, $C_{\mathrm{p,K}}$ and $C_{\mathrm{Q,K}}$ are found to deviate from their corresponding Poisson baselines. The effects of resonance decays on these correlations are explored by comparing the experimental measurements with Thermal-FIST model [19] calculations, both with and without the inclusion of resonance decay contributions. The model calculations use the canonical ensemble (CE) formalism, which enforces exact conservation of B, Q, and S in a correlation volume of $V_c = 3\mathrm{d}V/\mathrm{d}y$ ($\mathrm{d}V/\mathrm{d}y$ represents volume per unit rapidity), with model parameters sourced from Ref. [20]. The correlations $C_{\mathrm{p,K}}$ and $C_{\mathrm{Q,K}}$ are notably enhanced across all centralities when including resonance decays, and the model results incorporating these decay contributions show better alignment with the experimental data.

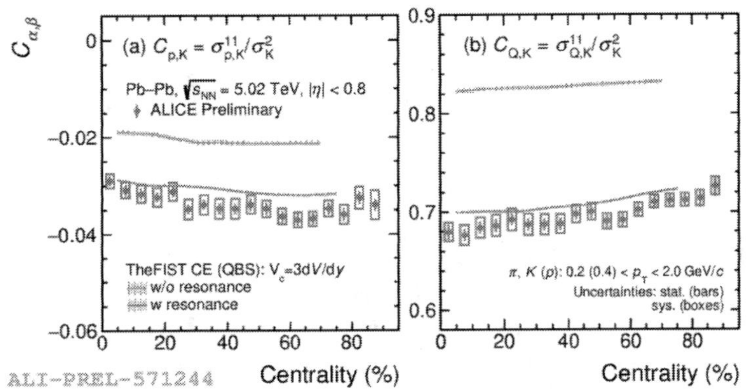

Figure 1. Centrality dependence of $C_{\mathrm{p,K}}$ (a) and $C_{\mathrm{Q,K}}$ (b) in Pb–Pb collisions at $\sqrt{s_{\mathrm{NN}}} = 5.02$ TeV, compared to Thermal-FIST (TheFIST) results [19] using the canonical ensemble (CE) formalism, enforcing exact conservation of baryon number, charge, and strangeness (B, Q, S) within a correlation volume of $V_c = 3\mathrm{d}V/\mathrm{d}y$. ALICE data are shown as solid red circles, while colored bands represent Thermal-FIST calculations with (w) and without (w/o) resonance contributions. The statistical (systematic) uncertainties are represented by vertical bars (boxes).

The conservation laws for Q, B, and S significantly influence the observed correlations in high-energy nuclear collisions. These laws constrain the types of particles produced, leading to distinct patterns in final-state particle correlations. The effect of different charge-conservation scenarios, investigated in the framework of Thermal-FIST model is reported here. In the left (right) panel of Fig. 2, the experimental results for $C_{\mathrm{p,K}}$ ($C_{\mathrm{Q,K}}$) are compared with model predictions imposing the conservation of electric-charge-only, strangeness-only, baryon-number-only, both strangeness and baryon number (electric charge), and all three combined. $C_{\mathrm{Q,K}}$ is found to be mainly sensitive to conservation of both electric charge and strangeness. For $C_{\mathrm{p,K}}$, conservation of electric-charge-only, and both strangeness and baryon number have a similar effect. The mesaurements for both the correlations are best described by model calculations that account for the conservation of all three quantum numbers.

4 Conclusions

In conclusion, the measured correlations between net-proton and net-kaon ($C_{\mathrm{p,K}}$) and between net-charge and net-kaon ($C_{\mathrm{Q,K}}$) in Pb–Pb collisions at $\sqrt{s_{\mathrm{NN}}} = 5.02$ TeV deviate significantly from their respective Poisson baseline values across all centralities. Resonance decays play a major role in enhancing these correlations, as demonstrated by the close agreement between

© The Authors, published by EDP Sciences. This is an open access article distributed under the terms of the Creative Commons Attribution License 4.0 (https://creativecommons.org/licenses/by/4.0/).

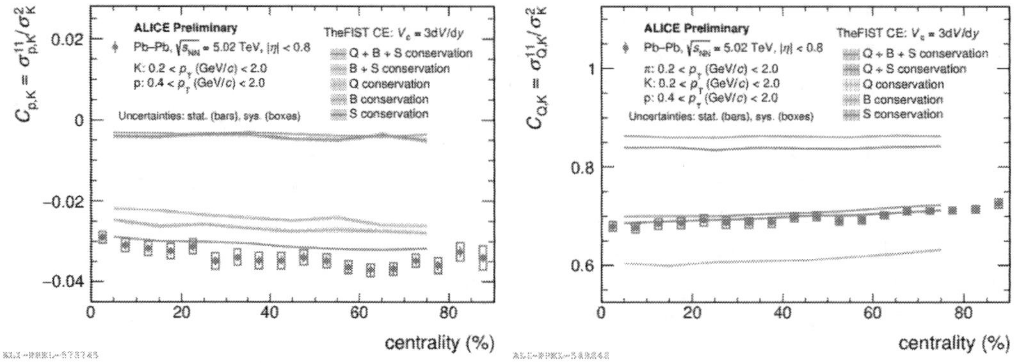

Figure 2. Centrality dependence of $C_{p,K}$ (left) and $C_{Q,K}$ (right) in Pb–Pb collisions at $\sqrt{s_{NN}} = 5.02$ TeV. ALICE measurements (solid red circles) are compared to predictions from the Thermal-FIST (TheFIST) model [19] in the canonical ensemble (CE), accounting for different conservation effects of charge (Q), baryon number (B), and strangeness (S). The statistical (systematic) uncertainties are represented by vertical bars (boxes).

experimental data and model calculations incorporating such contributions. The analysis using the Thermal-FIST model, which enforces the exact conservation of baryon number, electric charge, and strangeness, highlights the significant influence of these conservation laws on the observed correlations. Overall, the presented observables are best described when the model incorporates all three quantum number conservation laws.

References

[1] E. V. Shuryak, *Phys. Rept.* **61** (1980) 71.
[2] E. V. Shuryak, *Phys. Lett. B* **423** (1998) 9-14.
[3] M. A. Stephanov et al., *Phys. Rev. D* **60** (1999) 114028.
[4] B. Friman et al., *Eur. Phys. J. C* **71** (2011) 1694.
[5] F. Karsch, K. Redlich, *Phys. Lett. B* **695** (2011) 136-142.
[6] V. Vovchenko and V. Koch, *Phys. Rev. C* **103** (2021) 044903.
[7] X. Luo et al., *J. Phys. G* **40** (2013) 105104.
[8] P. Braun-Munzinger et al., *Nucl. Phys. A* **960** (2017) 114-130.
[9] P. Garg et al., *Phys. Lett. B* **726** (2013) 691-696.
[10] D. K. Mishra et al., *Phys. Rev. C* **94** (2016) 014905.
[11] S. Gupta et al., *Science* **332** (2011) 1525-1528.
[12] D. Bollweg et al., *Phys. Rev. D* **110** (2024) 054519.
[13] V. Koch et al., *Phys. Rev. Lett.* **95** (2005) 182301.
[14] ALICE Collaboration, K. Aamodt et al., *JINST* **3** (2008) S08002.
[15] ALICE Collaboration, B. Abelev et al., *Int. J. Mod. Phys. A* **29** (2014) 1430044.
[16] T. Nonaka et al., *Phys. Rev. C* **95** (2017) 064912.
[17] X.-N. Wang and M. Gyulassy, *Phys. Rev. D* **44** (1991) 3501-3516.
[18] R. Brun et al., *CERN – W5013* (10, 1994).
[19] V. Vovchenko and H. Stoecker, *Comput. Phys. Commun.* **244** (2019) 295-310.
[20] V. Vovchenko et al., *Phys. Rev. C* **100** (2019) 054906.

© The Authors, published by EDP Sciences. This is an open access article distributed under the terms of the Creative Commons Attribution License 4.0 (https://creativecommons.org/licenses/by/4.0/).

Measurements of Kaon Femtoscopy in Au+Au Collisions at $\sqrt{s_{NN}}$ = 3.0 - 4.5 GeV by the STAR Experiment

Bijun Fan[1,*] (for the STAR collaboration)

[1]Central China Normal University, Wuhan 430079, China

Abstract. In these proceedings, we present the measurements of charged K^+-K^+ and neutral $K_s^0-K_s^0$ correlation functions from Au+Au fixed-target collisions at $\sqrt{s_{NN}}$ = 3.0, 3.2, 3.5, 3.9 and 4.5 GeV at STAR. This is the first such systematic measurement of correlation functions involving strangeness in the high baryon density region. The source size values do not exhibit a clear energy dependence, and the transverse mass dependence of source size for kaons does not align with the trend observed for pions. Parameters extracted from UrQMD transport model calculations qualitatively capture the measured values.

1 Introduction

In heavy-ion collisions, the moment when the last scattering occurs among hadrons is referred to as "kinetic-freeze-out". Femtoscopy is a technique used to unravel the space-time and dynamical characteristics of the particle-emitting source at kinetic freeze-out. Kaons, in comparison to pions, experience reduced contributions from long-lived resonance decays and exhibit smaller rescattering cross-sections with hadronic matter, offering a more direct insight into the source. Theoretical momentum correlations between two identical particles are defined as the first part:

$$C(p_1, p_2) = \int S(\vec{r^*})|\Psi(q_{\rm inv}, \vec{r^*})|^2 \, d^3\vec{r^*} = \frac{\mathcal{N}(N_{\rm same}(q_{\rm inv}))}{N_{\rm mixed}(q_{\rm inv})} \tag{1}$$

where the $S(\vec{r^*})$ is emission function and $\Psi(q_{\rm inv}, \vec{r^*})$ is the two-particle wave function, $q_{\rm inv} = \sqrt{(\vec{p_1} - \vec{p_2})^2 - (E_1 - E_2)^2}$ means relative invariant momentum of the pair, and $\vec{r^*}$ represents relative distance of the emitters in the Pair Rest Frame(PRF). The experimental method to calculate the correlation function can be presented as the second part, in which $N_{same}(q_{inv})$ and $N_{mixed}(q_{inv})$ is the distribution of q_{inv} formed by the same event and separate events but with same multiplicity and position of primary vertex, the factor $\mathcal{N}$ used to normalize the correlation function to unity at large relative momentum. For charged kaons and pions, Sinyukov-Bowler [1] method is used to parameterize the correlation function

$$C(q_{inv}) = N[(1 - \lambda) + \lambda K_{Coul}(q_{inv}, R_G)(1 + e^{-R_G^2 q_{inv}^2})] \tag{2}$$

where R_G is the source size, λ is correlation strength, and $K_{Coul}(q_{inv}, R_G)$ is Coulomb factor. For neutral kaons, it is Lednicky-Lyuboshitz [2] approach shown as,

$$C(q_{inv}) = 1 + e^{-q_{inv}^2 R_G^2} + \frac{1-\epsilon^2}{2}\left[\left|\frac{f(k^*)}{R_G}\right|^2 + \frac{4\mathcal{R}f(k^*)}{\sqrt{\pi}R_G}F_1(q_{inv}R_G) - \frac{2\mathcal{I}f(k^*)}{R_G}F_2(q_{inv}R_G)\right] \tag{3}$$

*e-mail: bjfan@mails.ccnu.edu.cn

© The Authors, published by EDP Sciences. This is an open access article distributed under the terms of the Creative Commons Attribution License 4.0 (https://creativecommons.org/licenses/by/4.0/).

where $F_1(z) = \int_0^z dx \frac{e^{x^2-z^2}}{z}$ and $F_2 = \frac{1-e^{-z^2}}{z}$. Here, the ϵ is the kaon abundance asymmetry, it can be calculate by yield of kaons: $\epsilon = \frac{K-\bar{K}}{K+\bar{K}}$, and $f(k^*)$ is the s-wave scattering amplitude [3], where $k^* = \frac{1}{2}q_{inv}$ for identcal particles.

2 Data sets and Analysis strategy

The $\sqrt{s_{NN}}$ = 3.0 - 4.5 GeV Au + Au collision data were collected by STAR. The primary vertex position of each event along the beam direction, V_z, is limited to be 198 cm < V_z < 202 cm from the center of TPC (Time Projection Chamber). To eliminate possible beam interactions with the vacuum pipe, the vertex along the radial direction, V_r, is selected to be smaller than 2 cm (for π^+ and K^+) and 1.5 cm (for K_s^0). TPC and TOF (Time of flight) were used to identify the π^+ and K^+, and K_s^0 were reconstructed by Kalman Filter [4] Particle package via $\pi^+\pi^-$ chanel.

3 Results and Discussions

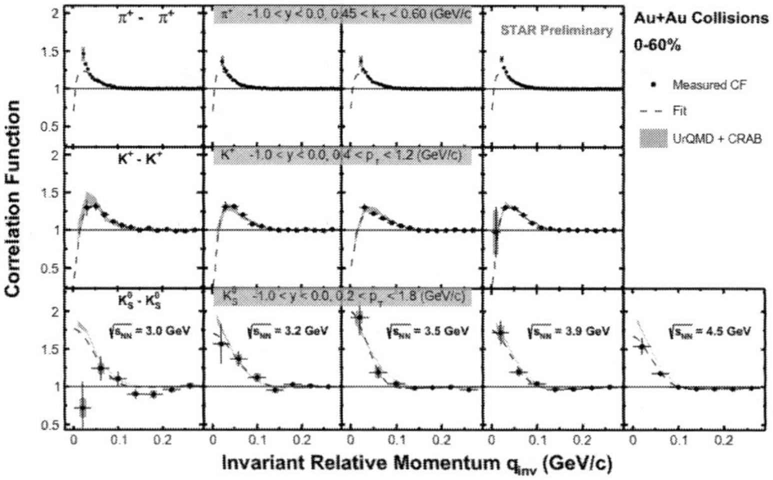

Figure 1. One dimensional correlation functions of $\pi^+ - \pi^+$ (upper panel), $K^+ - K^+$ (middle panel), $K_s^0 - K_s^0$ (bottom panel) in 0-60% Au + Au collisions at $\sqrt{s_{NN}}$ = 3.0 GeV - 4.5 GeV. Vertical lines and grey vertical band represent the statistical and systematic uncertainties, respectively.

Figure 1 shows the one dimensional correlation functions of $\pi^+ - \pi^+$, $K^+ - K^+$ and $K_s^0 - K_s^0$ from $\sqrt{s_{NN}}$ = 3.0 - 4.5 GeV Au + Au collisions at 0-60% centrality. The purple bands are correlation functions calculated from transport model UrQMD [5] plus CRAB (Correlation Afterburner) [6].

Figure 2 shows the energy dependence of extracted R_G and λ of $\pi^+ - \pi^+$, $K^+ - K^+$, $K_s^0 - K_s^0$ pairs in Au + Au 0-60% collisions at $\sqrt{s_{NN}}$ = 3.0 - 4.5 GeV. The grey vertical band represents systematical uncertainties, while the purple band is the result from UrQMD including CRAB to account for femtoscopic correlations, they are consistent with the data within uncertainty. There is no clear energy dependence of source radii observed in this energy region. The λ parameter of pions is smaller than that of kaons, reflecting the fact that kaons experience less contribution from resonance decays. The R_G and λ parameters of charged kaons are consistent with those of neutral kaons within the uncertainties.

© The Authors, published by EDP Sciences. This is an open access article distributed under the terms of the Creative Commons Attribution License 4.0 (https://creativecommons.org/licenses/by/4.0/).

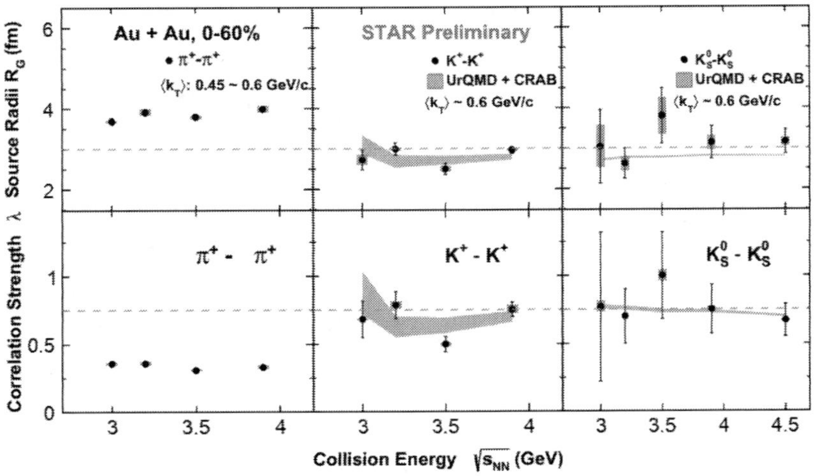

Figure 2. Energy dependence of extracted parameters of $\pi^+ - \pi^+$ (left panel), $K^+ - K^+$ (middle panel), $K_s^0 - K_s^0$ (right panel) in 0-60% Au + Au collisions at $\sqrt{s_{NN}}$ = 3.0 - 4.5 GeV.

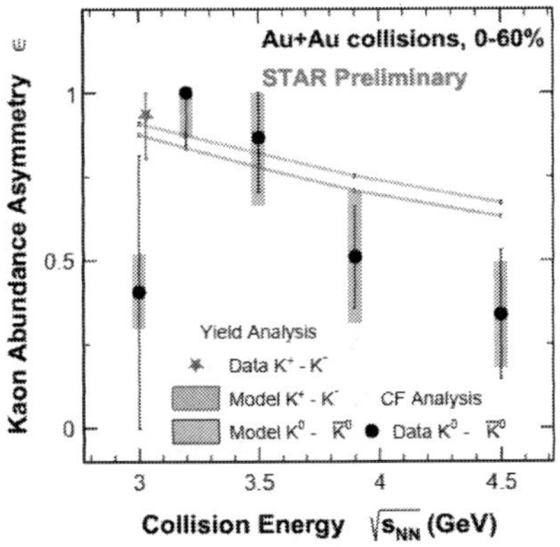

Figure 3. Energy dependence of kaon abundance asymmetry of charged kaons (red point) and neutral kaons (black points) in Au + Au collisions at $\sqrt{s_{NN}}$ = 3.0 GeV - 4.5 GeV.

The $K_s^0 - K_s^0$ correlation function provides insights into the asymmetry of neutral kaon abundance, as K_s^0 is a mixture of s and $\bar{s}$ quarks. In Fig. 3, the black solid points represent the neutral kaon abundance asymmetry extracted from $K_s^0 - K_s^0$ correlation functions, while the red point shows the charged kaon abundance asymmetry obtained from yield analysis based on STAR preliminary particle ratio results [7]. The UrQMD calculations are depicted as black and red bands for neutral and charged kaons, respectively. These calculations are derived from yield analysis of the model. This study marks the first measurement of neutral kaon abundance asymmetry in heavy-ion collisions. The results indicate a decrease in abundance asymmetry with increasing collision energy. The model results align with the data trend, suggesting that associated production dominates in the high baryon density region, while pair production becomes more significant at higher collision energies.

The average transverse mass dependence of source radii can also be examined. In Fig. 4, the black dashed line corresponds to the power-law fitting of all data points from pions. The

© The Authors, published by EDP Sciences. This is an open access article distributed under the terms of the Creative Commons Attribution License 4.0 (https://creativecommons.org/licenses/by/4.0/).

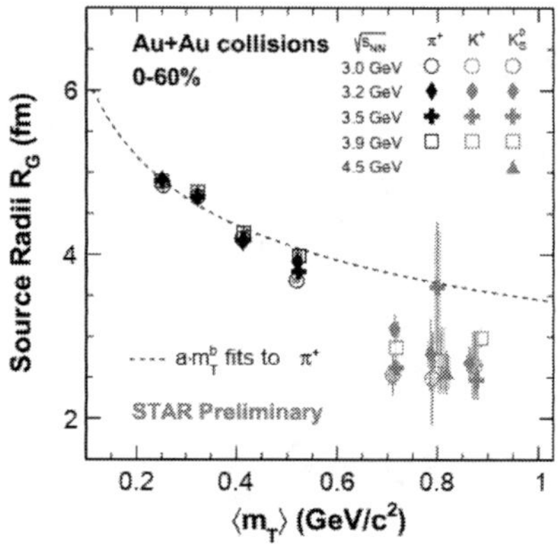

Figure 4. The average transverse mass dependence of source radii R_G of $\pi^+ - \pi^+$, $K^+ - K^+$, $K^0_s - K^0_s$ in 0-60% Au + Au collisions at $\sqrt{s_{NN}}$ = 3.0 GeV - 3.9 GeV. The R_G of $\pi^+ - \pi^+$ is represented by black points, while the $K^+ - K^+$ and $K^0_s - K^0_s$ are represented by green and brown points, respectively. The black dashed line represents a power-law fit to all the data points for pions, with the power-law exponent b is -0.25.

radii of pions are expected to follow the same scaling due to the universal collective flow predicted by hydrodynamics [8] at higher energies, while the R_G of kaons does not follow the trend of pions, suggesting a lack of equilibrium between pions and kaons in this energy range.

4 Summary

We reported the first measurements of kaon femtoscopy in Au+Au collisions in the high baryon density region, and the source parameters are extracted. However, R_G parameter of kaons does not follow the m_T scaling from pion correlation functions which may indicate there is no equilibrium amongst pions and kaons at these collisions. Meanwhile, the abundance asymmetry parameter of neutral kaon is determined and they are decreasing as a function of the collision energy, it indicates that the pair production becomes more important at higher collision energies. The UrQMD model reproduced all the above observations for kaons within uncertainties.

Acknowledgements This work is supported in part by the National Key Research and Development Program of China under Contract No. 2022YFA1604900; the National Natural Science Foundation of China (NSFC) under contract No. 12175084.

References

[1] Yu. M. Sinyukov, R. Lednicky, Phys. Lett. B **432**, 248 (1998).
[2] R. Lednicky, V. L. Lyuboshits, Yad. Fiz. **35**, 1316 (1981).
[3] S. Berkeley, R. Lednicky, Braz. J. Phys. **37**, 994 (2007).
[4] A. Banerjee, I. Kisel, M. Zyzak, Int. J. Mod. Phys. A **35**, 2043003 (2020).
[5] S. A. Bass et al., Prog. Part. Nucl. Phys. **41**, 225 (1998).
[6] https://web.pa.msu.edu/people/pratts/freecodes/crab/home.html.
[7] B. Kimelman, Acta Phys. Polon. Supp. **16**, 1-A49 (2023).
[8] B. Tomášik, U. A. Wiedemann, U. Heinz, Nucl. Phys. A **663**, 753c (2000).

© The Authors, published by EDP Sciences. This is an open access article distributed under the terms of the Creative Commons Attribution License 4.0 (https://creativecommons.org/licenses/by/4.0/).

Measurement of strange particle femtoscopic correlations at the CMS experiment

Raghunath Pradhan for the CMS Collaboration[1,*]

[1]University of Illinois, Chicago, USA, 60607

Abstract. The two particle correlations as a function of relative momenta of identified hadrons involving K^0_S and $\Lambda/\overline{\Lambda}$ are measured in PbPb collisions at $\sqrt{s_{NN}} = 5.02$ TeV with the data samples collected by the CMS experiment at the LHC. Such correlations are sensitive to the quantum statistics and possible final state interactions between the particles. The source radii are extracted from $K^0_S K^0_S$ correlations in different centrality regions and found to decrease from central to peripheral collisions. The strong interaction scattering parameters are extracted from $\Lambda K^0_S \oplus \overline{\Lambda} K^0_S$ and $\Lambda\Lambda \oplus \overline{\Lambda}\overline{\Lambda}$ correlations using the Lednicky-Lyuboshits model, and compared with other experimental and theoretical results.

1 Introduction

Two-particle correlations in relative momentum, so-called femtoscopic correlations, arising from relativistic heavy ion collisions provide a powerful tool for studying the spatiotemporal characteristics of the particle emitting sources created in these collisions and the subsequent interactions of the emitted particles [1]. The identical particle correlations are sensitive to quantum statistics (QS) and to possible final-state interactions (FSI), while nonidentical particle correlations are only sensitive to final-state interactions. The correlations among the neutral K^0_S, Λ, and $\overline{\Lambda}$ particles, collectively referred to as V^0 particles, are of particular interest. Because of their relatively large mass, femtoscopy based on K^0_S particles is less affected by resonance decays than is the case with more commonly studied pion and charged kaon pairs [2]. The results from $K^0_S K^0_S$, $\Lambda K^0_S \oplus \overline{\Lambda} K^0_S$, and $\Lambda\Lambda \oplus \overline{\Lambda}\overline{\Lambda}$ correlation studies can be used to determine the strong-interaction scattering parameters, i.e., the scattering length and the effective range, that are impossible to obtain from currently achievable scattering experiments [3–7]. This information can be used to constrain hadron-hadron correlation models that are used, for example, in modeling the composition of neutron stars [8, 9].

This note presents $K^0_S K^0_S$, $\Lambda K^0_S \oplus \overline{\Lambda} K^0_S$, and $\Lambda\Lambda \oplus \overline{\Lambda}\overline{\Lambda}$ femtoscopic correlations as a function of relative momentum in PbPb collisions at $\sqrt{s_{NN}} = 5.02$ TeV, using data recorded by the CMS experiment during the 2018 LHC run. The $K^0_S K^0_S$ correlations were measured in the extended range of centrality bin (0–60%), where centrality refers to the percentage of the total inelastic hadronic nucleus-nucleus cross section, and 0% corresponds to a maximum overlap of the colliding nuclei. The $\Lambda K^0_S \oplus \overline{\Lambda} K^0_S$ and $\Lambda\Lambda \oplus \overline{\Lambda}\overline{\Lambda}$ correlations were measured in the centrality range of 0–80%. This is the first measurement of the $\Lambda\Lambda \oplus \overline{\Lambda}\overline{\Lambda}$ femtoscopic correlation in

*e-mail: raghunath.pradhan@cern.ch

© The Authors, published by EDP Sciences. This is an open access article distributed under the terms of the Creative Commons Attribution License 4.0 (https://creativecommons.org/licenses/by/4.0/).

PbPb collisions at the LHC. The source size and strong interaction parameters are determined using the Lednický–Lyuboshitz (LL) model [11] and compared with theoretical calculations and results from other experiments.

2 $K^0_S K^0_S$, $\Lambda K^0_S \oplus \overline{\Lambda} K^0_S$, and $\Lambda\Lambda \oplus \overline{\Lambda\Lambda}$ femtoscopic correlations

The $K^0_S K^0_S$ correlations as a function of relative momenta of the particle pair q_{inv} [1, 12], shown in Fig. 1, measured in the six centrality bins with $0 < k_T < 2$ GeV, where $k_T \equiv |\vec{p}_{T,1} + \vec{p}_{T,2}|/2$ is the average transverse momentum of the particle pair [12]. The size of the particle-emitting source R_{inv} extracted from $K^0_S K^0_S$ correlation using the LL fit [11] together with nonfemtoscopic background [12] for different centrality ranges and plotted in the left panel of figure 2. It can be seen that R_{inv} decreases from central (0–10%) to peripheral (50–60%) events, as expected from a simple geometric picture of the collisions. The right panel of figure 2 shows CMS and ALICE results for R_{inv} as a function of transverse mass m_T in three centrality bins. The transverse mass can be calculated as $m_T = \sqrt{(m_{\text{inv}}/2)^2 + k_T^2}$, where m_{inv} is the invariant mass of the two-particle system [12]. Extrapolations of the m_T dependence found for the ALICE results at $\sqrt{s_{\text{NN}}} = 2.76$ TeV [13] to the m_T values of the CMS results at $\sqrt{s_{\text{NN}}} = 5.02$ TeV shows consistent results for the two experiments.

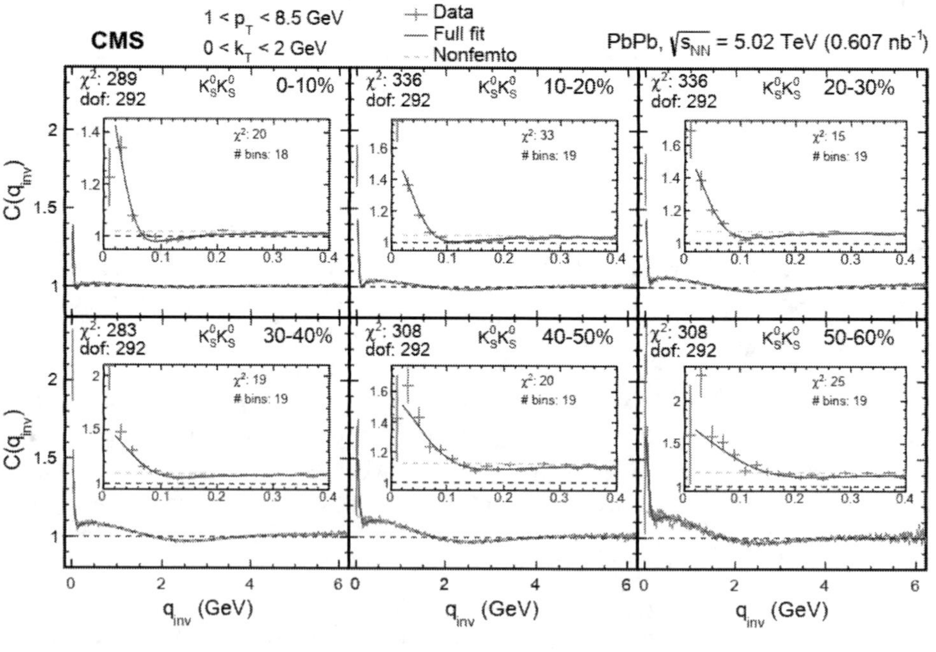

Figure 1. The correlation distributions and fits for $K^0_S K^0_S$ pairs in different centrality ranges, starting from 0–10% centrality to 50–60% centrality, with $0 < k_T < 2$ GeV. In each plot, the red circles represent the data, the blue solid line is the fit, and the green dashed line is the nonfemtoscopic background. The χ^2 and dof values are for the full q_{inv} range. The insert plots show the data and the fit for the $q_{\text{inv}} < 0.4$ GeV region, with the χ^2 and number of bins evaluated in that region.

Figure 3 shows the $\Lambda K^0_S \oplus \overline{\Lambda} K^0_S$ (left) and $\Lambda\Lambda \oplus \overline{\Lambda\Lambda}$ (right) correlation measurements in 0–80% centrality with no restriction on k_T [12]. The strong interaction scattering parameters: real scattering length ($\Re f_0$), imaginary scattering length ($\Im f_0$), and effective range (d_0), were extracted from the $\Lambda K^0_S \oplus \overline{\Lambda} K^0_S$ and $\Lambda\Lambda \oplus \overline{\Lambda\Lambda}$ correlations using LL fit [11] together with nonfemtoscopic background [12]. Figure 4 shows d_0 versus $\Re f_0$ (left) and $\Im f_0$ versus $\Re f_0$

© The Authors, published by EDP Sciences. This is an open access article distributed under the terms of the Creative Commons Attribution License 4.0 (https://creativecommons.org/licenses/by/4.0/).

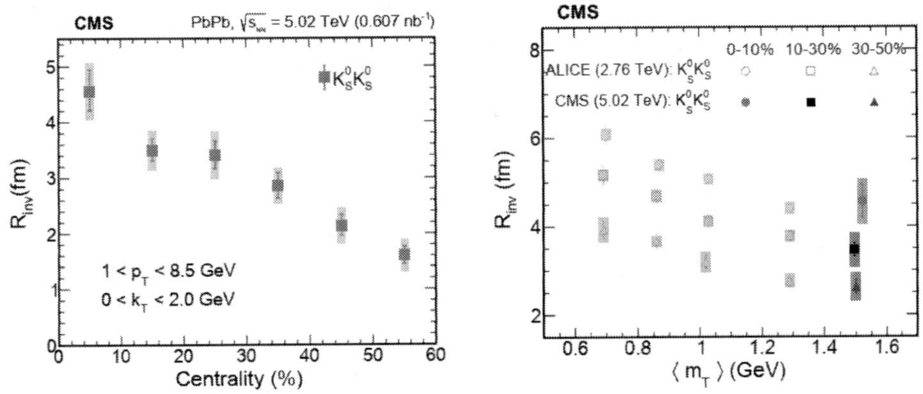

Figure 2. Left: The R_{inv} (left) as a function of centrality. Right: The R_{inv} (left) as a function of $\langle m_{\mathrm{T}} \rangle$ for the three centrality bins (0–10%, 10–30%, and 30–50%) from this analysis (dark filled symbols) and from the ALICE experiment (light open symbols) [13]. For each data point, the lines and the shaded areas indicate the statistical and systematic uncertainties, respectively.

(right). Comparisons to theoretical calculations and results from other experiments are also shown [12]. The negative value of $\Re f_0$ in $\Lambda K^0_S \oplus \overline{\Lambda} K^0_S$ correlations suggests that the ΛK^0_S ($\overline{\Lambda} K^0_S$) interaction is repulsive while $\Im f_0$ is consistent with zero within uncertainty, precluding any definitive claims regarding possible inelastic processes [12]. A positive $\Re f_0$ value for the $\Lambda\Lambda \oplus \overline{\Lambda\Lambda}$ correlations indicates that the $\Lambda\Lambda$ ($\overline{\Lambda\Lambda}$) interaction is attractive.

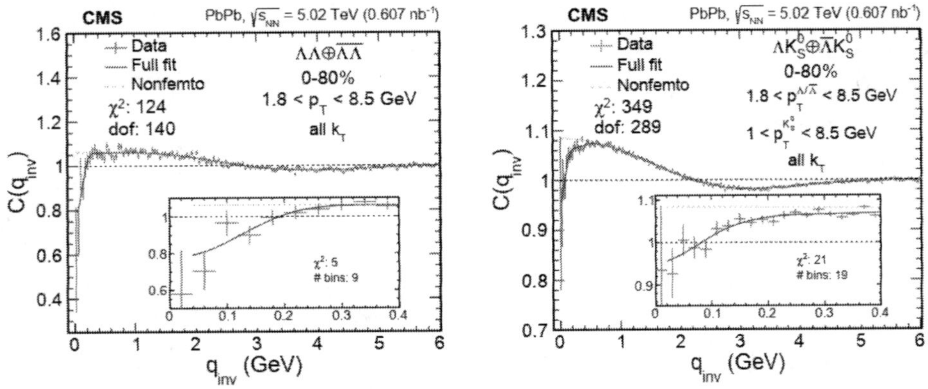

Figure 3. The correlation distributions and fits for $\Lambda K^0_S \oplus \overline{\Lambda} K^0_S$ (left) and $\Lambda\Lambda \oplus \overline{\Lambda\Lambda}$ (right) pairs with 0–80% centrality and no restriction on k_{T}. the red circles represent the data, the blue solid line is the fit, and the green dashed line is the non-fetoscopic background. The χ^2 and dof values are for the full q_{inv} range. The insert plots show the data and the fit for the $q_{\mathrm{inv}} < 0.4$ GeV region, with the χ^2 and number of bins evaluated in that region.

3 Summary

The $K^0_S K^0_S$, $\Lambda K^0_S \oplus \overline{\Lambda} K^0_S$, and $\Lambda\Lambda \oplus \overline{\Lambda\Lambda}$ femtoscopic correlations are studied using lead-lead (PbPb) collision data at a center-of-mass energy per nucleon pair of $\sqrt{s_{_{\mathrm{NN}}}} = 5.02$ TeV, collected by the CMS Collaboration. This is the first report on $\Lambda\Lambda \oplus \overline{\Lambda\Lambda}$ correlations in PbPb collisions at the CERN LHC. The source size R_{inv} extracted from $K^0_S K^0_S$ correlations in six centrality bins covering the 0–60% range. Along with the R_{inv}, the strong interaction scattering parameters, i.e., the complex scattering length and effective range, were extracted from

© The Authors, published by EDP Sciences. This is an open access article distributed under the terms of the Creative Commons Attribution License 4.0 (https://creativecommons.org/licenses/by/4.0/).

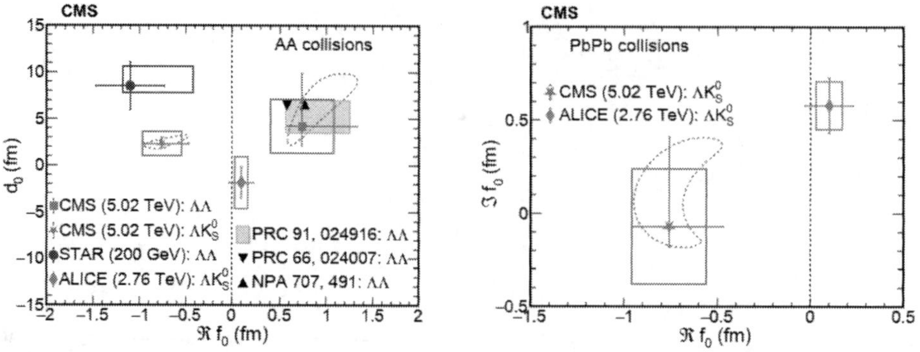

Figure 4. The measured values of d_0 versus $\Re f_0$ (left) and $\Im f_0$ versus $\Re f_0$ (right) from this analysis along with other measurements and predictions. For each data point, the lines and the boxes indicate the one-dimensional statistical and systematic uncertainties, respectively. For the CMS points, the dotted lines show two-dimensional contours corresponding to a 39% confidence level with just the statistical uncertainties included.

$\Lambda K^0_S \oplus \overline{\Lambda} K^0_S$ and $\Lambda\Lambda \oplus \overline{\Lambda\Lambda}$ correlations in the 0–80% centrality range. These scattering parameters indicate that the $\Lambda K^0_S \oplus \overline{\Lambda} K^0_S$ interaction is repulsive and that the $\Lambda\Lambda \oplus \overline{\Lambda\Lambda}$ interaction is attractive.

References

[1] M. A. Lisa, S. Pratt, R. Soltz, and U. Wiedeman, Ann. Rev. Nucl. Part. Sci. **55**, 357 (2005). https://doi.org/10.1146/annurev.nucl.55.090704.151533

[2] CMS Collaboration, Phys. Rev. C. **97**, 064912 (2018). https://doi.org/10.1103/PhysRevC.97.064912

[3] V. G. J. Stoks, R. A. M. Klomp, M. C. M. Rentmeester, and J. J. de Swart, Phys. Rev. C. **48**, 792, (1993). https://doi.org/10.1103/PhysRevC.48.792

[4] J.J de Swart and C Dullemond, Anna. Phys. **19**, 458 (1962). https://doi.org/10.1016/0003-4916(62)90185-9

[5] R. Engelmann, H. Filthuth, V. Hepp, and E. Kluge, Phys. Lett. **21**, 587 (1966). https://doi.org/10.1016/0031-9163(66)91310-2

[6] F. Eisele and H. Filthuth and W. Föhlisch and V. Hepp and G. Zech, Phys. Lett. B. **37**, 204 (1971). https://doi.org/10.1016/0370-2693(71)90053-0

[7] B. Sechi-Zorn, B. Kehoe, J. Twitty, and R. A. Burnstein, Phys. Rev. **175**, 1735 (1968). https://doi.org/10.1103/PhysRev.175.1735

[8] D. B. Kaplan and A. E. Nelson, Nucl. Phys. A. **479**, 273 (1988). https://doi.org/10.1016/0375-9474(88)90442-3

[9] Jürgen Schaffner-Bielich, Matthias Hanauske, Horst Stöcker, and Walter Greiner, Phys. Rev. Lett. **89**, 171101 (2002). https://doi.org/10.1103/PhysRevLett.89.171101

[10] CMS Collaboration, JINST **03** S08004 (2008). https://doi.org/10.1088/1748-0221/3/08/S08004

[11] R. Lednicky and V. Lyuboshits, Sov. J. Nucl. Phys. **35**, 770 (1982). https://inspirehep.net/literature/167537

[12] CMS Collaboration, CMS-PAS-HIN-21-006, https://arxiv.org/abs/2301.05290

[13] ALICE Collaboration, Phys. Rev. C. **92** 054908 (2015). https://doi.org/10.1103/PhysRevC.92.054908

© The Authors, published by EDP Sciences. This is an open access article distributed under the terms of the Creative Commons Attribution License 4.0 (https://creativecommons.org/licenses/by/4.0/).

Measurements of global and local spin polarization of Λ and $\bar{\Lambda}$ in Au+Au collisions from the RHIC Beam Energy Scan

Qiang Hu[1,2,*] for the STAR Collaboration

[1]Institute of Modern Physics, Chinese Academy of Sciences, Lanzhou 730000, China
[2]School of Nuclear Science and Technology, University of Chinese Academy of Sciences, Beijing 100049, China

Abstract. We report the measurements of Λ hyperons' global and local spin polarization from second phase of the RHIC Beam Energy Scan (BES-II) in Au+Au collisions at $\sqrt{s_{NN}}$= 7.7–27 GeV. Global polarization measurements of $\bar{\Lambda}$ and Λ show no significant differences, offering insights into the late-stage evolution of the magnetic field. The new measurements of the local polarization of Λ perpendicular to the reaction plane ($\langle P_{2,y}\rangle$) shows a monotonic increase with decreasing collision energy, while the component along the beam direction ($\langle P_{2,z}\rangle$) for both Λ and $\bar{\Lambda}$ is small in magnitude with no strong energy dependence. The net local polarization observable, $\langle P^{net}_{2,y}\rangle = \langle P_{2,y}(\Lambda)\rangle - \langle P_{2,y}(\bar{\Lambda})\rangle$ and $\langle P^{net}_{2,z}\rangle = \langle P_{2,z}(\Lambda)\rangle - \langle P_{2,z}(\bar{\Lambda})\rangle$ designed to probe baryonic spin Hall effect, is consistent with zero with large uncertainty.

1 Introduction

Relativistic heavy-ion collisions provide an excellent opportunity to examine the property of the quark-gluon plasma (QGP) in the laboratory. In non-central heavy-ion collisions, the system carries large initial orbital angular momentum. Subsequently, the quarks and the final state hadrons with non-zero spin could be polarized along global angular momentum due to spin-orbit coupling [1, 2], which is known as global polarization. Due to parity violation in the decay of a Λ hyperon, the daughter proton tends to emit along the spin direction of its parent, making Λs excellent candidates for measuring polarization in heavy-ion collisions.

The global polarization is determined by $\left\langle P_y\right\rangle = \frac{8}{\pi\alpha_\Lambda}\frac{1}{R^{(1)}_{EP}}\left\langle \sin(\Psi_1 - \phi^*_p)\right\rangle$, where α_Λ is the decay parameter, Ψ_1 is the first-order of event plane angle and $R^{(1)}_{EP}$ its resolution, and ϕ^*_p is the azimuthal angle of baryon (proton) in Λ's rest frame. The STAR Collaboration carried out the polarization measurement of Λ hyperons in Au+Au collisions at $\sqrt{s_{NN}}$ = 62.4 and 200 GeV, where the signal was consistent with zero within statistical uncertainties [3]. However, the data from first phase of the RHIC Beam Energy Scan (BES-I) provided the first evidence of non-zero global polarization for Λ hyperons [4, 5]. It supports the presence of vortices inside QGP and is considered as one of the important milestones in heavy-ion collisions. The BES-I results also show hints of a difference in global polarization between $\bar{\Lambda}$ and Λ which is expected from the effect of the late-stage magnetic field sustained by the QGP. Subsequently,

* e-mail: qianghu@impcas.ac.cn

© The Authors, published by EDP Sciences. This is an open access article distributed under the terms of the Creative Commons Attribution License 4.0 (https://creativecommons.org/licenses/by/4.0/).

global polarization has been measured by different collaborations in different systems from a few GeV to TeV.

The distribution of polarization as a function of azimuthal angle is referred to as local polarization. Local polarization can be measured with the polarization axis perpendicular to the reaction plane (along the out-of-plane, $\left\langle P_{2,y}\right\rangle$) and along the reaction plane (along in-plane or beam-axis, $\langle P_{2,z}\rangle$), these components are written as $\left\langle P_{2,y}\right\rangle = \frac{8}{\pi\alpha_\Lambda}\frac{1}{R_{EP}^{(1)}}\left\langle \sin(\Psi_1 - \phi_p^*)\cos(2\phi_\Lambda - 2\Psi_2)\right\rangle$ and $\langle P_{2,z}\rangle = \frac{\langle\cos\theta_p^*\sin(2\phi_\Lambda-2\Psi_2)\rangle}{\alpha_\Lambda\langle(\cos\theta_p^*)^2\rangle}$, where Ψ_2 is the second-order of event plane angle, θ_p^* is the polar angle of daughter proton in the Λ rest frame relative to the beam direction, ϕ_Λ is the azimuthal angle of Λ.

The recently predicted baryonic spin Hall effect (SHE) [6–8] in heavy-ion collisions describes the splitting of particles with spin up and spin down, driven by the gradient of the baryon chemical potential, similar to how an electric field induces the traditional spin Hall effect in condensed matter physics. The net local polarization, defined as the polarization difference between Λ and $\bar{\Lambda}$, denoted by $\langle P_{2,y}^{net}\rangle = \langle P_{2,y}(\Lambda)\rangle - \langle P_{2,y}(\bar{\Lambda})\rangle$ and $\langle P_{2,z}^{net}\rangle = \langle P_{2,z}(\Lambda)\rangle - \langle P_{2,z}(\bar{\Lambda})\rangle$, is predicted to be a sensitive probe for SHE.

2 Analysis

In these proceedings, we report the global and local polarization of Λ and $\bar{\Lambda}$ in Au+Au collisions at $\sqrt{s_{NN}}$ = 7.7, 9.2, 11.5, 14.6, 19.6, and 27 GeV, based on datasets from the second phase of the RHIC Beam Energy Scan (BES-II). The first-order event plane, Ψ_1 is reconstructed with the upgraded Event Plane Detector (EPD). The second-order event plane, Ψ_2, is determined by the Time Projection Chamber (TPC) detector. The particle identification is done with the TPC and Time-of-Flight (ToF) detector. The invariant mass spectra of Λ and $\bar{\Lambda}$ are reconstructed with the decay channels $\Lambda \longrightarrow p+\pi^-$ and $\bar{\Lambda} \longrightarrow \bar{p}+\pi^+$. The background is evaluated by fitting the side band with the second order polynomial function. Then the ratio of signal to background (f^{sig}) is obtained. The distribution of $\langle\sin(\Psi_1-\phi_p^*)\rangle$ and $\langle P_z\sin(2\Delta\phi)\rangle$ (where $\Delta\phi = \phi_\Lambda - \Psi_2$, $\langle P_z\rangle = \frac{\langle\cos\theta_p^*\rangle}{\alpha_\Lambda\langle(\cos\theta_p^*)^2\rangle}$) as functions of Λ invariant mass ($M_{p\pi^-}$) as shown in Fig. 1, are fitted with the expression $\langle P_H\rangle^{obs} = f^{Sig}(M_{p\pi^{-1}})\langle P_H\rangle^{Sig} + (1 - f^{Sig}(M_{p\pi^{-1}}))\langle P_H\rangle^{Bg}$ (P_H is hyperon's polarization, $\langle P_H\rangle^{Bg}$ is linear background in the considered invariant mass window) to extract the signal of polarization after event plane resolution correction.

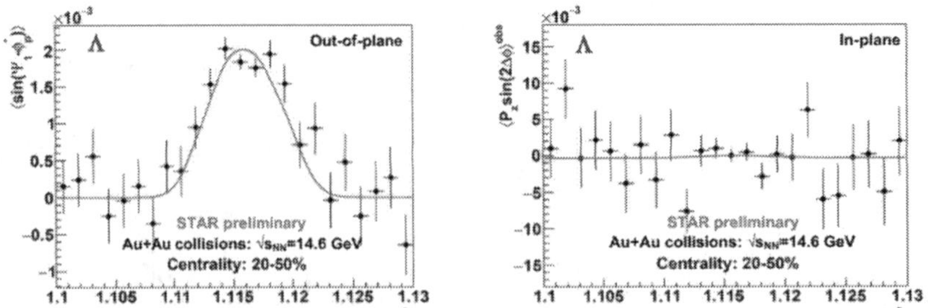

Figure 1: $\langle\sin(\Psi_1 - \phi_p^*)\rangle$ and $\langle P_z\sin(2\Delta\phi)\rangle^{obs}$ as functions of invariant mass of Λ in Au+Au collisions at $\sqrt{s_{NN}}$ = 14.6 GeV. Only statistical uncertainties are shown in the plots.

3 Results and discussion

Left and right panels of Fig. 2 present the comparison of Λ and $\bar{\Lambda}$ global polarization and their difference, respectively. The global polarization increases as the energy decreases, and the

© The Authors, published by EDP Sciences. This is an open access article distributed under the terms of the Creative Commons Attribution License 4.0 (https://creativecommons.org/licenses/by/4.0/).

results from BES-I and BES-II are consistent within uncertainties. The high-precision BES-II results show no significant splitting between Λ and $\bar{\Lambda}$. Upper limits on the late-stage magnetic field at $\sqrt{s_{NN}}$ = 19.6 and 27 GeV, are estimated to be $B < 9.2 \times 10^{12}$ T and $B < 1.4 \times 10^{13}$ T, respectively [9]. The left and right panels of Fig. 3 present the local polarization $\langle P_{2,y} \rangle$ and $\langle P_{2,z} \rangle$ for Λ and $\bar{\Lambda}$ as a function of energy. For $\langle P_{2,y} \rangle$, the polarization of Λ hyperons shows a clear increase with decreasing energy, while no distinct trend is observed for $\bar{\Lambda}$ within the current uncertainties. For $\langle P_{2,z} \rangle$, both Λ and $\bar{\Lambda}$ exhibit small magnitudes and no energy dependence. Left and right panels of Fig. 4 present the net Λ polarization, $\langle -P^{net}_{2,y} \rangle$ and $\langle P^{net}_{2,z} \rangle$, as a function of collision energy, an observable proposed to probe the SHE [1]. No significant energy dependence has been observed for either $\langle -P^{net}_{2,y} \rangle$ or $\langle P^{net}_{2,z} \rangle$. However, $\langle P^{net}_{2,z} \rangle$ shows non-trivial trends with hints of a sign change, albeit with large uncertainties. In previous studies, global polarization could be explained by thermal vorticity, and the sign of local polarization along the beam direction could be captured by incorporating shear-induced polarization into the models [7, 10].

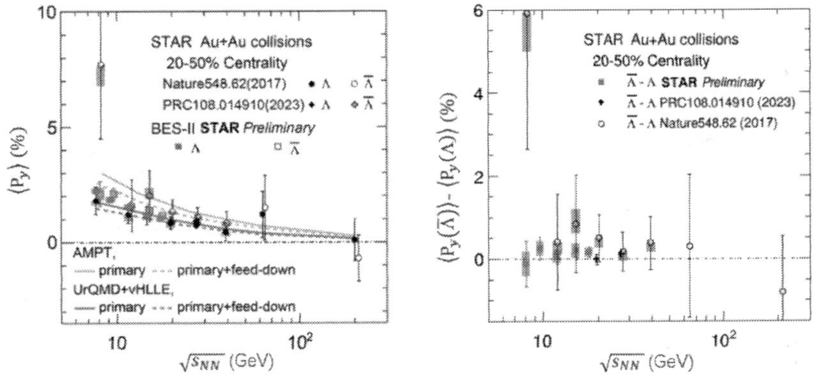

Figure 2: (Colour online) The Λ and $\bar{\Lambda}$ global polarization $\langle P_y \rangle$ (left panel) and their difference $\langle P_y(\bar{\Lambda}) \rangle - \langle P_y(\Lambda) \rangle$ (right panel) as a function of energy in Au+Au collisions. The points are data. Vertical lines and shaded boxes are statistical and systematic uncertainties.

4 Summary

In a summary, we report the collision energy dependence of global and local spin polarization of Λ and $\bar{\Lambda}$ hyperons, as well as their difference in Au+Au collisions at 7.7, 9.2, 11.5, 14.6, 19.6, and 27 GeV from the second phase of RHIC beam energy scan. With the high precision measurements, no splitting of global polarization between Λ and $\bar{\Lambda}$ has been observed. The upper limit of late stage magnetic field at 19.6 and 27 GeV have been estimated. The local polarization of Λ and $\bar{\Lambda}$ ($\langle P_{2,y} \rangle$, $\langle P_{2,z} \rangle$) are measured for the first time at RHIC BES energies. A monotonic energy dependence of Λ local polarization along out-of-plane direction has been observed. The net local polarization perpendicular to the reaction plane and along the beam directions have been measured. The newly proposed mechanism of the baryonic spin Hall effect has been probed using the local polarization difference between Λ and $\bar{\Lambda}$, and no indication of SHE is observed within the current precision. However, it brings challenges to the hydrodynamic models to describe $\langle P_{2,y} \rangle$ and $\langle P_{2,z} \rangle$ simultaneously with or without SHE. More theoretical input is needed to understand Λ polarization at high baryon densities.

[1] In the presentation at SQM2024, the results for $\langle P^{net}_{2,y} \rangle = \langle P_{2,y}(\Lambda) \rangle - \langle P_{2,y}(\bar{\Lambda}) \rangle$ were incorrectly plotted with a factor of -1 multiplied to the data. It is corrected in these proceedings.

© The Authors, published by EDP Sciences. This is an open access article distributed under the terms of the Creative Commons Attribution License 4.0 (https://creativecommons.org/licenses/by/4.0/).

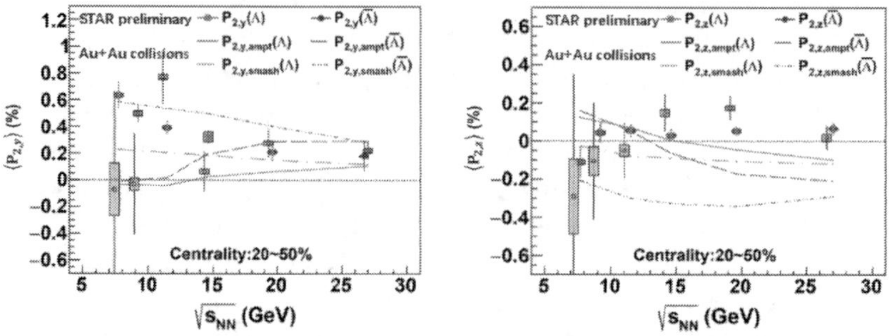

Figure 3: (Colour online) The Λ and $\bar{\Lambda}$ local polarization $\langle P_{2,y} \rangle$ (left pannel) and $\langle P_{2,z} \rangle$ (right panel) as a function of energy in Au+Au collisions. The red and blue points are data of Λ and $\bar{\Lambda}$, respectively. Vertical lines and shaded boxes are statistical and systematic uncertainties.

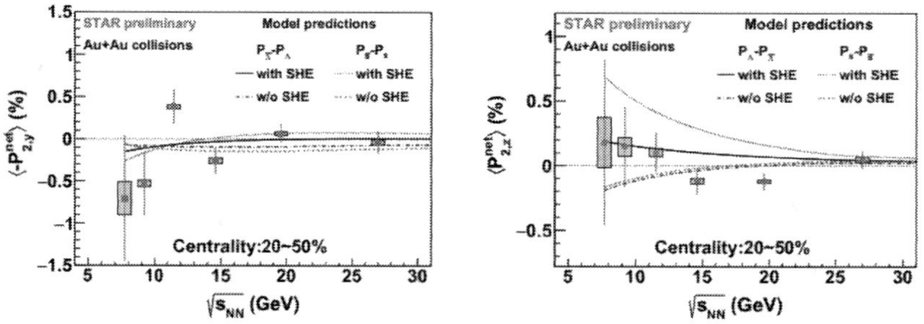

Figure 4: (Colour online) The net local polarization $\langle -P^{net}_{2,y} \rangle$ (left panel) and $\langle P^{net}_{2,z} \rangle$ (right panel) as a function of energy in Au+Au collisions. The red solid squares are data. Vertical lines and shaded boxes are statistical and systematic uncertainties.

Acknowledgments

This work is supported in part by the National Key Research and Development Program of China under Contract No. 2022YFA1604900.

References

[1] Z.-T. Liang and X.-N. Wang, Phys. Rev. Lett. **94**, 102301 (2005).
[2] S. A. Voloshin, arXiv:nucl-th/0410089.
[3] STAR Collaboration, Phys. Rev. C **76**, 024915 (2007).
[4] STAR Collaboration, Nature **548**. 62 (2017).
[5] J. Chen et al., arXiv:2407.02935[nucl-ex].
[6] S. Y. F. Liu et al., Phys. Rev. D **104**, 054043 (2021).
[7] B. Fu, S. Liu et al., Phys. Rev. Lett. **127**, 142301 (2021).
[8] B. Fu et al., arXiv:2201.12970.
[9] STAR Collaboration, Phys. Rev. C **108**, 014910 (2023).
[10] F. Becattini et al., Phys. Rev. Lett. **127**, 272302 (2021).

© The Authors, published by EDP Sciences. This is an open access article distributed under the terms of the Creative Commons Attribution License 4.0 (https://creativecommons.org/licenses/by/4.0/).

EPJ Web of Conferences **316**, 06014 (2025) https://doi.org/10.1051/epjconf/202531606014
SQM 2024

Differential measurements of ϕ-meson global spin alignment in Au+Au collisions at STAR

Gavin Wilks[1,*], for the STAR Collaboration

[1]University of Illinois at Chicago, 1200 W Harrison, Chicago, IL 60607, USA

Abstract. The STAR collaboration observed a significant global spin alignment (ρ_{00}) signal for ϕ-mesons in Au+Au collisions using the data from the BES-I [1] which cannot be explained by conventional mechanisms, but may be attributable to the influence of a ϕ-meson force field [2–6]. In this talk, we present differential measurements of ϕ-meson global spin alignment using the STAR detector in Au+Au collisions at $\sqrt{s_{NN}}$ = 14.6 and 19.6 GeV from the second phase of the Beam Energy Scan at RHIC (BES-II). The first rapidity (y) dependent ρ_{00} results for ϕ-mesons will be shown, alongside new centrality and transverse momentum (p_T) dependent measurements. The results presented in these proceedings will help understand the potential link of global spin alignment to vector meson fields and their roles in the evolution of nuclear matter.

1 Introduction

In non-central heavy-ion collisions, a large orbital angular momentum is produced, generating vorticity of the Quark Gluon Plasma (QGP) along the orbital angular momentum axis. Through spin-orbit couplings [7], a particle's spin can be polarized, known as global polarization. The global polarization of $\Lambda(\bar{\Lambda})$ hyperon can be used as a probe of $s(\bar{s})$ global polarization since, according to the flavor-spin wave function, the $\Lambda(\bar{\Lambda})$ polarization is carried solely by the $s(\bar{s})$ [8]. Recent measurements from the STAR collaboration show a significant signal of $\Lambda(\bar{\Lambda})$ hyperon global polarization [9].

Following the quark coalescence model, the production of ϕ-mesons ($s\bar{s}$) should be affected by the global polarization of the $s(\bar{s})$ [2]. In $\Lambda(\bar{\Lambda})$ hyperon decay, the global polarization can be measured since they decay through the weak force with parity violation, where decay products are preferentially emitted along the spin direction. However, for ϕ-mesons ($s\bar{s}$) which decay through the strong force with parity conservation, a direct measurement of spin polarization is not feasible. Rather, we can indirectly study global polarization using global spin alignment, measured by the ρ_{00} diagonal element of the spin density matrix. For the decay $\phi \rightarrow K^+K^-$, the K^+ or K^- daughter's polar angle (with respect to the spin-quantization axis) (θ^*) distribution within the ϕ-meson rest frame is given by:

$$\frac{dN}{d(\cos\theta^*)} \propto (1-\rho_{00}) + (3\rho_{00}-1)\cos^2\theta^*. \quad (1)$$

If $\rho_{00} \neq 1/3$, then the ϕ-meson spin states are not equally probable, leading to an anisotropic polar angle distribution of the daughter kaon. In the case of $\rho_{00} > 1/3$, the occupation of the

*e-mail: gwilks3@uic.edu

© The Authors, published by EDP Sciences. This is an open access article distributed under the terms of the Creative Commons Attribution License 4.0 (https://creativecommons.org/licenses/by/4.0/).

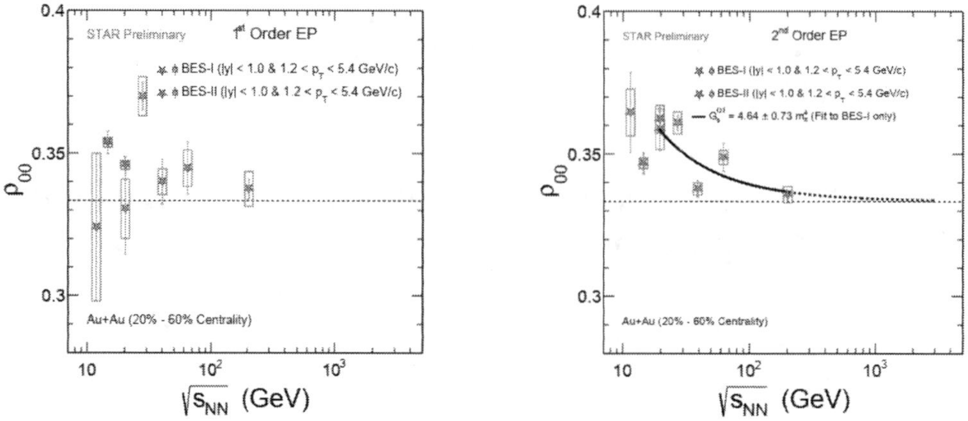

Figure 1. $\sqrt{s_{NN}}$ dependent ϕ-meson ρ_{00} for 20-60% centrality Au+Au collisions. Details about the fit can be found in [1].

spin-0 state is more probable, which corresponds to an alignment of the ϕ-meson's average polarization along the spin-quantization axis [10, 11].

Recent STAR measurements [1] reported a significant signal ($\rho_{00} > 1/3$) for ϕ-mesons in mid-central Au+Au collisions from the first phase of the Beam Energy Scan at RHIC (BES-I) at energies below $\sqrt{s_{NN}} = 62.4$ GeV. No conventional physical mechanism can explain the scale of the measured ρ_{00} signal, but it was proposed that fluctuations of the theoretical ϕ-meson strong force field could accommodate this signal [2]. If fully established, this possible connection between the ϕ-meson field and global spin alignment could allow us to probe previously inaccessible features of this field. In these proceedings, we report measurements of ϕ-meson ρ_{00} in Au+Au collisions at $\sqrt{s_{NN}} = 14.6$ and 19.6 GeV from the second phase of the Beam Energy Scan at RHIC (BES-II). BES-II provides significantly larger data samples, allowing for more precise measurements and differential studies not previously possible with BES-I data. Our precision is further increased by the many upgrades made to the STAR detector prior to BES-II, including the Event Plane Detector (EPD) [12] and an improved inner Time Projection Chamber (iTPC) [13].

2 Results and discussion

In this analysis, we use the orbital angular momentum direction in each event as the spin-quantization axis. To identify the orbital angular momentum axis, we use the normal of the reaction plane spanned by the impact parameter vector and the direction of the incoming colliding nuclei. To estimate the reaction plane, we can calculate the first- and second-order harmonic event planes through the methods detailed in [14]. The TPC ($|\eta| < 1.5$) is used to extract the second order event plane, and the EPD ($2.1 < |\eta| < 5.1$) is used for the first order event plane. We used both the first- and second-order harmonic event planes separately to calculate the θ^* angle of the K^+ daughter, providing two measurements of the ϕ-meson global spin alignment.

The collision energy dependence of ϕ-meson ρ_{00} is shown in Figure 1 for mid-central (20-60% centrality) Au+Au collisions from BES-II. We show new results for $\sqrt{s_{NN}} = 14.6$ GeV with $\rho_{00} > 1/3$ and consistency with BES-I for $\sqrt{s_{NN}} = 19.6$ GeV for measurements using both the first- and second-order event planes. The increased sample size increases our precision leading to a more significant measurement of $\rho_{00} > 1/3$ at 19.6 GeV. The solid line fit is derived from the ϕ-meson strong force field model. Details can be found in [1, 2, 6].

© The Authors, published by EDP Sciences. This is an open access article distributed under the terms of the Creative Commons Attribution License 4.0 (https://creativecommons.org/licenses/by/4.0/).

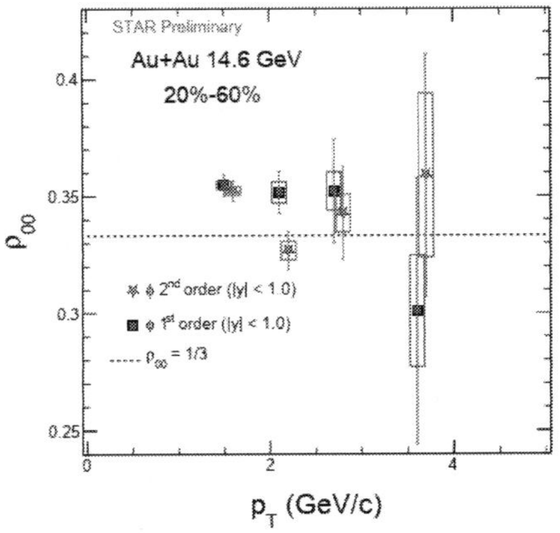

Figure 2. p_T dependent ϕ-meson ρ_{00} for 20-60% centrality BES-II Au+Au collisions at $\sqrt{s_{\mathrm{NN}}}$ = 14.6 GeV.

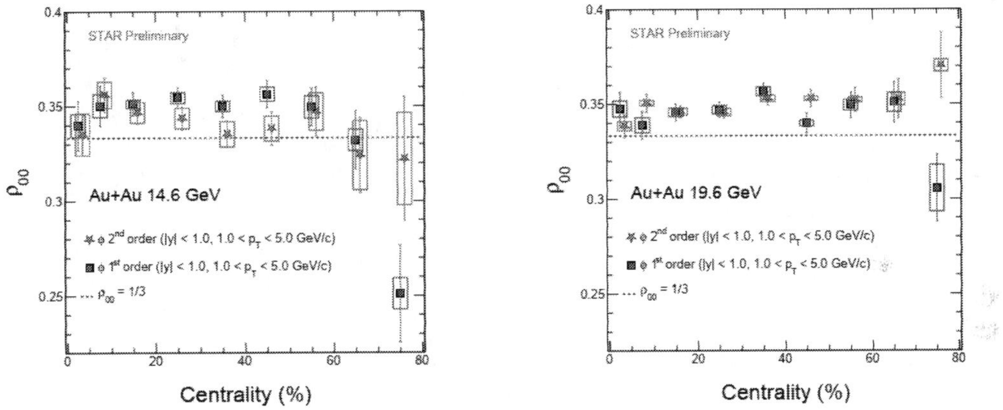

Figure 3. Centrality dependent ϕ-meson ρ_{00} for BES-II Au+Au collisions at $\sqrt{s_{\mathrm{NN}}}$ = 14.6 GeV and 19.6 GeV.

Figure 2 shows the p_T dependent ϕ-meson global spin alignment where no significant p_T dependence is observed. The centrality dependent ρ_{00} is shown in Figure 3 and there is no significant centrality dependence for 14.6 or 19.6 GeV. This contrasts to results of $\Lambda(\bar{\Lambda})$ polarization where a strong dependence on centrality is seen [15, 16]. This is unsurprising since the physical mechanisms driving $\Lambda(\bar{\Lambda})$ polarization are not expected to have significant contributions to ϕ-meson ρ_{00}; therefore, we do not necessarily expect the same centrality dependence [2]. Figure 4 shows the first rapidity dependent ϕ-meson ρ_{00} measurements. In both 14.6 GeV and 19.6 GeV collisions we observe $\rho_{00} = 1/3$ at mid-rapidity ($|y| = 0$) and an increase as $|y|$ approaches 1. Predictions of the $|y|$ dependence were made in [4] following directly from the ϕ-meson strong force field model discussed in [2, 6]. These predictions well describe the data for $|y| > 0.5$, but fail to do so for $|y| < 0.5$. In this model, the current understanding is that anisotropies in the ϕ-meson field fluctuations could cause ρ_{00} to deviate from 1/3 and these anisotropies in the ϕ-meson rest frame are affected by the motion of the ϕ-meson in the global frame.

© The Authors, published by EDP Sciences. This is an open access article distributed under the terms of the Creative Commons Attribution License 4.0 (https://creativecommons.org/licenses/by/4.0/).

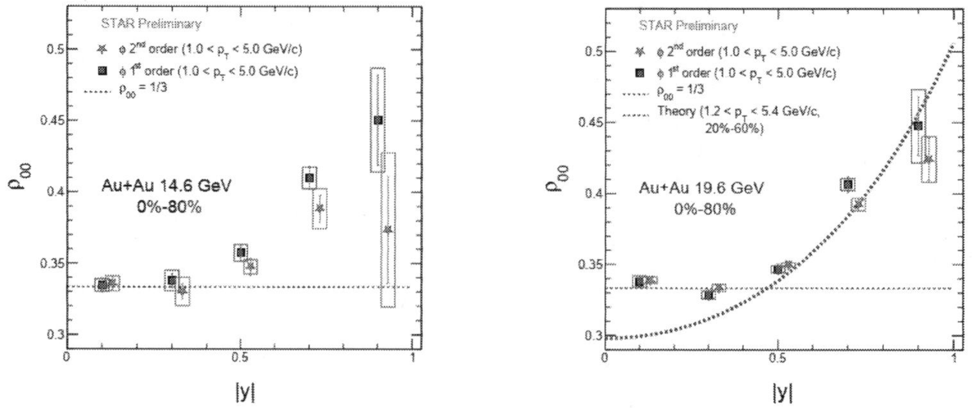

Figure 4. Rapidity dependent ϕ-meson ρ_{00} for 0-80% centrality BES-II Au+Au collisions at $\sqrt{s_{\mathrm{NN}}}$ = 14.6 GeV and 19.6 GeV. The dashed blue line represents the prediction from the ϕ-meson strong force field model in [4].

3 Conclusion

In these proceedings, we presented ϕ-meson ρ_{00} measurements with respect to the first- and second-order event planes for Au+Au collisions at $\sqrt{s_{\mathrm{NN}}}$ = 14.6 and 19.6 GeV from BES-II. We have new results for 14.6 GeV and our integrated values of ϕ-meson ρ_{00} for 19.6 GeV are consistent between BES-I and BES-II data. We observe no significant dependence of ρ_{00} on p_T or the collision centrality. Regarding the first measurements of the rapidity dependent ϕ-meson ρ_{00}, we observe an increasing ρ_{00} for rapidity ($|y| > 0.5$), consistent with the current ϕ-meson strong force field model [4]. We also see no significant difference between the first- and second-order event plane methods. In further studies, we will include more collisions energies from BES-II and expand the $|\eta|$ acceptance.

References

[1] STAR Collaboration, Nature **614** 244-248 (2023).
[2] Sheng et al., Phys. Rev. D **101**, 096005 (2020).
[3] Sheng et al., Phys. Rev. D **105**, 099903 (2022).
[4] Sheng et al., Phys. Rev. C **108**, 054902 (2023).
[5] X.L. Sheng et al., Physical Review Letters **131**, 042304 (2023).
[6] Sheng et al., Phys. Rev. D **102**, 056013 (2020).
[7] Liang et al., Phys. Lett. B **629**, 20–26 (2005).
[8] Close, *An Introduction to Quarks and Partons*, Academic Press (1979).
[9] STAR Collaboration, Nature **548**, 62–65 (2017).
[10] Schilling et al., Nucl. Phys. B **18**, 332 (1970).
[11] Chen et al., Science Bulletin **68** 874–877 (2023).
[12] STAR Collaboration, STAR Note 619, https://drupal.star.bnl.gov/STAR/starnotes/public/sn0619
[13] STAR Collaboration, STAR Note 666, https://drupal.star.bnl.gov/STAR/starnotes/public/sn0666
[14] Poskanzer et al., Phys. Rev. C **58**, 1671–1678 (1998).
[15] STAR Collaboration, Phys. Rev. C **98**, 014910 (2018).
[16] STAR Collaboration, arXiv:2305.08705 (2023).

© The Authors, published by EDP Sciences. This is an open access article distributed under the terms of the Creative Commons Attribution License 4.0 (https://creativecommons.org/licenses/by/4.0/).

Magnetic Field Effects on Hadron Yields and Fluctuations

Volodymyr Vovchenko[1,*]

[1]Physics Department, University of Houston, Box 351550, Houston, TX 77204, USA

Abstract. We explore the impact of an external magnetic field on hadron yields and fluctuations in a thermalized system within the hadron resonance gas (HRG) model using an expanded `Thermal-FIST` package. The magnetic field sizably enhances the final proton-to-pion (p/π) ratio due to $\Delta(1232)^{++}$ decay feeddown indicating that this ratio may serve as a potential magnetometer for freeze-out conditions. At the same time, magnetic field does not generate additional dynamical fluctuations of hadron numbers. This suggests that fluctuations in heavy-ion collisions provide limited additional information about the magnetic field beyond what is already inferred from mean multiplicities.

1 Introduction

The properties of hot QCD matter change under the presence of an external magnetic field eB [1–3]. Lattice QCD simulations indicate inverse magnetic catalysis of the chiral transition temperature at and a possibility of a critical point in temperature-magnetic field plane [4–7]. Recent lattice QCD simulations evaluate the effect of magnetic field on conserved charge susceptibilities [8], with a sizable influence of baryon-charge correlator. It has been argued that this quantity may serve as a magnetometer in heavy-ion collisions [9], where significant magnetic fields are generated in the initial state, and may potentially survive to later stages of the collision [10, 11]. Here, we re-evaluate the possible effect of the magnetic field on hadron yields and fluctuations by employing the hadron resonance model.

2 Hadron resonance gas in external magnetic field

The hadron resonance gas (HRG) model offers a well-established theoretical framework for studying the thermodynamics of hadronic matter. By incorporating external magnetic field into this model, one can investigate how it affects hadronic observables, such as yield ratios and conserved charge fluctuations. The HRG model in an external magnetic field eB has been formulated in Ref. [12], and this formulation is utilized here. The presence of magnetic field leads to the modification of the energy levels and discretization of transverse momenta for charged hadrons. It also leads to vacuum contribution to thermodynamics, which is, however, independent of chemical potentials and thus omitted in the present work. The partial pressure of charged hadron species i reads

$$p_c^i = \frac{|q_i|BT}{2\pi^2} \sum_{s_z=-s}^{s} \sum_{l=0}^{\infty} \varepsilon_c \sum_{k=1}^{\infty} (-\eta_i)^{k+1} \frac{e^{k\mu_i/T}}{k} K_1(k\varepsilon_c/T). \quad (1)$$

*e-mail: vvovchenko@uh.edu

© The Authors, published by EDP Sciences. This is an open access article distributed under the terms of the Creative Commons Attribution License 4.0 (https://creativecommons.org/licenses/by/4.0/).

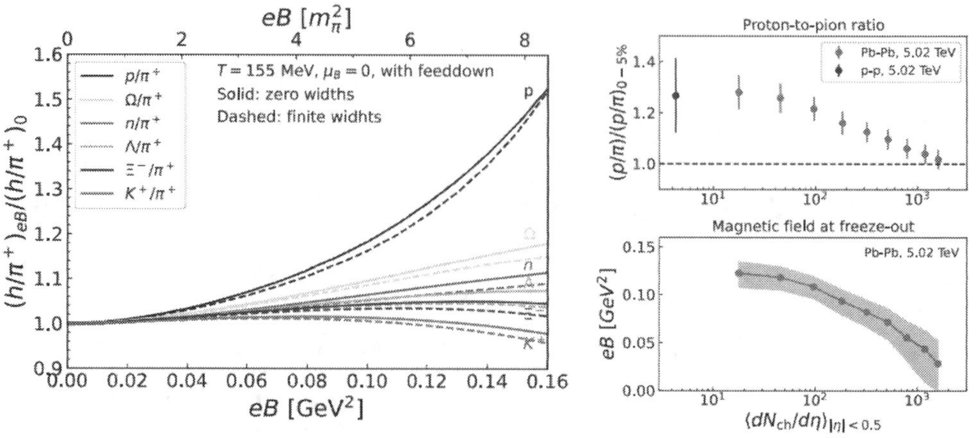

Figure 1. *Left:* Dependence of various final hadron-to-pion yield ratios on the external magnetic field eB in HRG model at temperature $T = 155$ MeV and vanishing chemical potentials, with (solid) and without (dashed) finite resonance widths. The ratios are normalized by their values at $eB = 0$. *Right:* Centrality dependence of the proton-to-pion ratio in Pb-Pb collisions at the LHC [17] and the extracted magnetic field necessary to describe the experimental data.

Here $\varepsilon_c = \sqrt{m^2 + 2|q|B(l + 1/2 - s_z)}$, s_z is the spin of hadron species i, m_i is its mass, $\mu_i = b_i\mu_B + q_i\mu_Q + s_i\mu_S$ is its chemical potential, and $\eta_i = \pm 1$ corresponds to Fermi (Bose) statistics. The hadron number density is calculated as $n_i = (\partial p_c^i/\partial \mu_i)$ while the susceptibilities of conserved charges are given by $\chi_{lmn}^{BQS} = \frac{\partial^{l+m+n}(p/T^4)}{\partial(\mu_B/T)^l\,\partial(\mu_Q/T)^m\,\partial(\mu_S/T)^n}$.

The effect of the magnetic field on HRG thermodynamics at finite temperature has been implemented in the open-source `Thermal-FIST` package [13] since version 1.5 (see [14] for details). The present implementation is based on Ref. [12] and does not include additional expected magnetic field effects such as modification of hadron masses (including neutral hadrons [15]) and decay rates [16]. It also neglects the internal structure of hadrons.

3 Hadron yields

Enhancement of the densities. Primordial hadron number densities are computed as chemical potential derivatives of the pressure in Eq. (1), which includes the effect of the magnetic field. In heavy-ion collisions, the final yields contain feeddown contributions from strong and electromagnetic decays. These are computed here assuming that magnetic field leaves decay branching ratios unaffected. Figure 1 depicts the dependence of various hadron-to-pion ratios computed at LHC conditions ($T = 155$ MeV, $\mu_B = 0$) as a function of the magnetic field eB. The calculations incorporate feeddown from strong and electromagnetic decays, with (dashed) or without (solid) finite resonance widths through energy-dependent Breit-Wigner prescription [18]. The effect of finite widths is mild. The main observation is the strong enhancement of the p/π ratio. This can be attributed to the feeddown from $\Delta(1232)^{++}$, which is strongly enhanced by the magnetic field due to its large spin (3/2) and charge ($q = 2$). Treatment of high-spin states should be improved, however, as the analysis of vacuum contributions to the pressure within the present framework indicates that $S = 3/2$ (and above) states may make the system unstable [12].

Proton-to-pion ratio as a magnetometer. Taking the calculated strong effect of magnetic field on the p/π ratio at face value, one can expect the magnetic field to induce sizable

© The Authors, published by EDP Sciences. This is an open access article distributed under the terms of the Creative Commons Attribution License 4.0 (https://creativecommons.org/licenses/by/4.0/).

centrality dependence of this ratio, with larger enhancements observed in peripheral collisions where the magnetic field is strongest. Attributing the observed enhancement of the p/π ratio as a function of centrality reported by the ALICE Collaboration for 5.02 TeV Pb-Pb collisions [17] entirely to the change of magnetic field, one can estimate the strength of the magnetic field at chemical freeze-out. This is shown in right panel of Fig. 1, indicating the maximum magnetic field in peripheral collisions of $eB \simeq (0.13 \pm 0.01)\text{GeV}^2 \simeq (6.6 \pm 0.6)m_\pi^2$.

However, several important caveats need to be considered in this interpretation. The assumption of a constant magnetic field throughout the fireball at freeze-out may not be entirely accurate, as large event-by-event fluctuations of the magnetic field are expected [19]. Perhaps even more importantly, other effects, such as baryon annihilation in the hadronic phase, can also induce centrality dependence of the p/π ratio [20], and must be carefully disentangled from the magnetic field's influence.

Isospin-symmetry breaking and neutron yields. In addition to the p/π ratio, other hadron yield ratios are also affected by the magnetic field. Neutron yields are particularly interesting. In the isospin-symmetric scenario, the yields of neutrons and protons are expected to be identical. On the other hand, magnetic field does induce the breaking of the isospin symmetry. This is reflected in Fig. 1 by a much weaker enhancement of the n/π ratio compared to p/π due to magnetic field. Thus, the corresponding measurements of n/p ratio and its possible deviations from unity as a function of centrality provide a more unambiguous probe of the magnetic field compared to the p/π ratio, motivating its challenging experimental measurement in the future.

4 Fluctuations and correlations

Comparing HRG model predictions for susceptibilities of conserved charges with lattice QCD allows one to probe the validity range of HRG. Figure 2 depicts the magnetic field dependence of χ_{11}^{BQ}, χ_2^B, and χ_2^Q at a temperature $T = 145$ MeV calculated using the HRG model with different hadron spectra and baryon excluded volume effects, and compared to lattice QCD simulations [9]. The comparison indicates qualitative agreement of the HRG model predictions with lattice QCD, which both predict the enhancement of χ's with eB, but also the need for quantitative improvement of the model, especially at $eB \gtrsim 0.08$ GeV^2.

HRG model calculations indicate that hadron number densities entirely drive the enhancement of susceptibilities with the magnetic field. When the corresponding fluctuations are normalized by the means – the Skellam baseline – the effect of magnetic field is essentially canceled out. This is illustrated by the middle panel in Fig. 2, where the normalized variance $\kappa_2[N_+ - N_-]/\langle N_+ + N_- \rangle$ of net-proton (solid black), net-kaon (dash-dotted blue), and net-pion (dotted red) numbers is shown as a function of eB. These quantities exhibit essentially flat dependence on the magnetic field, and the observed deviations from unity are driven by correlations from resonance decays rather than the magnetic field.

Finally, the normalized proton-charge correlator, χ_{11}^{pQ}/χ_2^Q can be considered as a proxy for the ratio χ_{11}^{BQ}/χ_2^Q that is enhanced by the magnetic field. This ratio is shown in Fig. 2 as a function of eB together with the p/π ratio. One can see that χ_{11}^{pQ}/χ_2^Q is indeed enhanced by the magnetic field, and it is primarily driven by the enhancement of the p/π ratio, given that χ_{11}^{pQ} and χ_2^Q are dominated by the proton and pion self-correlation terms, respectively. The relative enhancement of χ_{11}^{pQ}/χ_2^Q is stronger than that of the p/π ratio, which can be attributed to the $\Delta^{++} \to p + \pi^+$ decays. Thus, the relevance of the magnetic field could potentially be probed by precision measurements of the difference between χ_{11}^{pQ}/χ_2^Q and p/π^+ ratios.

© The Authors, published by EDP Sciences. This is an open access article distributed under the terms of the Creative Commons Attribution License 4.0 (https://creativecommons.org/licenses/by/4.0/).

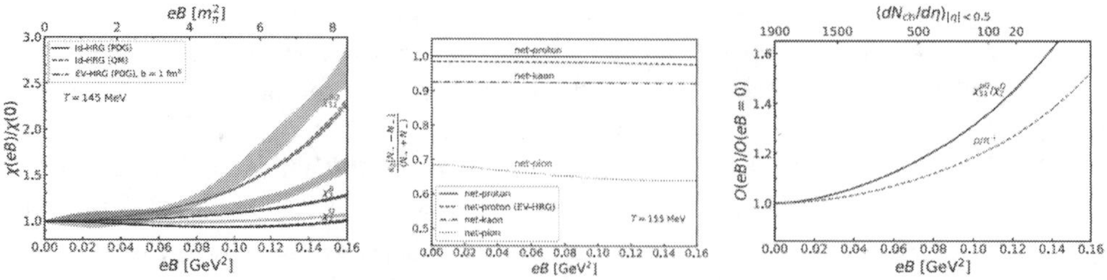

Figure 2. *Left:* Dependence of second-order susceptibilities of conserved charges on eB within different versions of the HRG model and comparison with lattice QCD at $T = 145$ MeV. *Middle:* Normalized variance of final net-proton, net-kaon, and net-pion numbers as a function of eB at $T = 155$ MeV. *Right:* Normalized proton-charge correlator, χ_{11}^{pQ}/χ_2^Q, and the p/π ratio as a function of the eB at $T = 155$ MeV.

5 Summary and Outlook

We have investigated the effects of magnetic fields on hadron yields and fluctuations in the hadron resonance gas model. Our results indicate that magnetic fields significantly influence select hadron yields, particularly by enhancing the final proton-to-pion ratio via the decay feeddown of $\Delta(1232)^{++}$, which can potentially serve as a magnetometer in heavy-ion collisions. The neutron-to-proton ratio can be a clearer probe of the magnetic field due to its isospin symmetry-breaking nature, though it is more challenging to measure experimentally. While magnetic fields have a substantial impact on hadron yields, their effect on normalized fluctuations of hadron yields is minimal. This suggests that hadron yield ratios may serve as better probes of the magnetic field in heavy-ion collisions than fluctuations.

References

[1] J. O. Andersen, W. R. Naylor and A. Tranberg, Rev. Mod. Phys. **88** (2016), 025001.
[2] V. A. Miransky and I. A. Shovkovy, Phys. Rept. **576** (2015), 1-209.
[3] G. Endrodi, [arXiv:2406.19780 [hep-lat]].
[4] G. S. Bali et al., JHEP **02** (2012), 044.
[5] G. S. Bali et al., Phys. Rev. D **86** (2012), 071502.
[6] G. Endrodi, JHEP **07** (2015), 173.
[7] M. D'Elia, L. Maio, F. Sanfilippo and A. Stanzione, Phys. Rev. D **105** (2022) 034511.
[8] H. T. Ding, S. T. Li, Q. Shi and X. D. Wang, Eur. Phys. J. A **57** (2021) 202.
[9] H. T. Ding, J. B. Gu, A. Kumar, S. T. Li, J. H. Liu, Phys. Rev. Lett. **132** (2024) 201903.
[10] D. E. Kharzeev, L. D. McLerran and H. J. Warringa, Nucl. Phys. A **803** (2008), 227-253.
[11] K. Tuchin, Adv. High Energy Phys. **2013** (2013), 490495.
[12] G. Endrödi, JHEP **04** (2013), 023.
[13] V. Vovchenko and H. Stoecker, Comput. Phys. Commun. **244** (2019), 295-310.
[14] V. Vovchenko, Phys. Rev. C **110** (2024) 034914.
[15] M. Coppola et al., Phys. Rev. D **100** (2019) 054014.
[16] M. Marczenko et al., [arXiv:2405.15745 [hep-ph]].
[17] S. Acharya *et al.* [ALICE], Phys. Rev. C **101** (2020) 044907.
[18] V. Vovchenko, M. I. Gorenstein and H. Stoecker, Phys. Rev. C **98** (2018) 034906.
[19] A. Bzdak and V. Skokov, Phys. Lett. B **710** (2012), 171-174.
[20] V. Vovchenko and V. Koch, Phys. Lett. B **835** (2022), 137577.

© The Authors, published by EDP Sciences. This is an open access article distributed under the terms of the Creative Commons Attribution License 4.0 (https://creativecommons.org/licenses/by/4.0/).

EPJ Web of Conferences **316**, 06016 (2025) https://doi.org/10.1051/epjconf/202531606016
SQM 2024

MIT-CTP/5662

Equilibrium expectations for non-Gaussian fluctuations near a QCD critical point

Jamie M. Karthein[1,*], *Maneesha* Pradeep[2], *Krishna* Rajagopal[1], *Mikhail* Stephanov[3], and *Yi* Yin[4]

[1]Center for Theoretical Physics, Massachusetts Institute of Technology, Cambridge, MA 02139, USA
[2]Department of Physics, University of Maryland, College Park, Maryland 20742, USA
[3]Department of Physics, University of Illinois, Chicago, Illinois 60607, USA
[4]School of Science and Engineering, The Chinese University of Hong Kong, Shenzhen, Guangdong, 518172, China

Abstract. With the highly anticipated results from the Beam Energy Scan II program at RHIC being recently revealed, an understanding of particle-number fluctuations and their significance as a potential signature of a possible QCD critical point is crucial. Early works that embarked on this endeavor sought to estimate the fluctuations due to the presence of a critical point assuming they stay in equilibrium. From these results came the proposal to focus efforts on higher, non-Gaussian, moments of the event-by-event distributions, in particular of the number of protons. These non-Gaussian moments are especially sensitive to critical fluctuations, as their magnitudes are proportional to high powers of the critical correlation length. As the equation of state provides key input for hydrodynamical simulations of heavy-ion collisions, we estimate equilibrium fluctuations from the BEST equation of state (EoS) that includes critical features from the 3D Ising Model. In particular, the proton factorial cumulants and their dependence on non-universal mapping parameters is investigated within the BEST EoS. Furthermore, the correlation length, as a central quantity for the assessment of fluctuations in the vicinity of a critical point, is also calculated in a consistent manner with the scaling equation of state. An understanding of the equilibrium estimates of proton factorial cumulants will be useful for further comparison to estimates of out-of-equilibrium fluctuations in order to determine the magnitude of the observable fluctuations to be expected in heavy-ion collision experiments, in which the time spent near a critical point is short.

1 Introduction

The phase structure of QCD matter at finite temperature and density remains an open question in nuclear science. While first-principles lattice QCD calculations are well-suited for vanishing density, finite density results are limited to $\mu_B/T \sim 3.5$ [1]. Experimental measurements are, therefore, the only modern tool that can access further reaches of the phase diagram. The Beam Energy Scan II (BESII) program at RHIC, which had the goal of mapping out the phase diagram, recently completed its runs and revealed measurements of the proton fluctuations for a range of μ_B [2]. The moments of the proton distribution are sensitive to the presence

*e-mail: jmkar@mit.edu

© The Authors, published by EDP Sciences. This is an open access article distributed under the terms of the Creative Commons Attribution License 4.0 (https://creativecommons.org/licenses/by/4.0/).

of a critical point in the phase diagram with higher orders diverging more strongly due to higher powers of the correlation length, a quantity which diverges at the critical point. (For a review see Ref. [3].) These results show no evidence of a critical point in the region of the phase diagram with $\mu_B \lesssim 420$ MeV where experimental measurements have focused to date. However, theoretical models of the noncritical baseline expected for these fluctuations do not appear to describe the data at all energies. Therefore, it is crucial to understand the critical point contribution to proton fluctuations in order to help interpret these results.

2 Scaling Equation of State Mapped to QCD

Studies of the effect of a critical point on the equation of state for QCD have been underway with the advent of the BEST equation of state in the original work of Parotto et al [4]. This equation of state incorporates critical features based on the universality principle of scaling near a critical point. The universality class for QCD has been shown to be that of the 3D Ising model, as discussed in the review Ref. [3]. The 3D Ising model scaling equation of state is given by the parametric form:

$$P_{\text{Ising}} = -G(R, \theta) = h_0 M_0 R^{2-\alpha}(\theta \tilde{h}(\theta) - g(\theta)), \tag{1}$$

where h_0, M_0 are normalization constants for the magnetic field and magnetization, $\alpha = 0.11$ is a 3D Ising critical exponent, and $\tilde{h}(\theta)$ and $g(\theta)$ are polynomials in θ [4, 5]. In order to express $P(T, \mu_B)$, we must utilize a map between Ising and QCD variables, the equations for which will be presented subsequently. The critical fluctuations are calculated as derivatives of the equation of state. For example the n-point correlations of baryon density are given by the kth derivative of pressure with respect to baryon chemical potential, μ_B at fixed temperature:

$$\langle(\delta n)^k\rangle = \frac{T^{k-1}}{V^{k-1}} \frac{\partial^k P}{\partial \mu^k} \tag{2}$$

Higher order fluctuations diverge more strongly at the critical point due to their scaling with higher powers of the correlation length. The correlation length itself may also be written the parametric form in the Ising model in terms of the parametric variables (R, θ) as [6]:

$$\xi^2(M, t) = R^{-2\nu} x_\xi(\theta). \tag{3}$$

We present the θ-dependence of the correlation length as calculated in the ϵ-expansion to $O(\epsilon^2)$:

$$x_\xi = x_\xi(0)\left(1 - \frac{5}{18}\epsilon\theta^2 + \left[\frac{1}{972}(24I - 25)\theta^2 + \frac{1}{324}(4I + 41)\theta^4\right]\epsilon^2\right), \tag{4}$$

where $I \equiv \int_0^1 \frac{\ln[x(1-x)]}{1-x(1-x)} dx \approx -2.3439$. This updated correlation length at $O(\epsilon^2)$ is, furthermore, consistent with the scaling equation of state valid to the same order in the ϵ-expansion.

In order to map the critical features of the 3D Ising model onto the QCD phase diagram, we employ a quadratic mapping as first introduced in Ref. [7]. This is an improvement over the original work of the BEST collaboration where a linear mapping allowed for the Ising transition line to be mapped onto the tangent to the chiral phase transition line from lattice QCD. This quadratic mapping is combined with an alternative expansion scheme, known as the T'-expansion, introduced in [1], which is a essentially a reshuffling of the Taylor expansion. The QCD variables $\mu_B^2 - \mu_{B,c}^2$ and $T'(\mu, T) - T_0$, are related to the Ising variables (r, h) near the critical point as:

$$\frac{T'(\mu_B, T) - T_0}{T'_T T_c} = -hw \frac{\sin(\alpha_1 - \alpha_2)}{\cos \alpha_1}\,; \quad \frac{\mu_B^2 - \mu_{B,c}^2}{2\mu_{B,c} T_c} = -w(r\rho \cos \alpha_1 + h \cos \alpha_2)\,. \tag{5}$$

© The Authors, published by EDP Sciences. This is an open access article distributed under the terms of the Creative Commons Attribution License 4.0 (https://creativecommons.org/licenses/by/4.0/).

where the mapping parameters T_c, $\mu_{B,c}$, $\alpha_{1,2}$, w and ρ were introduced in Ref.[4]. The location of the critical point in the phase diagram is given by $(\mu_{B,c}, T_c)$ while the size and shape of the critical region are controlled by the remaining parameters. However, by fixing the critical point to lie along the chiral phase transition line as in Refs. [4, 5, 7], T_c and α_1 are fixed by the choice of $\mu_{B,c}$, leaving 4 free parameters.

In order to calculate the critical contribution to the proton factorial cumulants, $\hat{\Delta}\omega_p^k$, from the fluctuations of the thermodynamic densities we utilize the maximum entropy approach from Ref. [8].

3 A Selection of Results, and a Look Ahead

We show the equilibrium results for the critical contribution to the second, third and fourth order fluctuations, $\langle(\delta n)^k\rangle$, in the QCD phase diagram in Fig. 1 for a given choice of the mapping parameters: $\mu_{B,c} = 600$ MeV, w = 5, $\rho = 1$, $\alpha_2 = 0$. Due to the constraint that the critical point should lie along the chiral phase transition line, the choice of $\mu_{B,c} = 600$ MeV fixes the remaining two parameters, $T_c = 90$ MeV, $\alpha_1 = 16.6^{\circ}$. Given this choice, we see that within the critical region the contours for these fluctuations obtain the typical features as expected from Ref. [3], including the strengthening of fluctuations toward the critical point, a sign change for the third cumulant from negative (red) to positive (blue) across the transition line, and a double-lobed structure for the fourth cumulant. Furthermore, due to the quadratic mapping between Ising and QCD variables as presented in Ref. [7] and shown here in Eq. (5), we see that these features follow the curve of the chiral phase transition line rather than a line tangent to the critical point as in Refs. [4, 5]. This in turn gives rise to the behavior of the factorial cumulants as shown in Fig. 2 along freeze-out lines in the phase diagram given by a shift ΔT from the transition line, which follow curves 4, 6, and 9 MeV below the critical point itself. Along these freeze-out lines, we see a peak in all three cumulants near the critical point with the magnitude of the peak increasing with increasing order of the fluctuations. Furthermore, a sign change can be seen in third and fourth cumulants. As expected, the magnitude of the fluctuations depends strongly on the distance ΔT between the freeze-out and the critical point, decreasing with increasing ΔT. We note that the choice of parameters we present here provides an example of what we can expect for equilibrium fluctuations. The features of these curves as manifest from the size and shape of the critical region are highly dependent on the choice of parameters. In fact, because the proton factorial cumulants are so sensitive to the parameters, these parameters can be constrained by the experimental measurements. In a forthcoming work, we will more fully explore the effect of the mapping parameters on the proton factorial cumulants. Furthermore, so far we have focused on equilibrium results by utilizing the equation of state, however, we will also study the non-equilibrium effects in order to provide a more direct comparison with experiment.

4 Acknowledgements

JMK is supported by an MPS Ascend Fellowship from the National Science Foundation under Award No. 2138063. This work is supported by the U.S. Department of Energy, Office of Science, Office of Nuclear Physics grants DE-SC0011090, DE-FG02-01ER41195 and DE-FG02-93ER40762.

References

[1] S. Borsányi, Z. Fodor, J.N. Guenther, R. Kara, S.D. Katz, P. Parotto, A. Pásztor, C. Ratti, K.K. Szabó, Lattice QCD equation of state at finite chemical potential from an alternative

© The Authors, published by EDP Sciences. This is an open access article distributed under the terms of the Creative Commons Attribution License 4.0 (https://creativecommons.org/licenses/by/4.0/).

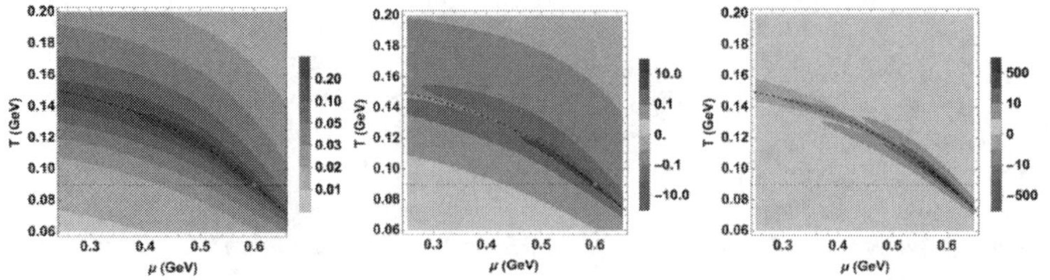

Figure 1. Contours of the critical contribution to $V^{k-1}T_c^{-3}\langle(\delta n)^k\rangle$ for $k = 2, 3, 4$ with the choice of $\mu_{B,c} = 600, \text{MeV}$, $\alpha_2 = 0°$, $w = 5$ and $\rho = 1$

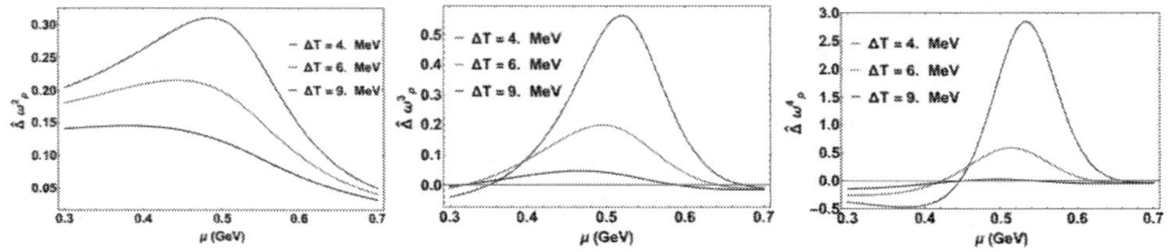

Figure 2. $\hat{\Delta}\omega_p^k$ plotted for $k = 2, 3, 4$ with the choice of $\mu_{B,c} = 600, \text{MeV}$, $\alpha_2 = 0°$, $w = 5$ and $\rho = 1$ along different freeze-out trajectories characterized by ΔT.

expansion scheme, Phys. Rev. Lett. **126**, 232001 (2021), 2102.06660. 10.1103/PhysRevLett.126.232001

[2] A. Pandav (STAR collaboration), https://conferences.lbl.gov/event/1376/contributions/8772/

[3] A. Bzdak, S. Esumi, V. Koch, J. Liao, M. Stephanov, N. Xu, Mapping the Phases of Quantum Chromodynamics with Beam Energy Scan, Phys. Rept. **853**, 1 (2020), 1906.00936. 10.1016/j.physrep.2020.01.005

[4] P. Parotto, M. Bluhm, D. Mroczek, M. Nahrgang, J. Noronha-Hostler, K. Rajagopal, C. Ratti, T. Schäfer, M. Stephanov, QCD equation of state matched to lattice data and exhibiting a critical point singularity, Phys. Rev. C **101**, 034901 (2020), 1805.05249. 10.1103/PhysRevC.101.034901

[5] J.M. Karthein, D. Mroczek, A.R. Nava Acuna, J. Noronha-Hostler, P. Parotto, D.R.P. Price, C. Ratti, Strangeness-neutral equation of state for QCD with a critical point, Eur. Phys. J. Plus **136**, 621 (2021), 2103.08146. 10.1140/epjp/s13360-021-01615-5

[6] J. Zinn-Justin, Quantum Field Theory and Critical Phenomena (Oxford University Press, 2002), ISBN 9780198509233, https://doi.org/10.1093/acprof:oso/9780198509233.001.0001

[7] M. Kahangirwe, S.A. Bass, E. Bratkovskaya, J. Jahan, P. Moreau, P. Parotto, D. Price, C. Ratti, O. Soloveva, M. Stephanov, Finite density QCD equation of state: Critical point and lattice-based T' expansion, Phys. Rev. D **109**, 094046 (2024), 2402.08636. 10.1103/PhysRevD.109.094046

[8] M.S. Pradeep, M. Stephanov, Maximum Entropy Freeze-Out of Hydrodynamic Fluctuations, Phys. Rev. Lett. **130**, 162301 (2023), 2211.09142. 10.1103/PhysRevLett.130.162301

© The Authors, published by EDP Sciences. This is an open access article distributed under the terms of the Creative Commons Attribution License 4.0 (https://creativecommons.org/licenses/by/4.0/).

The NA60+ experiment at the CERN SPS

Sabyasachi Siddhanta[1,*] for the NA60+ Collaboration

[1]Istituto Nazionale di Fisica Nucleare (INFN), Cagliari, Italy

Abstract. NA60+ is a new experiment designed to study the phase diagram of the strongly interacting matter at high baryochemical potential from 200 to 550 MeV at the CERN SPS. It is focused on precision studies of thermal dimuons, heavy quark and strangeness production in Pb–Pb collisions at center of mass energies ranging from 6 to 17 GeV. This contribution is focussing on the experimental apparatus, including the technical aspects and the R&D status, as well as the physics program and its competitiveness and complementarity to other experiments.

1 Introduction

Heavy ion collisions at low energies in the range 6 to 17 GeV is a tool to investigate the QCD phase diagram at large baryo-chemical potential (μ_B). This is in contrast to studies at top RHIC energies and at the LHC which are characterised by a high initial temperature and close to zero μ_B. The low energy studies could give insights to open questions at large μ_B like a first order phase transition with the presence of a critical point separating the hadronic phase with the QGP [1]. The low energy range is available at the CERN SPS and we propose a new fixed target experiment, presently denoted as the NA60+.

A rich physics program is foreseen, which includes the search for chiral symmetry restoration effects through the $\rho - a_1$ mixing, the study of the order of the phase transition at large baryochemical potential through the measurement of a caloric curve, the onset of the deconfinement through the measurement of J/ψ suppression. The measurement of the transport properties of the medium via open charm states and the study of hadrochemistry via detection of strange hadrons and hypernuclei are also part of the physics program.

2 Experimental apparatus

The proposed NA60+ apparatus, as shown in Figure 1, is inspired by the former NA60 apparatus[3]. It is composed of a muon spectrometer with six tracking stations, a large-acceptance toroidal magnet and a two-stage absorber system to filter out muons. This is complemented by a vertex telescope located close to the target and sitting before the absorber for precise measurement of muon kinematics. For the beam energy scan, the z-position can be adjusted to cover roughly one unit of rapidity in the forward direction at all energies.

The vertex telescope will be hosted inside a dipole magnet, a few centimeters after a lead target system comprising of five 1.5 mm thick Pb disks (bottom right image of Figure 2). An

*e-mail: sabyasahi.siddhanta@cern.ch

© The Authors, published by EDP Sciences. This is an open access article distributed under the terms of the Creative Commons Attribution License 4.0 (https://creativecommons.org/licenses/by/4.0/).

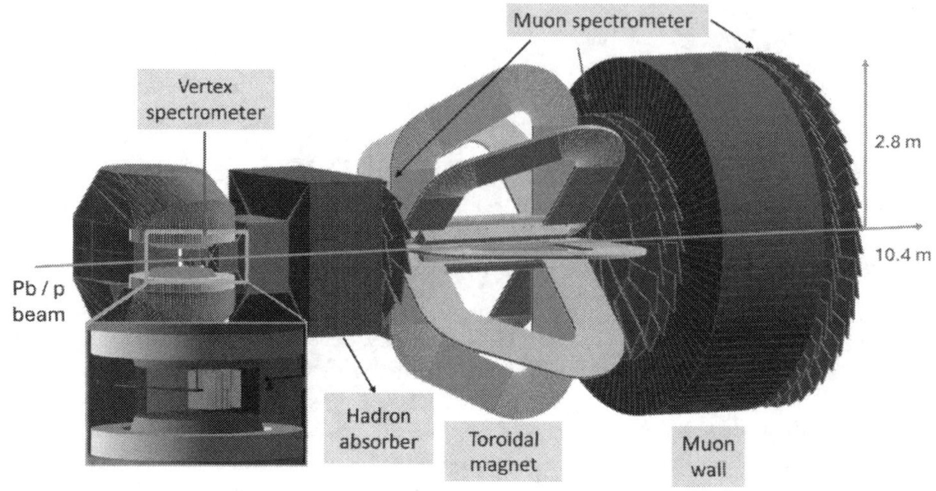

Figure 1. The NA60+ apparatus.

existing magnet - MEP48 which is stored at CERN is proposed. It provides 1.5 T over a 40 cm gap. The vertex telescope will consist of several planes of ultra-thin, large area Monolithic Active Pixel sensors (MAPS). For proton runs, which is needed to provide a reference to Pb-Pb runs, several target species from beryllium to tungsten will be exposed to the beam.

For the pixel sensor we plan to use large area stitched sensors, based on the R&D carried out by ALICE for ITS3[2], which has the same timeline for the installation as NA60+. The sensor is based on 25 mm sub-units called Repeated Sub-Units (RSU), which are replicated up to 14 cm for NA60+ through a stitching technology provided by the foundry. These 14 cm strips are then replicated vertically in the silicon wafer to obtain 14x14 cm^2 sensors. Each plane will be formed by four of such sensors, covering approximately an area of 28x28 cm^2 as shown in the top right image in Figure 2. The material budget will be less than 0.1 % X_0 and the spatial resolution better than 5 μm. A dedicated R&D for mechanics and cooling, based on airflow and water, is presently ongoing

The muon spectrometer will utilize large area gaseous detectors for muon tracking and a large-acceptance toroidal magnet based on a new light-weight and general-purpose concept. The tracking is based on six stations, with increasing size, with a modular structure exploiting a trapezoidal unit, arranged in concentric rings. An absorber comprising a BeO and a graphite section will be placed after the vertex telescope to reduce the background in the muon spectrometer. A second absorber will be placed before the last two stations to stop the residual background. The warm toroidal magnet with low material budget in the acceptance area will be placed between the second and third station to measure the muon momentum. The overall scheme is shown in the left image of Figure 2. The muon spectrometer will be placed on rails, so as to adjust its length to better cover the mid-rapidity region at different energies. For higher energies, the graphite secion of the first absorber will be extended. The rates after the absorber, evaluated with dedicated Fluka simulations, do not exceed few kHz/cm^2 for 10^6 Pb ions per second. This would mean that both Multi Wire Proportional Chambers (MWPC) and/or Gas Electron Multiplier (GEM) detectors can match the requirements.

First MWPC prototypes have been built and beam tests were performed in 2023. It is based on double strip readout providing a spatial resolution around 100 μm. There are on-

© The Authors, published by EDP Sciences. This is an open access article distributed under the terms of the Creative Commons Attribution License 4.0 (https://creativecommons.org/licenses/by/4.0/).

 https://doi.org/10.1051/epjconf/202531607001

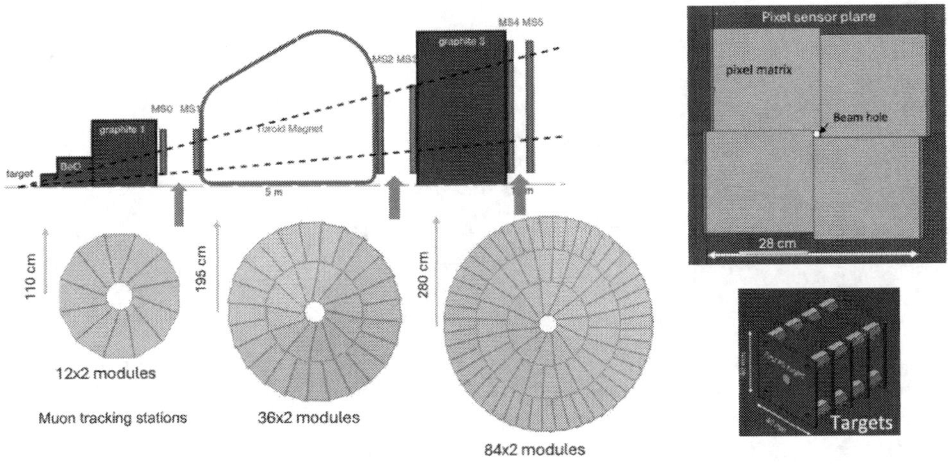

Figure 2. Left: Schematic layout of the muon spectrometer. The setup is adapted for a beam energy of $\sqrt{s_{NN}} = 6.3$ GeV. For higher beam energies, the graphite absorber will be extended. Right (top): One plane of the vertex telescope consisting of four large area (14 x 14 cm^2) MAPS sensors. Right (bottom): Sketch of the NA60+ target system (Pb–Pb collisions set-up).

going discussions for the final setup, with the possibility to use GEMs for the first two stations, and MWPC for the most downstream stations. Discussions for the overall readout electronics are also ongoing.

A complete feasibility study has been performed to install NA60+ in the EHN1-PP138 area, where CERES-NA45 was previously installed. This includes rails needed to shift the spectrometer longitudinally, a floor excavation to cope with the vertical position of the beam line and a rather massive shielding needed for complying with radio-protection safety rules.

3 Physics performance studies

Several new and unique measurements are foreseen with NA60+ in the energy range 6 - 17 GeV related to thermal dimuons, quarkonia, open heavy flavor and strangeness. The NA60+ physics program includes, among others, the measurement of a caloric curve of temperature vs. collision energy, investigation of chiral symmetry restoration close to the phase boundary, charmonium melting in the QGP and open heavy flavor to extract QGP transport coefficients and investigate hadronization [4].

Detailed performance studies were carried out with different observables for the 5% most central Pb–Pb collisions at $\sqrt{s_{NN}} = 6.3$, 8.8 and 17.3 GeV[4].

The left plot of Figure 3 shows the NA60+ performance on the measurement of a caloric curve which displays the temperature evolution as a function of collision energy. Thermal dileptons provide a precise thermometer by measuring the invariant mass slope in the mass region 1.5 - 2.5 GeV. This slope is an average temperature of the early stage of the system. There are only two precision measurements of the temperature till date, done by NA60 and HADES as shown in Figure 3, which also shows the evolution of T_{slope} vs $\sqrt{s_{NN}}$ from theory. In the energy range covered by SPS, NA60+ would extract information in the region close to the deconfinement temperature, with a possible signal of a first order phase transition.

© The Authors, published by EDP Sciences. This is an open access article distributed under the terms of the Creative Commons Attribution License 4.0 (https://creativecommons.org/licenses/by/4.0/).

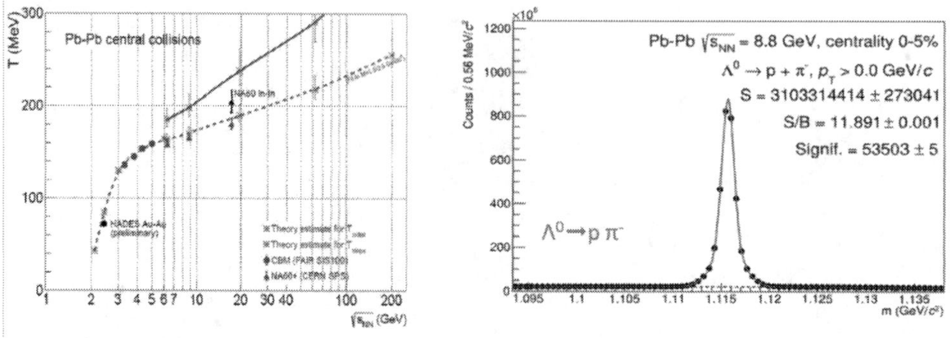

Figure 3. Left: Caloric curve - medium temperature evolution vs $\sqrt{s_{NN}}$ in central Pb–Pb collisions. Blue triangles are the expected performance from NA60+. Right: Projection for the invariant-mass distribution of Λ^0 candidates (right).

The hadronic decays of strange hadrons can be studied with the Vertex Telescope of the NA60+ apparatus. Several decal channels involving hadronic decays of $\Lambda, \Xi, \Omega, K_s^0$ and ϕ were studied. For example, the right plot of Figure 3 shows the projection for the invariant-mass distribution of Λ^0 candidates. The analysis was performed on 10^7 simulated Pb-Pb collisions in the 0-5% centrality class at $\sqrt{s_{NN}} = 8.8$ GeV.

4 Conclusion and outlook

The Letter of Intent for the NA60+ experiment[4] was submitted to the SPS Committee (SPSC) in 2022 and was discussed in February 2023 with a favorable feedback. The project is mentioned in US 2023 Long Range Plan for Nuclear Science and in the NUPECC Long Range Plan 2024. Significant progress was made in the detector and toroidal magnet R&D and beam optics studies in 2023 and 2024. Presently, preparation and consolidation work is ongoing for the Technical Proposal which will be submitted to the SPSC in early 2025.

NA60+ is planned to be commissioned during the Long Shutdown 3 (LS3) of the LHC (2026-28). The data taking will start from 2029 and will continue over a period of seven years. Data taking for one energy point will be performed including one month of Pb-Pb and a few weeks of p-A runs each year.

References

[1] K. Fukushima and T. Hatsuda, The phase diagram of dense QCD. Reports on Progress in Physics **74.1** (2010): 014001

[2] ALICE Collaboration, Technical Design Report for the ALICE Inner Tracking System 3 - ITS3. CERN-LHCC-2024-003/ALICE-TDR-021

[3] NA60 Collaboration, Study of prompt dimuon and charm production with proton and heavy ion beams at the CERN SPS. CERN-SPSC-2000-010

[4] NA60+ Collaboration, Letter of Intent: the NA60+ experiment. CERN-SPSC-2022-036; SPSC-I-259

© The Authors, published by EDP Sciences. This is an open access article distributed under the terms of the Creative Commons Attribution License 4.0 (https://creativecommons.org/licenses/by/4.0/).

https://doi.org/10.1051/epjconf/202531607002

The ITS3 detector and physics reach of the LS3 ALICE Upgrade

Chun-Zheng Wang (for the ALICE Collaboration)[1,2,*]

[1]Key Laboratory of Nuclear Physics and Ion-beam Application (MOE), Institute of Modern Physics, Fudan University, 200433, Shanghai, China
[2]Key Laboratory of Quark & Lepton Physics (MOE) and Institute of Particle Physics, Central China Normal University, 430079, Wuhan, China

Abstract. During Large Hadron Collider (LHC) Long Shutdown 3 (LS3) (2026-30), the ALICE experiment is replacing its inner-most three tracking layers by a new detector, Inner Tracking System 3. It will be based on newly developed wafer-scale monolithic active pixel sensors, which are bent into truly cylindrical layers and held in place by light mechanics made from carbon foam. Unprecedented low values of material budget (per layer) and closeness to interaction point (19 mm) lead to a factor two improvement in pointing resolutions from very low $p_{\rm T}$ (O(100MeV/c)), achieving, for example, 20 μm and 15 μm in the transversal and longitudinal directions, respectively, for 1 GeV/c primary charged pions. After a successful R&D phase 2019-2023, which demonstrated the feasibility of this innovational detector, the final sensor and mechanics are being developed. This contribution briefly reviews the conceptual design and the main R&D achievements, as well as the current activities and road to completion and installation. It concludes with a projection of the improved physics performance, in particular for heavy-flavour hadrons, as well as for thermal di-electrons, that will come into reach with this new detector installed.

1 Introduction

ALICE (A Large Ion Collider Experiment) is a detector at LHC specifically designed to investigate high-energy heavy-ion collisions, and to study the formation of the quark–gluon plasma, a colour-deconfined state of the matter. To achieve this, ALICE is equipped with multiple detectors, including the ITS, which plays a crucial role in determining the interaction and decay vertices and precisely tracking particles with low momentum and angular resolution in conjunction with the TPC and other detectors. The ITS was upgraded during LHC LS2 to the ITS2, which is constructed using ALPIDE chips based on 180 nm CMOS technology [1].

With advances in chip manufacturing technology, wafer-scale chips can now be produced using stitching techniques. This development has opened new avenues for the next upgrade of the ITS. After a successful R&D phase 2019-23, a next-generation ITS3 design has been proposed by ALICE [2], aimed at replacing the innermost three tracking layers of ITS2 during LHC LS3 (2026-30). ITS3 features three layers of wafer-scale chips manufactured using stitching technology, and these chips will be bent into a half-cylinder shape. The chips,

*e-mail: chunzheng.wang@cern.ch

© The Authors, published by EDP Sciences. This is an open access article distributed under the terms of the Creative Commons Attribution License 4.0 (https://creativecommons.org/licenses/by/4.0/).

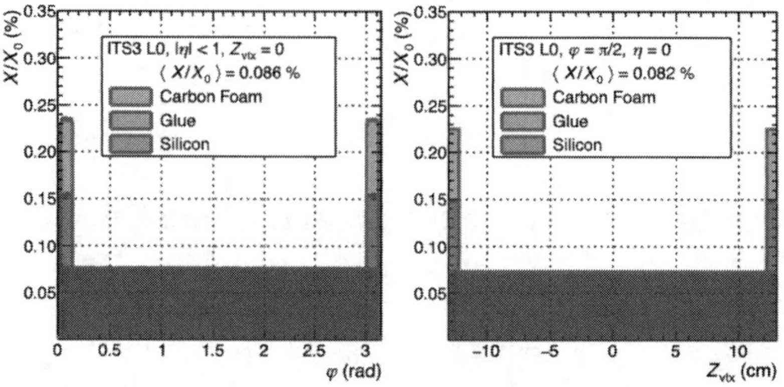

Figure 1. Material budget for tracks of particles originating from the interaction point as a function of the azimuthal angle ϕ (left panel) and the coordinate of the interaction vertex along the beam axis Z_{vtx} (right panel)

based on 65 nm CMOS MAPS technology, with a power density below 40 mW/cm^2, allow for air cooling to replace the water cooling system used in ITS2. These improvements will significantly reduce the material budget from 0.35% X_0 to 0.07% X_0 per layer from ITS2 to ITS3. Additionally, the material distribution will become quasi uniform as Fig. 1 illustrates. Furthermore, the beam pipe's radius will be reduced from 18.2 mm to 16.0 mm, leading to a decrease in the radius of layer 0 of ITS3 from approximately 24.0 mm to 19.0 mm.

2 Chip design and characterization

Supporting this ambitious design, the chip development for ITS3 involves multiple phases. In the first phase, the MLR1 phase, includes three types of chips—Analogue Pixel Test Structure (APTS), Digital Pixel Test Structure (DPTS), and Circuit Exploratoire 65 nm (CE65). Through this phase, the 65 nm process chips were thoroughly studied, including the confirmation of radiation hardness. Even when subjected to 100 times the radiation dose required for ITS3 (10 kGy and 10^{13} 1MeV n_{eq} cm^2), the detection efficiency of the chips remained sufficiently high, and the spatial resolution was almost unaffected. For more detailed results, please refer to Ref. [3].

In the ER1 phase, ALICE designed two types of wafer-scale chips: MOnolithic Stitched sensor (MOSS) and MOnolithic Stitched sensor with Timing (MOST). The MOSS chip is the first stitched chip used for high-energy physics, with a length of 259 mm and a width of 14 mm. The MOSS chip is segmented into a left end cap (LEC), a right cap (REC), and 10 repeated sensor units (RSU), with each RSU containing a top and a bottom half-unit. The top half-unit is divided into four pixel regions, where each pixel has a width of 22.5μm, and there are 256 × 256 pixels per region, with slight design variations among the regions. The bottom half-unit is also divided into four regions, each with a pixel width of 18 μm and 320 × 320 pixels. Fig. 2 shows the beam test results for the MOSS chip at CERN PS, which indicate that the MOSS chip achieves a detection efficiency of over 99% with a very low fake-hit rate, below 10^{-6} pixel^{-1} event^{-1} for region 2 when operated at an optimal working point. More detailed description can be found in the ITS3 TDR[2].

The successful design of the MOSS chip has confirmed the effectiveness of using wafer-scale chips for particle detection, and provided the guidance for the design and fabrication of ITS3 chips in the subsequent ER2 phase.

© The Authors, published by EDP Sciences. This is an open access article distributed under the terms of the Creative Commons Attribution License 4.0 (https://creativecommons.org/licenses/by/4.0/).

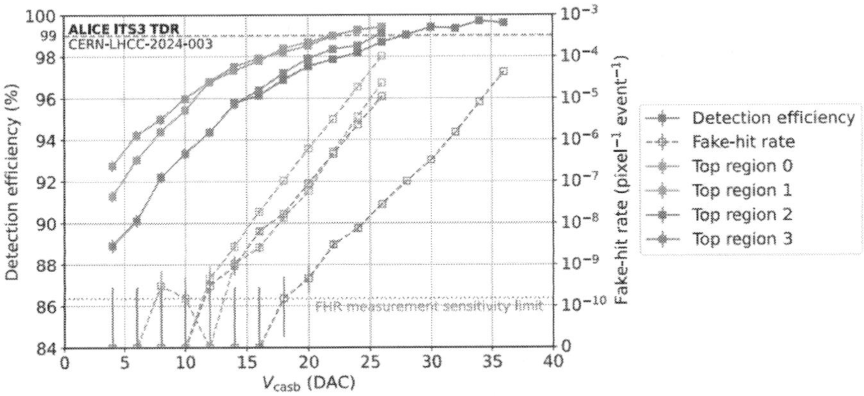

Figure 2. Detection efficiency and fake-hit rate as a function of V_{casb} (inversely proportional to the threshold in number of electrons) measured for MOSS at the CERN PS.

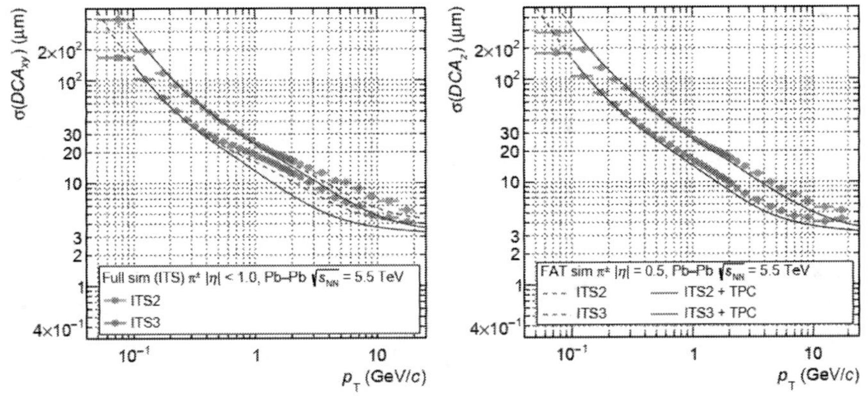

Figure 3. Impact parameter resolution in the transverse (left panel) and longitudinal (right panel) direction for primary charged pions with $|\eta| < 1$ as a function of the track p_T for ITS2 and ITS3 detectors from the O^2 simulation

3 Physics performance and physics reach of ITS3

Using the O^2 framework [4] developed by ALICE for the reconstruction, calibration, and simulation of the ALICE experiment for LHC Run 3 and 4, the impact parameter resolution of ITS3 has been estimated using two simulation methods: full simulation and fast analytic tool (FAT), both yielding consistent results. Compared to the ITS2, the ITS3 is expected to achieve a twofold improvement in impact parameter resolution [5]. This will greatly benefit numerous analyses, such as studies of the participation of heavy-flavour particles in the collective motions of the system created in heavy-ion collisions

The production of heavy-flavor particles in the QGP mainly arises from two mechanisms: fragmentation, where the transition from heavy quark to hadron occurs via the emission of a parton shower, resulting in a hadron that carries a fraction of the original parton momentum, and coalescence, where partons close in phase space can recombine. These two mechanisms contribute differently to the elliptic flow coefficient [6] of heavy-flavor particles. The performance of the ITS3 in reconstructing heavy-flavor particles is expected to be significantly better compared to the ITS2. As shown in the left panel of Fig. 4, although the dead zones have a minor impact, the reconstruction remains extremely precise, leading to accurate flow calculations. The right panel of Fig. 4 further demonstrates, according to the TAMU model [7], that in the $p_T > 4$ GeV/c region, there will be a significant difference in the flow of Λ_c^+

© The Authors, published by EDP Sciences. This is an open access article distributed under the terms of the Creative Commons Attribution License 4.0 (https://creativecommons.org/licenses/by/4.0/).

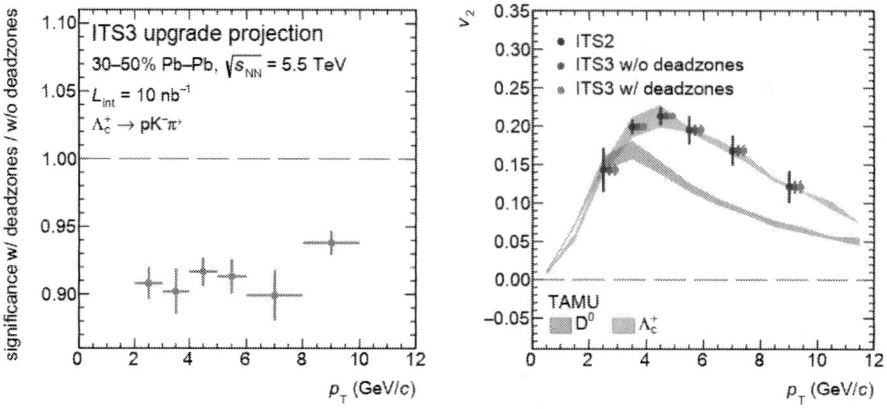

Figure 4. Left: Ratio between the expected significance with or without deadzones. Right panel: Comparison of the expected performance for the measurement of the Λ_c elliptic flow with the ITS2 and the ITS3 with and without deadzones.

and D^0 particles. With ITS3, the statistical uncertainty in the v_2 measurement for Λ_c^+ can be reduced by up to a factor of 4. This study utilizing ITS3 will further constrain the modeling of charm diffusion and hadronization in the QGP. More details on the physics performance can be found in Ref. [5].

4 Summary

The ITS3, a bent wafer-scale monolithic pixel detector, is set to replace the ITS2 inner barrel during LHC LS3. Characterization results from each phase of chip design and production have confirmed that the bent 65 nm process wafer-scale chips fully meet the requirements of ITS3, including radiation hardness, spatial resolution, and detection efficiency. The ITS3 project remains on track for installation as planned. This upgrade, offering a twofold improvement in spatial resolution compared to ITS2, will significantly enhance various analyses, such as heavy-flavor collectivity and thermal dielectrons measurement.

The author is supported in part by the National Key Research and Development Program of China (Nos. 2022YFA1602103, 2018YFE0104600), the National Natural Science Foundation of China (No. 12322508) and the Science and Technology Commission of Shanghai Municipality (No. 23590780100).

References

[1] M. Mager, on behalf of the ALICE collaboration, NIM-A, 824, 434 (2016). https://doi.org/10.1016/j.nima.2015.09.057

[2] ALICE collaboration, Technical Design Report for the ALICE Inner Tracking System 3 (2024). https://cds.cern.ch/record/2890181

[3] G. A. Rinella, and et al., arXiv:2403.08952. https://doi.org/10.48550/arXiv.2403.08952

[4] M. Richter, and et al., AliceO2Group/AliceO2. https://doi.org/10.5281/zenodo.1493334

[5] ALICE collaboration, Upgrade of the ALICE Inner Tracking System during LS3: study of physics performance (2023). https://cds.cern.ch/record/2868015

[6] ALICE Collaboration, PRL, 116, 132302 (2016). https://doi.org/10.1103/PhysRevLett.116.132302

[7] M. He and R. Rapp, PRL, 124, 042301 (2020). https://doi.org/10.1103/PhysRevLett.124.042301

© The Authors, published by EDP Sciences. This is an open access article distributed under the terms of the Creative Commons Attribution License 4.0 (https://creativecommons.org/licenses/by/4.0/).

Fast timing silicon R&D for the future Electron-Ion Collider

Xuan Li[1,*], *Eric* Renner[1], *Ming* Liu[1], *Walter* Sondheim[1], *Carlos* Solans Sanchez[2], *Marcos* Vazquez Nuñez[3], *Vicente* Gonzalez[3], and *Yasser* Corrales Morales[4]

[1]Los Alamos National Laboratory, USA
[2]CERN, Switzerland
[3]University of Valencia, Spain
[4]Massachusetts Institute of Technology, USA

Abstract. The proposed Electron-Ion Collider (EIC) will utilize high-luminosity high-energy electron+proton ($e + p$) and electron+nucleus ($e + A$) collisions to solve several fundamental questions including searching for gluon saturation and studying the proton/nuclear structure. Complementary to the ongoing EIC project detector technical prototype carried out by the ePIC collaboration, a Depleted Monolithic Active Pixel Sensor (i.e., MALTA2) based fast timing silicon tracking detector (FMT) has been proposed to provide additional hits for track reconstruction in the forward region at the EIC to improve the overall track reconstruction quality. The fast timing resolution of the MALTA2 technology will help reject background events at the EIC as well. Progress of latest MALTA2 R&D, the development of a new MALTA2 quad-sensor prototype module and impacts of the proposed FMT in EIC physics studies will be discussed.

1 Introduction

The future Electron-Ion Collider (EIC) is expected to start construction at Brookhaven National Laboratory in 2025. It will support two interaction points (IP6 and IP8), where the detector located at IP6 is referred to as the EIC project detector and the one at IP8 is named as detector II. A series of electron+proton ($e + p$) and electron+nucleus ($e + A$) collisions (A=2-238) at the center of mass energies of 20-141 GeV will be operated at the EIC [1]. The bunch crossing period at the EIC is around 10.2 ns with a 25 mrad beam crossing angle is introduced to IP6 to reduce impacts from beam backgrounds. The current technical design of the EIC project detector at IP6 led by the ePIC collaboration utilizes the 65 nm Monolithic Active Pixel Sensor (MAPS) [2], the Micro Pattern Gaseous Detector (MPGD) [3] and the AC-coupled Low Gain Avalanche Diode (AC-LGAD) [4] to form its vertex and tracking subsystem. In addition to ongoing EIC detector development, a Depleted MAPS (i.e., MALTA2 [5]) based Fast MAPS Tracker (FMT) had been proposed as a new subsystem for either the ePIC detector upgrade or the EIC detector II. The design and R&D for the proposed FMT is underway supported by the EIC generic R&D program.

*e-mail: xuanli@lanl.gov

© The Authors, published by EDP Sciences. This is an open access article distributed under the terms of the Creative Commons Attribution License 4.0 (https://creativecommons.org/licenses/by/4.0/).

2 MALTA2 R&D progress

The MALTA2 sensor fabricated with the Tower 180 nm CMOS technology consists of a 224 by 512 pixel array on a 1 cm by 2 cm active area (see the left panel of figure 1) [5]. The average digital power consumption of the MALTA2 sensor is around 10 mW/cm^2 and its analog power consumption can reach around 70 mW/cm^2. Beam tests have been performed at different facilities to characterize the MALTA2 sensor and the right panel of Figure 1 shows the test setup for a MALTA2 sensor placed in the middle of a 6-layer MALTA telescope at CERN SPS. Test results of MALTA2 sensor with 100 μm thickness [6] are shown in Figure 2. The average hit spatial resolution is around 4.1 μm and the timing resolution is around 2.1 ns for individual MALTA2 sensors [6]. Minimal performance loss has been found on MALTA2 sensors at the irradiation dose around 10^{15} 1MeV n_{eq}/cm^2 [7], which is significantly higher than the expected irradiation level at the EIC.

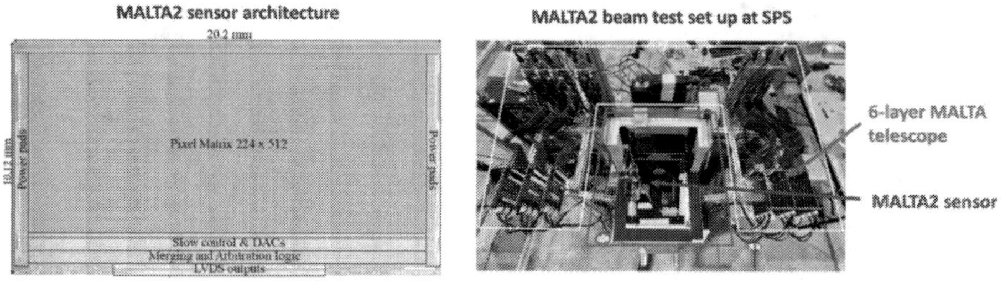

Figure 1. The architecture of the MALTA2 sensor (left) and the CERN SPS beam test setup to characterize the MALTA2 sensor (right).

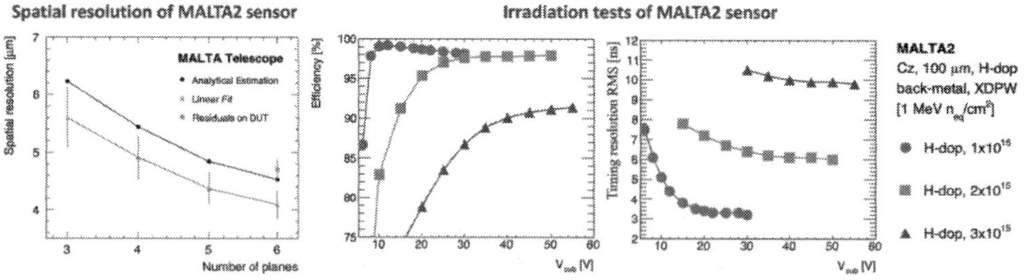

Figure 2. Spatial resolution of the MALTA2 sensor from the CERN SPS beam tests [6] (left). The bias voltage dependent efficiency (middle) and timing RMS (right) of the MALTA2 sensor with the irradiation dose at 1-3 x 10^{15} 1MeV n_{eq}/cm^2 and 100 μm wafer thickness [7].

3 The proposed Fast MAPS Tracker (FMT) design

The MALTA2 based Fast MAPS Tracker (FMT) is proposed to consist of 2 disks in the forward region ($\eta > 2.5$) near the edge of the EIC central magnet. Another candidate location is to place the FMT disks in the backward region ($\eta < -2.5$). As a starting point, the EIC IP6 beam pipe and three 65 nm MAPS (the ePIC default technology) disks with the pseudorapidity coverage up to 2.5 have been used to develop the FMT design. Several optimization iterations have been applied for the FMT geometry and the left panel of Figure 3 presents the

© The Authors, published by EDP Sciences. This is an open access article distributed under the terms of the Creative Commons Attribution License 4.0 (https://creativecommons.org/licenses/by/4.0/).

Table 1. The geometry of the proposed FMT in the forward pseudorapidity region

Disk index	Inner radius	Outer radius	z location	Material budget	Hit efficiency
1	7.014 cm	23.095 cm	145 cm	$0.74\%X/X_0$	98%
2	7.014 cm	23.095 cm	165 cm	$0.74\%X/X_0$	98%

optimized FMT geometry in the forward region and the corresponding geometry parameters are summarized in table 1. The backward geometry optimization will be performed later on.

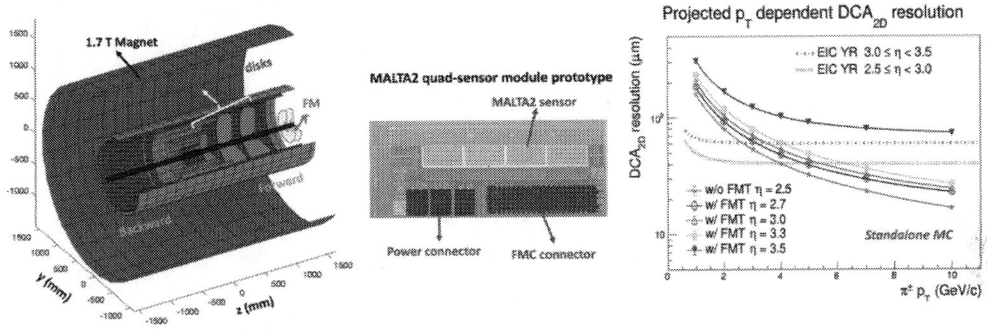

Figure 3. The proposed FMT geometry implemented in fast simulation (left). The first design of the MALTA2 quad-sensor prototype module (middle). The transverse momentum dependent track DCA_{2d} resolution in different pseudorapidity regions with FMT evaluated in standalone simulation (right).

The proposed FMT will utilize either groups of MALTA2 quad-sensor staves/modules or MALTA2 octuple-sensor staves/modules to build the disk. The middle panel of Figure 3 shows the first prototype design of the MALTA2 quad-sensor module carrier board. Production of the MALTA2 quad-sensor is underway and performance characterization has been planned using lab bench and beam tests at CERN. Further development of the MALTA2 octuple-sensor module is underway. Tracking performance such as spatial and momentum resolutions with the proposed FMT geometry (see Table 1) has been evaluated in standalone single particle simulation. The right panel of Figure 3 shows the track p_T dependent transverse Distance of Closest Approach (DCA_{2D}) resolution with and without the FMT in comparison with the EIC yellow report detector requirement [1]. The proposed FMT can extend the track reconstruction of the 3 MAPS disks to the forward pseudorapidity region of $2.5 < \eta < 3.3$, and meet the EIC detector requirements for tracks with $p_T > 2$ GeV/c.

4 Physics impacts of the proposed FMT

Fast timing is another advanced feature of the proposed FMT. Recent beam tests have demonstrated that the MALTA2 sensor can accomplish data acquisition within around 2 ns timing window [6]. The estimated collision rate at the EIC is around 500 kHz, which is equivalent to a 2 μs data collection window. The FMT will be able to carry out physics measurements for individual collisions at the EIC, which can significantly reject pileup events. A PYTHIA8 [9] based simulation framework has been setup to evaluate the impact on forward heavy flavor hadron reconstruction using different readout time windows. Figure 4 compares the invariant mass of reconstructed D^0 ($\overline{D^0}$) ($D^0/\overline{D^0} \to K^{\mp}\pi^{\pm}$) with the projected FMT spatial resolutions in the $2.5 < \eta \leq 3.5$ region using two different readout timing windows: 2 ns (left) and 10 μs

© The Authors, published by EDP Sciences. This is an open access article distributed under the terms of the Creative Commons Attribution License 4.0 (https://creativecommons.org/licenses/by/4.0/).

(right). The proposed FMT with around 2 ns readout can significantly reduce the combinatorial background contributions to the forward heavy flavor reconstruction at the EIC, which has a great sensitivity in exploring the heavy quark hadronization process under different medium conditions [8].

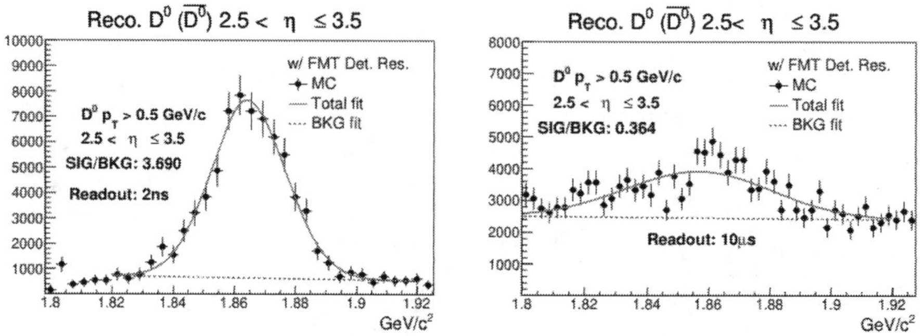

Figure 4. Invariant mass spectrum of reconstructed D^0 $(\overline{D^0})$ in the pseudorapidity regions of $2.5 < \eta \leq 3.5$ with 2 ns readout window (left) and with 10 μs readout window (right).

5 Summary and Outlook

Latest bench and beam tests have demonstrated good spatial and timing resolutions for MALTA2 sensors. As evaluated in recent simulation studies, the proposed MALTA2 based FMT will help reduce pileup and combinatorial background contributions for forward heavy flavor reconstruction at the EIC. Further ongoing R&D for the FMT, which includes design and production of a MALTA2 quad-sensor prototype module, will provide more technical details to validate the overall detector performance.

References

[1] R. Abdul Khalek et al., *Science Requirements and Detector Concepts for the Electron-Ion Collider: EIC Yellow Report*, Nucl. Phys. A 1026 (2022) 122447.

[2] ALICE collaboration, *Letter of intent for an ALICE ITS upgrade in LS3*, CERN-LHCC-2019-018, LHCC-I-034 (2019).

[3] L. Shekhtman *Micro-pattern gaseous detectors*, Nuclear Instruments and Methods in Physics Research Section A: Accelerators, Spectrometers, Detectors and Associated Equipment 494 (2002) 1-3.

[4] G. Giacomini et al., *Fabrication and performance of AC-coupled LGADs*, JINST **14** (2019) P09004.

[5] F. Piro et al., *A 1-µW Radiation-Hard Front-End in a 0.18-µm CMOS Process for the MALTA2 Monolithic Sensor* IEEE Trans. Nucl. Sci., vol. 69, no. 6, pp. 1299-1309, 2022.

[6] M. van Rijnback et al., *Performance of the MALTA telescope*, Eur. Phys. J. C **83** (2023) 581.

[7] M. van Rijnback et al., *Radiation hardness of MALTA2 monolithic CMOS imaging sensors on Czochralski substrates*, Eur. Phys. J. C **84** (2024) 251.

[8] X. Li *Exploration of hadronization through heavy flavor production at the future Electron-Ion Collider*, EPJ Web of Conferences 296, 16001 (2024).

[9] T. Sjöstrand et al., *An introduction to PYTHIA 8.2*, Comput. Phys. Commun. 191 (2015) 159-177.

© The Authors, published by EDP Sciences. This is an open access article distributed under the terms of the Creative Commons Attribution License 4.0 (https://creativecommons.org/licenses/by/4.0/).

Multiplicity and Rapidity Dependent Study of (Multi)-strange Hadrons in d+Au collisions using the STAR detector

Ishu Aggarwal (For the STAR Collaboration)[1]*

[1]Department of Physics, Panjab University, Chandigarh - India

Abstract. Strangeness enhancement has long been considered as a signature of the quark-gluon plasma formation in heavy-ion collisions. Strangeness enhancement has also been observed in small systems at the Large Hadron Collider (LHC), but the underlying physics still needs to be fully understood. This motivates us to study the strange hadron production in small systems at RHIC. We present new measurements of (multi-)strange hadrons (K^0_S, Λ, Ξ and Ω) in d+Au collisions at $\sqrt{s_{NN}}$ = 200 GeV, collected by STAR in 2016. We report the multiplicity and rapidity dependence of strange hadron transverse momentum (p_T) spectra, p_T-integrated yields dN/dy, average transverse momentum ($\langle p_T \rangle$), yield ratios of these strange hadrons to pions, nuclear modification factors and rapidity asymmetry for these particles in d+Au collisions.

1 Introduction

The production of strange hadrons in high-energy hadronic interactions provides a way to investigate the properties of quantum chromodynamics (QCD), the theory of strongly interacting matter. Unlike up (u) and down (d) quarks, strange (s) quarks are not present as valence quarks in the initial state and are created during the collisions. Strangeness enhancement in heavy ion collisions with respect to proton-proton (p+p) collisions has been suggested as a signature of quark-gluon plasma (QGP) formation [1]. However, the creation of QGP in small systems is still under intense debate.

Asymmetric small collision systems like proton-nucleus (p–A) and deuteron-nucleus (d–A) can be considered as control experiments where the formation of an extended QGP phase is not expected. These collision systems are used for baseline measurements to study the possible effects of cold nuclear matter and to disentangle them from hot dense matter effects signifying the QGP formation in heavy-ion collisions. Cold nuclear matter effects can be evident through Cronin enhancement at intermediate p_T due to the transverse momentum broadening. The process of generating hadrons can be affected by various factors that include modifications in parton distribution functions within nuclei, the possibility of parton saturation, multiple scatterings of the partons traversing the nucleus, and the radial flow. It is anticipated that the magnitude of these effects may vary with the rapidity of the produced particles.

*e-mail: iaggarwal@rcf.rhic.bnl.gov

© The Authors, published by EDP Sciences. This is an open access article distributed under the terms of the Creative Commons Attribution License 4.0 (https://creativecommons.org/licenses/by/4.0/).

The abundances of strange particles relative to pions in heavy-ion collisions for top RHIC and LHC energies do not show a significant dependence on the initial volume (collision centrality) [2].

Nuclear effects can be quantified using variables such as nuclear modification factor and rapidity asymmetry. The nuclear modification factor is defined as the ratio of the yield of a particle in heavy-ion collisions to its yield in *p*+*p* collisions, scaled by the number of binary nucleon-nucleon inelastic collisions [3].

$$R_{AB}(p_T) = \frac{Yield_{AB}}{\langle N_{bin} \rangle Yield_{pp}} \quad (1)$$

where $\langle N_{bin} \rangle$ is the average number of binary nucleon-nucleon collisions. A comparative study of particle production in forward and backward rapidity regions uses rapidity asymmetry (Y_{Asym}) [4]. Y_{Asym} is defined as

$$Y_{Asym}(p_T) = \frac{Y_B(p_T)}{Y_F(p_T)} \quad (2)$$

where Y_B and Y_F are particle yields in backward and forward rapidity regions, respectively.

2 Analysis Details

A successful run of *d*+Au collisions at $\sqrt{s_{NN}}$ = 200 GeV was carried out in 2016 at RHIC. A total of approximately 100 million good events have been selected in STAR for the reconstruction of K^0_s, $\Lambda(\bar{\Lambda})$, $\Xi(\bar{\Xi})$, $\Omega(\bar{\Omega})$. K^0_s, $\Lambda(\bar{\Lambda})$, $\Xi(\bar{\Xi})$, $\Omega(\bar{\Omega})$ are weakly decaying particles, and they travel a certain distance before decaying into daughter particles ($K^0_S \to \pi^+ + \pi^-$, $\Lambda(\bar{\Lambda}) \to p(\bar{p}) + \pi^-(\pi^+)$, $\Xi^-(\bar{\Xi}^+) \to \Lambda(\bar{\Lambda}) + \pi^-(\pi^+)$, $\Omega^-(\bar{\Omega}^+) \to \Lambda(\bar{\Lambda}) + K^-(K^+)$). The daughter particles (π, K and p) are identified via measuring $\langle dE/dx \rangle$ using Time Projection Chamber (TPC).

The K^0_s, $\Lambda(\bar{\Lambda})$, $\Xi(\bar{\Xi})$, $\Omega(\bar{\Omega})$ signals are extracted by reconstructing the invariant mass of the decay daughter pairs. The decay topology can be used to reconstruct and suppress the background. We have used double Gaussian and second order polynomial function to describe the signal and background invariant mass distributions respectively. Raw yield is estimated using functional fitting and bin counting method (which is considered in systematic uncertainty). Raw yield is determined within the mass window of $M_0 \pm 3\sigma$, where M_0 is mass of K^0_S (or $\Lambda/\Xi/\Omega$) and σ is the fitted width. Raw yield for each p_T interval is corrected for branching ratio, acceptence and efficiency to obtain corrected p_T spectra. Weak decay feed down correction from Ξ is applied to Λ.

3 Results and Discussions

Fig. 1 (left) shows the integrated yield particle-to-pion ratio for K^0_S, Λ, Ξ and Ω as a function of multiplicity for *d*+Au collisions at $\sqrt{s_{NN}}$ = 200 GeV. A smooth transition of these ratios from *p*+*p* to heavy ion collisions is observed. Data from different collision systems are consistent with each other at similar multiplicities [2] and yields of particles with more strangeness content decrease faster as we move from high to low multiplicities.

In Fig. 1 (right), nuclear modification factors for K^0_S, Λ and Ξ at mid rapidity ($|y| < 0.5$) for *d*+Au collision is presented, and Cronin-like enhancement is observed, which is stronger for baryons (Ξ, Λ, p) as compared to mesons (K^0_S, π).

In Fig. 2 (left), integrated yield and $\langle p_T \rangle$ as function of multiplicity for mid-rapidity in *p*+*p*, *d*+Au, Cu+Cu and Au+Au collisions at $\sqrt{s_{NN}}$ = 200 GeV are shown. Integrated yield increases as a function of multiplicity, and we have also observed that particle production is

© The Authors, published by EDP Sciences. This is an open access article distributed under the terms of the Creative Commons Attribution License 4.0 (https://creativecommons.org/licenses/by/4.0/).

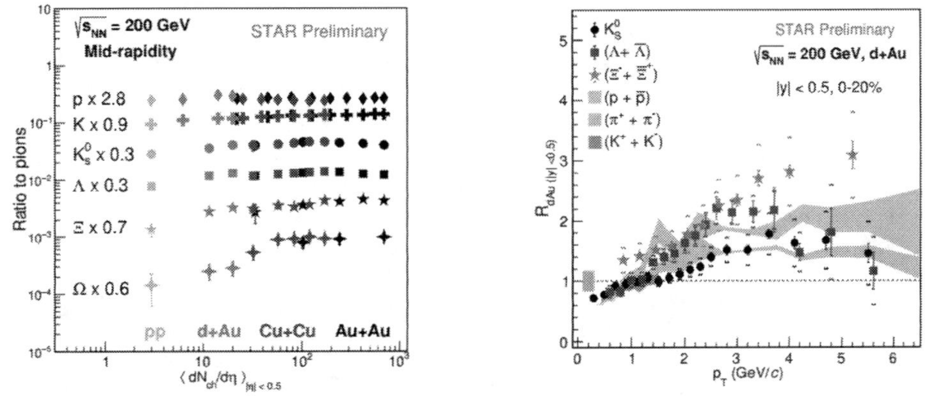

Figure 1: Integrated yield particle-to-pion ratio (left) for strange particles (K^0_S, Λ, Ξ and Ω) as a function of multiplicity for *p*+*p*, *d*+Au, Cu+Cu and Au+Au collisions and Nuclear modification factors (R_{dAu}) (right) for strange particles (K^0_S, Λ and Ξ) in *d*+Au collisions at $\sqrt{s_{NN}}$ = 200 GeV.

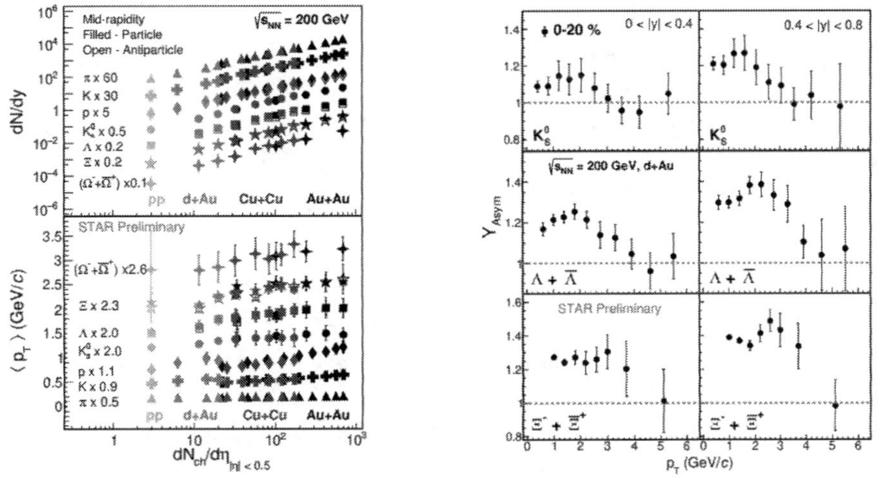

Figure 2: Integrated yield (dN/dy) and mean transverse momentum $\langle p_T \rangle$ (left) for strange particles (K^0_S, Λ, Ξ and Ω) as a function of multiplicity for *p*+*p*, *d*+Au, Cu+Cu and Au+Au collisions and rapidity asymmetry (Y_{Asym}) as a function of p_T (right) for strange particles (K^0_S, Λ and Ξ) in *d*+Au collisions at $\sqrt{s_{NN}}$ = 200 GeV.

mainly driven by multiplicity and not by collision system. A hint of an increase in $\langle p_T \rangle$ is observed for heavier particles, which supports the picture of radial flow.

We have also measured the transverse momentum dependence of Y_{Asym} for K^0_s, Λ and Ξ in two rapidity intervals (0 < |y| < 0.4, 0.4 < |y| < 0.8) in *d*+Au collision at $\sqrt{s_{NN}}$ = 200 GeV. Y_{Asym} values deviate from unity at low p_T (< 3 GeV/c), suggesting the presence of a rapidity dependence in the nuclear effects. Y_{Asym} is consistent with unity at high p_T, hinting that nuclear effects become weaker at high p_T as shown in Fig. 2 (right). Asymmetry is more prominent for larger rapidity interval and heavier particles.

© The Authors, published by EDP Sciences. This is an open access article distributed under the terms of the Creative Commons Attribution License 4.0 (https://creativecommons.org/licenses/by/4.0/).

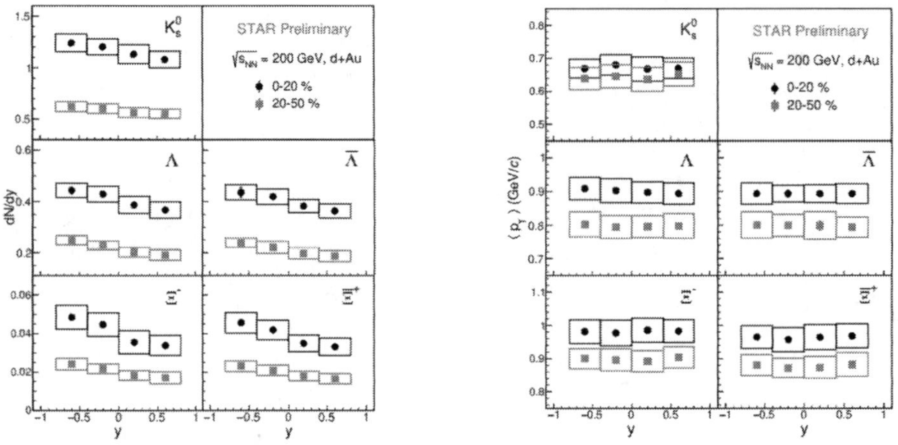

Figure 3: dN/dy (left) and ⟨p_T⟩ (Right) of strange particles (K^0_s, $\Lambda(\bar{\Lambda})$ and $\Xi(\bar{\Xi})$) as a function of rapidity in d+Au collisions at $\sqrt{s_{NN}}$ = 200 GeV for multiplicity classes 0-20% and 20-50%.

In Fig. 3, integrated yield (dN/dy) and mean transverse momentum (⟨p_T⟩) are calculated for different rapiditiy intervals ($0 < |y| < 0.4$, $0.4 < |y| < 0.8$). For strange particles K^0_s, $\Lambda(\bar{\Lambda})$ and $\Xi(\bar{\Xi})$, integrated yield (dN/dy) decreases slightly from negative (Au going side) to positive (d going side) rapidity but mean transverse momentum (⟨p_T⟩) is constant in the measured rapidity range which suggests a similar radial flow for rapidity ranging from $-0.8 < y < 0.8$.

4 Summary

We present new measurements of multiplicity and rapidity dependence of (multi-)strange hadrons (K^0_s, $\Lambda(\bar{\Lambda})$, $\Xi(\bar{\Xi})$ and $\Omega(\bar{\Omega})$) production in d+Au collisions at $\sqrt{s_{NN}}$ = 200 GeV with the STAR experiment. p_T-integrated yields of strange and multi-strange particles relative to pions increase significantly with multiplicity. Results suggest that strange particle production is independent of the collision system and mainly driven by multiplicity. The observed enhancement increases with strangeness content rather than with mass or baryon number of the hadron. Hint of Cronin-like enhancement is observed for 0-20% centrality for K^0_S, Λ and Ξ. $Y_{Asym} > 1$ at low p_T, indicating the presence of the nuclear effects. The Y_{Asym} is more prominent for heavier mass and larger rapidity interval.

References

[1] G. Agakishiev et al. (STAR Collaboration) Phys. Rev. Lett. **108**, 072301 (2012).
[2] J. Adam et al. (ALICE Collaboration) Nature Phys **13**, 535–539 (2017).
[3] J. Adams et al. (STAR Collaboration) Phys. Lett. B **637**, 161-169 (2006).
[4] B. I. Abelev et al. (STAR Collaboration) Phys. Rev. C **76**, 054903 (2007).
[5] B. I. Abelev et al. (STAR Collaboration) Phys. Rev. C **75**, 064901 (2007).
[6] B. I. Abelev et al. (STAR Collaboration) Phys. Rev. C **79**, 034909 (2009).
[7] B. I. Abelev et al. (STAR Collaboration) Phys.Lett.B **616**, 8-16 (2008).

© The Authors, published by EDP Sciences. This is an open access article distributed under the terms of the Creative Commons Attribution License 4.0 (https://creativecommons.org/licenses/by/4.0/).

[8] M. M. Aggarwal et al. (STAR Collaboration) Phys. Rev. C **83**, 034910 (2011).

© The Authors, published by EDP Sciences. This is an open access article distributed under the terms of the Creative Commons Attribution License 4.0 (https://creativecommons.org/licenses/by/4.0/).

Signals of initial state quantum entanglement in relativistic particle collision

Rene Bellwied[1,*] and *Alek* Hutson[1]

[1]Physics Department, University of Houston, Houston TX 77204

Abstract. We present thermodynamic entropy calculations based on charged particle multiplicity data from proton-proton collisions measured by ALICE at the LHC in comparison to entanglement entropy calculations based on initial state parton distributions. The relative agreement of these distributions can be quantified by studying their higher order cumulants. The commonalities between the initial and final state entropy suggest that entanglement could be a possible source for the seemingly thermal and collective behavior in small systems.

1 Introduction

The deconfined phase of strongly interacting matter, generated in particle collisions at the RHIC and LHC, can be described by a combination of hydrodynamical and statistical models that trace the evolution of collisions from an initially thermalized system, even for elementary particle collisions [1], to the final state. While these models have successfully predicted particle yields with statistically significant accuracy, they rely on the assumption that the system reaches partial thermal equilibrium at very early times (less than 1 fm/c), which is highly improbable through particle collisions, such as e^+e^- or pp. Furthermore, this type of phenomenological approach fails to capture all the quantum dynamics of the initial system. Standard MC event generators, such as PYTHIA, describe an elementary proton-proton collision using single parton-parton interactions, thus ignoring the impact that the non-interacting partons, described by the proton wave function, might have on the reduced density matrix that captures the number of available final states. These deficiencies in the current models have led physicists to take new approaches relying on fundamental quantum principles to give a more accurate and complete description [2–4]. These approaches are supported by empirical evidence of quantum thermalization through entanglement found initially in condensed matter experiments at ultra-cold temperatures [5], but more recently also at higher temperatures and over finite volume [6].

2 Initial State Entropy

In a quantum system the degree of entanglement, or the information shared between a subsystem and the remainder of the system, can be quantified using the von Neumann entropy, defined for a subsystem A as:

$$S(A) = -\mathrm{Tr}(\rho_A \ln(\rho_A)), \tag{1}$$

*e-mail: bellwied@uh.edu

© The Authors, published by EDP Sciences. This is an open access article distributed under the terms of the Creative Commons Attribution License 4.0 (https://creativecommons.org/licenses/by/4.0/).

where ρ_A is the reduced density matrix of subsystem A. For a pure state, described by a single wavefunction, the von Neumann entropy of the complementary subsystem (the remainder of the system outside A) is equal in magnitude and opposite in sign, resulting in a total entropy of zero for the composite system. This entropy also known as entanglement entropy quantifies the level of disorder in the system and provides a quantum analog to thermodynamic entropy.

In proton-proton collisions, entropy in the initially pure system originates from a break in entanglement between the interacting parton and the remaining partonic spectator region. By tracing over the degrees of freedom in the reduced density matrix of the system, one can calculate the entanglement entropy between different regions. This calculation can be challenging due to the complex nature of the density matrix, however, in the limit of small momentum fractions ($x < 10^{-3}$), where the proton is described as a highly dense system of indistinguishable gluons, the relevant degrees of freedom can be effectively reduced to the parton number. In this scenario, the density matrix simplifies to:

$$\rho_A \approx \frac{1}{N} I_{N\times N} \tag{2}$$

where N represents the number of partons, and $I_{N\times N}$ represents the identity matrix in N×N dimensions. Consequently, by restricting the kinematics of our collision system to low x, we can simplify the calculation of the entanglement entropy, which reduces to:

$$S(A) \approx \ln(N) \tag{3}$$

Here, N is determined by integrating over the known parton probability distribution, with integration limits based on the spatial region measured in the final state. PDF's are provided in terms of the parton momentum fraction (or Bjorken-x) and momentum transfer Q^2. Our present analysis is limited to the central rapidity region covered by the Time Projection Chamber (TPC) in ALICE, which limits our x-range to values around 10^{-4}, corresponding to Q^2 values around unity. It is important to note here that in this low Q^2 regime higher order expansions of the QCD coupling constant give unreliable results for gluon distributions. Therefore, we only consider leading order (LO) parton distribution functions (PDFs)[7].

Prior to the collision we have a Lorentz contracted proton moving near the speed of light. The partons exist in a complete coherent state (fully entangled) which contains zero von-Neumann entropy. At the onset of the collision there is a break of entanglement which is encoded in the reduced density matrix that defines all quantum states of constituent partons within the overlap region. After the initial quench of the wave function, the combined system of interacting partons interacts through flux tube or string generation along the beam axis, thereby generating an extensive volume [8].

We can begin to calculate an entropy in the low-x limit based on the number of interacting gluons calculated by summing over PDF's, although the need to extend this formalism to sea quark contributions at low Q^2 and moderate x was shown by Hentschinski et al. [9].

The authors also added a factor of 2/3 to account for unmeasured neutral states in the final distribution. The entropy equation then becomes:

$$S_{EE} = ln(2/3(N_{gluons} + N_{quarks})) = ln(N_{gluons} + N_{quarks}) + ln(2/3) \tag{4}$$

Finally, the entanglement entropy calculated this way is only taking into account the diagonal elements of the reduced density matrix and not the non-diagonal terms of the infinite density matrix. In the literature the inclusive calculation is known as the entropy of ignorance. Using an initial CGC model, theorists have been able to determine the ratio between this entropy of ignorance and the entanglement entropy [10]. In the calculation it is observed that at low Q^2 this ratio is quite large with an upper limit of about 1.4. For the mid-rapidity

© The Authors, published by EDP Sciences. This is an open access article distributed under the terms of the Creative Commons Attribution License 4.0 (https://creativecommons.org/licenses/by/4.0/).

ALICE measurements the ratio reaches around 1.24 for $\sqrt{s}$ = 0.9 TeV [10]. We apply this correction factor to the measured initial state entropy.

3 Final State Entropy

Since the fractional momentum and momentum transfer scale are not directly measurable in proton-proton collisions, an alternative way of comparing entropies has to be employed. In the final-state we can calculate a thermodynamic entropy using the probability distribution of produced particles over any range in pseudo-rapidity as long as we can define a mapping between η in the final state and the fractional momentum x in the initial state. Within the CGC model a common approximation is given by $\ln(1/x) = y_{\text{initial proton}} - y_{\text{hadron}}$.

This relation might break down at large x [11], but we are mostly interested in the low x, high gluon density region. ALICE has published mid-rapidity charged particle multiplicity data for four collision energies at $\sqrt{s}$ = 0.9, 2.76, 7, and 8 TeV [12]. As examples, the measured multiplicity distributions for 0.9 and 8 TeV are shown in Fig.1, and are well described by a Negative Binomial Distribution (NBD). Since both colliding protons contribute equally to the final state, the distribution is halved to obtain a meaningful comparison to single proton PDFs [4] and calculate the Shannon entropy of a single proton.

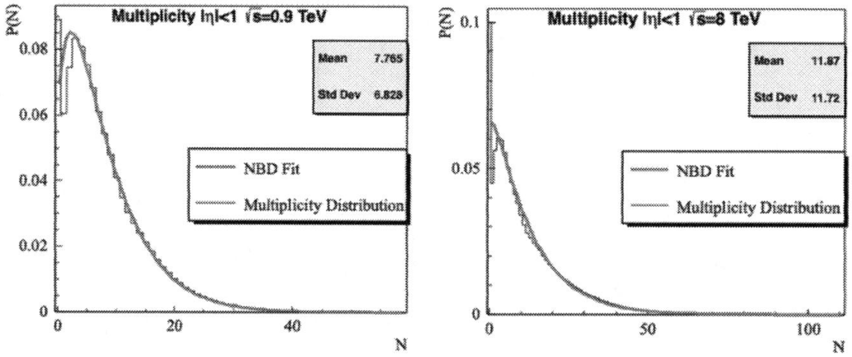

Figure 1. Measured hadron multiplicity distributions from ALICE data [12]

4 Results and Conclusions

Higher moments of the NBD distribution are used to compare a 1+1 toy model of non-linear QCD evolution of the BK equation as suggested by Kharzeev and Levin [2]. They demonstrated that one can construct a generating function that captures the non-linear interactions leading to dipole formation, in an entangled system, which sets an upper limit on the moments of the final state NBD distribution. We observe that the final state multiplicity distribution approaches the limit expected from a fully entangled state, see Fig.2.

Fig.3a shows final state entropy shown as red points compared with the initial-state entropy shown as bands, corrected for additional quark contributions, neutral particle contributions and the aforementioned entropy of ignorance factor. Entropies from two different corrected leading order parton distribution functions at the relevant x, Q^2 and α_s values are shown.

The final particle production can also be modeled by the simple fragmentation of strings, but as expected, even the most recent PYTHIA tunes underestimate the final state entropy

© The Authors, published by EDP Sciences. This is an open access article distributed under the terms of the Creative Commons Attribution License 4.0 (https://creativecommons.org/licenses/by/4.0/).

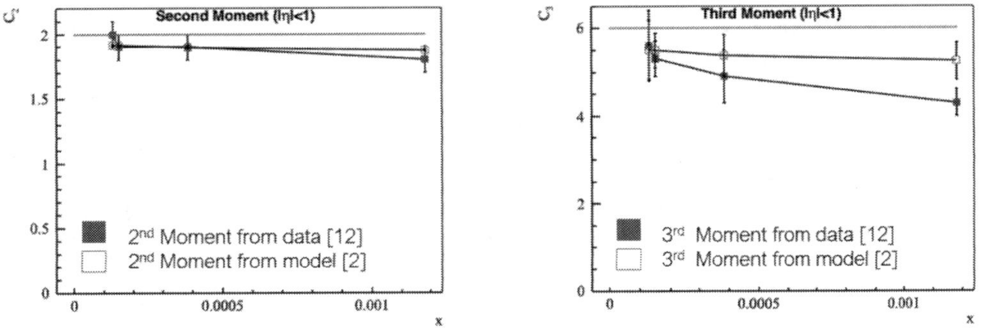

Figure 2. Comparison of theoretical moments from 1+1 toy model [2] final state moments of the multiplicity distribution [12]. Red lines represent the upper bound on entangled cumulants in the limit of $\overline{n}$ approaching infinity.

significantly when not taking into account some type of quantum effects. Lately these effects have been approximated via two distinct PYTHIA parameters and modes, namely the level of multi-parton interactions (MPI) and the color reconnection mode (CR). We show that these modes indeed help to bring the simple string fragmentation multiplicities to a level similar to the entangled initial state. Fig.3b shows the comparison to various PYTHIA modes.

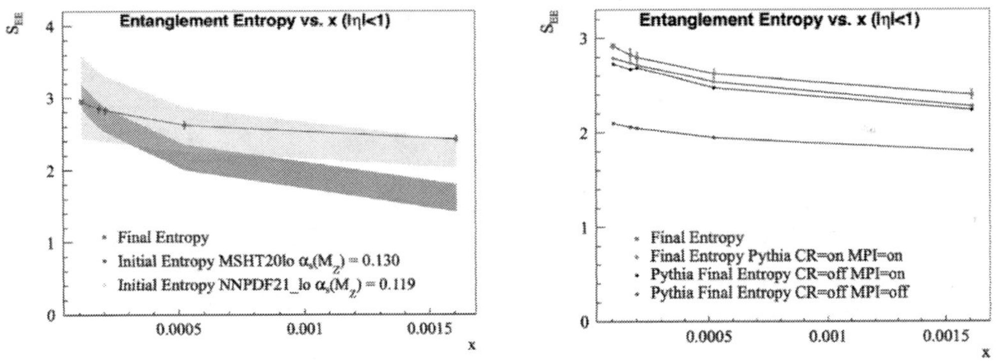

Figure 3. Left: Final state entropy including all corrections in red, and the initial state entropy shown by colored bands for two models (NNPDF [7] and MHST [13]), Right: Final state entropy calculated from data to PYTHIA using three different tunes.

We have shown that the entropies calculated from hadron multiplicity in the final state and parton multiplicity in the initial state approach one another at low-x. We observe better agreement between the initial state and final state for a lower strong coupling constant of 0.119. We have not yet seen any effect of gluon saturation in the covered x-range, but future measurements at forward rapidity at the LHC and over the full phase space at the EIC, will be highly relevant, since it is predicted that the entropy generation should follow the gluon saturation curve at very low x.

Overall our measurements give support to the theory that the quantum entangled state in the initial phase of a relativistic hadron collision could maintain its coherence throughout the early evolution and lead to a seemingly thermal final state, where the observed particle multiplicities are governed by the partition function of the initial parton states.

© The Authors, published by EDP Sciences. This is an open access article distributed under the terms of the Creative Commons Attribution License 4.0 (https://creativecommons.org/licenses/by/4.0/).

Acknowledgments: This material is based upon work supported by the U.S. Department of Energy, Office of Science, under Award Number DE-FG02-07ER41521.

References

[1] F. Becattini, P. Castorina, A. Milov, H. Satz, Eur. Phys. J. C **66**, 377 (2010).
[2] D.E. Kharzeev, E.M. Levin, Phys. Rev. D **95**, 114008 (2017).
[3] R. Bellwied, J. Phys. Conf. Ser. **1070**, 012001 (2018).
[4] Z. Tu, D.E. Kharzeev, T. Ullrich, Phys. Rev. Lett. **124**, 062001 (2020).
[5] A.M. Kaufman, M.E. Tai, A. Lukin, M. Rispoli, R. Schittko, P.M. Preiss, M. Greiner, Science **353** (2016).
[6] J. Kong, R. Jiménez-Martínez, C. Troullinou, V.G. Lucivero, G. Tóth, M.W. Michell, Nat Commun **11** (2020).
[7] R.D. Ball, V. Bertone, F. Cerutti, L. Del Debbio, S. Forte, A. Guffanti, J.I. Latorre, J. Rojo, M. Ubiali (NNPDF), Nucl. Phys. B **855**, 153 (2012).
[8] J. Berges, S. Floerchinger, R. Venugopalan, JHEP **04**, 145 (2018).
[9] M. Hentschinski, K. Kutak, R. Straka, Eur. Phys. J. C **82**, 1147 (2022).
[10] H. Duan, C. Akkaya, A. Kovner, V.V. Skokov, Phys. Rev. D **101**, 036017 (2020).
[11] J. Jalilian-Marian, Nucl. Phys. A **1005**, 121943 (2021).
[12] J. Adam et al. (ALICE), Eur. Phys. J. C **77**, 33 (2017).
[13] S. Bailey, T. Cridge, L.A. Harland-Lang, A.D. Martin, R.S. Thorne, Eur. Phys. J. C **81**, 341 (2021).

© The Authors, published by EDP Sciences. This is an open access article distributed under the terms of the Creative Commons Attribution License 4.0 (https://creativecommons.org/licenses/by/4.0/).

Probing Hydrodynamics in PbPb Collisions at 5.02 TeV with Higher-order Cumulants

Aryaa Dattamunsi[1] for the CMS Collaboration

[1]Indian Institute of Technology Madras, Chennai, India

Abstract. The elliptic flow of charged hadrons is studied through multiparticle correlations in PbPb collisions at $\sqrt{s_{NN}}$ = 5.02 TeV, corresponding to an integrated luminosity of 0.607 nb^{-1}. $v_2\{2k\}$ are measured up to the tenth order (k = 5) as functions of the collision centrality. They are shown to follow the ordering : $v_2\{2\} > v_2\{4\} \gtrsim v_2\{6\} \gtrsim v_2\{8\} \gtrsim v_2\{10\}$. The centrality-dependent moments - skewness, kurtosis and, for the first time, superskewness, are measured for the fluctuation-driven event-by-event v_2 distribution. These moments can be good probes of the initial-state geometry in high-energy nucleus-nucleus collisions, assuming a hydrodynamic expansion.

1 Introduction

The quark-gluon plasma (QGP) is a strongly interacting state of highly energetic and dense matter which has been achieved in experiments [1–3] by colliding heavy nuclei at ultra-relativistic speeds. The QGP is known to undergo collective hydrodynamic expansion. The initial spatial anisotropy in the overlap region of the colliding nuclei leads to anistropic pressure gradients, which in turn leads to the azimuthally anisotropic flow of outgoing particles. The distribution of the anisotropic flow of particles can be expressed in terms of the Fourier series [4, 5] :

$$P(\phi) = \frac{1}{2\pi}\left[1 + 2\sum_{n=1}^{\infty} v_n \cos n(\phi - \psi_n)\right] \tag{1}$$

where ϕ is the particle azimuthal angle, ψ_n is the nth order flow symmetry plane, and v_n are the nth order Fourier coefficients called the flow harmonics. v_2 refers to elliptic flow, and is the leading term in this Fourier expansion.

Event-by-event fluctuations of the v_2 distribution lead to non-zero values of its higher-order moments, like skewness, kurtosis and superskewness. Skewness refers to the overall asymmetry of the distribution, kurtosis measures the peakedness or flatness, while superskewness quantifies the asymmetry of the tails. Assuming hydrodynamic expansion, initial-state eccentricity (ϵ_2) fluctuations directly lead to non-Gaussianities in the v_2 distribution, giving us insight into the initial-state geometry and structure. These have been measured using multi-particle cumulants [6].

In this paper [7], v_2 from 10 particle correlations, $v_2\{10\}$, has been measured for the first time in PbPb collisions. Two hydrodynamic probes - one introduced in Ref.[8] and another

© The Authors, published by EDP Sciences. This is an open access article distributed under the terms of the Creative Commons Attribution License 4.0 (https://creativecommons.org/licenses/by/4.0/).

one using $v_2\{10\}$ and all lower-order $v_2\{2k\}$, have been derived using these higher-order moments. The higher-order moments have been defined in terms of $v_2\{2k\}$. Introduction of these higher-order moments have been shown to describe the centrality dependence of the hydrodynamic probes very accurately.

2 Analysis Technique

In order to calculate $v_2\{2k\}$ and the two hydrodynamic probes, the Q-cumulant method [9] has been used. With this method, any arbitrary $2k^{th}$ order cumulant can be constructed from the recursion relation [10] :

$$c_n\{2k\} = \langle\langle 2k \rangle\rangle - \sum_{m=1}^{k-1} \binom{k}{m}\binom{k-1}{m} \langle\langle 2m \rangle\rangle c_n\{2k-2m\} \tag{2}$$

From Eq. (2), we can get the $v_n\{2k\}$ values as :

$$v_n\{2k\} = \sqrt[2k]{a_{2k}^{-1} c_n\{2k\}} \tag{3}$$

where

$$a_{2k} = 1 - \sum_{m=1}^{k-1} \binom{k}{m}\binom{k-1}{m} a_{2k-2m} \tag{4}$$

Writing the higher-order moments in terms of $v_2\{2k\}$, the two hydrodynamic probes have been defined as :

$$h_1 = \frac{v_2\{6\} - v_2\{8\}}{v_2\{4\} - v_2\{6\}} \approx h_1^{Taylor} = \frac{1}{11} - \frac{1}{11}\frac{v_2\{4\}^2 - 12v_2\{6\}^2 + 11v_2\{8\}^2}{v_2\{4\}^2 - v_2\{6\}^2} \tag{5}$$

$$h_2 = \frac{v_2\{8\} - v_2\{10\}}{v_2\{6\} - v_2\{8\}} \approx h_2^{Taylor} = \frac{3}{19} - \frac{1}{19}\frac{3v_2\{6\}^2 - 22v_2\{8\}^2 + 19v_2\{10\}^2}{v_2\{6\}^2 - v_2\{8\}^2} \tag{6}$$

Finally, the "corrected" skewness, kurtosis and superskewness have been constructed [7], making them independent of the other moments. This has been done by assuming an elliptic power distribution [11], setting the initial eccentricity parameter $\epsilon_0 \leq 0.15$. This shows the utility of these observables to constrain the initial state.

3 Results

The measured $v_2\{2k\}$ ($k = 1, .., 5$) are shown in Figure 1 as functions of collision centrality in PbPb collisions at $\sqrt{s_{NN}} = 5.02$ TeV. Charged particles having pseudorapidity values $|\eta| < 2.4$ and transverse momentum $0.5 < p_T < 3.0$ GeV/c have been used.

It is observed that there is a clear splitting between $v_2\{2\}$ and the higher-order cumulant based $v_2\{2k\}$ values. The difference is due to flow fluctuations, with $v_2\{2\}^2 \approx v_2\{2k\}^2 + 2\sigma_v^2$ for $k > 1$ [8], where σ_v^2 is the v_2 variance. It is also seen that the flow fluctuations become larger going to more peripheral collisions (higher centrality percentages).

The cumulants of different orders have been used to calculate the hydrodynamic probes given by the left-hand sides of Eqs. (5) and (6). Figure 1 displays these distributions with closed symbols. Open symbols in the same figure show the right-hand sides of Eqs. (5) and (6). Without taking the higher-order moments into consideration, the distributions are represented by the constant values of $\frac{1}{11}$ and $\frac{3}{19}$, represented by the blue and red lines for the

© The Authors, published by EDP Sciences. This is an open access article distributed under the terms of the Creative Commons Attribution License 4.0 (https://creativecommons.org/licenses/by/4.0/).

https://doi.org/10.1051/epjconf/202531607006

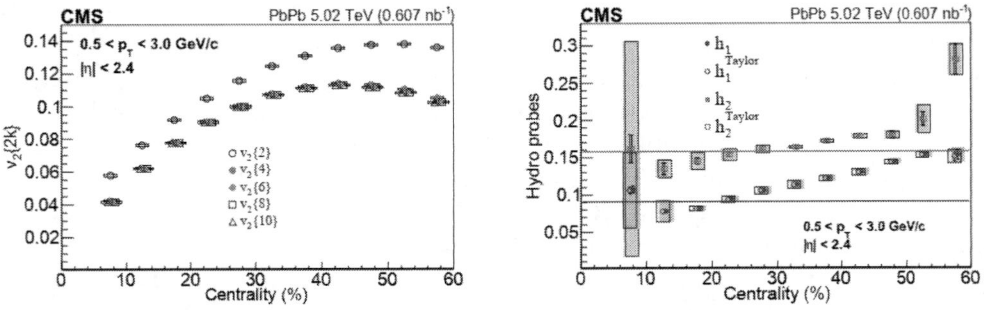

Figure 1. Left : $v_2\{2k\}$ ($k = 1, .., 5$) as functions of centrality in PbPb collisions at $\sqrt{s_{NN}} = 5.02$ TeV. Right : The two hydrodynamic probes as functions of centrality. The vertical sizes of the open boxes denote the systematic uncertainties. Statistical uncertainties are negligible compared to the marker size. Points are plotted at the center of the respective centrality ranges. The markers are displaced horizontally for better visibility.

two probes respectively in the right-hand side plot of Figure 1. However, it is clear that the values of these probes are centrality-dependent and are very well described by the right-hand side expansions of Eqs. (5) and (6).

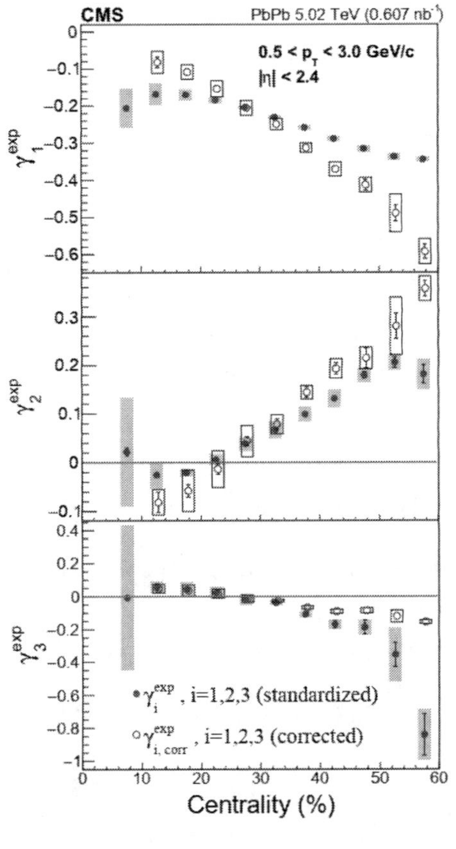

Figure 2. The skewness, kurtosis and superskewness and their corrected values as functions of centrality in PbPb collisions at $\sqrt{s_{NN}} = 5.02$ TeV.

© The Authors, published by EDP Sciences. This is an open access article distributed under the terms of the Creative Commons Attribution License 4.0 (https://creativecommons.org/licenses/by/4.0/).

Finally, the standardized higher-order moments - skewness, kurtosis and superskewness (full circles) and their corrected versions (open circles) are plotted as functions of centrality in Figure 2. Both the standardized and corrected skewness are negative for the whole centrality range. This indicates that the $v_{2,x}$ distribution has a long tail on the low $v_{2,x}$ side (as shown in Figure 1 of [8]). This measurement confirms the prediction that the skewness becomes more negative as the centrality percentage increases. The corrected kurtosis values are positive for the whole centrality range except 10-20%, which qualitatively agrees with the prediction in [12], as it is driven by the mean eccentricity. Except for centrality 5-25%, superskewness is negative, with its absolute magnitude increasing towards more peripheral collisions. This is the first measurement of this moment.

4 Summary

The different orders of multiparticle cumulants, $v_2\{2k\}$ ($k = 1, .., 5$), are determined as functions of centrality in PbPb collisions at 5.02 TeV, with an integrated luminosity of 0.607 nb^{-1}. A splitting is observed between $v_2\{2\}$ and higher-order $v_2\{2k\}$ due to flow fluctuations. Two hydrodynamic probes have been defined in terms of $v_2\{2k\}$, which are well-explained with the inclusion of higher-order moments of the v_2 distribution. The centrality-dependent values of standardized and corrected skewness, kurtosis and superskewness are measured. These results can provide basic input for a precision test of models that assume a hydrodynamic expansion of the QGP.

References

[1] BRAHMS collaboration, Nucl.Phys.A **757**, 1 (2005). https://doi.org/10.48550/arXiv.nucl-ex/0410020

[2] PHOBOS collaboration, Nucl.Phys.A **757**, 28 (2005). https://doi.org/10.48550/arXiv.nucl-ex/0410022

[3] PHENIX collaboration, Nucl.Phys.A **757**, 184 (2005). https://doi.org/10.48550/arXiv.nucl-ex/0410003

[4] J.-Y. Ollitrault, Phys.Rev.D. **46**, 229 (1992). https://doi.org/10.1103/PhysRevD.46.229

[5] S. Voloshin and Y. Zhang, Z.Phys.C. **70**, 665 (1996). https://doi.org/10.1007/s002880050141

[6] N. Borghini, P.M. Dinh and J.-Y. Ollitrault, Phys.Rev.C. **63**, 054906 (2001). https://doi.org/10.48550/arXiv.nucl-th/0007063

[7] CMS collaboration, J. High Energ. Phys. **2024**, 106 (2024). https://doi.org/10.48550/arXiv.2311.11370

[8] G. Giacalone, L. Yan, J. Noronha-Hostler and J.-Y. Ollitrault, Phys.Rev.C. **95**, 014913 (2017). https://doi.org/10.1103/PhysRevC.95.014913

[9] A. Bilandzic, R. Snellings and S. Voloshin, Phys.Rev.C. **83**, 044913 (2011). https://doi.org/10.48550/arXiv.1010.0233

[10] L. Nadderd, J. Milosevic and F. Wang, Phys.Rev.C. **104**, 034906 (2021). https://doi.org/10.48550/arXiv.2104.00588

[11] L. Yan, J.-Y. Ollitrault, A. M. Poskanzer, Phys.Rev.C **90**, 024903 (2014). https://doi.org/10.48550/arXiv.1405.6595

[12] R. S. Bhalerao, G. Giacalone, J.-Y. Ollitrault, Phys.Rev.C **99**, 014907 (2019). https://doi.org/10.48550/arXiv.1811.00837

© The Authors, published by EDP Sciences. This is an open access article distributed under the terms of the Creative Commons Attribution License 4.0 (https://creativecommons.org/licenses/by/4.0/).

Quarkyonic solution to the hyperon puzzle

Yuki Fujimoto[1,*], *Toru* Kojo[2], and *Larry* McLerran[1]

[1]Institute for Nuclear Theory, University of Washington, Box 351550, Seattle, WA 98195, USA
[2]Department of Physics, Tohoku University, Sendai 980-8578, Japan

Abstract. We show that Quarkyonic Matter can mitigate the hyperon puzzle. The key observation is that the hyperon threshold is shifted to a higher density by a factor of constituent strange quark mass. We illustrate this effect by using the ideal dual Quarkyonic (IdylliQ) model with multiple flavors.

1 Introduction

The existence of hyperons in neutron star cores remains as an open question. Hyperons are expected to appear since it is at some point more likely to add a particle with different flavor due to Pauli blocking than adding another nucleon. The onset of hyperons leads to the problematic reduction of the maximum mass of a neutron star, which is at odds with the presence of heavy pulsars. This problem is known as *hyperon puzzle*.

Meanwhile, we believe that nuclear matter undergoes deconfinement and eventually turns into quark matter with increasing baryon density. It is natural to expect so since nucleons in nuclear matter overlap at a few times the saturation density, so quarks become relevant degrees of freedom. The liberation of quarks at high density is led by screening of the confinement potential by quarks in the medium, which is based on the weak-coupling analysis [1].

However, as was pointed out in Ref. [2] based on the large-N_c QCD, the deconfinement at high density may not be as simple as this conventional picture. Namely, in the large-N_c limit of QCD, the screening mass for gluons from quark loops is suppressed by the factor $1/N_c$. This should be contrasted to QCD at finite temperatures in which the thermal screening of the confinement potential is efficient enough to form the deconfined quark-gluon plasma.

Such an observation in large-N_c QCD at high density led to the notion of Quarkyonic matter [2]: The fundamental degrees of freedom in dense large-N_c QCD matter are confined baryons because the confinement interaction is never screened from the argument above. At the same time, matter can be described in terms of quarks in the weakly-coupled regime because the fundamental degrees of freedom in the weak-coupling calculations are quarks. This can be interpreted as the duality between baryonic and quark degrees of freedom.

Quarkyonic matter has successfully been applied to reproduce the semi-quantitative feature of the neutron star equation of state [3]. The key feature of Quarkyonic matter is the nontrivial occupation number in the phase space. This has been recently derived in Ref. [4] by using the ideal dual Quarkyonic (IdylliQ) model, which explicitly takes into account the dual aspect of Quarkyonic matter.

In this contribution, we show that the onset of hyperons in the IdylliQ matter is shifted to a higher density compared to the conventional picture, so the hyperon puzzle is alleviated.

*e-mail: yfujimoto@berkeley.edu

© The Authors, published by EDP Sciences. This is an open access article distributed under the terms of the Creative Commons Attribution License 4.0 (https://creativecommons.org/licenses/by/4.0/).

2 Ideal dual Quarkyonic (IdylliQ) model

In this section, we briefly sketch the IdylliQ model of the symmetric nuclear matter developed in [4, 5] and discuss its solution [4].

From the dual nature of Quarkyonic matter, we expect the thermodynamic quantities to be described simultaneously in terms of baryons and quarks. Here, we specifically consider the baryon density n_B and the energy density ε. We introduce the phase-space density for baryons $f_B(k)$ and quarks $f_q(q)$. The quark distribution f_q is defined for a fixed color, $f_q \equiv f_q^R = f_q^G = f_q^B$. In such a setup, the baryon distribution is given by

$$n_B = 4\int_k f_B(k) = 4\int_q f_q(q)\,, \tag{1}$$

where the factor 4 arises from the spin and isospin degeneracy, and $\int_k$ and $\int_q$ are the shorthand notation for the integration over the momentum $\int \frac{d^3\boldsymbol{k}}{(2\pi)^3}$ and $\int \frac{d^3\boldsymbol{q}}{(2\pi)^3}$, respectively. The energy density is

$$\varepsilon_B[f_B] = 4\int_k E_B(k) f_B(k)\,, \qquad \varepsilon_q[f_q] = 4\int_q E_q(q)[N_c f_q(q)]\,, \tag{2}$$

where we fixed the dispersion relation for a baryon as $E_B(k) = \sqrt{M_N^2 + k^2}$ and M_N is the nucleon mass. Duality means $\varepsilon = \varepsilon_B[f_B] = \varepsilon_q[f_q]$. Each description is related through the following relation, which effectively takes into account the quark confinement inside baryons:

$$f_q(q) = \int_k \varphi\left(\boldsymbol{q} - \frac{\boldsymbol{k}}{N_c}\right) f_B(k)\,. \tag{3}$$

The function φ describes a momentum distribution of a single quark in a single baryon state, so the physical requirement for φ is to have a peak at $\boldsymbol{q} - \boldsymbol{k}/N_c = 0$ with the finite width to account for the quark confinement effect. Here, for φ, we choose

$$\varphi(\boldsymbol{q}) = \frac{2\pi^2}{\Lambda^3}\frac{e^{-q/\Lambda}}{q/\Lambda}\,, \tag{4}$$

where the parameter Λ is an effective confinement scale. This specific choice (4) was made so that the model becomes analytically soluble while fulfilling the physical requirement.

Now, we find the optimal solutions of f_B and f_q by minimizing the ε at a given n_B and fulfilling the constraints $0 \le f_B \le 1$ and $0 \le f_q \le 1$. The minimum energy solution is $f_B(k) = \Theta(k_F - k)$ as in the ordinary Fermi gas model when the corresponding quark distribution obtained through (3) is less than the unity. When $f_q(q) \ge 1$, the Fermi-Dirac distribution $f_B(k) = \Theta(k_F - k)$ has to be modified. We find the minimum energy solution [4]

$$f_B(k) = \frac{1}{N_c^3}\Theta(k_{\rm bu} - k) + \Theta(k_{\rm sh} - k)\Theta(k - k_{\rm bu})\,, \tag{5}$$

where $k_{\rm bu}$ and $k_{\rm sh}$ refer to the edges of the bulk and the shell part, respectively. The corresponding quark distribution is

$$f_q(q) = \Theta(q_{\rm bu} - q) + f_q^{f_B=1}(q)\Theta(q_{\rm sh} - q)\Theta(q - q_{\rm bu}) + f_q^{f_B=0}(q)\Theta(q - q_{\rm sh})\,, \tag{6}$$

where $q_{\rm bu} = k_{\rm bu}/N_c$ and $q_{\rm sh} = k_{\rm sh}/N_c$. Also, $f_q^{f_B=1}(q) = N_c^3 + d_+\frac{e^{q/\Lambda}}{q/\Lambda} + d_-\frac{e^{-q/\Lambda}}{q/\Lambda}$ and $f_q^{f_B=0}(q) = c_-\frac{e^{-q/\Lambda}}{q/\Lambda}$. The coefficients c_- and $d_\pm$ are determined by the matching conditions at $q = q_{\rm bu}$ and $q_{\rm sh}$. One can find the relation between $q_{\rm bu}$ and $q_{\rm sh}$ from the minimum energy condition:

$$\frac{\Lambda + q_{\rm bu}}{\Lambda + q_{\rm bu} - (\Lambda + q_{\rm sh})e^{-(q_{\rm sh} - q_{\rm bu})/\Lambda}} = N_c^3\,. \tag{7}$$

© The Authors, published by EDP Sciences. This is an open access article distributed under the terms of the Creative Commons Attribution License 4.0 (https://creativecommons.org/licenses/by/4.0/).

3 Hyperons in IdylliQ model

To address the hyperon puzzle, we consider including a hyperon on top of the pure neutron matter. For simplicity, we neglect the lepton contributions and limit ourselves to the charge-neutral hadron species. Now we extend the above model to the multi-flavor case [6]. In the above, we consider the symmetric nuclear matter, and the baryons are described by a single distribution f_B. Here, we consider the momentum distributions for neutrons f_n and for the strangeness -1 hyperons f_Y. To be more specific, we consider Λ^0 and Σ^0 hyperons and neglect their mass differences. The number density and energy density are modified as

$$n_n = 2\int_k f_n(k)\,, \qquad n_Y = 4\int_k f_Y(k)\,, \tag{8}$$

$$\varepsilon_B[f_n, f_Y] = 2\int_k \left[E_N(k)f_n(k) + 2E_Y(k)f_Y(k)\right]\,, \tag{9}$$

where the dispersion relations for neutrons and hyperons are $E_N(k) = \sqrt{k^2 + M_N^2}$ and $E_Y(k) = \sqrt{k^2 + M_Y^2}$, respectively. The baryon density is given by $n_B = n_n + n_Y$. In the flavor asymmetric case, the relation (3) is modified as

$$f_d(q) = \int_k \varphi\left(\boldsymbol{q} - \frac{\boldsymbol{k}}{N_c}\right)\left[\frac{2}{3}f_n(k) + \frac{2}{3}f_Y(k)\right]\,, \tag{10}$$

where the factor in front of f_n originates from the baryon number content $2/3$ in a neutron, and the factor in front of f_Y from the baryon number content $1/3$ in a hyperon times the mass degeneracy 2 for Λ^0 and Σ^0. One can write down the similar relation for u and s quarks.

As we will see below, since the system is d-quark abundant, only the d-quark distribution is saturated. Thus, the d-quark distribution is given by Eq. (6), and from the relation (10), the distribution at $k < k_{\rm bu}$ must fulfill the condition:

$$\frac{2}{3}f_n(k) + \frac{2}{3}f_Y(k) = \frac{1}{N_c^3}\,. \tag{11}$$

From this condition, we find that the optimal distributions for neutrons and hyperons:

$$f_n(k) = \frac{3}{2N_c^3}\Theta(k - k_Y)\Theta(k_{\rm bu} - k) + \Theta(k - k_{\rm bu})\Theta(k_{\rm sh} - k)\,, \tag{12}$$

$$f_Y(k) = \frac{3}{2N_c^3}\Theta(k_Y - k)\Theta(k)\,. \tag{13}$$

In Fig. 1, we plot f_n and f_Y, and the corresponding quark distributions, f_d, f_u, and f_s.

Now, from the relations above, we can compute the onset density of hyperons. This can be computed from the β-equilibrium condition, i.e., the chemical potential for neutrons μ_n and hyperons μ_Y should be the same. They can be computed from the thermodynamic relations $\mu_n = (\partial\varepsilon/\partial n_n)_{n_Y}$ and $\mu_Y = (\partial\varepsilon/\partial n_Y)_{n_n}$. We find that the onset of hyperons is at [6]

$$\mu_B = 2M_Y - M_N\,, \tag{14}$$

where μ_B is the baryon chemical potential and is equal to μ_n. The hyperon onset in the non-interacting baryon gas is at $\mu_B = M_Y$. Our estimate is shifted to much higher μ_B by $M_Y - M_N \simeq 0.18\,\text{GeV}$.

© The Authors, published by EDP Sciences. This is an open access article distributed under the terms of the Creative Commons Attribution License 4.0 (https://creativecommons.org/licenses/by/4.0/).

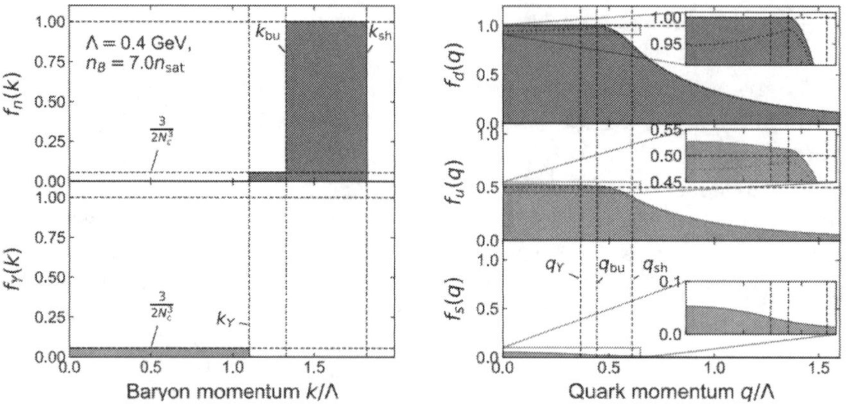

Figure 1. Phase-space density in the IdylliQ model. Left: The momentum distributions for neutrons (top) and hyperons (bottom). Right: The momentum distributions for d (top), u (middle), and s quarks.

4 Conclusions

We presented that the $S = -1$ hyperon threshold is shifted from $\mu_B = M_Y$ to higher density $2M_Y - M_N$ because of the Pauli exclusion effect of the quark substructure of hyperons. Since d-quark states are already filled by nucleons, hyperons that contains d quarks are less likely to appear. This makes the hyperon contribution to a neutron star less important and thus works toward the resolution of the hyperon puzzle; the shifted threshold is high enough so that it may not greatly affect the maximum mass of neutron stars. The $S = -2$ hyperon such as Ξ^0 can contribute at μ_B slightly above the $S = -1$ hyperon threshold; we could not mention it due to the limitation of the space. See Ref. [6] for further discussions.

Acknowledgment

This research was supported in part by the INT's U.S. Department of Energy grant No. DE-FG02-00ER41132.

References

[1] J.C. Collins, M.J. Perry, Superdense Matter: Neutrons Or Asymptotically Free Quarks?, Phys. Rev. Lett. **34**, 1353 (1975). 10.1103/PhysRevLett.34.1353

[2] L. McLerran, R.D. Pisarski, Phases of cold, dense quarks at large N(c), Nucl. Phys. A **796**, 83 (2007), 0706.2191. 10.1016/j.nuclphysa.2007.08.013

[3] L. McLerran, S. Reddy, Quarkyonic Matter and Neutron Stars, Phys. Rev. Lett. **122**, 122701 (2019), 1811.12503. 10.1103/PhysRevLett.122.122701

[4] Y. Fujimoto, T. Kojo, L.D. McLerran, Momentum Shell in Quarkyonic Matter from Explicit Duality: A Dual Model for Cold, Dense QCD, Phys. Rev. Lett. **132**, 112701 (2024), 2306.04304. 10.1103/PhysRevLett.132.112701

[5] T. Kojo, Stiffening of matter in quark-hadron continuity, Phys. Rev. D **104**, 074005 (2021), 2106.06687. 10.1103/PhysRevD.104.074005

[6] Y. Fujimoto, T. Kojo, L. McLerran, Quarkyonic matter pieces together the hyperon puzzle (2024), 2410.22758.

© The Authors, published by EDP Sciences. This is an open access article distributed under the terms of the Creative Commons Attribution License 4.0 (https://creativecommons.org/licenses/by/4.0/).

Directed Flow of Λ, ${}^{3}_{\Lambda}\mathrm{H}$ and ${}^{4}_{\Lambda}\mathrm{H}$ in Au+Au collisions at $\sqrt{s_{NN}}$ = 3.2, 3.5, 3.9 and 4.5 GeV at RHIC

Junyi Han[1,2,*] (for the STAR collaboration)

[1]Central China Normal University
[2]Heidelberg University

Abstract. Studying hyper-nuclei yields and their collectivity can shed light on their production mechanism as well as the hyperon-nucleon interactions. Heavy-ion collisions from the RHIC beam energy scan phase II (BES-II) provide an unique opportunity to understand these at high baryon densities.
In these proceedings, we present a systematic study on energy dependence of the directed flow (v_1) for Λ and hyper-nuclei (${}^{3}_{\Lambda}\mathrm{H}$, ${}^{4}_{\Lambda}\mathrm{H}$) from mid-central Au+Au collisions at $\sqrt{s_{NN}}$ = 3.2, 3.5, 3.9 and 4.5 GeV, collected by the STAR experiment with the fixed-target mode during BES-II. The rapidity (y) dependence of the hyper-nuclei v_1 is studied in mid-central collisions. The extracted v_1 slopes ($dv_1/dy|_{y=0}$) of the hyper-nuclei are positive and decrease gradually as the collision energy increases. These hyper-nuclei results are compared to that of light-nuclei including p, d, t/^{3}He and ^{4}He. Finally, these results are compared with a hadronic transport model including coalescence after-burner.

1 Introduction

Heavy-ion collisions (HICs) are a powerful tool to explore the properties of strongly interacting QCD matter. An extremely high temperature and high density matter is produced during HICs. The interations between the constituents produced in HICs will lead to the formation of bound states. Hypernuclei are bound states of nucleons and hyperons. In astrophysics, the "hyperon puzzle" shows it is difficult to reconcile the measured masses of neutron stars with the presence of hyperons in their interiors. Hypernuclei provide a good opprotunity to study the hyperon-nucleon(Y-N) interactions in the QCD equation of state at high baryon density. Understanding hyperon-nucleon(Y-N) interaction in high density region is essential for solving the hyperon puzzle. The production mechanism of hypernuclei is also a topic of importance to understand.

From the thermal model calculations of light-nuclei and hyper-nuclei, their production are enhanced at high baryon density region [1, 2]. In the second phase of the RHIC Beam Energy Scan program [3, 4], the collision energy for fixed-target Au+Au collisions is extended down to $\sqrt{s_{NN}}$ = 3 GeV. It gives us a good opportunity to study nucleon-nucleon(N-N) and hyperon-nucleon(Y-N) interactions by analyzing light-nuclei and hyper-nuclei. Collective flow of hyper-nuclei can help us to explore its production mechanism and hyperon interactions in the medium.

*e-mail: JYHan@mails.ccnu.edu.cn

© The Authors, published by EDP Sciences. This is an open access article distributed under the terms of the Creative Commons Attribution License 4.0 (https://creativecommons.org/licenses/by/4.0/).

2 Datasets and Particle Reconstruction

The datasets used in this analysis were collected by the upgraded STAR experiment at RHIC with the fixed-target mode setup at $\sqrt{s_{NN}}$ = 3.2, 3.5, 3.9 and 4.5 GeV. STAR has upgraded the inner Time Projection Chamber (iTPC), endcap Time-of-Flight (eTOF) and Event Plane Detector (EPD) [5] in recent years. These upgrades bring larger acceptance, better particle identification ability with uniform efficiency and higher event plane resolution.

The reaction plane is estimated via the reconstruction of the event plane in the EPD [6]. Three sub-event method is used to determined the EPD event plane resolution. In 5-40% centrality, we can both have high hyper-nuclei yield and high event plane resolution. The event plane resolution can reach over 70% at 3.2 GeV.

In order to reconstruct hyper-nuclei, we need the π^-, p, d, ^{3}He and ^{4}He. These charged particles can be selected based on the ionization energy loss (dE/dx) measured in the TPC [7] as a function of rigidity. For the short-lived particle like hyper-nuclei, they are reconstructed with KFParticle package based on Kalman filter method to improve signal significance [8]. The following decay channels are used: $\Lambda \rightarrow p + \pi^-$, ${}^4_\Lambda H \rightarrow {}^4He + \pi^-$, ${}^3_\Lambda H \rightarrow {}^3He + \pi^-$ and ${}^3_\Lambda H \rightarrow p + d + \pi^-$. Combinatorial backgrounds for hyper-nuclei is estimated by rotating one of the daughter tracks by random angle in transverse plane.

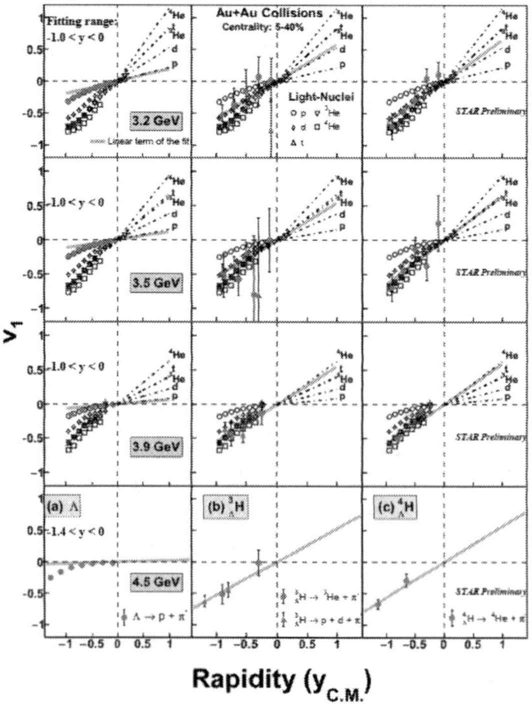

Figure 1. Light-nuclei and hyper-nuclei directed flow v_1 as a function of center-of-mass rapidity at 3.2, 3.5, 3.9 and 4.5 GeV with 5-40% centrality Au+Au collisions [9]. The rapidity dependence of v_1 for Λ on the left, ${}^3_\Lambda H$ 2-body and 3-body decay in the middel, ${}^4_\Lambda H$ results on the right represented by solid markers. The linear terms of the fitting for Λ, ${}^3_\Lambda H$ and ${}^4_\Lambda H$ are represented as brown lines. For comparison, the rapidity dependence of v_1 for p, d, t, ^{3}He and ^{4}He are also shown as open markers, and the linear terms of the fitting results are represented as dashed lines in the positive rapidity region.

Systematic uncertainties of the measured collective flow mainly come from the event plane resolution, reconstruction efficiency, particle identification cuts and topological variable

© The Authors, published by EDP Sciences. This is an open access article distributed under the terms of the Creative Commons Attribution License 4.0 (https://creativecommons.org/licenses/by/4.0/).

cuts. In case of hyper-nuclei, the dominant source of systematic uncertainty arises from the topological cuts (used for their selection).

3 Results and Discussion

The v_1(y) of Λ hyperon and hyper-nuclei are extracted with event plane method [9]. The distributions of light-nuclei and hyper-nuclei $v_1(y)$ from 5-40% mid-central Au+Au collisions at 3.2, 3.5, 3.9 and 4.5 GeV are shown in the Fig.1. Since collective flow depends on p_T, we select a similar p_T/A range (approximately $0.4 < p_T/A < 0.8$ GeV/c) for comparison for a given particle. As we can see, the Λ hyperon $v_1(y)$ distribution is close to that of the proton, and hyper-nuclei $v_1(y)$ distributions are also close to those light-nuclei with the same mass numbers. For the Λ hyperon and light-nuclei, we use the third-order polynomial $v_1(y) = p_0 y + p_1 y^3$ to extract the v_1 slope ($p_0 = (v_1)_{slope} = dv_1/dy|_{y=0}$). Due to poor statistics for ${}^3_\Lambda$H and ${}^4_\Lambda$H, we use the first order polynomial v_1(y) $= p_0 y$ to fit, the fitting functions are shown by the solid brown lines.

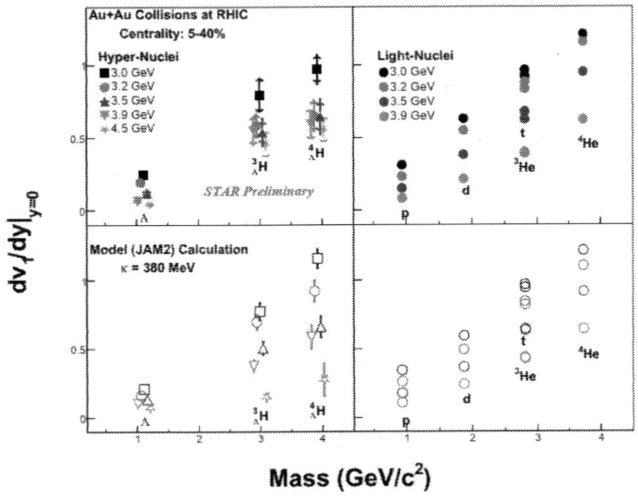

Figure 2. Particle mass dependence of the mid-rapidity v_1 slope for light-nuclei and hyper-nuclei at 3.2, 3.5, 3.9 and 4.5 GeV with 5-40% centrality Au+Au collisions. The top results are from experimental data for hyper-nuclei(left) and light-nuclei(right), and the bottom results with open markers are corresponding results from hadronic transport model (JAM2) with coalescence afterburner calculation.

The results of the mid-rapidity v_1 slope as a function of particle mass for light-nuclei and hyper-nuclei with 5-40% centrality are shown in Fig.2. The top panels present results from hyper-nuclei and light-nuclei, and the bottom panels with the open markers present the hadronic transport model (JAM2) [10] with coalescence afterburner calculation results. As we can see, at given energy, for both light-nuclei and hyper-nuclei, the mid-rapidity v_1 slopes are scaled with particle mass, implying that coalescence is the major process for the light-nuclei and hyper-nuclei production. And the feature is also reproduced by JAM2 model calculations.

The collision energy dependence of the mass scaled mid-rapidity v_1 slope for light-nuclei and hyper-nuclei with 5-40% are shown in Fig.3. From the results, we observe that as the collision energy increases, the mid-rapidity v_1 slope of light-nuclei and hyper-nuclei have a downward trend. For comparison, hadronic transport model calculations with a coalescence afterburner are also included and are found consistent with the observed energy dependence in the data.

© The Authors, published by EDP Sciences. This is an open access article distributed under the terms of the Creative Commons Attribution License 4.0 (https://creativecommons.org/licenses/by/4.0/).

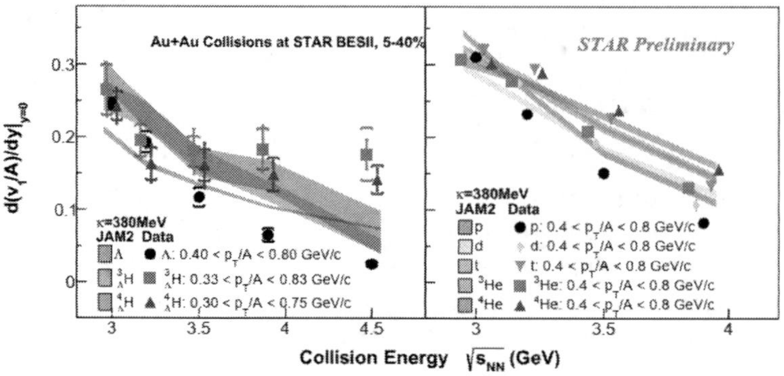

Figure 3. Collision energy dependence of the mid-rapidity v_1 slope for light-nuclei and hyper-nuclei at 3.2, 3.5, 3.9 and 4.5 GeV with 5-40% centrality Au+Au collisions. The left results are from hyper-nuclei and the right are light-nuclei. The JAM2 model calculation results are represented by colored bands both for light-nuclei and hyper-nuclei.

4 Summary and conclusions

In this analysis, we presented the particle mass and collision energy dependence of the directed flow v_1 for light-nuclei and hyper-nuclei in $\sqrt{s_{NN}}$ = 3.2, 3.5, 3.9 and 4.5 GeV Au+Au collisions at RHIC-STAR. The distribution of particle mass and collision energy dependence of the mass scaled mid-rapidity v_1 slope of light- and hyper-nuclei shows that as the collision energy increases, the baryon number scale breaks down, indicating that the coalescence process dominates the production mechanism of the light clusters becoming weaker. Additionally, the hadronic transport model calculations are consistent with the observed mass and energy dependence. STAR has collected 2 billion events from 3.0 GeV Au+Au collisions. Analysis of this data will provide improved precision, leading to stronger constraints on the coalescence parameters for both light-nuclei and hyper-nuclei.

Acknowledgement: This work was supported in part by the National Natural Science Foundation of China (Grant No. 12375134), the National Key Research and Development Program of China under Grant No. 2020YFE0202002, and the Fundamental Research Funds for the Central Universities (Grant No. CCNU22QN005).

References

[1] A. Andronic et al, Phys. Lett. B **697**, 203 (2011)
[2] J. Steinheimer et al, Phys. Lett. B **714**, 85 (2012)
[3] J.H. Chen et al, arXiv:2407.02935 [nucl-ex].
[4] F. Liu et al, Sci Sin-Phys Mech & Astron **53**, 290003 (2023)
[5] C. Yang, EPJ Web of Conferences **182**, 02130 (2018)
[6] Sergei A. Voloshin et al, Nuclear Experiment (nucl-ex), arXiv:0809.2949 [nucl-ex].
[7] M. Anderson et al, Nucl. Instrum. Meth. A **499**, 659–678 (2003)
[8] KF Particle Finder: M. Zyzak, Dissertation thesis, Goethe University of Frankfurt (2016)
[9] H. Masui et al, Nucl. Instrum. Meth. A **833**, 181–185 (2016)
[10] Y. Nara et al, Phys. Rev. C **106**, 044902 (2022)

© The Authors, published by EDP Sciences. This is an open access article distributed under the terms of the Creative Commons Attribution License 4.0 (https://creativecommons.org/licenses/by/4.0/).

AUTHOR INDEX

AUTHOR INDEX